5th International Workshop on Top Quark Physics

(TOP2012)

Journal of Physics: Conference Series Volume 452

Winchester, United Kingdom
16-21 September 2012

ISBN: 978-1-62993-090-9
ISSN: 1742-6588

Printed from e-media with permission by:

Curran Associates, Inc.
57 Morehouse Lane
Red Hook, NY 12571

proceedings.com

Some format issues inherent in the e-media version may also appear in this print version.

Printed by Curran Associates, Inc. (2013)

For permission requests, please contact the Institute of Physics
at the address below.

Institute of Physics
Dirac House, Temple Back
Bristol BS1 6BE UK

Phone: 44 1 17 929 7481
Fax: 44 1 17 920 0979

techtracking@iop.org

Additional copies of this publication are available from:

Curran Associates, Inc.
57 Morehouse Lane
Red Hook, NY 12571 USA
Phone: 845-758-0400
Fax: 845-758-2634
Email: curran@proceedings.com
Web: www.proceedings.com

5th International Workshop on Top Quark Physics

(TOP2012)

Journal of Physics: Conference Series Volume 452

Winchester, United Kingdom
16-21 September 2012

ISBN: 978-1-62993-090-9
ISSN: 1742-6588

IOP Institute of Physics

TABLE OF CONTENTS

ANGULAR DISTRIBUTIONS AND ASYMMETRIES

INTERPLAY WITH OTHER SECTORS

NEW PHYSICS IN THE TOP QUARK SECTOR

OBJECTS, PERFORMANCE AND BACKGROUNDS

SINGLE TOP QUARK PRODUCTION

TOP QUARK SIGNAL MODELING UNCERTAINTIES

SUMMARY TALKS

POSTER PAPERS

Preface

The 5th International Workshop on Top Quark Physics (TOP 2012) took place in Winchester, UK, from the 16–21 September. It gathered students as well as people active in the top quark sector and provided a framework to highlight the newest results and matters related to top quark physics. Discovered in 1995, the top quark is the sixth and heaviest of all quarks, and it is the only one with a lifetime short enough to be observed 'naked'. This makes it an important testing ground in the search for new physics. In fact, the fact of its mass being so much larger than the other quarks, hints at its special role in the Higgs mechanism. For the same reason, in many models of New Physics, new heavy resonances are expected to couple mostly with top quarks. Even if no new particles are observed, the direct correlation between its angular momentum and that of its detectable decay products allows us to probe indirectly New Physics in action when top quarks are created.

In this edition of the TOP conference series, for the first time, the agenda was equally balanced between 'traditional' measurements and the now vast number of searches for physics BSM in the top quark sector, thanks mostly to the amount of data collected at the LHC in its Run I. New results were presented by both the Tevatron and the LHC collaborations: improved ttbar and single top cross-section measurements, refined techniques to measure the top quark mass and a large number of results on properties such as spin correlation and W boson polarization in top quark decays were shown. More technical discussions on the experimental issues, both from the detector and the simulation side also took place, drawing together experimentalists and theorists. Reviews of the latest results on ttbar asymmetry both from CDF and D0 and from ATLAS and CMS were shown, and theorists active in the field made some interesting points on this hot topic. Additionally, results on the search for fourth generation fermions and new resonances in the boosted top regime were also provided. Finally, a set of dedicated talks on the interplay between the top sector and other hot subjects, like the Higgs and SUSY, were given both at the theory and experimental level.

Furthermore, *ad hoc* student sessions were organized to allow younger colleagues to pose questions to the senior experts in the field and contribute with their more recent studies.

The conference has been a definitive success, not just scientifically: about 130 participants from all over the world created a collegiate spirit which culminated in the social events at Winchester Hall below King Arthur's table; and in a cosy 16th century barn for the social dinner.

The Local Organizing Committee would like to thank all participants, and in particular the speakers, for their high level contributions to TOP 2012 and for making this a very fruitful and pleasant time together.

We conclude by wishing the Organizing Committee of TOP 2013 all the best for a successful conference. We look forward to seeing everyone in Germany in 2013.

Giuseppe Salamanna
Local Organizing Committee
London, June 2013

Journal of Physics: Conference Series **452** (2013) 011001 doi:10.1088/1742-6596/452/1/011001

Local Organising Committee

Veronique Boisvert (Chair, *Royal Holloway, University of London*)
Lucio Cerrito (*Queen Mary, University of London*)
Akram Khan (*Brunel University, London*)
Stefano Moretti (*University of Southampton*)
Mark Owen (*University of Manchester*)
Giuseppe Salamanna (*Queen Mary, University of London*)
Christian Schwanenberger (*University of Manchester*)

International Advisory Committee

Roberto Tenchini (*INFN, Pisa*)
Martine Bosman (*IFAE, Barcelona*)
Michelangelo Mangano (*CERN*)
Scott Willenbrock (*University of Illinois, Urbana*)
Werner Bernreuther (*RWTH, Aachen*)
Jorgen D'Hondt (*VUB, Brussels*)
Antonio Onofre (*LIP, University Minho*)
Fabio Maltoni (*UCL, Louvain*)
Eric Laenen (*NIKHEF*)
Fabrizio Margaroli (*INFN, Roma 1*)
Juan Antonio Aguilar Saavedra (*University of Granada*)
Yvonne Peters (*University of Manchester*)
Roberto Chierici (*CERN*)
Markus Cristinziani (*University of Bonn*)

Peer review statement

All papers published in this volume of *Journal of Physics: Conference Series* have been peer reviewed through processes administered by the proceedings Editors. Reviews were conducted by expert referees to the professional and scientific standards expected of a proceedings journal published by IOP Publishing.

Journal of Physics: Conference Series **452** (2013) 012001 doi:10.1088/1742-6596/452/1/012001

Angular distributions at the Tevatron

A Harel, on behalf of the D0 and CDF collaborations.

University of Rochester, Rochester, New York 14627, USA

E-mail: Amnon.Harel@cern.ch

Abstract. We review measurements of angular distributions in top pair events by the CDF and D0 collaborations. The helicity of W bosons produced in top decays was measured, as was the correlation between the top quark and antiquark spins in $p\bar{p} \to t\bar{t}$ production. We also review the information available on top quark polarization in $p\bar{p} \to t\bar{t}$ production.

1. Introduction

As the heaviest of the known elementary quanta of the standard model of particle physics (SM), the top quark offers many opportunities to search for new phenomena beyond the SM. Given the precise SM predictions for polarizations in the process $p\bar{p} \to t\bar{t} \to (W^+b)\left(W^-\bar{b}\right)$, we seek to identify such new phenomena by measuring these polarizations.

Details on event selection, the estimation of sample composition, and the modeling of signal and background processes can be found in other proceedings from this conference. In the spirit of the papers being reviewed here, we concentrate on the unique aspects of the measurements of angular distributions.

2. Measurements of W helicity in top decays

Data supports the SM prediction of $B\,(t \to Wb) \approx 100\%$ [1]. The SM further predicts the fractions of decay into each of the W boson helicity states: $f_- = 30.3\%$ for left handed helicity, $f_0 = 69.6\%$ for longitudinally polarized W bosons, and $f_+ = 0.1\%$ for right handed helicity, whose production is suppressed by the $V - A$ structure of the weak force [2]. The uncertainties on these predictions are smaller than the experimental uncertainties. Nevertheless, data may indicate new physics, for example, with an observation of sizable f_+.

We select events in the "lepton+jets" channel, which corresponds to $t\bar{t} \to (W^+b)\left(W^-\bar{b}\right)$ decays where one W boson decays hadronically and the other decays leptonically into an electron or a muon and one or more neutrinos, possibly through an intermediate τ lepton. We also select events in the "dilepton" channel which corresponds to decays where both W bosons decay leptonically. Leptons are required to be well isolated in the detector, and to have transverse momentum $p_T > 20\,\text{GeV}$. We use only events where at least one of the selected jets is tagged as arising from b decay. The tagging is based primarily on tracks and secondary vertices that are displaced from the primary collision vertex. Transverse momentum imbalance is also used to suppress background, as it is expected for signal events due to the undetected neutrino(s).

We distinguish between helicity states by reconstructing the angle between the up-type decay product and the incoming top quark in the W boson's rest frame, θ^* (see figure 1). In the lepton+jets channel we perform a kinematic fit that varies the four-momenta of the detected objects within their experimental resolutions and minimizes a χ^2 statistic within the constraints $M_W = 80.4\,\text{GeV}$ and $m_t = 172.5\,\text{GeV}$; the χ^2 statistic is used to assign the observed jets to the final state quarks from top decay. $\cos\theta^*$ is reconstructed both for the leptonic decay and for the hadronic decay. For the latter, the up-type quark is not identified, and due to this ambiguity we only measure $|\cos\theta^*|$, which is useful to constrain f_0. In the dilepton channels D0 [3] accounts for the resolutions using a statistical procedure that considers both possible jet-quark assignments.

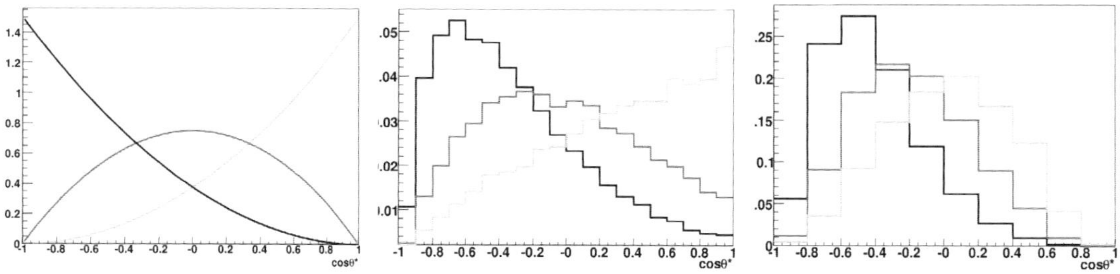

Figure 1. $\cos\theta^*$ distributions for left-handed (black), longitudinal (red), and right-handed (green) W bosons, shown at parton level (left), and after reconstruction in the lepton+jets (middle) and $e\mu$ (right) channels.

The preliminary CDF dilepton measurement [4] uses a different technique, calculating the likelihood of the helicity fractions given the data without choosing a particular reconstruction. Assuming a certain set of helicity fractions, the probabilities of the four-vectors of the final state partons are predicted according to the SM matrix elements and the proton's parton distribution functions. The probabilities for the observed final state four-vectors are then calculated according to the experimental resolutions, taking into account all jet-quark assignments. Background contributions are similarly included using the W+jets matrix elements from VECBOS [5]. Unlike previous CDF measurements, this "matrix element" technique uses the hadronic decay and all of the jet kinematics to constrain the helicity fractions, thus improving the precision by $\approx 20\%$. Due to effects that are not modeled by the likelihood, such as higher order diagrams, events without a correct jet-quark assignment, and other backgrounds, we expect the fitted fractions to be biased. The fits are calibrated using fully simulated events. We find that the measured fractions depend linearly on the simulated fractions, which simplifies the interpretation of the fit results.

The CDF and D0 collaborations released a preliminary combined measurement of W helicity in top decays [7]. The combined measurement found $f_0 = 0.722 \pm 0.081$ and $f_+ = -0.033 \pm 0.046$. Preliminary CDF lepton+jets results [6] supersede earlier work included in the combination. They are $f_0 = 0.726 \pm 0.066 \pm 0.067$ and $f_+ = -0.045 \pm 0.043 \pm 0.058$, where the first quoted uncertainty is statistical and the second one systematic. These results are shown in figures 2 and 3. Taken together, these results are in excellent agreement with the SM.

3. Top quark polarization

The lifetime of the top quark is much shorter than the time scale for its spin flip [8], so its production and decay polarizations are almost the same. The production mechanisms at the Large Hadron Collider and the Fermilab Tevatron Collider differ significantly, and so, there is interest in measuring the top quark polarization at both accelerators. Unfortunately, no dedicated measurements of top quark polarization in $p\bar{p} \to t\bar{t}$ production were published by the Tevatron collaborations. However, several relevant data distributions were published, and these will be reviewed here.

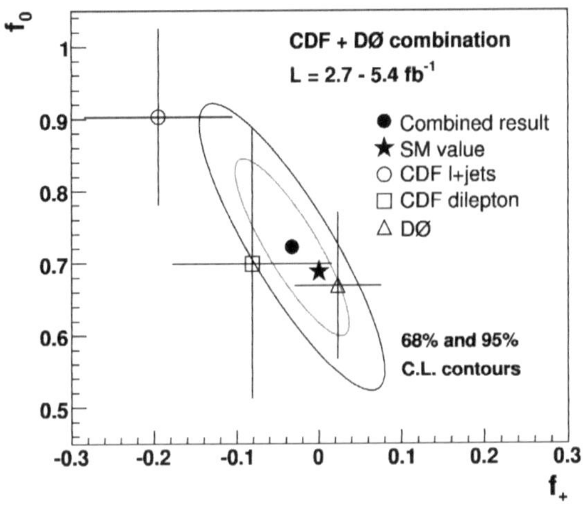

Figure 2. Combined measurement of W helicity fractions and its inputs.

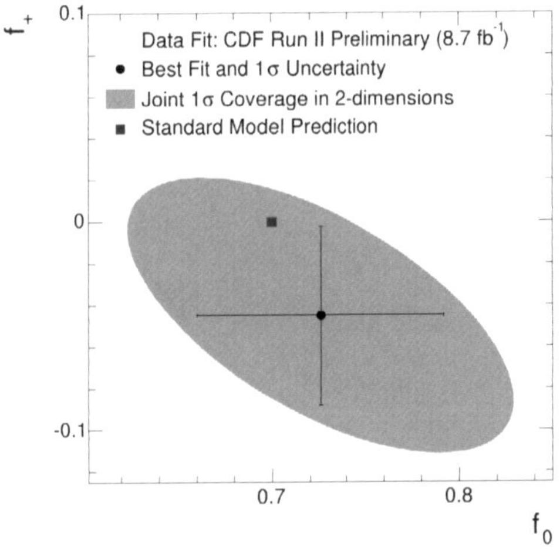

Figure 3. Updated CDF lepton+jets measurement of W helicity fractions.

In the top quark's rest frame we define the angle θ between the direction of flight of a certain decay product and the chosen polarization axis. The differential cross section is then:

$$\frac{1}{\sigma}\frac{d\sigma}{d\cos\theta} = \frac{1}{2}\left(1 + \left|\vec{P}\right|\alpha\cos\theta\right), \tag{1}$$

where $\vec{P}$ is the polarization vector and α is the analyzing strength for that decay product. We analyze the spin of the top quark by looking at the direction of its decay lepton, as for this decay α is essentially one [9] and the lepton is easily identified.

Figure 4, from reference [10], shows distributions of the cosine of the angle between the flight directions of the lepton and its progenitor top quark, $\cos\theta^{l,t}$. This angle measures the top quark polarization in the helicity basis. From CP invariance we expect the distributions for positive and for negative leptons to be the same, and they are combined to increase statistical strength. The SM prediction is contrasted with production through a heavy Z-like boson whose couplings are identical to those of the Z boson, resulting in highly polarized top quark production. The reconstruction in the dilepton channel of reference [10] uses the "neutrino-weighting" method, which samples possible neutrino pseudorapidity (η) values, weighting each set by its compatibility with the measured momentum imbalance given the experimental resolutions. For each set of neutrino ηs, the full decay chain is derived analytically using M_W and m_t constraints.

Figure 5, from reference [11], shows distributions of the cosine of the angle between the flight directions of the lepton and of the same-sign colliding proton, $\cos\theta_{\pm}$. Again, CP invariance predicts identical distributions for positive and negative leptons, which are combined in the plot. To conclude, the published Tevatron data hints at SM-like, unpolarized top pair production.

4. Measurements of spin correlations in top pair production

The spins of the top quark-antiquark pair produced by the SM process $p\bar{p} \to t\bar{t}$ are highly correlated. This correlation leads to the doubly differential cross section:

$$\frac{1}{\sigma}\frac{d^2\sigma}{d\cos\theta_1 d\cos\theta_2} = \frac{1}{4}\left(1 - A\alpha_1\alpha_2\cos\theta_1\cos\theta_2\right), \tag{2}$$

Journal of Physics: Conference Series **452** (2013) 012001 doi:10.1088/1742-6596/452/1/012001

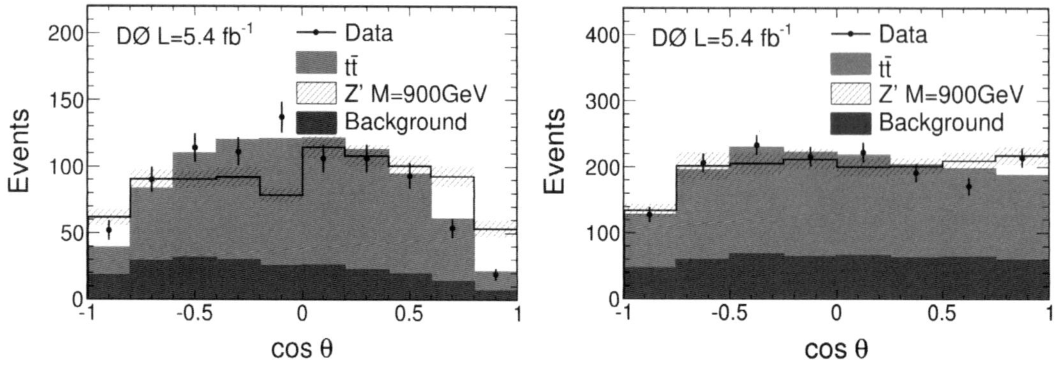

Figure 4. $\cos\theta^{l,t}$ distributions for the dilepton (left) and lepton+jets (right) channels.

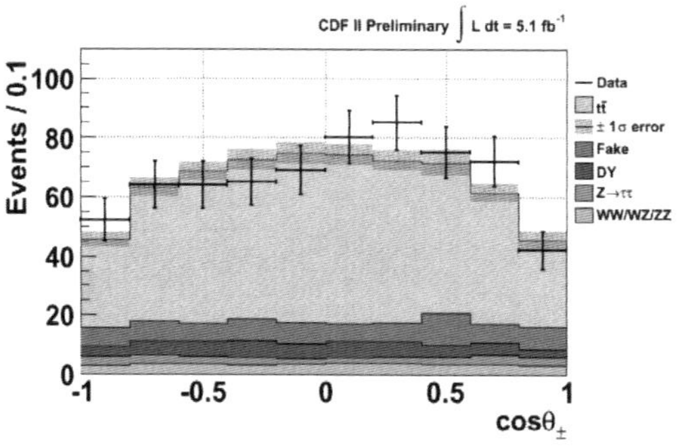

Figure 5. $\cos\theta_\pm$ distributions.

where the θs are leptonic decay angles along some axis, and A is the strength of the corresponding spin correlation. In the beam axis, the SM prediction is $A_{SM}^{\mathrm{beam}} = 0.78$, calculated to next-to-leading order in quantum chromodynamics. Only one lepton, with $\alpha \approx 1$, is available in the lepton+jets channel. The spin of the other top must be inferred from the angle of one of the jets arising from its decay products. For the b-jet, $\alpha = -0.41$, and for the jet arising from the down-type quark, $\alpha = 0.97$. Identifying the latter jet is difficult, making the dilepton channel particularly attractive for this measurement.

A preliminary measurement of spin correlations by the CDF collaboration in the lepton+jets channel [11] was reported in the previous workshop. Here we report on a new measurement in the dilepton channel [12].

To reconstruct $t\bar{t}$ events in the dilepton channel, we perform a fit for the energies of the b-jets, the transverse momentum imbalance ($\not{E}_T$) due to the neutrinos, and the best jet-quark assignment. For each set of values assumed in the fit, the six unknown neutrino momentum components are solved for analytically from the four mass constraints and the two neutrino-$\not{E}_T$ values. The fit maximizes a likelihood with χ^2 terms for each parameter and with a-priori probability density functions for $p_z^{t\bar{t}}$, $p_T^{t\bar{t}}$, and $m_{t\bar{t}}$. These density functions are simulated by the PYTHIA event generator.

The spin correlation coefficient, $\kappa = -A\alpha_1\alpha_2$, is measured for the beam axis, defined here using both proton and anti-proton directions which results in the minus sign. It is extracted from a template fit to the observed two-dimensional distributions of $(\cos\theta_+, \cos\theta_-)$ and of $(\cos\theta_b, \cos\theta_{\bar{b}})$. The expected sample composition is assumed during this fit. The templates are produced by reweighting $t\bar{t}$ events generated by PYTHIA without spin correlations. To model a given correlation strength, $\kappa \in (-1, -0.8, \dots 1)$, we use weights of $1 + \kappa \cos\theta_+ \cos\theta_-$, where the θs are the generator-level decay angles of the down-type fermion. Templates for the diboson and Drell-Yan background processes were simulated using PYTHIA and ALPGEN (respectively). Templates for the multijet background were derived from control data.

The fit procedure was checked for biases using ensembles of simulated pseudo datasets. No calibration is needed, and the measured value is $\kappa = 0.042^{+0.563}_{-0.562}$.

In references [13] and [14], the D0 Collaboration reports measurements of spin correlations in the dilepton channel and lepton+jets channels, respectively. Both analyses measure the fraction f of $t\bar{t}$ events whose spins are correlated as in the SM predictions (while the rest have uncorrelated spins) using a discriminant that summarizes efficiently all event kinematics:

$$R(x) = \frac{P_c(x)}{P_c(x) + P_u(x)}, \tag{3}$$

where x is the observed event, P_c is its probability assuming SM production, and P_u is its probability assuming uncorrelated spins. The Ps are calculated as in the matrix element technique described above. This discriminant was suggested in reference [15].

The observed distributions of R in the dilepton channel, and in the best of the four sub-channels of the lepton+jets analysis, are shown in figure 6. In the dilepton channel we measure $f = 0.74 \pm 0.41$, in the lepton+jets channel $f = 1.15 \pm 0.43$, and combined $f = 0.85 \pm 0.29$. Deriving frequentist limits on the physical value of f with a C.L. of 99.7%, we find it is positive. This is the first evidence for spin correlations in top pair production.

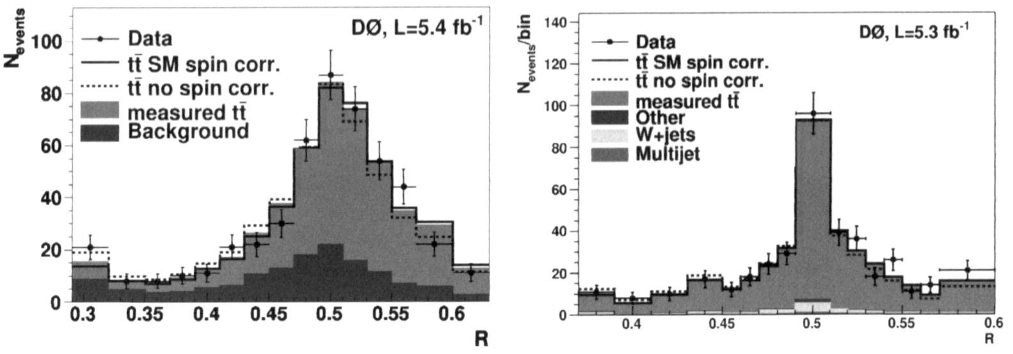

Figure 6. R distributions from the dilepton (left) and lepton+jets (right) channels.

[1] Abazov V M *et al* 2011 *Phys. Rev. Lett.* **107** 121802
CDF Collaborations 2012 *Preprint* CDF Public Note 10887
[2] Fischer M *et al* 2001 *Phys. Rev.* D **63** 031501
[3] Abazov V M *et al* 2011 *Phys. Rev.* D **83** 032009
[4] CDF Collaboration 2012 *Preprint* arXiv:1205.0354 [hep-ex]
[5] Berends F A, Kuijf H, Tausk B and Giele W T 1991 *Nucl.Phys.* B **357** 32-64
[6] CDF Collaboration 2012 *Preprint* arXiv:1211.4523 [hep-ex]
[7] Aaltonen T *et al* 2012 *Phys. Rev.* D **85** 071106
[8] Willenbrock S and Stelzer T 1996 *Phys. Lett.* B **374** 169-172
[9] Czarnecki A, Jezabek M and Kuhn H 1991 *Nucl. Phys.* B **351** 70
[10] D0 Collaboration 2012 *Preprint* arXiv:1207.0364 [hep-ex]
[11] CDF Collaboration 2012 *Preprint* CDF Public Note 10719
[12] CDF Collaboration 2010 *Preprint* CDF Public Note 10211
[13] Abazov V M *et al* 2012 *Phys. Rev. Lett.* **108** 032004
[14] Abazov V M *et al* 2011 *Phys. Rev. Lett.* **107** 032001
[15] Melnikov K and Schulze M 2011 *Phys. Lett.* B **700** 17-20

Journal of Physics: Conference Series **452** (2013) 012002 doi:10.1088/1742-6596/452/1/012002

The role of charge and spin asymmetries in profiling $Z' \to t\bar{t}$ events

L. Basso[1], **K. Mimasu**[2] **and S. Moretti**[2,3]

[1]Physikalisches Institut, Albert-Ludwigs-Universität Freiburg
D-79104 Freiburg, Germany
[2]School of Physics & Astronomy, University of Southampton,
Highfield, Southampton, SO17 1BJ, UK
[3]Particle Physics Department
Rutherford Appleton Laboratory
Chilton, Didcot, Oxon OX11 0QX, UK

E-mail: `lorenzo.basso@physik.uni-freiburg.de`

E-mail: `k.mimasu@soton.ac.uk`

E-mail: `s.moretti@soton.ac.uk`

Abstract. We study the sensitivity of top pair production at the Large Hadron Collider (LHC) to the nature of an underlying Z' boson, including full tree level standard model background effects and interferences while assuming realistic final state reconstruction efficiencies. We demonstrate that exploiting asymmetry observables represents a promising way to distinguish between a selection of benchmark Z' models due to their unique dependences on the chiral couplings of the new gauge boson.

1. Introduction

Z' bosons are a ubiquitous feature of theories beyond the Standard Model (SM) arising from various BSM scenarios such as $U(1)$ gauge extensions of the SM motivated by supersymmetry or grand unified theories, Kaluza-Klein excitations of SM gauge fields or excitations of composite exotic vector mesons in technicolor theories to name a few.

Typically, such resonances are searched for at hadron colliders via the Drell-Yan (DY) production of a lepton pair, i.e., $pp(\bar{p}) \to (\gamma, Z, Z') \to \ell^+\ell^-$, where $\ell = e, \mu$. The Tevatron places limits on the Z' mass, $M_{Z'}$, at around 1 TeV [1] (for a sequential Z') while the latest LHC limits lie around 2.3 TeV [2] from this channel. Several phenomenological studies on how to measure the Z' properties and couplings to SM particles in this clean DY channel have been performed.

These proceedings summarise a recently published paper [3] addressing the use of the top-antitop final state, i.e., $pp(\bar{p}) \to (\gamma, Z, Z') \to t\bar{t}$, to probe these Z' properties. While it may not have as much 'discovery' scope as the DY channel, owing to the large QCD background combined with the complex six-body final state and the associated poor reconstruction efficiency, it remains important to extract the couplings of new physics to the top quark. Furthermore, the fact that the top decays before hadronising, transmitting spin information to its decay products, allows for the definition of spin asymmetry observables which provide an extra handle on Z' couplings not present in non-decaying final states.

We study the scope of the LHC to profile a Z' boson mediating $t\bar{t}$ production, in both standard kinematic variables as well as spatial/spin asymmetries, by adopting some benchmark scenarios for several realisations of the sequential, Left-Right symmetric and E_6 based Z' models (specifically, the same as those in [4]). Specifically, the issue of distinguishability of various models using these observables is addressed.

2. Asymmetries and Z' couplings

We define the asymmetry observables considered with the aim of determining their power to discriminate between Z's. We refer the reader to our paper for a more detailed discussion on these as well as the selection of benchmark models, statistical uncertainties and definitions of significance. This study investigated charge (spatial) and spin asymmetries and their dependence on top couplings to profile and distinguish the models considered.

A selection of charge asymmetry variables were investigated with the most sensitive found to be A_{RFB}, defined by the rapidity difference of the top and antitop, $\Delta y = |y_t| - |y_{\bar{t}}|$, while also cutting on the boost of the $t\bar{t}$ system. This increases the contribution from the $q\bar{q}$ initial state by probing regions of higher partonic momentum fraction, x, where its parton luminosity is more important:

$$A_{RFB} = \left.\frac{N(\Delta y > 0) - N(\Delta y < 0)}{N(\Delta y > 0) + N(\Delta y < 0)}\right|_{|y_{t\bar{t}}|>|y_{t\bar{t}}^{cut}|} . \tag{1}$$

The two spin asymmetries considered, termed double (LL) and single (L), are defined as follows:

$$A_{LL} = \frac{N(+,+) + N(-,-) - N(+,-) - N(-,+)}{N_{Total}} \quad ; \quad A_L = \frac{N(-,-) + N(-,+) - N(+,+) - N(+,-)}{N_{Total}} \tag{2}$$

where N denotes the number of observed events and its first(second) argument corresponds to the helicity of the top (anti)quark. These observables are alternatively known as the spin correlation and spin polarisation asymmetries respectively and can be extracted as coefficients in differential angular distributions of the top decay products.

The dependence of the asymmetries on the vector and axial couplings g_V and g_A or alternatively, the left and right handed couplings g_L and g_R of a generic neutral current are shown in [3]. The charge asymmetry depends on the product of the vector and axial couplings of both the initial and final state particle and can only be generated when all of these are non-vanishing. This is equivalent to measuring the relative magnitudes of their right and left handed couplings of the Z'. For the spin asymmetries, A_{LL} depends on the couplings in a similar way to the total cross section, while A_L is only non-vanishing for non-zero vector and axial couplings of the final state tops. It is additionally sensitive to their relative sign i.e. it measures their relative handedness as with the charge asymmetry but only for the final state.

With these unique coupling dependences, it is our aim to show that asymmetries can provide extra information to distinguish Z' models and ultimately contribute to extracting the couplings of an observed neutral resonance.

3. Results

We present a selection of results profiling the spatial and spin asymmetry distributions of the benchmark Z' models compared to the SM including interference effects. The set of benchmarks are split into two categories: those with a vanishing vector or axial coupling (the E_6 models with the 'B-L' generalised left-right symmetric model) are classed as the 'E_6' type while the rest, with both couplings non-zero, are referred to as the 'generalised' models. The variables described in section 2 were computed as a function of the $t\bar{t}$ invariant mass within $\Delta M_{t\bar{t}} = |M_{Z'} - M_{t\bar{t}}| < 500$ GeV and compared to the tree-level SM predictions. The code exploited for our study is based on helicity amplitudes, defined through the HELAS subroutines [8], and built up by means of MadGraph [9]. CTEQ6L1 [10] Parton Distribution Functions (PDFs) were used, with the factorisation/renormalisation scale at $Q = \mu = 2m_t$. VEGAS [11] was used for numerical integration.

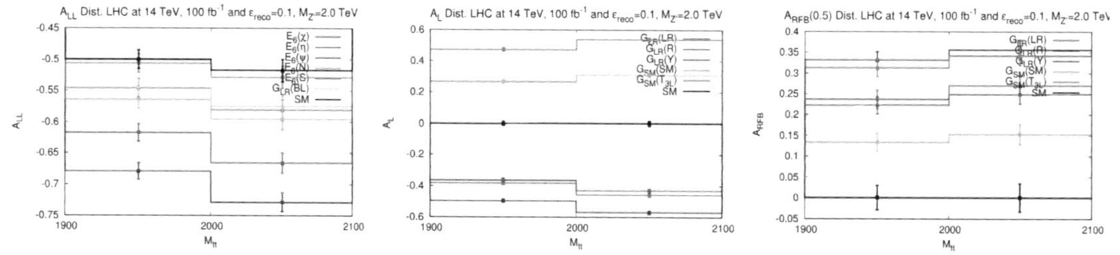

Figure 1. A_{LL} for the E_6-type models and A_L and $A_{RFB}(|y_{t\bar{t}}| > 0.5)$ for the generalised models binned in $M_{t\bar{t}}$ 100 GeV either side of $M_{Z'}$ =2 TeV for the LHC at 14 TeV assuming 100 fb^{-1} of integrated luminosity.

The asymmetry profiles shown in figure 1 include statistical uncertainties and fold in an estimated 10% reconstruction efficiency of the $t\bar{t}$ pair assuming the use of all possible decay channels. The figures show that the majority of benchmark models can be distinguished from one another using these variables, noting the sensitivity of A_L to the relative sign of the vector and axial couplings that allows for a clear distinction between the G_{SM} (sequential) and G_{LR} (left-right symmetric).

A_{LL} depends on the couplings in the same way as the total cross section and therefore models that cannot be distinguished in the invariant mass spectrum will remain so in this observable. It is clear that A_L is the most powerful observable in that it provides the best distinguishing power along with the extra feature of being sensitive to the handedness of the top couplings. A_{RFB} provides some distinction but the handedness sensitivity is not present for reasons discussed in the previous section. Increasing the Z' mass increases the statistical uncertainties but also slightly raises the central values, both as a consequence of the lower SM background. Table 1 shows the significance of A_L between benchmarks assuming 100 fb^{-1} of integrated luminosity. In almost all cases it is effective in disentangling the 'generalised' models with the discrimination decreasing slightly for higher masses. Although not shown in these proceedings, a similar statistical analysis was performed in determining how much integrated luminosity would be required to achieve a significance of 3σ between models using asymmetries. We showed that in most cases, the models can be distinguished at the relatively early stages of the LHC ($\sim$ 100 fb^{-1} at 14 TeV) even for the higher mass of 2.5 TeV. The cases where this is not possible reflect mostly instances where the couplings are too similar.

4. Conclusion

We have presented an overview of a phenomenological study on classes of Z' models in both spin and spatial asymmetries of $t\bar{t}$ production and showed that there is much scope to observe deviations from the SM and even distinguish between various models, particularly for spin asymmetries. This suggests that the $t\bar{t}$ channel may be a useful complement to the DY channel in the aim of profiling a Z' resonance should one be observed in the near future.

Table 1. Significance for A_L values around the Z' peak of generalised models, for the LHC at 14 TeV only. Upper triangle for $M_{Z'} = 2.0$ TeV and lower triangle for $M_{Z'}$=2.5 TeV. Figures refer to $\Delta M_{t\bar{t}} < 100(500)$ GeV.

A_L	SM	$GLR(LR)$	$GLR(R)$	$GLR(Y)$	$GSM(SM)$	$GSM(T_{3L})$
SM	-	31.9(11.1)	40.6(18.3)	30.1(11.2)	22.1(9.8)	38.7(22.5)
$GLR(LR)$	16.9(7.7)	-	10.0(6.6)	2.0(0.1)	62.2(21.7)	81.3(34.5)
$GLR(R)$	21.3(11.5)	4.6(4.0)	-	12.0(6.5)	72.2(30.4)	91.3(44.1)
$GLR(Y)$	16.3(7.8)	1.0(0.1)	5.8(3.9)	-	60.2(21.8)	79.3(34.6)
$GSM(SM)$	11.8(6.3)	33.1(14.8)	38.8(18.8)	33.0(14.9)	-	19.1(13.7)
$GSM(T_{3L})$	20.1(13.9)	42.5(23.0)	48.5(27.2)	42.7(23.2)	9.7(7.8)	-

We note that the benchmarks considered are put forward to set bounds on Z' masses and are best probed in the di-lepton channels. Other models could be better suited to the $t\bar{t}$ channel, such as leptophobic/top-phillic Z's occurring in composite/multi-site and extra-dimensional models. The profiling techniques discussed in this study would be even more applicable in these scenarios.

Finally, although not addressed in the paper to which these proceedings refer, one can ask whether more can be done in profiling an observed resonance with the view of extracting its fermionic couplings. A previous study attempting to do this in the light lepton sector [12] finds a degeneracy in determining the quark and lepton couplings which can be solved by considering asymmetries in an alternate final state. In a more recent paper [13], we show that asymmetry observables in the $t\bar{t}$ provide independent information which would break this degeneracy and allow for a fit to all couplings in the 'minimal' framework of 5 independent couplings used in many benchmarks.

Acknowledgments

The work of KM and SM is partially supported through the NExT Institute. LB is supported by the Deutsche Forschungsgemeinschaft through the Research Training Group grant GRK 1102 *Physics of Hadron Accelerators*. We would like to thank E. Alvarez for pointing out the discussion on systematic uncertainties. KM would additionally like to thank the organisers of TOP2012 for their financial support.

References

[1] T. Aaltonen *et al.* [CDF Collaboration], *Phys. Rev. Lett.* **102** (2009) 091805 and *Phys. Rev. Lett.* **102** (2009) 031801 ; V. M. Abazov *et al.* [DØ Collaboration], **4375-Conf** and **4577-Conf**.
[2] G. Aad *et al.* [ATLAS Collaboration], *Phys. Rev. Lett.* **107** (2011) 272002 and ATLAS-CONF-2012-007; S. Chatrchyan *et al.* [CMS Collaboration], *JHEP* **1105** (2011) 093 and *Phys. Lett.* B **714** (2012) 158.
[3] L. Basso, K. Mimasu and S. Moretti, *JHEP* **1209** (2012) 024.
[4] E. Accomando, A. Belyaev, L. Fedeli, S. F. King and C. Shepherd-Themistocleous, *Phys. Rev.* D **83** (2011) 075012.
[5] R. W. Brown, D. Sahdev and K. O. Mikaelian, *Phys. Rev. Lett.* **43** (1979) 1069; J. H. Kuhn and G. Rodrigo, *Charge asymmetry of heavy quarks at hadron colliders*, *Phys. Rev.* D **59** (1999) 054017.
[6] M. Arai, N. Okada, K. Smolek and V. Simak, *Acta Phys. Polon.* B **40** (2009) 93.
[7] K. Hagiwara and D. Zeppenfeld, *Nucl. Phys.* B **274** (1986) 1.
[8] H. Murayama, I. Watanabe and K. Hagiwara, KEK-91-11.
[9] T. Stelzer and W. F. Long, *Comput. Phys. Commun.* **81** (1994) 357.
[10] J. Pumplin, D. R. Stump, J. Huston, H. L. Lai, P. M. Nadolsky and W. K. Tung, *New generation of parton distributions with uncertainties from global QCD analysis*, *JHEP* **0207** (2002) 012.
[11] preprint CLNS-80/447, March 1980].
[12] F. Petriello and S. Quackenbush, *Measuring Z′ couplings at the CERN LHC*, *Phys. Rev.* D **77** (2008) 115004.
[13] L. Basso, K. Mimasu and S. Moretti, arXiv:1208.0019 [hep-ph].

Top Quark Angular Distributions - Theory

Stephen J. Parke[1]

Theoretical Physics Dept., Fermi National Accelerator Laboratory, Batavia IL 60510, USA

E-mail: parke@fnal.gov

Abstract. The theoretical aspects of top quark angular distributions will be reviewed with special emphasis on spin correlations in top quark pair production.

1. W-boson Helicity fractions in Top Quark decay

In the Standard Model, the W-bosons from top quark decay are primarily longitudinal which imparts particular angular distributions for the decay products of the W-boson, as given by

$$\frac{1}{\Gamma_T}\frac{d\Gamma}{d\cos\chi_l} = \frac{3}{4}\frac{(m_t^2 \sin^2\chi_l \; + \; 2m_W^2(1-\cos\chi_l)^2/2)}{(m_t^2 + 2M_W^2)} \tag{1}$$

where the angle, χ_l is defines in Fig. 1. Thus, the W-boson helicity fractions are

$$\text{W Helicity Fractions} = \begin{cases} \text{longitudinal} & \sim m_t^2 \quad \approx 70\% \\ \text{transverse(L)} & \sim 2m_W^2 \approx 30\% \\ \text{transverse(R)} & \mathcal{O}(m_b^2) \ll 1\%. \end{cases} \tag{2}$$

The W-boson helicity fractions effect the p_T distributions of the decay products and thus these distributions can we used to measure the helicity fractions.

2. Top Quarks decay before Spin Decorrelation

The top quark decay width ($G_F m_t^3 \sim 1$ GeV) is much larger than the QCD hadronization scale ($\Lambda_{QCD} \sim 0.1$ GeV) and much larger than the spin decorrelation scale ($\Lambda_{QCD}^2/m_t \sim 0.1$ MeV). Therefore, spin correlations in top quark pair production are reflected in angular correlations of the decay products, see [1] and [2].

3. Top Quark Pair Production

3.1. Quark-antiquark annihilation or unlike helicity gluon fusion

For top quark pair production via quark-antiquark annihilation or unlike helicity gluon fusion, there exists a spin axis such that the top quarks are produced in only the up-down or down-up configuration, i.e. parallel, since the spin axes are back to back,

$$q_L\bar{q}_R,\; q_L\bar{q}_R,\; g_L g_R,\; g_R g_L \rightarrow t_U\bar{t}_D + t_D\bar{t}_U. \tag{3}$$

t

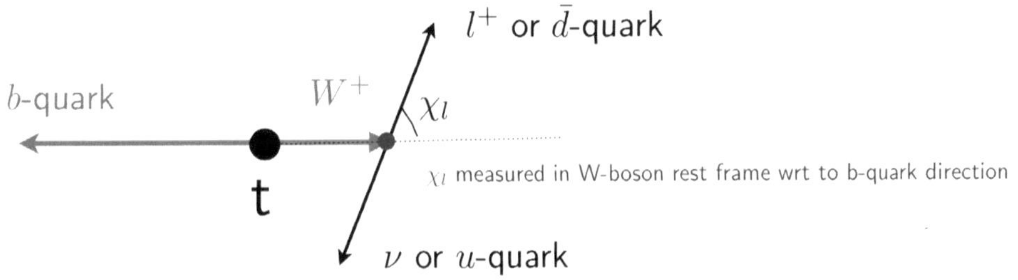

Figure 1. The definition of the χ_l is the W-boson rest frame.

No combinations $t_U\bar{t}_U$ or $t_D\bar{t}_D$ are produced, see Fig. 2. This spin basis is known as the Off-Diagonal basis, see [3] and [4], and the spin axis makes an angle Ω wrt to the top quark momentum direction in the ZM frame. This angle is given by

$$\tan\Omega = (1-\beta^2)\tan\theta = \frac{1}{\gamma^2}\tan\theta. \tag{4}$$

Where the speed and the scattering angle of the top quark are given by β and θ, respectively. Note, at threshold, $\Omega = \theta$ at the spin axis is aligned along the beamline, whereas at ultra-high energies, $\Omega = 0$ and the spin axis is aligned along the direction of motion of the top quark (ZMF helicity).

3.2. Like helicity gluon fusion

For top quark pair production via like helicity gluon fusion in the helicity basis, the top quarks are produced in only the left-left or right-right configuration, i.e. anti-parallel,

$$g_Lg_L,\; g_Rg_R \to t_L\bar{t}_L + t_R\bar{t}_R. \tag{5}$$

No combinations $t_L\bar{t}_R$ or $t_R\bar{t}_L$ are produced in this process, see Fig. 3.

4. Polarized top quark decay

The decay products from a polarized top quark have their moment vectors correlated with the top quark spin axis as follows:

$$\frac{1}{\Gamma_T}\frac{d\Gamma}{d\cos\chi_i} = (1+\alpha_i\cos\chi_i)/2, \quad \text{where} \quad \alpha_i = \begin{cases} +1.0 & (+0.998) & l^+ \\ +1.0 & (+0.966) & \bar{d}\text{-quark} \\ -0.31 & (-0.314) & \bar{\nu} \\ -0.31 & (-0.317) & u\text{-quark} \\ -0.41 & (-0.393) & b\text{-quark} \end{cases} \tag{6}$$

at LO with the NLO results, see [5], in parenthesis. χ_i is the angle between the decay product and the spin axis in the top quark rest frame.

5. Understanding the Full Correlated Matrix Element Squared

In top quark rest frame, denote the polar and azimuthal angles of the charged lepton from top quark decay wrt to the top spin axis by χ and ϕ respectively. Whereas for the anti-top quark these angles wrt to the anti-top quark spin axis by $\bar{\chi}$ and $\bar{\phi}$ respectively. ϕ and $\bar{\phi}$ are determined

t

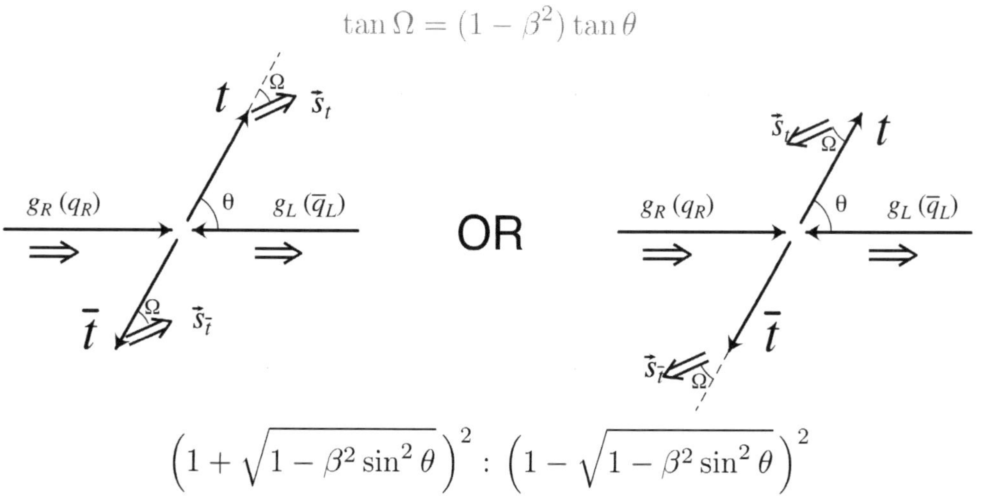

Figure 2. Spin correlations for the production of top quark pairs via unlike helicity gluon fusion or quark-antiquark annihilation.

wrt to the scattering plane. Then for quark-antiquark annihilation, the correlated part of the full matrix element squared, apart from over all factor $(b \cdot \nu)(\bar{b} \cdot \bar{\nu})$, is given by

$$|\mathcal{A}|^2_{LR} + |\mathcal{A}|^2_{RL} \sim [(2 - \beta^2 s^2_\theta)(1 + c_\chi c_{\bar{\chi}}) + \beta^2 s^2_\theta s_\chi s_{\bar{\chi}} \cos(\phi + \bar{\phi})] \tag{7}$$

in the Off-Diagonal bases. Also for unlike helicity gluon fusion

$$|\mathcal{A}|^2_{LR} + |\mathcal{A}|^2_{RL} \sim \frac{(7 + 9\beta^2 c^2_\theta)}{(1 - \beta^2 c^2_\theta)^2} \, \beta^2 s^2_\theta \, [(2 - \beta^2 s^2_\theta)(1 + c_\chi c_{\bar{\chi}}) + \beta^2 s^2_\theta s_\chi s_{\bar{\chi}} \cos(\phi + \bar{\phi})] \tag{8}$$

again in the Off-Diagonal bases. Notice that the part of these two expressions between $[\cdots]$ are identical, so that these two expressions only differ by an overall factor.

Whereas for like helicity gluon fusion in the helicity bases, we have

$$|\mathcal{A}|^2_{LL} + |\mathcal{A}|^2_{RR} \sim \frac{(7 + 9\beta^2 c^2_\theta)}{(1 - \beta^2 c^2_\theta)^2} \, (1 - \beta^2)[(1 + \beta^2)(1 - c_\chi c_{\bar{\chi}}) - (1 - \beta^2) s_\chi s_{\bar{\chi}} \cos(\phi - \bar{\phi})] \tag{9}$$

in the helicity basis. So the full correlation is determined by the angles of the charge leptons (or down type quarks) wrt the spin axis and momentum conservation. The term proportional to $\cos(\phi - \bar{\phi})$ ($\cos(\phi + \bar{\phi})$) is the interference terms between the $t_L \bar{t}_L$ & $t_R \bar{t}_R$ ($t_U \bar{t}_D$ & $t_D \bar{t}_U$) for like helicity gluon fusion (unlike helicity gluon fusion or quark-antiquark annihilation). Eqn (5), (6) and (7) can be easily derived from Eqn (B1), (47) and (48) of ref. [4] respectively.

For the uncorrelated case, just set $c_\chi c_{\bar{\chi}}$ and $\cos(\phi \pm \bar{\phi})$ equal to their uncorrelated average values, zero! The expressions for arbitrary spin axis can be found in the slides presented at the workshop.

b

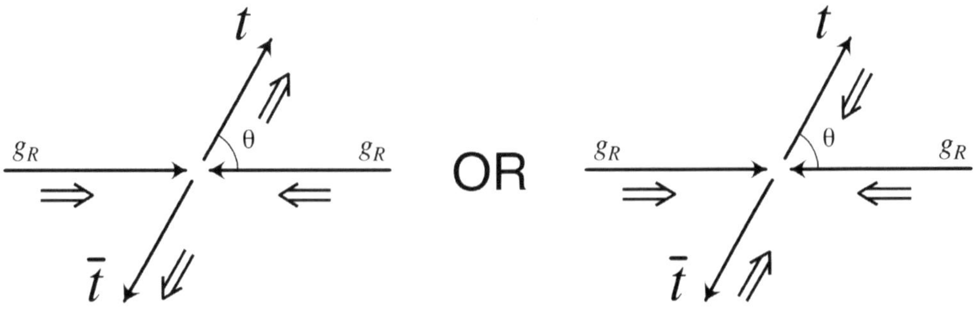

$$(1+\beta)^2 : (1-\beta)^2$$

Figure 3. Spin correlations for the production of top quark pairs via like helicity gluon fusion.

6. Observables

6.1. The Effects of Spin Correlations

The dominant effect of the spin correlations is to correlated the angles of the decay products between the top quark and antitop quark, i.e. between χ_i and $\bar{\chi}_{\bar{i}}$. This correlation is given by

$$\frac{1}{\sigma_T}\frac{d^2\sigma}{d\cos\chi_i \; d\cos\bar{\chi}_{\bar{i}}} = \frac{1}{4}(1 + C_{t\bar{t}}\;\alpha_i\alpha_{\bar{i}}\;\cos\chi_i\cos\bar{\chi}_{\bar{i}}) \tag{10}$$

where

$$C_{t\bar{t}} \equiv \frac{\sigma_{\uparrow\uparrow}+\sigma_{\downarrow\downarrow}-\sigma_{\uparrow\downarrow}-\sigma_{\downarrow\uparrow}}{\sigma_{\uparrow\uparrow}+\sigma_{\downarrow\downarrow}+\sigma_{\uparrow\downarrow}+\sigma_{\downarrow\uparrow}} = \begin{cases} -0.456 & (-0.389) & \text{Helicity at Tevatron} \\ +0.910 & (+0.806) & \text{Beamline at Tevatron} \\ +0.918 & (+0.913) & \text{Off}-\text{Diagonal at Tevatron} \\ +0.305 & (+0.311) & \text{Helicity at LHC(14 TeV)}, \end{cases}$$

at LO with NLO in parenthesis, see [5]. At the LHC, the coefficient $C_{t\bar{t}}$ in the off-diagonal and beamline bases is small, < 0.10.

6.2. Azimuthal Correlations

However, there are interference effects between the various spin components of the top-antitop system, e.g. between $t_L\bar{t}_L$ & $t_R\bar{t}_R$ for like helicity gluon fusion, which leads to azimuthal correlations between the decay products,

$$\frac{1}{\sigma_T}\frac{d\sigma}{d\Delta\phi} = \frac{1}{2}(1 - D\cos\Delta\phi) \tag{11}$$

where in the ZM frame, the azimuthal correlations along the production axis are given by

$$D = \begin{cases} +0.132 & \text{Tevatron} \\ -0.353 & \text{LHC(14 TeV)} \end{cases} \tag{12}$$

from [6]. The azimuthal correlations about the beam axis in the Laboratory frame for the dilepton events are discussed in [4], see Fig. 4. These azimuthal correlations are easier to

t

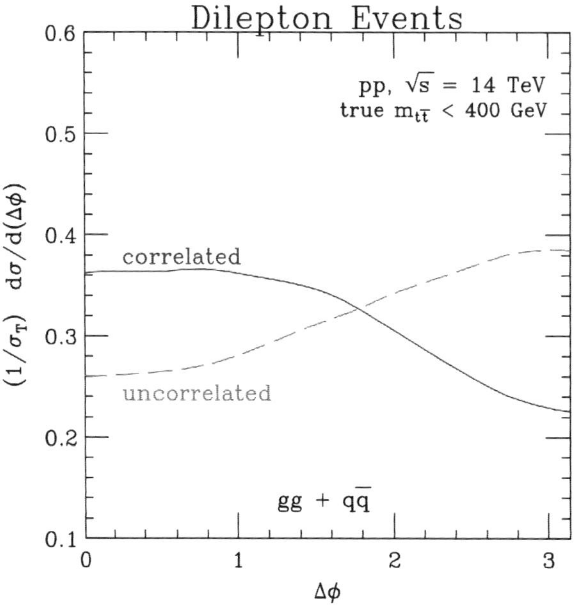

Figure 4. The $\Delta\Phi$ distribution of di-lepton events assuming the true $m_{t\bar{t}} < 400$ GeV.

observe than the other angular correlations. In fact, they have been observed in the last year by both the Atlas [7] and CMS [8] experiments at the LHC. See also relevant talks by ATLAS and CMS in these proceedings.

Acknowledgments

Fermilab is operated by the Fermi Research Alliance under contract no. DE-AC02-07CH11359 with the U.S. Department of Energy. I wish to thank the organizers of Top 2012 for a wonderful, stimulating workshop near the beautiful historic city of Winchester, England.

References

[1] G. Mahlon and S. J. Parke, Phys. Rev. D **53**, 4886 (1996) [hep-ph/9512264].
[2] T. Stelzer and S. Willenbrock, Phys. Lett. B **374**, 169 (1996) [hep-ph/9512292].
[3] S. J. Parke and Y. Shadmi, Phys. Lett. B **387**, 199 (1996) [hep-ph/9606419] and G. Mahlon and S. J. Parke, Phys. Lett. B **411**, 173 (1997) [hep-ph/9706304].
[4] G. Mahlon and S. J. Parke, Phys. Rev. D **81**, 074024 (2010) [arXiv:1001.3422 [hep-ph]].
[5] W. Bernreuther, A. Brandenburg, Z. G. Si and P. Uwer, Phys. Rev. Lett. **87**, 242002 (2001) [hep-ph/0107086].
[6] W. Bernreuther and Z. -G. Si, Nucl. Phys. B **837**, 90 (2010) [arXiv:1003.3926 [hep-ph]].
[7] G. Aad *et al.* [ATLAS Collaboration], Phys. Rev. Lett. **108**, 212001 (2012) [arXiv:1203.4081 [hep-ex]].
[8] [CMS Collaboration], CMS-PAS-TOP-12-004.

Journal of Physics: Conference Series **452** (2013) 012004 doi:10.1088/1742-6596/452/1/012004

Top quark angular distributions at the LHC

Thorsten Wengler on behalf of the ATLAS and CMS Collaborations
CERN, PH-Department, 1211 Geneva 23, Switzerland
E-mail: Thorsten.Wengler@cern.ch

Abstract.
Results from the CMS and ATLAS Collaborations are presented on angular distributions in top quark decays. The studies presented here are based on proton-proton collisions at a center of mass energy of 7 TeV collected by the two experiments at the LHC in 2011. The angular distributions of the top quark decay products are used to probe several fundamental properties of the top quark and its couplings. The polarization of produced top quarks is investigated as well as the correlation of the spin of pair-produced top and anti-top quarks. The polarization states of the W-boson originating from the decay of the top quark are determined and used to constrain possible contributions of new physics processes to the Wtb vertex.

1. Introduction

Since its discovery at the Tevatron in 1995 [1, 2] the top quark[1] has been the object of an intense program of study. The fact that it still provokes so much interest today is due to several reasons. To start with, the top quark is by far the heaviest known elementary particle and hence an ideal probe of the electro-weak symmetry breaking sector. The top quark can also be used to look for signs of new physics both in its production and decay, for example via searches for resonance production or flavor changing neutral currents. In addition the top quark decays before it has time to hadronize, thereby giving access to its properties for example through the angular distributions of its decay products. It is this latter characteristic that will be exploited here to study several fundamental properties of the top quark. Last but not least the decay products of the top quark provide a striking experimental signature. At the LHC, top quarks are mainly produced in pairs. They decay almost exclusively to a W boson and a bottom quark. $t\bar{t}$ events are usually classified according to the decay mode of the two produced W bosons, which may decay either hadronically or into a charged lepton and a neutrino. The studies by the ATLAS [3] and CMS [4] Collaborations presented here make use of the single lepton and dilepton topologies, considering only electrons and muons, including those from leptonic τ decays. The data used have been collected from proton-proton collisions at a center of mass energy of 7 TeV during 2011.

2. W boson polarization

In the Standard Model, the top quark nearly always decays via a Wtb vertex defined by the electroweak interaction. The W boson is produced as a real particle in top decays and can hence assume longitudinal, left-handed or right-handed polarization states. The fractions of W bosons

[1] Unless explicitly stated, top quark refers to both the top and the anti-top quark.

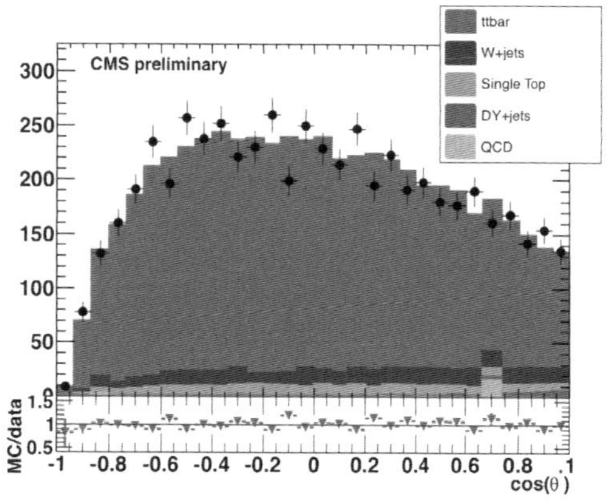

Figure 1. The CMS [6] $\cos\theta^\star$ distribution, from where the W helicity fractions are obtained.

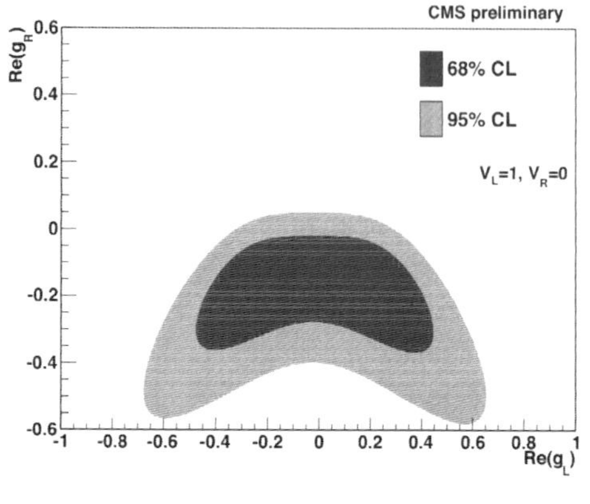

Figure 2. CMS [6] limits on the real components of the anomalous couplings g_L, g_R at 68% and 95% confidence level.

produced in each state are predicted in next-to-next-to-leading-order (NNLO) calculations as $F_0 = 0.687 \pm 0.005$, $F_\mathrm{L} = 0.311 \pm 0.005$ and $F_\mathrm{R} = 0.0017 \pm 0.0001$ [5]. Experimentally the helicity fractions F_0, F_R, and F_L can be determined from the angular distribution of the top decay products. In particular, ATLAS and CMS use the distribution of the angle $\theta^\star$ between the charged lepton in the W boson rest frame and the W boson momentum in the top quark rest frame, which can be expressed in terms of the helicity fractions as [5]:

$$\frac{1}{\sigma}\frac{\mathrm{d}\sigma}{\mathrm{d}\cos\theta^\star} = \frac{3}{4}(1-\cos^2\theta^\star)F_0 + \frac{3}{8}(1-\cos\theta^\star)^2 F_\mathrm{L} + \frac{3}{8}(1+\cos\theta^\star)^2 F_\mathrm{R}.$$

CMS [6] has determined the W boson helicity fractions in the single lepton muon channel using a data sample of 2.2 fb^{-1}. The $\cos\theta^\star$ distribution obtained from the data is shown in Figure 1, compared to the sum of simulated signal and background processes, normalized to the data luminosity. The $t\bar{t}$ sample is additionally scaled by 10%. The main result is obtained by a template fit to the $\cos\theta^\star$ distribution using the three free parameters F_0, F_L, and $F_{t\bar{t}}$ (F_R is constrained to be $1 - F_0 - F_\mathrm{L}$). $F_{t\bar{t}}$ is a normalization factor of the $t\bar{t}$ component in the number of expected events entering the likelihood function. The helicity fractions obtained in this way are $F_0 = 0.567 \pm 0.074(\mathrm{stat.}) \pm 0.047(\mathrm{syst.})$, $F_\mathrm{L} = 0.393 \pm 0.045(\mathrm{stat.}) \pm 0.029(\mathrm{syst.})$ and $F_\mathrm{R} = 0.040 \pm 0.035(\mathrm{stat.}) \pm 0.044(\mathrm{syst.})$, in agreement with the Standard Model prediction. CMS has performed a further fit also constraining F_R to be zero, which leads to consistent results. The measurement of the helicity fractions can be used to set limits on new physics contributions to the Wtb vertex. Figure 2 shows the derived limits on the anomalous couplings g_L and g_R at 68% and 95% confidence level. ATLAS [7] has studied both electron and muon signatures in the single lepton and dilepton channel in a 1.04 fb^{-1} data sample to determine the helicity fractions. Both a template fit to the observed $\cos\theta^\star$ distributions and angular asymmetries as calculated from the $\cos\theta^\star$ distributions unfolded for detector effects are used. The asymmetries $A_\pm$ are defined as the difference in numbers of events below and above a cutoff $z = \pm(1 - 2^{2/3})$ in the $\cos\theta^\star$ distribution, normalized to the sum. The results for both methods are shown in Figure 3 in comparison to the NNLO QCD theory calculations [5]. Combining the separate results yields the main result of $F_0 = 0.67 \pm 0.03(\mathrm{stat.}) \pm 0.06(\mathrm{syst.})$, $F_\mathrm{L} = 0.32 \pm 0.02(\mathrm{stat.}) \pm 0.03(\mathrm{syst.})$ and $F_\mathrm{R} = 0.01 \pm 0.01(\mathrm{stat.}) \pm 0.04(\mathrm{syst.})$. Consistent results are obtained when fixing $F_\mathrm{R} = 0$. The limits on the real parts of the Wtb vertex anomalous couplings g_L and g_R obtained by ATLAS using the helicity fractions are shown in Figure 4. The second minimum around $\mathrm{Re}(g_\mathrm{R}) = 0.8$

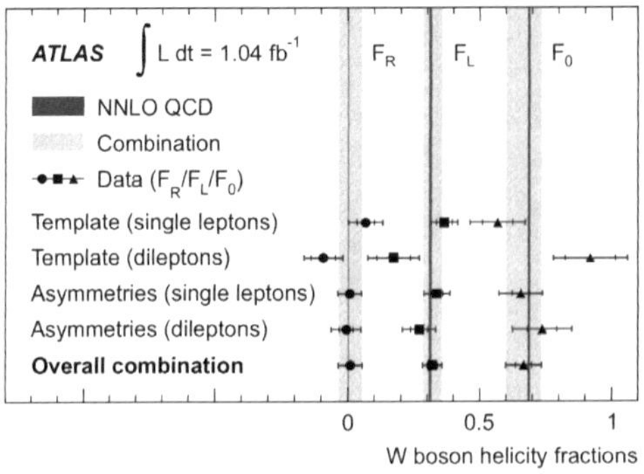

Figure 3. Overview of the four ATLAS [7] measurements of the W boson helicity fractions and the combined values. The error bars correspond to the statistical and total uncertainties.

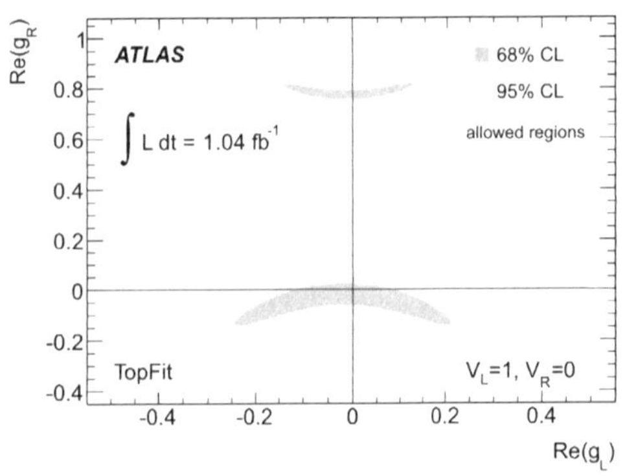

Figure 4. Allowed regions at 68% and 95% confidence level for the Wtb anomalous couplings $g_{\rm L}$ and $g_{\rm R}$ obtained by ATLAS [7].

(also present, but not shown, in the CMS analysis) is not consistent with measurements of the single top cross section.

3. $t\bar{t}$ spin correlations

The polarization of pair-produced t and $\bar{t}$ quarks is predicted to be very small, however their spins are predicted to be correlated. Since the top quarks decay before they can hadronize, this correlation is passed on to the decay products. For the studies presented here the correlation in azimuthal angle, $\Delta\phi$, between the two charged leptons in the dilepton channel is used to determine the amount of spin correlation present, as suggested in [8]. To extract the spin correlation from the $\Delta\phi$ distribution, a template fit is employed, using templates of correlated

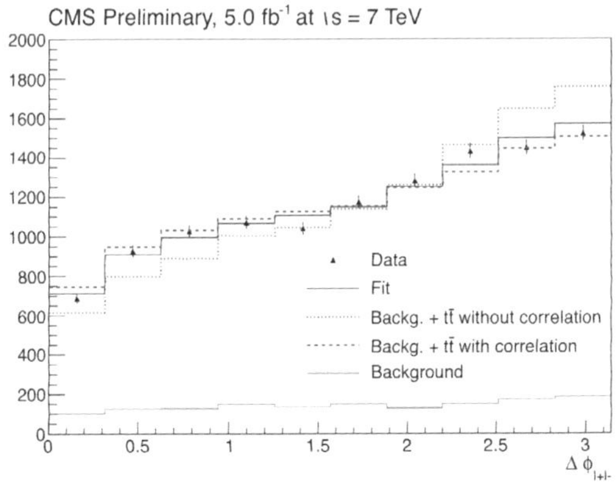

Figure 5. CMS [9] result of the spin correlation fit (solid line) performed on data (triangles) after the combination of the three channels. The data are also compared to the $\Delta\phi_{l^+l^-}$ distribution of $t\bar{t}$ events with and without spin correlation.

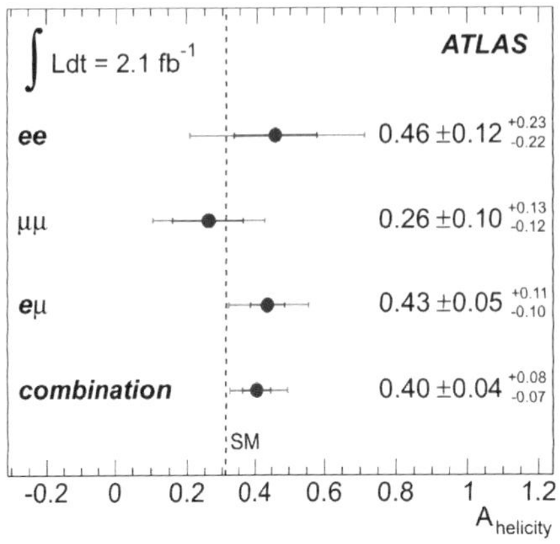

Figure 6. Value of $A_{\rm helicity}$ compared to the Standard Model prediction for the ee, $\mu\mu$ and $e\mu$ channels and the combination, as obtained by ATLAS [11].

and uncorrelated samples, as well as background contributions. CMS [9] has analyzed a data sample of 5 fb^{-1} in the dilepton channel. Figure 5 shows the $\Delta\phi$ distribution after combining the three dilepton channels, compared to the simulated sample with and without spin correlation. The data clearly prefer the slope with spin correlations, which is also evident in the fit result shown in the figure. Translating the fit result into a measured asymmetry in the helicity spin basis yields $A_{\rm hel}^{\rm meas} = 0.24 \pm 0.02(\rm stat.) \pm 0.08(\rm syst.)$, in agreement with the Standard Model value of $A_{\rm hel}^{\rm SM} = 0.31$ [10]. CMS [9] also studied asymmetrics of numbers of events above and below certain cut-offs in angular distributions sensitive to spin correlation effects. These asymmetries, investigated before and after unfolding for detector effects, yield results consistent with those obtained from the template fit. ATLAS [11] has studied a data sample of 2.1 fb^{-1} in the dilepton channel. The template fit is performed for the ee, $\mu\mu$ and $e\mu$ channels separately, and for all channels combined. The results are shown in Figure 6 and compared to the Standard Model value. The combined fit yields an asymmetry in the helicity spin basis of $A_{\rm helicity} = 0.40 \pm 0.04(\rm stat.)\ ^{+0.08}_{-0.07}(\rm syst.)$. The hypothesis of zero spin correlation is excluded by the ATLAS result with a significance of 5.1 standard deviations, amounting to the first observation of $t\bar{t}$ spin correlations.

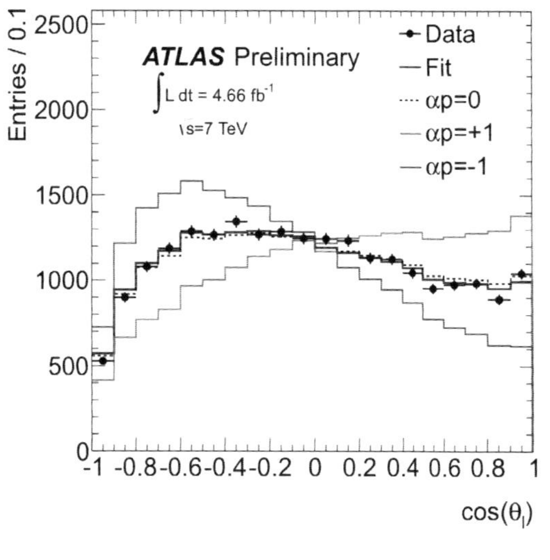

Figure 7. ATLAS [12] result of the top polarization fit to the electron channel compared to the data; the maximally polarized templates and the no polarization expectation are also shown.

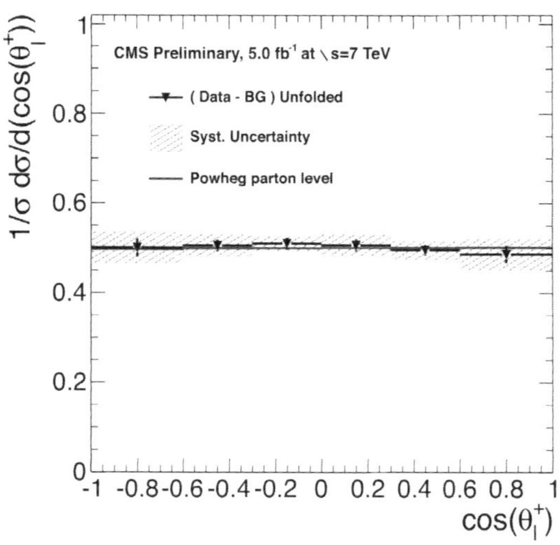

Figure 8. CMS [13] background-subtracted and unfolded $\cos\theta_l^+$ distribution used to determine the top quark polarization.

4. Top polarization

The polarization of t and $\bar{t}$ quarks can be obtained from the angular distribution of the decay products. The results presented here use leptonic decays of top quarks in the single lepton or dilepton decay channels of $t\bar{t}$ events. In particular the distribution of the angle θ_l between the lepton direction in the top rest frame and the top direction in the $t\bar{t}$ centre of mass frame is obtained through performing a full event reconstruction. ATLAS [12] performs a template fit to the $\cos\theta_l$ distribution, using fully positively and negatively polarized signal templates added to background expectations. The data set is obtained from a sample of 4.66 fb^{-1} in the single lepton channel. Figure 7 shows the $\cos\theta_l$ distribution in the electron channel together with the maximally polarized fit templates, the no polarization case, and the fit result. The combined result from electron and muon channels yields a degree of polarization of $-0.060 \pm 0.018(\text{stat.})\ ^{+0.046}_{-0.064}(\text{syst.})$, consistent with no polarization and the Standard Model expectation. CMS [13] studies the $\cos\theta_l$ distribution in dileptonic $t\bar{t}$ decays in a data sample of 5.0 fb^{-1}. The reconstructed distribution is unfolded for detector effects and compared to the Standard Model expectation in Figure 8 for positively charged leptons. The final measured polarization value is $-0.009 \pm 0.029(\text{stat.}) \pm 0.041(\text{syst.})$, again consistent with no polarization and the Standard Model expectation.

5. Conclusion

The ATLAS and CMS Collaborations have used angular distributions of leptons from the single lepton and dilepton decay channels of $t\bar{t}$ events to investigate several fundamental properties of the top quark. The data used in these studies have been collected from pp collisions at a center of mass energy of 7 TeV at the LHC. The structure of the Wtb vertex has been investigated via a determination of the helicity fractions of the W boson produced in the decay of the top quark. The polarization of the produced top quarks has been investigated as well as the spin correlation of top and anti-top quark for produced $t\bar{t}$ pairs. All obtained results are consistent with the Standard Model expectations.

References

[1] D0 Collaboration, Abachi S *et al.* 1995 *Phys. Rev. Lett.* **74** 2422.
[2] CDF Collaboration, Abe F *et al.* 1995 *Phys. Rev. Lett.* **74** 2626.
[3] ATLAS Collaboration 2008 *JINST* **3** S08003.
[4] CMS Collaboration 2008 *JINST* **3** S08004.
[5] Czarnecki A, Korner J G and Piclum J H 2010 *Phys. Rev.* D **81** 111503.
[6] CMS Collaboration 2012 Measurement of the W boson polarization in semileptonic top pair decays with the CMS detector at the LHC CMS-PAS-TOP-11-020 http://cds.cern.ch/record/1423370.
[7] ATLAS Collaboration 2012 *JHEP* **1206** 088.
[8] Mahlon G and Parke S 2010 *Phys. Rev.* D **81** 074024.
[9] CMS Collaboration 2012 Measurement of Spin Correlations in $t\bar{t}$ events in the dilepton channels in pp collisions at $\sqrt{s} = 7$ TeV CMS-PAS-TOP-12-004 http://cdsweb.cern.ch/record/1461788.
[10] Bernreuther W and Si Z G 2010 *Nucl. Phys.* B **837** 90.
[11] ATLAS Collaboration 2012 *Phys. Rev. Lett.* **108** 212001.
[12] ATLAS Collaboration 2012 Measurement of top quark polarisation in $t\bar{t}$ events with the ATLAS detector in proton-proton collisions at $\sqrt{s} = 7$ TeV ATLAS-CONF-2012-133 http://cds.cern.ch/record/1478373.
[13] CMS Collaboration 2012 Measurement of the top polarization in the dilepton final state CMS-PAS-TOP-12-016 http://cds.cern.ch/record/1460102.

5th International Workshop on Top Quark Physics (TOP2012) IOP Publishing
Journal of Physics: Conference Series **452** (2013) 012005 doi:10.1088/1742-6596/452/1/012005

Experimental interplay between the top quark and Supersymmetry at the LHC and Tevatron

Jean-François Arguin
On behalf of the ATLAS, CDF, CMS and DØ Collaborations
Département de physique, Université de Montréal, C.P. 6128, succ. centre-ville, Montréal (Québec), H3C 3J7, Canada

E-mail: arguin@lps.umontreal.ca

Abstract. The search for Supersymmetry is of highest priority for the physics programs of the LHC and Tevatron experiments. The top quark plays an important role in this quest, both as a dominant background and as a key signature for the signal, in particular in the decay of third generation squarks. We review the latest status of the interplay of top quark and SUSY at hadron colliders.

1. Introduction

Supersymmetry (SUSY) [1–9] is often considered the favorite extension of the Standard Model (SM). It solves several of its limitations such as fixing the fine-tuning problem of the Higgs boson, providing a viable dark matter candidate and unifying the coupling constants at the Grand Unification scale. Supersymmetry implies that each SM particle will have one or two[1] supersymmetric partners, with a spin differing from the associated SM particle by one half. For example the top quark will have two scalar top "squark" partners ($\tilde{t}_1$ and $\tilde{t}_2$).

Supersymmetry must be broken since the SUSY partners have not been observed to date. A subset of them need to be light to solve the fine-tuning problems of the SM Higgs and thus are expected to be accessible by the Tevatron and/or LHC experiments. In most searches R-parity conservation is assumed, which implies that the number of SUSY particles is conserved. The typical SUSY phenomenology involves strongly produced particles, such as gluinos ($\tilde{g}$, the partner of the gluon) and squarks, which produce a decay chain involving several jets and leptons of high transverse momentum (p_{T}), and missing transverse energy ($E_{\mathrm{T}}^{\mathrm{miss}}$) due to the lightest SUSY particle (LSP), which must be stable by R-parity conservation.

The top quark plays an important role in the search for SUSY at hadron colliders for two main reasons. Firstly its experimental signature is similar to the ones described above for SUSY. It typically constitutes a dominant background for these searches. Secondly its SUSY partners, the two top squarks, are required to be two of the lightest SUSY particles in order to solve the hierarchy problem. In this document we will explore these two aspects of the interplay between top quark and SUSY in more details. The presented analyses utilize the ATLAS [10], CDF, CMS [11] and DØ detectors, analysis selections and identification algorithms which are described in detail elsewhere in these TOP2012 workshop proceedings.

[1] Fermions have two scalar partners to conserve the number of degrees of freedom.

Journal of Physics: Conference Series **452** (2013) 012005 doi:10.1088/1742-6596/452/1/012005

2. The top quark as a background to SUSY searches

The pair production of top quarks tends to be a dominant background to SUSY searches because of a combination of factors. Firstly the top quark is the most massive SM particle, and thus closest in energy scale to the new particles that are presumably more massive. Secondly its pair production has a high cross-section since it is mediated by the strong interaction, e.g. $\sigma_{t\bar{t}} \approx 165$ pb at $\sqrt{s}$=7 TeV. Finally the decay products involve several high-p_{T} jets, lepton(s) and $E_{\mathrm{T}}^{\mathrm{miss}}$, and can thus mimic the typical decay chains from SUSY particles. In fact it constitutes the single largest background for several inclusive searches for strongly produced SUSY (a few recent examples can be found in Ref. [12–20]), in particular those involving leptons and/or b-jets.

The details of the method used to estimate the $t\bar{t}$ background are analysis-dependent but a few general principles apply. The estimates are purely or partially data-driven because the SUSY background consists of events in the high-p_{T} tails of the $t\bar{t}$ events that suffer from larger theoretical, experimental and MC statistical uncertainties than typical $t\bar{t}$ events. A top-enriched control region is employed, which, on one hand, is designed to be as free as possible of SUSY signal and other SM backgrounds. On the other hand it is devised to be kinematically as close as possible to the signal region(s) in order to reduce extrapolation uncertainties. An extrapolation to the signal region is then performed to predict the $t\bar{t}$ background, either to correct the MC prediction for the signal region, or by using only data for purely data-driven background estimates.

One example is a CMS analysis that uses a lepton p_{T} spectrum method to predict the semi-leptonic $t\bar{t}$ background in an inclusive SUSY search with one high-p_{T} lepton, for which the details of the analysis are provided in Ref. [13]. This method is almost entirely data-driven. It employs the fact that, in a $t\bar{t}$ event, the charged lepton p_{T} spectrum can be used to predict the neutrino momentum, and thus $E_{\mathrm{T}}^{\mathrm{miss}}$. Corrections need to be applied to account for the slightly different $l^{\pm}$ and ν spectrum from W boson decays (due to W polarization), the lepton p_{T} threshold in the selections and the different lepton p_{T} and $E_{\mathrm{T}}^{\mathrm{miss}}$ resolutions. Excellent agreement is obtained between the data and predicted $E_{\mathrm{T}}^{\mathrm{miss}}$ spectra as shown in Fig. 1.

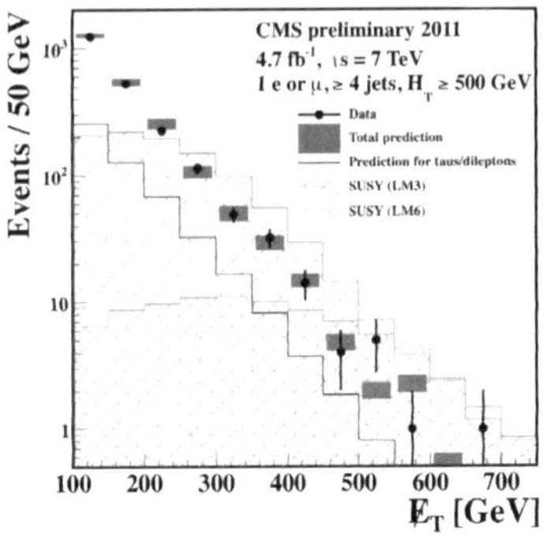

Figure 1. Data and predicted $E_{\mathrm{T}}^{\mathrm{miss}}$ distribution for the $t\bar{t}$ background of the CMS 1-lepton SUSY search described in Ref. [13].

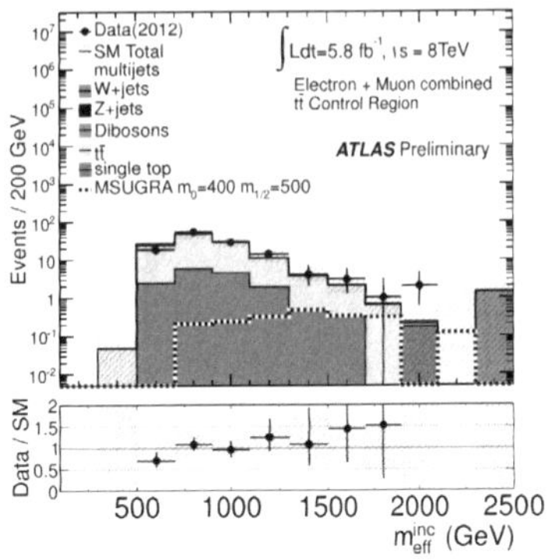

Figure 2. The $t\bar{t}$ control region for the ATLAS 1-lepton SUSY search described in Ref. [18].

In general the ATLAS collaboration uses a transfer factor method for estimating its $t\bar{t}$ background in SUSY searches. The signal region prediction is obtained by using a MC-based transfer factor translating the number of events from a top-enriched control region to the signal region. This transfer factor is applied to the number of events in the control region in data which is corrected for the number of background events. One example is a recent search requiring exactly one isolated lepton. The details of the analysis are provided in Ref. [18]. The signal region requires at least four jets with $p_T > 80$ GeV, a muon or an electron with $p_T > 25$ GeV, the transverse mass between the lepton and E_T^{miss}, m_T, greater than 100 GeV, as well as $E_T^{miss} > 250$ GeV. Also the inclusive transverse mass m_{eff}^{inc}, defined as the total transverse energy of the selected objects and of all jets with $p_T > 40$ GeV, is required to be greater than 800 GeV. The $t\bar{t}$ control region utilizes the same selections except that at least one jet is required to be b-tagged, $100 < E_T^{miss} < 180$ GeV and $500 < m_{eff}^{inc} < 1300$ GeV. The distribution of m_{eff}^{inc} for the $t\bar{t}$ control region is shown in Fig. 2 before applying the m_{eff}^{inc} cut. The region is pure in $t\bar{t}$ events and the signal contamination, illustrated by a low-mass SUSY point, is negligible.

3. Searches for third generation squarks

The ATLAS and CMS collaborations have increased their focus on more specific signatures as inclusive searches for SUSY have yet to uncover a signal. Only a few SUSY particles need to be relatively light for SUSY to cancel the large corrections to the Higgs boson mass [21, 22], and thus be "natural". At the one loop level, only $\tilde{t}_1$, $\tilde{t}_2$ and $\tilde{b}_1$ are required to be less than a TeV or so. The gluino, which appears at the two loop level in the Higgs boson mass corrections, should also be relatively light. A natural decay chain for SUSY thus involves the pair production of $\tilde{g}$, each decaying to the lightest top squark $\tilde{t}_1$, as illustrated in Fig. 3. Another example is the direct pair production of $\tilde{t}_1$ which is illustrated in Fig. 4. The inclusive searches discussed in Sec. 2 are generally not optimized to observe these processes. We describe in this section the search for third generation squarks at LHC and the Tevatron. For these proceedings to the TOP2012 workshop, we concentrate on searches with final states involving $\tilde{t}_1$ and top quarks. All quoted limits are at the 95% confidence level.

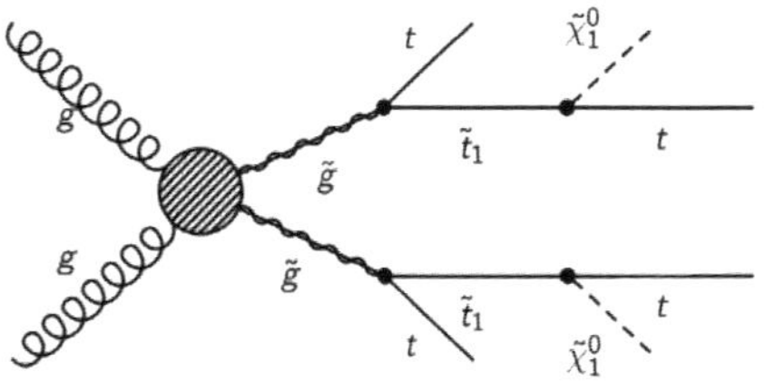

Figure 3. Feynman diagram of pair production of $\tilde{g}$, each decaying to $\tilde{t}_1$ and a top quark.

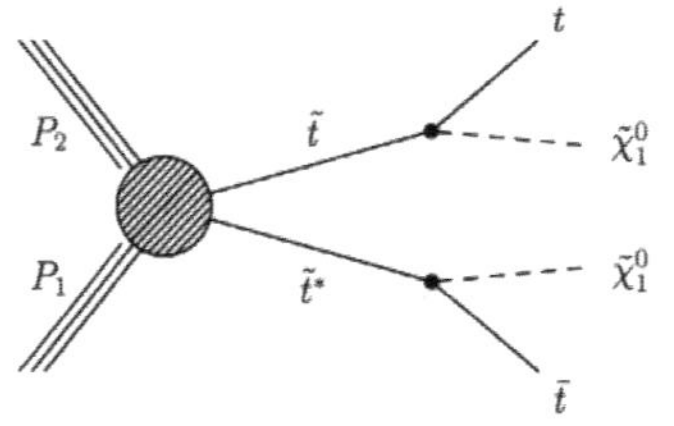

Figure 4. Feynman diagram of the direct production of $\tilde{t}_1$ decaying to a top quark and $\tilde{\chi}_1^0$.

3.1. Searches for the gluino-mediated production of top squarks

The process described in Fig. 3 has a spectacular signature consisting of four top quarks as well as two $\tilde{\chi}_1^0$'s, resulting in high E_T^{miss}. Since each top quark decays approximately 100% of the time to a W boson and a b-quark, each event will contain at least four b-jets, up to 12 jets from decays and up to four leptons from W boson decays. Several searches have been designed and we describe them below in the order defined by the increasing number of isolated leptons required. In this document we only quote limits on a four-top simplified model (FTS), where the $\tilde{t}_1$ is assumed to be heavier than the $\tilde{g}$, in which case each $\tilde{g}$ decays directly to $t\bar{t}$ (via a virtual $\tilde{t}_1$) and $\tilde{\chi}_1^0$. This model has the advantage that the limits depend solely on the $\tilde{g}$ and the $\tilde{\chi}_1^0$ masses. Limits on the full process in Fig. 3 can be found in the references provided. The exclusion limits obtained in the FTS model are summarized in Fig. 5 (CMS) and Fig. 6 (ATLAS).

ATLAS has searched for gluino-mediated top squarks production in events with zero isolated electrons or muons and at least three b-jets in 4.7 fb^{-1} of $\sqrt{s}$=7 TeV data [23]. Several signal regions are devised with varying requirements on the number of jets, E_T^{miss}, b-tagging and m_{eff}, where m_{eff} is defined as the total transverse energy of the selected objects. No excess is found and limits are obtained that span $\tilde{g}$ masses up to 0.94 TeV and $\tilde{\chi}_1^0$ masses up to 300 GeV in the FTS model. ATLAS has also searched for this process in events with zero leptons with no b-jet requirements but with high jet multiplicity requirements, i.e. between six and nine jets with $p_{\text{T}} > 55$ to 80 GeV [24]. The search is performed in 5.8 fb^{-1} of $\sqrt{s}$=8 TeV data and limits are set in the FTS model for $\tilde{g}$ ($\tilde{\chi}_1^0$) mass up to 1.0 (0.36) TeV. CMS performed a search in this channel that uses 3.9 fb^{-1}of $\sqrt{s}$=8 TeV data and exploits the α_T variable [12] to discriminate between events with real and fake $E_{\text{T}}^{\text{miss}}$. At least two jets are required with $p_{\text{T}} > 100$ GeV. Several signal regions are designed with varying requirements on the total transverse energy of selected objects (H_{T}) and the number of b-jets. Limits in the FTS model are obtained that span $\tilde{g}$ ($\tilde{\chi}_1^0$) mass up to 0.85 (0.25) TeV. Several other recent searches in the zero-lepton channel have been performed at CMS and interpreted as searches for gluino-mediated top squarks [25–29].

CMS also performed a search [30] in which one isolated lepton with $p_{\text{T}} > 20$ GeV is required in addition to at least three jets with $p_{\text{T}} > 40$ GeV, $H_{\text{T}} > 350$ GeV and $E_{\text{T}}^{\text{miss}} > 100$ GeV. Limits up to approximately 0.9 TeV and 200 GeV in the $\tilde{g}$ and $\tilde{\chi}_1^0$ mass, respectively, are obtained.

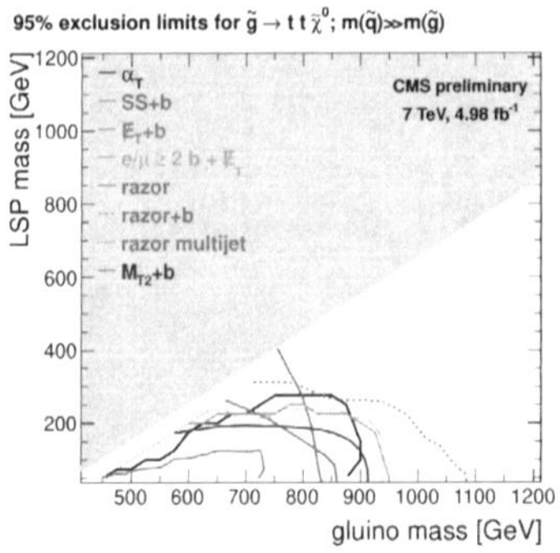

Figure 5. Exclusion limits in the $\tilde{g}$ and $\tilde{\chi}_1^0$ mass plane of the FTS model for CMS [31].

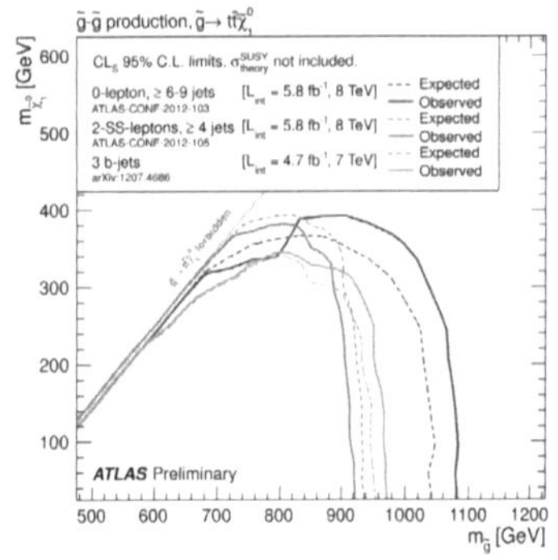

Figure 6. Exclusion limits in the $\tilde{g}$ and $\tilde{\chi}_1^0$ mass plane of the FTS model for ATLAS [32].

Since four top quarks are present, there is a significant probability that two leptons carry the same electric charge in the same event, a signature which suffers little background from SM processes. CMS has performed such a search in 3.95 fb^{-1} of $\sqrt{s}$=8 TeV data by requiring two same-sign leptons with $p_{\text{T}} > 20$ GeV, at least two jets with $p_{\text{T}} > 40$ GeV, of which at least two must be b-tagged [14]. Several signal regions are devised with varying requirements on $E_{\text{T}}^{\text{miss}}$, H_{T}, and the number of jets and b-jets. No excess is observed in data and limits are obtained on the $\tilde{g}$ and $\tilde{\chi}_1^0$ masses up to 0.9 TeV and 450 GeV, respectively. ATLAS has performed a similar search which sets comparable limits on gluino-mediated production of top squarks [33]. ATLAS also performed a search in which at least three isolated leptons ($p_{\text{T}} > 15$ GeV) are required [34].

Figures 5 and 6 summarize the limits in the $\tilde{g}$ and $\tilde{\chi}_1^0$ mass plane of the FTS model of a subset of the analyses presented above for the CMS ($\sqrt{s}$=7 TeV) and ATLAS experiments ($\sqrt{s}$=7 TeV and $\sqrt{s}$=8 TeV), respectively.

3.2. Searches for direct production of top squarks

Searches for the process in Fig. 4 are important to cover the possibility that the $\tilde{g}$ is too massive to be (yet) observable at the LHC. The experimental signature is the same as for $t\bar{t}$ events but with additional $E_{\text{T}}^{\text{miss}}$ due to the presence of two $\tilde{\chi}_1^0$'s.

At the time of this workshop, the CMS collaboration did not have public dedicated searches for this process. However the searches have been re-interpreted for direct stop production and yield limits on the $\tilde{t}_1$ mass between approximately 325 and 475 GeV for light $\tilde{\chi}_1^0$'s, and up to 150 GeV in $\tilde{\chi}_1^0$ [26, 35, 36].

ATLAS has performed a dedicated search in the zero-lepton channel [37], in which at least six high-p_{T} jets are required and $E_{\mathrm{T}}^{\mathrm{miss}} > 150$ GeV. The reconstruction of the two top quarks in the event is attempted and topological cuts are applied to reduce the multi-jets background. No excess is observed and limits are obtained on the $\tilde{t}_1$ mass between 370 and 465 GeV for massless $\tilde{\chi}_1^0$, and up to $m_{\tilde{\chi}_1^0} \approx 50$ GeV. An analysis is also performed in the one-lepton channel [38]. Stringent selections on $E_{\mathrm{T}}^{\mathrm{miss}}$, m_{T} and the reconstructed top quark mass are applied to reject the dominant $t\bar{t}$ background. The limits on $m_{\tilde{t}_1}$ span 230−440 GeV for massless $\tilde{\chi}_1^0$ and reach up to $m_{\tilde{\chi}_1^0} \approx 125$ GeV. Other dedicated searches for $\tilde{t}_1$ have been performed by ATLAS that include the other possible decay channel $\tilde{t}_1 \to b\tilde{\chi}_1^{\pm}$ [39–41]. The limits from ATLAS in the $\tilde{t}_1$ and $\tilde{\chi}_1^0$ mass plane are summarized in Fig. 7.

The top squark has also been searched for at the Tevatron. The searches have focused on low $m_{\tilde{t}_1}$ decays such as $\tilde{t}_1 \to c\tilde{\chi}_1^0$ [42] and $\tilde{t}_1 \to bl\tilde{\nu}$ [43, 44].

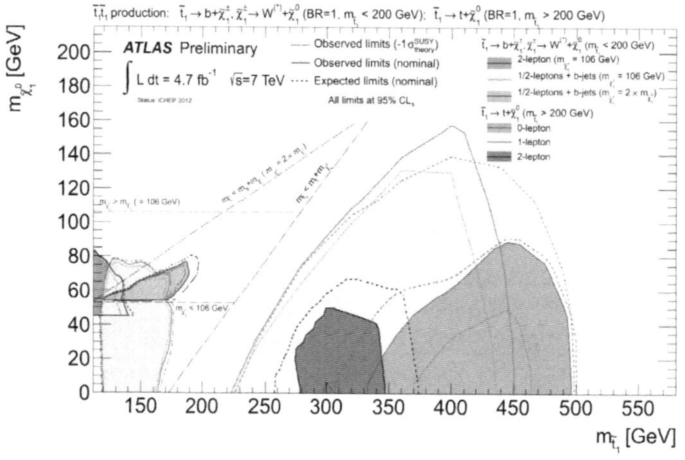

Figure 7. Exclusion limits in the $\tilde{t}_1$ and $\tilde{\chi}_1^0$ mass plane of the FTS model for ATLAS [32].

4. Conclusions

The top quark plays an important role in the search for Supersymmetry at hadron colliders. We have provided an overview of the techniques performed to estimate the $t\bar{t}$ background, which is often dominant. Top quarks can also be part of the signal, in particular as the decay product of third generation squarks. We have summarized the searches for the top squark. Increased luminosity and $\sqrt{s}$ will substantially improve the sensitivity of future searches.

References

[1] Miyazawa H 1966 *Prog. Theor. Phys.* **36 (6)** 1266–1276

[2] Ramond P 1971 *Phys. Rev.* **D3** 2415–2418

[3] Golfand Y A and Likhtman E P 1971 *JETP Lett.* **13** 323326 [Pisma Zh.Eksp.Teor.Fiz.13: 452-455,1971]

[4] Neveu A and Schwarz J H 1971 *Nucl. Phys.* **B31** 86–112

[5] Neveu A and Schwarz J H 1971 *Phys. Rev.* **D4** 1109–1111
[6] Gervais J and Sakita B 1971 *Nucl. Phys.* **B34** 632–639
[7] Volkov D V and Akulov V P 1973 *Phys. Lett.* **B46** 109–110
[8] Wess J and Zumino B 1974 *Phys. Lett.* **B49** 52
[9] Wess J and Zumino B 1974 *Nucl. Phys.* **B70** 39–50
[10] ATLAS Collaboration, 2008 *JINST* **3** S08003
[11] CMS Collaboration, 2008 *JINST* **3** S08004
[12] CMS Collaboration, CMS-PAS-SUS-12-016, `https://cdsweb.cern.ch/record/1460095`.
[13] CMS Collaboration, CMS-PAS-SUS-12-010, `https://cds.cern.ch/record/1445275`.
[14] CMS Collaboration, CMS-PAS-SUS-12-017, `http://cdsweb.cern.ch/record/1459811`.
[15] CMS Collaboration, Submitted to *Phys.Lett.* **B** (*Preprint* hep-ex/1206.3949)
[16] ATLAS Collaboration, ATLAS-CONF-2012-109, `http://cds.cern.ch/record/1472710`.
[17] ATLAS Collaboration, 2012 *Phys.Rev.* **D85** 112006 (*Preprint* hep-ex/1206.3949)
[18] ATLAS Collaboration, ATLAS-CONF-2012-104, `http://cds.cern.ch/record/1472673`.
[19] ATLAS Collaboration, ATLAS-CONF-2012-105, `http://cds.cern.ch/record/1472674`.
[20] ATLAS Collaboration, 2012 *Phys.Rev.* **D86** 092002 (*Preprint* hep-ex/1208.4688)
[21] Barbieri R and Giudice G 1988 *Nucl.Phys.* **B306** 63
[22] de Carlos B and Casas J 1993 *Phys.Lett.* **B309** 320–328 (*Preprint* `hep-ph/9303291`)
[23] ATLAS Collaboration, 2012 *Eur.Phys.J.* **C72** 2174 (*Preprint* hep-ex/1207.4686)
[24] ATLAS Collaboration, ATLAS-CONF-2012-103, `https://cds.cern.ch/record/1472672`.
[25] CMS Collaboration, CMS-PAS-SUS-11-024, `https://cds.cern.ch/record/1503581`.
[26] CMS Collaboration, CMS-PAS-SUS-12-005, `https://cds.cern.ch/record/1430715`.
[27] CMS Collaboration, CMS-PAS-SUS-12-009, `https://cds.cern.ch/record/1459812`.
[28] CMS Collaboration, 2012 *JHEP* **1210** 018 (*Preprint* hep-ex/1207.1798)
[29] ATLAS Collaboration, 2012 *Phys.Rev.* **D86** 072010 (*Preprint* hep-ex/1208.4859)
[30] CMS Collaboration, Submitted to *Phys.Rev.* **D** (*Preprint* hep-ex/1211.3143)
[31] CMS Collaboration, SUSY summary plots page,
`https://twiki.cern.ch/twiki/bin/view/CMSPublic/SUSYSMSSummaryPlots`.
[32] ATLAS Collaboration, SUSY public page,
`https://twiki.cern.ch/twiki/bin/view/AtlasPublic/SupersymmetryPublicResults`.
[33] ATLAS Collaboration, ATLAS-CONF-2012-105, `https://cds.cern.ch/record/1472674`.
[34] ATLAS Collaboration, ATLAS-CONF-2012-108, `https://cds.cern.ch/record/1472687`.
[35] CMS Collaboration, CMS-PAS-SUS-11-024, `https://cds.cern.ch/record/1459813`.
[36] CMS Collaboration, Submitted to *JHEP* (*Preprint* hep-ex/1210.8115)
[37] ATLAS Collaboration, 2012 *Phys.Rev.Lett.* **109** 211802 (*Preprint* hep-ex/1208.1447)
[38] ATLAS Collaboration, 2012 *Phys.Rev.Lett.* **109** 211803 (*Preprint* hep-ex/1208.2590)
[39] ATLAS Collaboration, 2012 *JHEP* **11** 094 (*Preprint* hep-ex/1209.4186)
[40] CMS Collaboration, Submitted to *Phys.Lett.* **B** (*Preprint* hep-ex/1209.2102)
[41] ATLAS Collaboration, 2012 *Eur.Phys.J.* **C72** 2237 (*Preprint* hep-ex/1208.4305)
[42] CDF Collaboration, 2012 *JHEP* **1210** 158 (*Preprint* hep-ex/1203.4171)
[43] DØ Collaboration, 2012 *Phys.Lett.* **B710** 578 (*Preprint* hep-ex/1202.1978)
[44] DØ Collaboration, 2011 *Phys.Lett.* **B696** 321 (*Preprint* hep-ex/1009.5950)

Journal of Physics: Conference Series **452** (2013) 012006 doi:10.1088/1742-6596/452/1/012006

Search for Higgs boson production in association with a top quark pair in pp collisions [1]

Andrew Brinkerhoff for the CMS Collaboration
University of Notre Dame
Notre Dame, IN

E-mail: andrewbrink@gmail.com

Abstract. We present a search for the standard model Higgs boson produced in association with a top quark pair in 5 fb^{-1} of 7 TeV pp collision data. We look for events where the Higgs boson decays to $b\bar{b}$ and the top quark pair decays to either lepton plus jets ($t\bar{t} \to \ell\nu jjb\bar{b}$) or dileptons ($t\bar{t} \to \ell\nu\ell\nu b\bar{b}$), where ℓ is an electron or muon. The major background to our signal is top pair production. We use artificial neural networks to discriminate between background and signal events. We perform a simultaneous fit for signal and background fractions using the neural network output distributions. Our expected limit on Higgs boson production cross section times $H \to b\bar{b}$ branching ratio for a Higgs boson mass of 125 GeV/c^2 is 4.6 times the standard model expectation, while the observed limit is 3.8 times the standard model expectation.

1. Introduction

In this note, we present a search for Higgs boson production in the $t\bar{t}H$ channel using data taken in 2011 by the CMS detector [2], where $H \to b\bar{b}$ and the top pair decays semi-leptonically ($t\bar{t} \to \ell\nu jjb\bar{b}$) or fully leptonically ($t\bar{t} \to \ell\nu\ell\nu b\bar{b}$). The $t\bar{t}H$ process is interesting for a number of reasons. The rate of this process depends on the couplings of the Higgs boson to both the top quark and the bottom quark, the two largest couplings in the SM. If a Higgs-like excess is observed in the data, measuring the values of these couplings will help to determine whether the object responsible for the excess has the properties expected from a SM Higgs boson. Because of the large mass of the top quark, $t\bar{t}H$ production represents the only opportunity to directly probe the $t\bar{t}H$ vertex without making assumptions on contributions beyond the SM. In addition, $t\bar{t}H$ production provides a probe of the Higgs boson coupling to the b quark that is complementary to the VH channel, with $t\bar{t}$+jets instead of W+jets providing the dominant background contribution. Should the Higgs boson be discovered, $t\bar{t}H$ production will play an important role in determining its properties.

2. Event reconstruction and selection

This analysis selects events consistent with the production of a top quark pair in association with a Higgs boson, where the top quark pair decays by either the lepton + jets ($t\bar{t} \to \ell\nu jjb\bar{b}$) or dilepton ($t\bar{t} \to \ell\nu\ell\nu b\bar{b}$) modes and the Higgs boson decays to a pair of bottom quarks. For the lepton + jets case, we select events containing an energetic, isolated, charged electron or muon, and at least four energetic jets, two or more of which should be identified as containing a bottom quark hadron (b-tagged). For the dilepton case, we require a pair of oppositely charged energetic leptons (two electrons, two muons or one electron and one muon) and two or more jets, with at least two of the jets being b-tagged. In both channels, a significant amount of missing transverse energy (MET) should be present as the result of one or more energetic neutrinos. For both channels, we use a common set of criteria for selecting individual objects (electrons, muons, and jets).

Both lepton + jets and dilepton events are required to have one tight muon or electron. The second charged lepton in dilepton events is allowed to be either tight or loose. For lepton + jets events, tight muons are required to have $p_T > 30$ GeV/c and $|\eta| < 2.1$. For dilepton events, tight muons are allowed with $p_T > 20$ GeV/c and $|\eta| < 2.4$. Loose muons must have $p_T > 10$ GeV/c and $|\eta| < 2.4$. The degree of isolation for the muon is assessed by comparing the sum of the transverse energy of particles in a cone of $\Delta R = \sqrt{\Delta\eta^2 + \Delta\phi^2} = 0.4$ around the muon direction, excluding the muon itself, to the p_T of the muon. For tight muons, the ratio of the neighboring energy deposits to the muon p_T is required to be less than 0.125. For loose muons, this ratio must be less than 0.2. Tight electrons in lepton + jets events are required to have $E_T > 30$ GeV/c^2, while in dilepton events they must have $E_T > 20$ GeV/c^2. Loose electrons must have $E_T > 15$ GeV/c^2. All electrons are required to have $|\eta| < 2.5$. Electrons that fall into the transition region between the barrel and endcap of the electromagnetic calorimeter ($1.442 < |\eta| < 1.566$) are rejected. Electron isolation is calculated in a similar manner to muon isolation. Tight electrons must have an isolation less than 0.1, while loose electrons may have an isolation less than 0.2.

Jets are required to have corrected $p_T > 30$ GeV/c and $|\eta| < 2.4$. However, in lepton + jets events, because of the jet requirement in the electron trigger, the leading three jets are required to have $p_T > 40$ GeV/c. For consistency, this requirement is also applied to lepton + jets events containing a muon, even though there is no jet component in the trigger for those events. Jets are identified as containing a b quark hadron using the Combined Secondary Vertex (CSV) algorithm. This algorithm combines information about the impact parameter of tracks and reconstructed secondary vertices within the jets in a multivariate algorithm to separate jets containing b quark hadrons from jets containing charm quarks (c), light quarks (uds), or gluons (g). Both categories of events use the medium operating point of the CSV algorithm. For this operating point, the b-tag efficiency is 70% (20%) for jets originating from a b (c) quark, and the probability of mis-tagging for jets originating from light quarks or gluons is 2%.

3. Artificial neural networks

The first step in separating signal from background is to divide the lepton + jets (dilepton) sample into 7 (2) categories based on the number of jets and tags in the event. Artificial neural networks (ANNs) are used to further improve signal-to-background sensitivity in all categories of the analysis. Separate ANNs are trained for different categories, utilizing information related to object kinematics, event shape, and the discriminant output from the b-tagging algorithm. The choice of variables is optimized for each category. In the lepton + jets (dilepton) categories, ten (five) variables are chosen from a pool of candidates based on initial separation between signal and background. In addition, the number of jets is employed as a variable in the dilepton channel for the category $\geq$3 jets + $\geq$3 b-tags.

In both lepton + jets and dilepton channels, the variables that discriminate best between signal and background directly involve b-tagging information, such as the average CSV output value for b-tagged jets. The variables that distinguish best between signal and $t\bar{t} + b\bar{b}$, a background very similar to the signal, are variables that include b-tagging information, kinematic information, and angular correlation information, such as the minimum ΔR between all pairs of b-tagged jets. The modeling of these variables were checked before using them as inputs to the ANN.

4. Results

Our background model matches data well for the event yields and for the shapes of the neural network input variables. We checked the modeling of our input variables in all categories in our search region. In general, we found that all of our input variables were well modeled within the statistics of the events in each category.

To search for the presence of signal in our data, we use a simultaneous maximum-likelihood fit to nine lepton+jets and dilepton neural network output distributions, some of which are shown in Figs. 1 and 2. These plots are input to the CMS standard "Higgs Combination" limit-setting package. The package uses a modified-frequentist approach. Our uncertainties are handled as nuisance parameters in the fit.

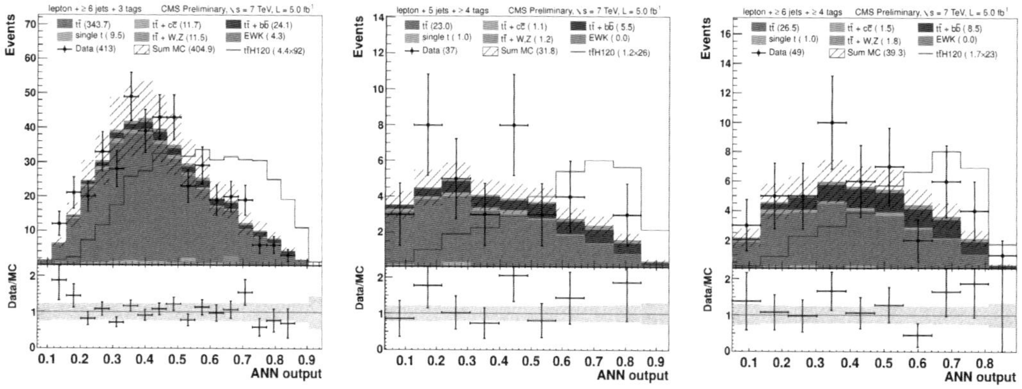

Figure 1. ANN output for lepton + jet events. Background-like events have a low neural network output value. Signal-like events have a high value. The uncertainty band includes statistical and systematic uncertainties that affect both the rate and shape of the background distributions. The plots read from left to right: ≥6 jets + 3 tags, 5 jets + ≥4 b-tags, and ≥6 jets + ≥4 b-tags. The $t\bar{t}H$ signal (m_H = 120 GeV/c^2) is normalized to the total background yield.

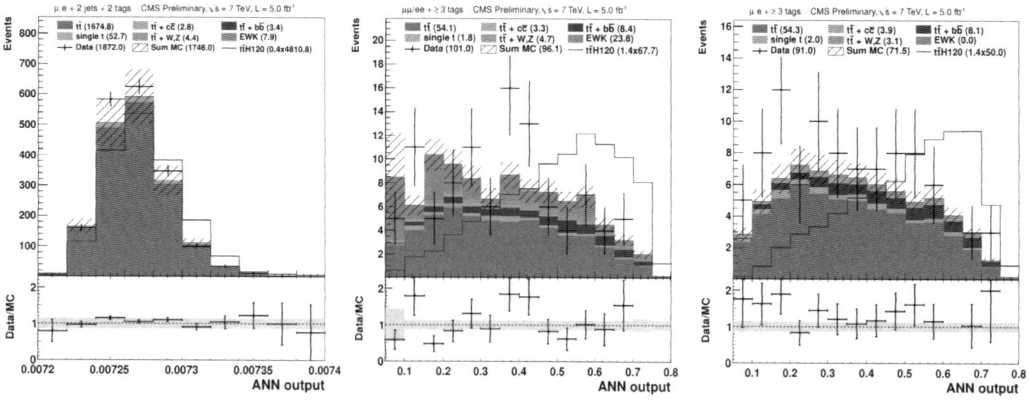

Figure 2. ANN output for dilepton events. The left plot is for $e\mu$ events in the 2 jets + 2 b-tags category. The right plots are for the ≥3 b-tag category, $\mu\mu$ or ee and $e\mu$ events, respectively.

m_H (GeV/c^2)	Obs limit	Median Exp limit	Expected range for 68% CL	Expected range for 95% CL
110	2.3	2.9	[2.1,4.0]	[1.6,5.4]
115	2.8	3.5	[2.4,4.7]	[1.8,6.2]
120	3.1	3.8	[2.8,5.3]	[2.1,7.1]
125	3.8	4.6	[3.3,6.4]	[2.5,8.5]
130	4.4	5.7	[4.1,7.9]	[3.1,10.5]
135	5.7	7.0	[5.1,9.8]	[3.8,13.0]
140	6.6	9.5	[6.8,13.2]	[5.1,17.5]

Table 1. Expected and observed upper limits for SM Higgs boson for the lepton + jets and dilepton channels combined.

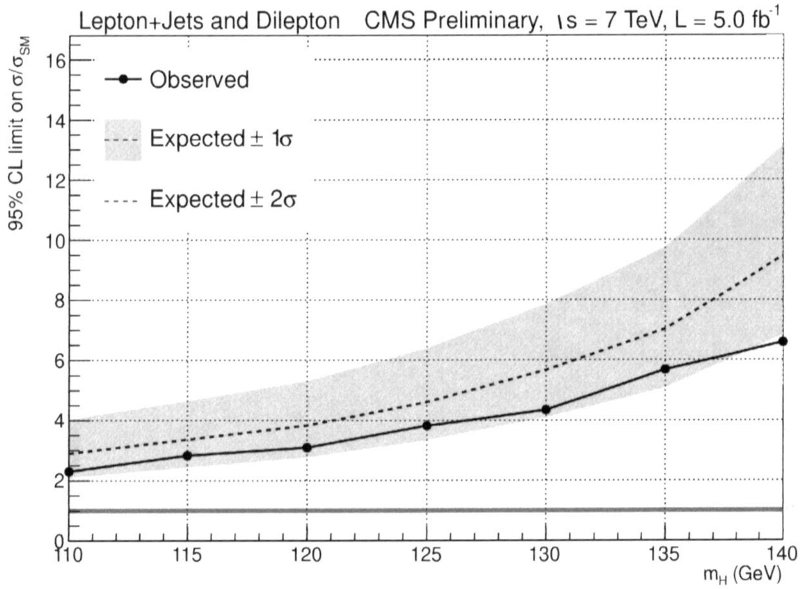

Figure 3. The expected and observed 95% CL upper limits on the signal strength parameter $\mu = \sigma/\sigma_{SM}$ for lepton + jets and dilepton channels combined.

5. Conclusions

The results presented in these proceedings represent the first CMS search for SM Higgs boson production in association with a top quark pair decaying semi-leptonically ($t\bar{t} \to \ell\nu jjb\bar{b}$) or fully leptonically ($t\bar{t} \to \ell\nu\ell\nu b\bar{b}$). Using the full 2011 dataset, we set limits on Higgs boson production comparable in sensitivity to the limits set in the CMS analysis searching for SM Higgs boson production in association with a vector boson [3]. For example, at a Higgs boson mass of 125 GeV/c^2, the expected limit is 4.6 times the SM expectation. The observed limit for $m_H = 125$ GeV/c^2 is 3.8 times the SM expectation.

References

[1] CMS Collaboration (CMS) 2012 Search for higgs boson production in association with a top quark pair in pp collisions CMS Physics Analysis Summary CMS-PAS-HIG-12-025 these TOP2012 conference proceedings are excerpted from the CMS PAS, which can be found at https://cds.cern.ch/record/1460423 URL `https://cds.cern.ch/record/1460423`

[2] CMS Collaboration (CMS) 2008 *JINST* **3** S08004 URL `http://dx.doi.org/10.1088/1748-0221/3/08/S08004`

[3] Chatrchyan S *et al.* (CMS) 2012 *Phys.Lett.* **B710** 284–306 (*Preprint* `1202.4195`)

Journal of Physics: Conference Series **452** (2013) 012007 doi:10.1088/1742-6596/452/1/012007

Interplay of Top Quark and Higgs Boson Measurements at the Tevatron and LHC

Sarah Boutle for the ATLAS, CMS and CDF collaborations
University of Virginia, Charlottesville VA, USA
E-mail: sboutle@cern.ch

Abstract. Given its large mass, the top quark plays an important role in the understanding of the mechanism for electroweak symmetry breaking (EWSB) in the Standard Model (SM). The recently observed new boson, consistent in many ways with the Higgs boson, could be the agent of EWSB. Studies of the production of this new particle in association with top-quark pairs will give important insight into the true nature of this particle. The current status of studies of $t\bar{t}H$ production at the LHC and Tevatron are described herein, as well as one of the most important background channels, $t\bar{t}+b\bar{b}$ production. Further, the top quark plays an important role in exotic Higgs models; one such search is also described, looking for $t \rightarrow H^{\pm}b$.

1. Introduction

In summer 2012, a new boson with a mass of approximately 125 GeV was observed [1, 2] at the Large Hadron Collider (LHC) by the CMS and ATLAS experiments [3, 4]. After several decades of searching for the Standard Model (SM) Higgs boson, this observation is a significant breakthrough. However the effort is far from complete, as hadron collider experiments must now turn their focus to the characterization of the particle to ascertain its consistency with the Standard Model (SM) Higgs boson. Greater than three standard deviation (σ) significance has been established for direct production in the decay to pairs of either photons, W bosons or Z bosons. The observations in these channels are consistent with the SM Higgs boson within the uncertainties but in order to build a complete understanding of the characteristics of the new boson, the rate of production across all decay channels must be measured. Currently, the least understood aspects of the new boson are the fermionic couplings.

The production of Higgs bosons in association with a pair of top quarks ($t\bar{t}H$) is becoming a particularly important channel in the characterisation of the new boson. It is the only production mode which is directly sensitive to the top-Higgs Yukawa coupling; the cross section is proportional to the coupling squared. Another fermionic coupling, to the b quark, is also accessible through the $t\bar{t}H$ channel. At a Higgs boson mass of $\sim$ 125 GeV, the dominant decay mode is to a pair of bottom quarks. In gluon fusion production, the dominant production mechanism at the LHC, this decay channel is not experimentally feasible to probe due to large backgrounds, but associated production mechanisms such as VH [5, 6, 7] and $t\bar{t}H$, despite having a much smaller cross section, give an experimental handle on the Higgs-b coupling. Therefore, $t\bar{t}H$ is a unique channel in that it can probe the Higgs coupling to top and bottom in the same process.

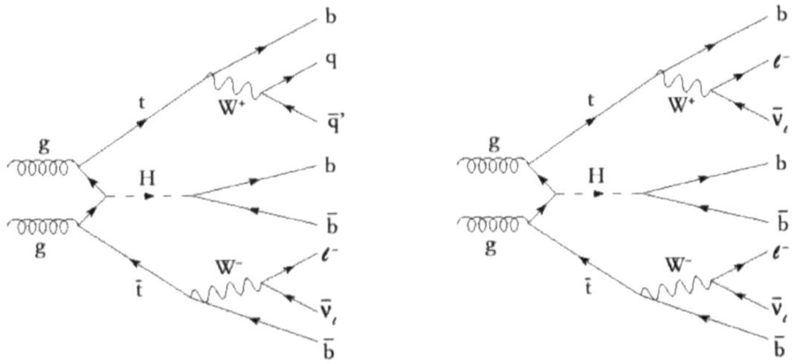

Figure 1. Feynman diagrams of the Lepton+Jets (left) and Dilepton (right) $t\bar{t}H$ channels.

The top quark may play a significant role in understanding EWSB. The large mass of the top quark means that it is predicted to have a large coupling to the SM Higgs boson. But in other models, the top quark could play a role in EWSB beyond that of the Higgs mechanism of the SM. Top quark studies also bring about sensitivity to non-SM Higgs bosons for example in the decay of a t quark to a b quark with a charged Higgs boson.

This paper discusses recent experimental results from the LHC and Tevatron which probe the interplay between the top quark and the Higgs boson, either SM or non-SM.

2. Search for SM Higgs boson in association with a top quark pair: $t\bar{t}H$

The SM Higgs boson produced in association with a pair of top quarks has so far been studied at three hadron collider experiments: CDF, CMS and ATLAS [8, 9, 10]. All three analyses select events in the "lepton+jets" channel, but CMS additionally selects "dilepton" events. Figure 1 shows the Feynman diagram for $t\bar{t}H$ production in these two final states which are defined in terms of the decay of the Ws originating from each top quark. Events where one W decays leptonically and the other hadronically are referred to as "lepton+jets" events, whereas in "dilepton" events both W bosons decay leptonically. In the ATLAS results described in this paper, only events with $H\rightarrow b\bar{b}$ are considered as signal and subsequently any limits being set are on the cross section times $H\rightarrow b\bar{b}$ branching ratio. The CMS and CDF analyses, although optimized for $H\rightarrow b\bar{b}$ decays, do not exclude events from other possible Higgs boson decay modes and as such set limits on the $t\bar{t}H$ production cross section.

The lepton+jets final state consists of a high p_T lepton, a large number of jets, some of which are b jets, and a significant amount of missing transverse energy (E_T^{miss}) from the presence of a neutrino. Therefore all three analyses start with a $t\bar{t}$-enriched data sample with all or most of these features. The CMS analysis selects events with an isolated lepton with $p_T \geq 30$ GeV, at least three jets with $p_T \geq 40$ GeV and another jet with $p_T \geq 30$ GeV. At least two jets are required to be tagged as b jets. The dilepton channel events have two leptons: tight ($p_T \geq 20$ GeV) and loose ($p_T \geq 15(10)$ GeV electron(muon)), at least with 2 jets $p_T \geq 30$ GeV and 2 b-tags. In general, $t\bar{t}H$ signal events tend to have a larger number of jets and a larger number of b-tagged jets than background processes, the largest of which is $t\bar{t}$+jets. This feature can be exploited to separate signal and background and so the events are split into 7 jet-tag categories for lepton+jets: ≥6 jets + 2 b-tags, 4 jets + 3 b-tags, 5 jets + 3 b-tags, ≥6 jets + 3 b-tags, 4 jets + ≥4 b-tags, 5 jets + ≥4 b-tags, ≥6 jets + ≥4 b-tags. The most sensitive category is ≥6 jets + ≥4 b-tags with $S/\sqrt{B}$ of 0.31 but categories with lower sensitivity are included to better constrain the background. For dilepton events, two categories are used: 2 jets + 2 b-tags, ≥3 jets + ≥3 b-tags.

Artificial Neural Networks (ANNs) are trained in each category to better separate signal and

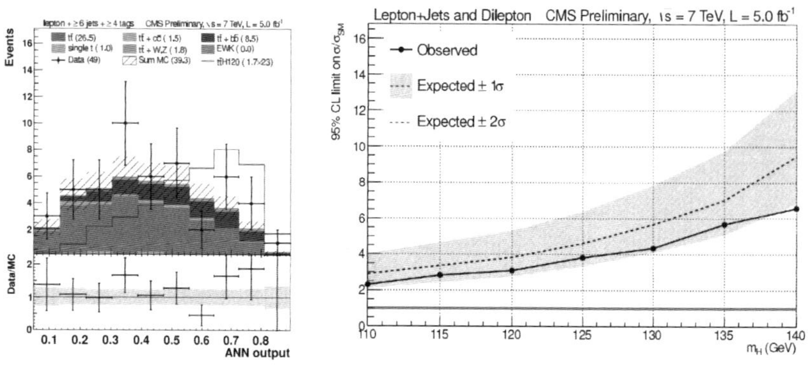

Figure 2. Left: ANN output distribution for the ≥6 jets + ≥4 b-tags of the CMS $t\bar{t}H$ analysis. Right: Lepton+jets and dilepton combined 95% C.L. upper limit $t\bar{t}H$ cross section with respect to the SM prediction as a function of m_H with $5.0 fb^{-1}$ of 7 TeV CMS data.

background. The input variables vary from category to category but in general can be described by four main classes of variables: basic single object kinematics (e.g. jet p_T), kinematics of pairs of jets (e.g. invariant mass of pairs of untagged jets), event shape variables (e.g. sphericity) and b-tag related variables (e.g. average b-tag discriminant value which is the most discriminating variable in the majority of jet-tag categories). Ten input variables are used in each lepton+jets jet-tag category and five or six in the dilepton categories. The ANN output gives better discriminating power than any one of the single input variables. The ANN output distribution is shown in Fig. 2 for the ≥6 jets + ≥4 b-tags comparing data and MC simulated background. The signal MC distribution is also shown, normalized to the area of the background for visual purposes. A maximum likelihood fit of the ANN output distributions from the nine jet-tag categories considered in the analysis. Both background and signal processes are modelled with Monte Carlo simulation. Currently no significant excess in the data is observed and so 95% confidence level (C.L.) upper limits are set as a function of m_H on the possible presence of a SM-like signal, as shown in Fig. 2. The median expected limit for a Higgs boson mass of 125GeV is 4.6 σ_{SM} while the observed limit is 3.8 σ_{SM}.

The CDF collaboration have previously performed an analysis which is very similar to the CMS lepton+jets search, fitting an ANN output distribution to separate signal from background. The 6 jet-tag categories used were 4, 5 and ≥6 jets with 2 or ≥3 b-tags. The data used corresponded to 9.45 fb^{-1} of 1.96 TeV $p\bar{p}$ collisions collected with the CDF detector. Again a 95% C.L. upper limit was set as a function of m_H on the possible presence of a SM-like signal: the median expected limit for a Higgs boson mass of 125GeV is 12.6 σ_{SM} while the observed limit is 20.5 σ_{SM}.

The ATLAS search also exploits the difference between signal and background in terms of the number of jets and the number of tags in events, by dividing into jet-tag categories. Events were selected with an isolated electron(muon) with $p_T \geq 25(20)$ GeV, at least four jets with $p_T \geq 25$ GeV. Additional cuts on the E_T^{miss} and the transverse mass of the leptonic W (M_T) are included to reduce the QCD multijet background contribution. The nine jet-tag categories used are divided into signal enriched: 5 jets (3, ≥4 b-tags) and ≥6 jets (3, ≥4 b-tags), and background enriched categories: 4 jets (0, 1, ≥2 b-tags), 5 jets + 2 b-tags and ≥6 jets + 2 b-tags. A discriminating variable is employed in each jet-tag category to further separate signal and background. For the first variable, a kinematic fit is used first to reconstruct the $t\bar{t}$ system from selected objects (jets, lepton and E_T^{miss}). The remaining two jets are then combined to

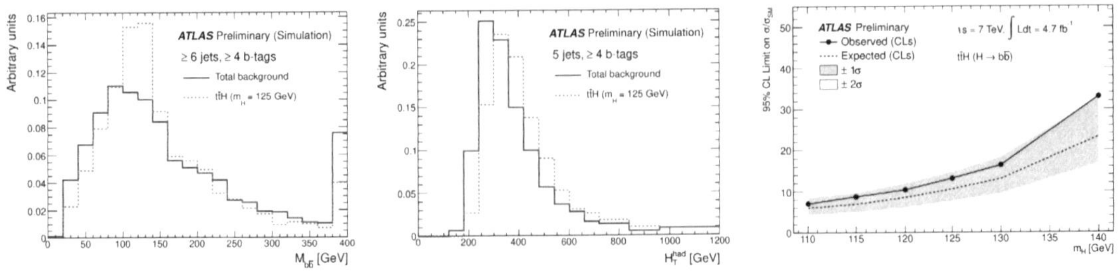

Figure 3. Comparison of $H_{\mathrm{T}}^{\mathrm{had}}$ (left) and $m_{\mathrm{b\bar{b}}}$ (middle) distributions for $\mathrm{t\bar{t}H}$ signal (m_{H} = 125 GeV) and background simulation. Right: 95% C.L. upper limit $\mathrm{t\bar{t}H}$ cross section with respect to the SM prediction as a function of $\mathrm{m_H}$ with 4.7 fb^{-1} of 7 TeV ATLAS data.

reconstruct the invariant mass of the $\mathrm{b\bar{b}}$ system ($m_{\mathrm{b\bar{b}}}$). This variable is used in a fit in the categories with ≥6 jets and shows a peak near the Higgs boson mass in signal MC. In all other categories, the scalar sum of jet p_{T}s ($H_{\mathrm{T}}^{\mathrm{had}}$) is used in the fit. Signal to background shape comparisons of the distributions of both these variables are shown in Fig. 3. The simultaneous fit to background-like and signal-like topologies acts to improve the background prediction and reduce background related uncertainties.

The signal and largest background process ($\mathrm{t\bar{t}}$+jets) are modelled using MC, but data are used in the construction of the W+jets and QCD multijet backgrounds. Observed and expected 95% C.L. upper limits were set on the Higgs boson production cross section times the branching ratio to a pair of b quarks ($\sigma(\mathrm{t\bar{t}H}) \times BR(\mathrm{H} \to \mathrm{b\bar{b}})$) as a function of m_{H}, as shown in Fig. 3. The median expected limit for a Higgs boson mass of 125GeV is 10.5 σ_{SM} while the observed limit is 13.1 σ_{SM}.

3. Measurement of $\sigma(\mathrm{t\bar{t}+b\bar{b}})/\sigma(\mathrm{t\bar{t}+jj})$

One of the main challenges in the search for $\mathrm{t\bar{t}H}$ is the $\mathrm{t\bar{t}+b\bar{b}}$ background which has the same signature as $\mathrm{t\bar{t}H}$. It has a very small cross section which makes it difficult to measure experimentally; while theoretical predictions, although available at NLO, suffer from larger scale uncertainties due to the presence of two very different energy scales (m_{t} and the jet p_{T} threshold). A more accessible quantity is the ratio of the $\mathrm{t\bar{t}+b\bar{b}}$ cross section to the total $\mathrm{t\bar{t}+jj}$ cross section. CMS carried out the first measurement [11] of this ratio is 5 fb^{-1} of 7 TeV data in the dilepton channel. Events were selected with 2 isolated leptons (ee, $e\mu$, $\mu\mu$) with p_{T} ≥20 GeV. These lepton pairs were required to have an invariant mass (m_{ll}) outside a 12 GeV window around the Z boson mass to reduce the large background from $\mathrm{Z} \to l^+l^-$ decays present in the dilepton channel. Events were required to have $E_{\mathrm{T}}^{\mathrm{miss}} \geq$ 30GeV in the $ee/\mu\mu$ channels to suppress any remaining QCD multijet background. Finally, only events with ≥4 jets and ≥2 (4) b-tags were used for the $\mathrm{t\bar{t}+jj}$ ($\mathrm{t\bar{t}+b\bar{b}}$) sample. The difference in the b-jet multiplicity spectrum between $\mathrm{t\bar{t}+b\bar{b}}$ and $\mathrm{t\bar{t}+lf}$ was exploited to extract $\mathrm{t\bar{t}+b\bar{b}}$ signal by performing a fit of this distribution to data. Correcting to the visible phase space, a value for the ratio of the cross sections was measured to be $\sigma(\mathrm{t\bar{t}+b\bar{b}})/\sigma(\mathrm{t\bar{t}+jj})$ = 3.6 ± 1.1 (stat.) ± 0.9 (sys.). The result cannot be directly compared to current NLO QCD calculations since the p_{T} thresholds differ. The result is somewhat larger than predictions from MadGraph and Powheg, which are 1.2% and 1.3% respectively.

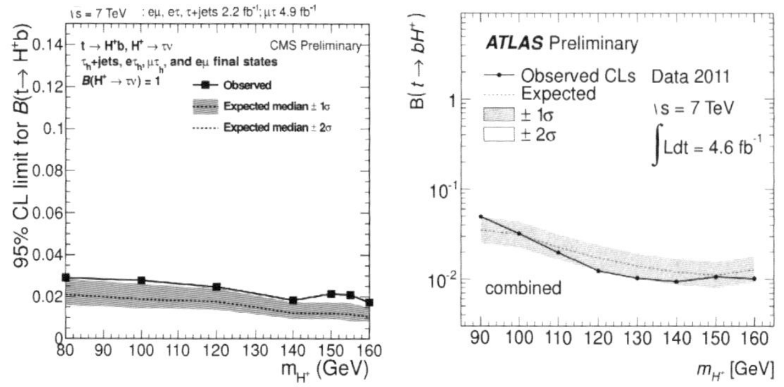

Figure 4. The upper limit on the branching fraction for t to a charged Higgs and a b quark for CMS (left) and ATLAS (right).

4. Direct Searches for Charged Higgs

Events containing top quarks can also be used to search for evidence of Higgs physics beyond the SM, such as MSSM charged Higgs production through $\mathrm{t} \rightarrow \mathrm{H}^{\pm}\mathrm{b}$. In the SM, $\mathrm{t\bar{t}}$ events are dominated by the Wb final states. The presence of a light charged Higgs, with mass smaller than that of the top quark, would result in a different distribution of final states than is expected in the SM. For low Higgs mass and high tanβ, the dominant decay mode of a charged Higgs is $\mathrm{H}^{\pm} \rightarrow \tau\nu$. The CMS and ATLAS collaborations have completed searches [12, 13] for charged Higgs bosons in the $\mathrm{H}^{\pm} \rightarrow \tau\nu$ decay in a variety of final states, each defined by (a) the decay of the τ from the Higgs (hadronic: τ_{h} or leptonic: "e/μ"), and (b) the decay of the W from the other top decay (hadronic: "jets" or leptonic: "e/μ").

The three channels used in the CMS analysis are τ_{h}+jets, τ_{h}+e/μ, and e+μ. In the τ_{h}+jets, the transverse mass of the τ_{h} and $E_{\mathrm{T}}^{\mathrm{miss}}$ (m_{T}) is constructed. This quantity is expected to be larger in signal where the τ and neutrino originate from a $\mathrm{H}^{\pm}$ as opposed to from a W, as in background. A binned maximum-likelihood fit to this distribution is used to extract the signal. In the τ_{h}+e/μ and e+μ channels, a counting experiment is performed in order to extract the final limits. In the analysis performed by ATLAS, the τ_{h}+jets and τ_{h}+e/μ channels and a third channel, e/μ+jets are used. A different discriminating variable is used in each channel to extract the signal: m_{T} in τ_{h}+jets events, $E_{\mathrm{T}}^{\mathrm{miss}}$ in τ_h+e/μ events and m_T^H (the transverse mass of the leptonically decaying $\mathrm{H}^{\pm}$, or W in the case of the background).

Both collaborations set 95% C.L. upper limits on the $BR(\mathrm{t} \rightarrow \mathrm{H}^{\pm}\mathrm{b})$ as a function of the mass of $\mathrm{H}^{\pm}$ using 2.3(4.6) fb^{-1} of CMS (ATLAS) data, as shown in Fig. 4. An upper limit of 2-3% for a mass range of 80-160 GeV was set by CMS and of 1-5% over a mass range of 90-160 GeV by ATLAS. The analyses from both collaborations also exclude a large region of $m_{\mathrm{H}^{\pm}}$ - tanβ parameter space.

5. Conclusion

The current status of searches for the SM Higgs boson in association with top quarks has been presented, including analyses from the LHC and Tevatron. These are the first searches in this production mode. The first measurement from CMS of the $\mathrm{t\bar{t}} + \mathrm{b\bar{b}}$ cross section ratio was also described, a large and important background to $\mathrm{t\bar{t}H}$ searches. Finally, two searches for MSSM charged light Higgs in the $\mathrm{H}^{\pm} \rightarrow \tau\nu$ decay channel were described. These analyses were carried out by the CMS and ATLAS collaborations and set limits on $BR(\mathrm{t} \rightarrow \mathrm{H}^{\pm}\mathrm{b})$.

References

[1] ATLAS Collaboration, "Observation of a new particle in the search for the Standard Model Higgs boson with the ATLAS detector at the LHC", *Phys. Lett.* **B716** (2012) 1-29, doi:10.1016/j.physletb.2012.08.020, arXiv:1207.7214.

[2] CMS Collaboration, "Observation of a new boson at a mass of 125 GeV with the CMS experiment at the LHC", *Phys. Lett.* **B716** (2012) 30-61, doi:10.1016/j.physletb.2012.08.021, arXiv:1207.7235.

[3] ATLAS Collaboration, "The ATLAS experiment at the CERN LHC", *JINST* **3** (2008) S08003.

[4] CMS Collaboration, "The CMS experiment at the CERN LHC", *JINST* **3** (2008) S08004.

[5] ATLAS Collaboration, "Search for the Standard Model Higgs boson produced in association with a vector boson and decaying to a b-quark pair with the ATLAS detector", *Phys. Lett.* **B718** (2012) 369-390, doi:10.1016/j.physletb.2012.10.061, arXiv:1207.0210

[6] CMS Collaboration, "Search for the standard model Higgs boson decaying to bottom quarks in pp collisions at sqrt(s)=7 TeV", *Phys. Lett.* **B710** (2012) 284-306, doi:10.1016/j.physletb.2012.02.085, arXiv:1202.4195.

[7] CDF Collaboration, D0 Collaboration, "Evidence for a particle produced in association with weak bosons and decaying to a bottom-antibottom quark pair in Higgs boson searches at the Tevatron", *Phys. Rev. Lett.* **109** (2012) 071804, doi:10.1103/PhysRevLett.109.071804, arXiv:1207.6436.

[8] CDF Collaboration, Search for the standard model Higgs boson produced in association with top quarks using the full CDF data set, *Phys. Rev. Lett.* **109** (2012) 181802, doi:10.1103/PhysRevLett.109.181802, arXiv:1208.2662.

[9] CMS Collaboration, "Search for Higgs boson production in association with a top quark pair in pp collisions", CMS-PAS-HIG-12-025.

[10] ATLAS Collaboration, "Search for the Standard Model Higgs boson produced in association with top quarks in proton-proton collisions at $\sqrt{s}$ = 7 TeV using the ATLAS detector", ATLAS-CONF-2012-135.

[11] CMS Collaboration, "First Measurement of the cross section ratio ttbb/ttjj in pp Collisions at $sqrts$ = 7TeV", CMS-PAS-TOP-12-024.

[12] CMS Collaboration, "Search for a light charged Higgs boson in top quark decays in pp collisions at $\sqrt{s}$=7TeV" *JHEP* **07** (2012) 143, doi:10.1007/JHEP07(2012)143, arXiv:1205.5736.

[13] ATLAS Collaboration, "Search for charged Higgs bosons decaying via $H^{\pm} \to \tau\nu$ in $t\bar{t}$ events using pp collision data at $\sqrt{s}$ = 7 TeV with the ATLAS detector", ATLAS-CONF-2012-011.

5th International Workshop on Top Quark Physics (TOP2012) IOP Publishing
Journal of Physics: Conference Series **452** (2013) 012008 doi:10.1088/1742-6596/452/1/012008

Physics of the Interplay Between the Top Quark and the Higgs Boson

Mikael Chala and José Santiago
CAFPE and Departamento de Física Teórica y del Cosmos,
University of Granada, 18071, Granada, Spain

E-mail: miki@ugr.es, jsantiago@ugr.es

Abstract. We discuss some aspects of the interplay between the top quark and the Higgs boson at the LHC. First we describe what indirect information on the top Yukawa coupling can be extracted from measurements in the Higgs sector. We then show that the study of processes involving $Ht\bar{t}$ and $Hb\bar{b}$ final states can give us information on the spectrum in models of strong electroweak symmetry breaking. In particular, we introduce a novel analysis of the $Hb\bar{b} \to b\bar{b}b\bar{b}$ channel with an excellent discovery potential at the LHC.

1. Introduction

With the discovery of the Higgs boson at the LHC, the quest to understand the mechanism of electroweak symmetry breaking (EWSB) has finally started. The top quark has an undeniable leading role in the physics of the Higgs boson, as it dominates, in the Standard Model (SM), the main production mechanism ($gg \to H$) and contributes in an important way to one of the most relevant decay channels ($H \to \gamma\gamma$). Hence, experimental tests of the Higgs sector can give us precious information on aspects of the top quark, like its Yukawa coupling or on the possible existence of top partners or even new vector resonances. At the same time the LHC is a top factory and many of the top quark properties will be measured to an unprecedented accuracy. Searches of processes involving the top quark or some of its possible partners can in turn give us non-trivial information on the sector of EWSB. In these proceedings we will discuss some aspects of this interplay between the top quark and the Higgs boson. In Section 2 we will describe what indirect information on the top quark Yukawa coupling we can infer from current results on Higgs searches and how that information might be combined with other searches to obtain a coherent picture of the EWSB sector. In Section 3 we show how searches involving the production of $Ht\bar{t}$ and $Hb\bar{b}$ can give us information on the spectrum in models of strong EWSB. In these models, new vector-like quarks can be singly produced in association with a SM quark, mediated by a color octet resonance. Subsequent decays of the heavy quark can result in $Ht\bar{t}$ and $Hb\bar{b}$ final states. We show that these channels can be measured at the LHC for a large region of parameter space. In particular we introduce a novel analysis of the $Hb\bar{b} \to b\bar{b}b\bar{b}$ channel that takes full advantage of the distinctive kinematics of the process, leading to an excellent reach potential at the LHC. We finally conclude in Section 4.

2. What is the Higgs telling about the top?

Now that the discovery of a Higgs-like boson has been settled all the efforts are directed towards measuring its main properties. Even though the errors are still large, a coherent picture is starting to emerge from the combination of all the different channels [1, 2]. The experimental data point to a new boson of mass $m_H \approx 126$ GeV with production and decay compatible with those of the standard model Higgs in all channels with the possible exception of the $H \to \gamma\gamma$ decay, which seems to be consistently larger than the SM one (but still compatible with it at the two standard deviation level). [1]

The $H \to \gamma\gamma$ decay is unique for two reasons. First, it occurs at the loop level in the SM, being *a priori* particularly sensitive to new physics. Second, this decay arises from interfering contributions with the top and the W boson running in the loop and it is therefore sensitive not only to the absolute value of the top Yukawa coupling but also to its sign. It turns out that although it is still possible to explain the data with new particles contributing at the loop level, it would not be easy to accommodate the current enhancement without modifying the production cross section through gluon fusion and without destabilizing the vacuum (see for instance [3]). If we assume instead that there is no significant contribution to the Higgs production and decay from new particles and the only effect is a possible modification of the couplings of the SM particles to the Higgs we find that apart from a region with SM-like couplings there is a second possible explanation to the current data. In this region the couplings of the gauge boson to the Higgs are slightly reduced whereas the top Yukawa coupling has the opposite sign than in the SM [4]. The "wrong sign" top Yukawa coupling turns the destructive interference in the $H \to \gamma\gamma$ decay into a constructive one thus explaining the observed enhancement.

Most physical processes are sensitive to the absolute value of the top Yukawa and not to its sign so it is not easy to test the hypothesis of a wrong sign Yukawa coupling. One possibility, recently proposed in [5], is to use t-channel single top production in association with a Higgs boson. This is a tree level process in which there is interference between diagrams involving the top Yukawa and a W, making it also sensitive to the sign of the top Yukawa coupling relative to the WWH coupling. The process is very small in the SM due to an almost perfect destructive interference. With the wrong sign top Yukawa coupling however it becomes observable at the LHC and it is therefore a potentially crucial test of this explanation of current Higgs data. It should be noted however that the information on the top Yukawa coupling is obtained from a loop process in one case and from a tree level process in the other. These two Yukawa couplings are in general different as new physics can affect differently the tree and loop processes. Therefore care should be exercised when interpreting the results of these searches.

3. New Higgs Production Mechanisms in Composite Higgs Models

We have seen in the previous section how Higgs searches can give us information on some properties of the top quark like its Yukawa coupling. In this section we will see how searches involving the production of $Ht\bar{t}$ and $Hb\bar{b}$ can give us information on the spectrum of new particles in models of strong EWSB. Many models of strong EWSB include in their spectrum light top partners (new vector-like quarks that mix strongly with the top and/or bottom quarks) together with massive color-octet vector resonances, that we will call heavy gluons. In some regions of parameter space the leading production mechanism of these top partners is single production mediated by the exchange of the heavy gluons. If the top partners have electric charge 2/3 or $-1/3$ they can result in $Ht\bar{t}$ and $Hb\bar{b}$ final states, respectively. The corresponding process is shown in Fig. 1. In the following we discuss how to use these channels to search for the top partners and even for the heavy gluons in these models of strong EWSB. In our studies we have

[1] Currently ATLAS also observes an incompatibility between the mass of the new boson as measured in the ZZ^* and $\gamma\gamma$ channels at the level of almost three standard deviations.

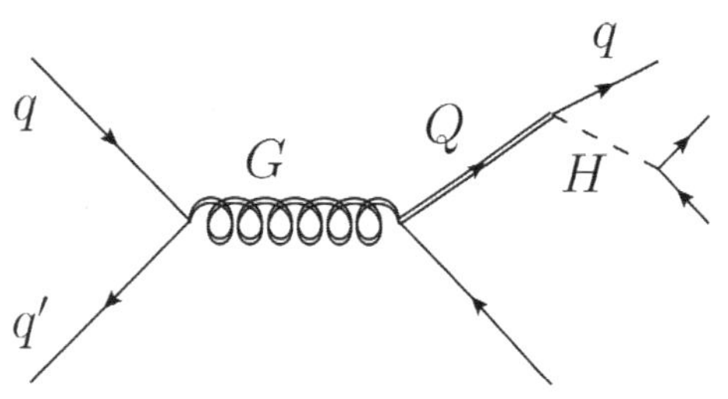

Figure 1. Main diagram contributing to $Ht\bar{t}$ and $Hb\bar{b}$ production in our model. Here Q (q) stands for T (t) or B (b).

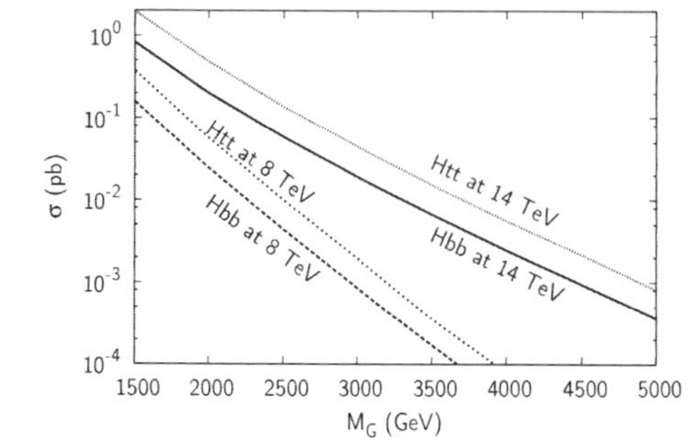

Figure 2. $Ht\bar{t}$ and $Hb\bar{b}$ production cross section as a function of M_G for $\sin(\phi_{t_R}) = \sin(\phi_{b_R}) = 0.6$.

used a simplified version of the minimal composite Higgs model [6] as described in [7, 8] (see also [9]). The most relevant parameter that we will consider is the degree of compositeness of the top or bottom quarks, denoted by $\sin(\phi_{t_R})$ and $\sin(\phi_{b_R})$, respectively. The $Ht\bar{t}$ and $Hb\bar{b}$ production cross sections, for $\sin(\phi_{t_R}) = \sin(\phi_{b_R}) = 0.6$ are reproduced in Fig. 2 at the LHC for $\sqrt{s} = 8$ and 14 TeV. We have simulated our signal and backgrounds using `MADGRAPH V4.5.0` [10] and `ALPGEN V.2.13` [11], respectively. We have then passed the events through `PYTHIA 6.4` [12] for hadronization and showering and `DELPHES V1.9` [13] for detector simulation. Further details on the implementation of the model and the simulations can be found in reference [8].

3.1. $Ht\bar{t}$ Channel

We consider the process shown in Fig. 1 with $Q = T$ and $q = t$ and the leading $H \to b\bar{b}$ decay channel of the Higgs and a semileptonic decay of the $t\bar{t}$ system. The specific process is

$$pp \to G \to T\bar{t} + \bar{T}t \to Ht\bar{t} \to 4b + 2j + l\not{E}_T. \tag{1}$$

A detailed analysis of this channel has been performed in Ref. [8] where all the relevant details can be found. Here we will just summarize the main results for the LHC running at a center of mass energy $\sqrt{s} = 14$ TeV. The heavy gluon masses that can be probed at this energy are large enough to make all the decay products quite hard. Also T is typically heavy so that the use of boosted techniques probed useful in these searches.

In order to take maximum advantages of the particular features of the signal we have implemented the following set of cuts

- At least 3 jets, with a minimum of 2 b tags.
- At least 1 isolated charged lepton.
- The two jets with the largest invariant masses, $j_{1,2}$, are required to have invariant masses close to the top and Higgs mass, respectively, $|m_{j_1} - m_t| \leq 40$ GeV and $|m_{j_2} - m_H| \leq 40$ GeV.
- A cut on S_T (the scalar sum of the p_T of the three hardest jets, the charged lepton and the missing transverse energy) that depends on the test M_G we are considering

$$S_T > 1.2,\ 1.5,\ 1.7,\ 2 \text{ TeV for } M_G = 2,\ 2.5,\ 3, \geq 3.5 \text{ TeV}. \tag{2}$$

We have considered all the relevant backgrounds in our analysis. The most important ones turn out to be $t\bar{t}$ and $t\bar{t}b\bar{b}$. The result of this analysis is summarized in Fig. 3 in which we show the

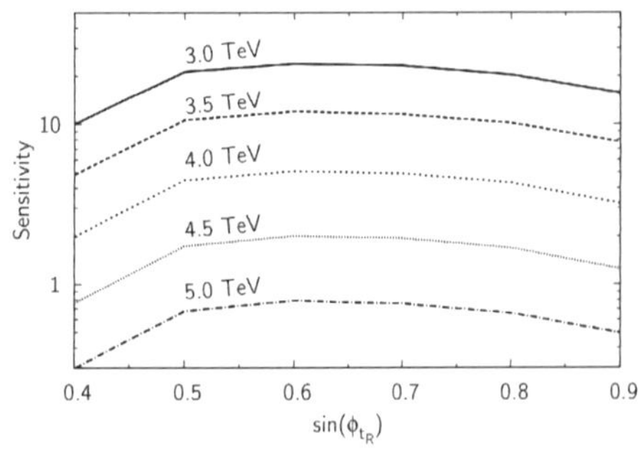

Figure 3. Sensitivity obtained in the $Ht\bar{t}$ channel for different values of the heavy gluon mass as a function of $\sin(\phi_{t_R})$ after an integrated luminosity of 100 fb^{-1}.

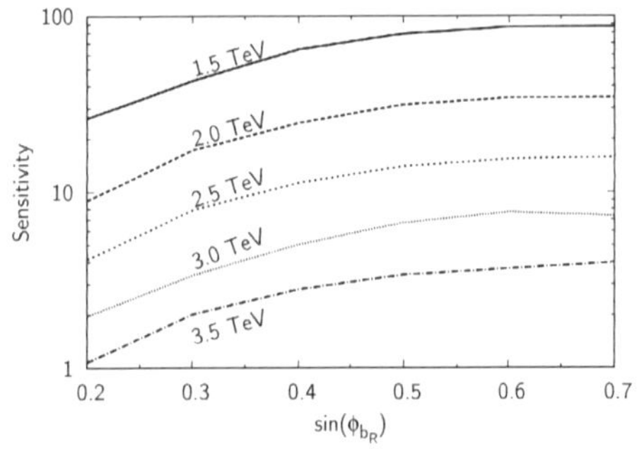

Figure 4. Sensitivity obtained in the $Hb\bar{b}$ channel for different values of the heavy gluon mass as a function of $\sin(\phi_{b_R})$ after an integrated luminosity of 100 fb^{-1}.

sensitivity that can be reached, for different values of M_G, with an integrated luminosity of 100 fb^{-1} as a function of $\sin(\phi_{t_R})$.

3.2. $Hb\bar{b}$ Channel

We now turn our attention to the case $Q = B$ and $q = b$. We will show that contrary to naive expectations, the $H \to b\bar{b}$ decay channel, resulting in a four b quark final state, can be quite competitive, allowing for the reconstruction not only of the heavy quark but also of the heavy gluon. The process we are interested in is

$$pp \to G \to B\bar{b} + \bar{B}b \to Hb\bar{b} \to 4b. \qquad (3)$$

In order to reduce the background to manageable levels we need to require all four b-quarks to be measured, well isolated and also all four of them to be quite hard. Once these cuts are imposed and the signal and background cross sections are comparable, we can take advantage of the particular kinematics of the signal in which the two b-jets coming from the decay of the Higgs are typically softer than the other two. Thus the requirement that the two softest b-jets reconstruct the Higgs mass is a good extra discriminator. Finally, we can reconstruct the B quark using the reconstructed Higgs together with the hardest b-jet and G using the invariant mass of all four b-jets. We use the latter invariant mass as a final discriminating variable. The proposed cuts are

- 4 b-tagged jets with $\Delta R(bb) > 0.7$ and $p_T(b) > 50$ GeV
- No isolated leptons
- $p_T(b_h) > 300$ GeV for the hardest b-jet
- $|M_{b_3b_4} - m_H| < 30$ GeV, where $b_{3,4}$ are the two softest b-jets
- $M_G - 1000 \text{ GeV} < M_{4b} < M_G + 500$ GeV.

Our results are summarized in Fig. 4 in which we show the sensitivity that can be reached with 100 fb^{-1} for different values of M_G as a function of $\sin(\phi_{b_R})$. Further details of the study will be presented elsewhere [14]. A sample of the reconstruction power for the heavy quark and the heavy gluon is shown in Fig. 5 for the particular case of $M_B = 1.25$ TeV and $M_G = 2.5$ TeV.

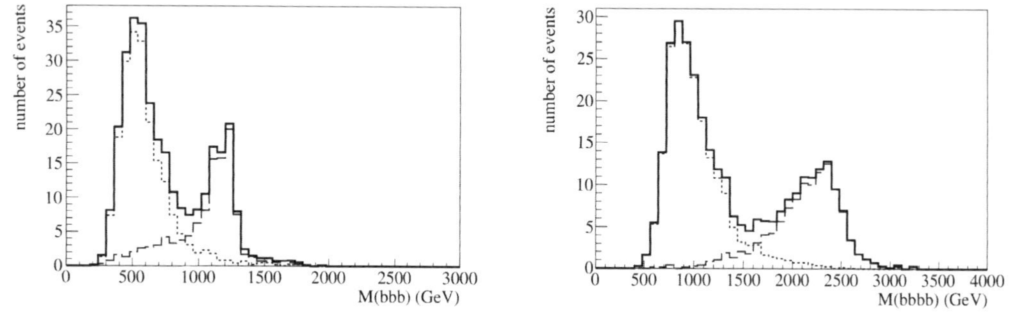

Figure 5. Reconstruction of the heavy quark (left) and the heavy gluon (right) with $m_B = 1250$ GeV and $m_G = 2500$ GeV with an integrated luminosity of 100 fb^{-1}. The dotted, dashed and solid lines correspond to the background, signal and sum of both, respectively.

4. Conclusions

We have discussed how searches involving the Higgs boson can give us valuable information on the top quark and other particles beyond the SM related to the top quark. Current experimental data on the Higgs boson might be pointing to a large correction to the top Yukawa coupling, effectively reversing its sign. This possibility might be tested through single top production in association with a Higgs boson although further studies are required before considering the results of such searches conclusive. We have also shown how searches involving the production of $Ht\bar{t}$ and $Hb\bar{b}$ can give us information on the spectrum of new particles in models of strong EWSB. In particular, they can be used to search for top partners, new vector-like quarks that mix strongly with the top and/or bottom quarks.

Acknowledgments

We would like to thank N. F. Castro and J. P. Araque for the useful discussions. This work has been supported by MICINN projects FPA2006-05294 and FPA2010-17915, through the FPU programme and by Junta de Andalucía projects FQM 101, FQM 03048 and FQM 6552.

References

[1] CMS Collaboration, CMS-PAS-HIG-12-045; ATLAS Collaboration, ATLAS-CONF-2012-170.
[2] Tevatron New Physics Higgs Working Group and CDF and D0 Collaborations, arXiv:1207.0449.
[3] M. Carena, I. Low and C. E. M. Wagner, JHEP **1208** (2012) 060 [arXiv:1206.1082]; N. Arkani-Hamed, K. Blum, R. T. D'Agnolo and J. Fan, JHEP **1301** (2013) 149 [arXiv:1207.4482]; M. Reece, arXiv:1208.1765.
[4] A. Azatov, R. Contino and J. Galloway, JHEP **1204** (2012) 127 [arXiv:1202.3415]; P. P. Giardino, K. Kannike, M. Raidal and A. Strumia, Phys. Lett. B **718** (2012) 469 [arXiv:1207.1347]; J. R. Espinosa, C. Grojean, M. Muhlleitner and M. Trott, JHEP **1212**, 045 (2012) [arXiv:1207.1717]; D. Carmi, A. Falkowski, E. Kuflik, T. Volansky and J. Zupan, JHEP **1210** (2012) 196 [arXiv:1207.1718].
[5] M. Farina, C. Grojean, F. Maltoni, E. Salvioni and A. Thamm, arXiv:1211.3736.
[6] K. Agashe, R. Contino and A. Pomarol, Nucl. Phys. B **719** (2005) 165 [hep-ph/0412089].
[7] C. Bini, R. Contino and N. Vignaroli, JHEP **1201** (2012) 157 [arXiv:1110.6058].
[8] A. Carmona, M. Chala and J. Santiago, JHEP **1207** (2012) 049 [arXiv:1205.2378].
[9] R. Barcelo, A. Carmona, M. Chala, M. Masip, J. Santiago, Nucl. Phys. B **857** (2012) 172 [arXiv:1110.5914].
[10] J. Alwall, P. Demin, S. de Visscher, R. Frederix, M. Herquet, F. Maltoni, T. Plehn and D. L. Rainwater *et al.*, JHEP **0709** (2007) 028 [arXiv:0706.2334].
[11] M. L. Mangano, M. Moretti, F. Piccinini, R. Pittau and A. D. Polosa, JHEP **0307** (2003) 001 [hep-ph/0206293].
[12] T. Sjostrand, S. Mrenna and P. Z. Skands, JHEP **0605** (2006) 026 [hep-ph/0603175].
[13] S. Ovyn, X. Rouby and V. Lemaitre, arXiv:0903.2225.
[14] M. Chala and J. Santiago, *in preparation.*

New physics in top decay

Celine Degrande
Department of Physics, University of Illinois at Urbana-Champaign
1110 W. Green Street, Urbana, IL 61801
E-mail: cdegrand@illinois.edu

Abstract. After a short introduction to effective field theories, most of their features are illustrated using the top decay. The effects of heavy new physics on the top decay are computed and the constraints on the coefficients of the dimension-six operators are derived from the available measurements.

1. Introduction

The outstanding performances of the LHC have led last summer to the discovery of a new particle compatible with the Higgs boson. However, the LHC has not been built only to find the last missing piece of the Standard Model (SM) but also to find new physics. If the new degrees of freedom are light enough to be produced by the LHC, the new physics would manifest itself as new resonances. On the contrary, new physics would appear as anomalous interactions between the known particles if the center of mass energy is lower than the scale of new physics. This paper focus on the second case. The theoretical framework will be introduced in section 2 and will be applied to top decay in section 3.

2. Effective field theories

Effective field theories (EFT) rely on the hierarchy between the energy reached in the experiment and the scale of new physics. Since their ratio can only be used as an expansion parameter if it is smaller than one, an EFT is only valid below the new physics scale, Λ. In the Lagrangian, the new interactions arise from higher dimensional operators suppressed by the new physics scale,

$$\mathcal{L} = \mathcal{L}_{SM} + \sum_{d>4} \sum_{i} \frac{c_i^d}{\Lambda^{d-4}} \mathcal{O}_i^d, \tag{1}$$

where d is the dimension of the operator $\mathcal{O}_i^d$ and c_i^d are coefficients which can be computed from the high energy theory. Far below the new physics scale, only the operators with the lowest dimension are required for a given precision. Their coefficients can be kept as free parameters to be model independent since the number of relevant operators is finite. However, the infinite sum of operators is needed as the new scale is reached by the experiment and the effective theory loses its predictive power.

The new physics at the Tevatron or at the LHC can only be described by an EFT if the associated scale is at or above one TeV. Therefore the low energy degrees of freedom are the SM fields including the Higgs field. We assume also that the SM gauge symmetries as well as the Baryon and Lepton numbers are preserved in agreement with the experimental data. Those requirements imply that only operators with an even dimension can be built [1]. Consequently, the largest new physics contributions are due to the dimension-six operators. While there are already 59 dimension-six operators for one generation of fermions [2, 3], only a few can affect a specific process [2, 4] and they can often be distinguished by their different effects. Due to the symmetries and the scale suppression, the model is more predictive than the anomalous couplings approach and therefore provides some guidance in the quest for new physics. In fact, anomalous couplings Lagrangians only satisfy Lorentz invariance and usually electromagnetic gauge invariance is imposed afterwards while the $SU(2)_L$ is ignored. Only the operators with the lowest number of derivatives are kept without any justification as the scale suppression is lacking. Moreover, the effective Lagrangian can be used for any process because gauge invariance is guaranteed by construction. Like in the SM, gauge invariance requires that vertices with different numbers of gauge bosons are related to each other. On the contrary, all the vertices from an anomalous couplings Lagrangian are independent. Since the effective field theory is renormalizable in the modern sense, it can be used in loop computation. Therefore the parameters can be constrained both by direct and indirect measumements. Finally, unitarity is satisfied in the EFT validity region, *i.e.* below the new scale and no form factors are needed [1]. However, this last point can hardly be illustrated for top decay as the energies of the decay products are bound by the top mass.

3. Top decay

We will focus here on the SM-like top decay, namely $t \to bW$ with the W decaying eventually in a pair of leptons or light quarks. Exotic decay channels can be found for example in Ref. [5].

3.1. The operators

In the massless limit for the b quark[1], only two operators affect the $t \to bW$ decay at the order Λ^{-2}[7],

$$\mathcal{O}_{\phi q}^{(3)} = i\left(\phi^\dagger \tau^i D_\mu \phi\right)\left(\bar{q}\gamma^\mu \tau^i q\right) + h.c. \qquad \text{and} \qquad \mathcal{O}_{tW} = \bar{q}\sigma_{\mu\nu}\tau^i t\tilde{\phi} W_i^{\mu\nu}. \tag{2}$$

Keeping only the terms with the top, the bottom and one W field and comparing to the anomalous Lagrangian [8],

$$\mathcal{L}_{AC} = \frac{g}{\sqrt{2}}\bar{b}\gamma^\mu V_{tb}\left(f_V^L \gamma_L + f_V^R \gamma_R\right) t W_\mu - \frac{ig}{\sqrt{2}m_W}\bar{b}\sigma^{\mu\nu} q_\nu V_{tb}\left(f_V^L \gamma_L + f_V^R \gamma_R\right) t W_\mu + h.c., \tag{3}$$

the prediction of the effective field theory for the anomalous couplings reads

$$\begin{array}{ll} f_V^L = \frac{C_{\phi q}^{(3)} v^2}{V_{tb}\Lambda^2} & f_V^R = 0 \\ f_T^L = 0 & f_T^R = -\sqrt{2}\frac{C_{tW} v^2}{V_{tb}\Lambda^2}. \end{array} \tag{4}$$

As already mentioned, the anonalous Lagrangian in Eq. (3) is not completely generic as further terms with extra momenta can be added. The effective field theory is more predictive since it has two parameters less than anomalous couplings for the Wtb vertex but also because it fixes some of the anomalous couplings of other vertices. For example, the vertices generated by $\mathcal{O}_{tW}$ are shown on Fig. 1. Those vertices are required in higher multiplicity processes like $t \to bW\gamma$ to insure gauge invariance. If only the anomalous tbW vertex from Eq. (3) was added to the SM Lagrangian for this process, the result would not be gauge invariant.

[1] Extra operators can contribute to the $t \to Wb$ decay if this assumption is removed [6].

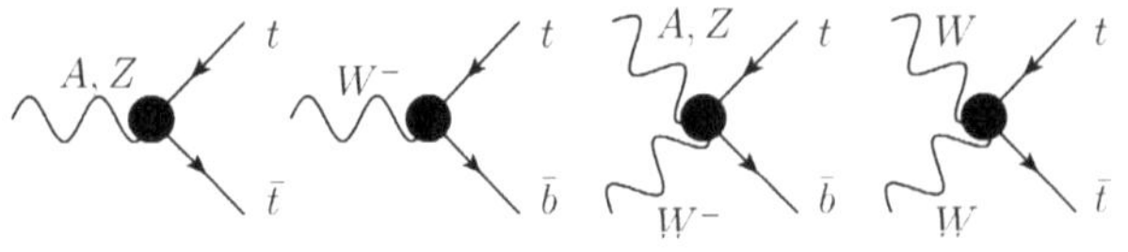

Figure 1. Vertices of the $\mathcal{O}_{tW}$ operators. For each vertex in this figure, there is one extra vertex with an additional Higgs line.

In addition to the operators in Eq. (2), one operator contribute to the $t \to bl\bar{\nu}_l$ decay at $\mathcal{O}\left(\Lambda^{-2}\right)$[5] without the exchange of a W boson[2],

$$\mathcal{O}_{ql}^{(3)} = \left(\bar{q}\gamma^\mu\tau^i q\right)\left(\bar{l}\gamma_\mu\tau^i l\right) \tag{5}$$

and a similar operator contribute to the $t \to bd\bar{u}$ decay. With all those dimension-six operators, the square matrix element for $t \to bl\bar{\nu}_l$ reads

$$\begin{aligned}\frac{1}{2}\Sigma|M|^2 &= \frac{V_{tb}^2 g^4 u(m_t^2-u)}{2(s-m_W^2)^2}\left(1+2\frac{C_{\phi q}^{(3)}v^2}{V_{tb}\Lambda^2}\right)+\frac{4\sqrt{2}\mathrm{Re}C_{tW}V_{tb}m_t m_W}{\Lambda^2}\frac{g^2 su}{(s-m_W^2)^2}\\ &+\frac{4C_{ql}^{(3)}}{\Lambda^2}\frac{g^2u(m_t^2-u)}{s-m_W^2}+\mathcal{O}\left(\Lambda^{-4}\right)\end{aligned} \tag{6}$$

where $s \equiv (p_t - p_b)^2$ and $u \equiv (p_t - p_e)^2$. The contribution of $\mathcal{O}_{\phi q}^{(3)}$ is proportional to the SM one because they generate the same Wtb vertex up to a global factor.

3.2. The width

The width is affected mainly by $\mathcal{O}_{\phi q}^{(3)}$ and $\mathcal{O}_{tW}$,

$$\frac{\Gamma\left(t \to be^+\nu_e\right)}{GeV} = 0.1541 + \left[0.019\frac{C_{\phi q}^{(3)}}{\Lambda^2} + 0.026\frac{C_{tW}}{\Lambda^2} + 0\frac{C_{ql}^{(3)}}{\Lambda^2}\right]\mathrm{TeV}^2. \tag{7}$$

The contribution of the four-fermion operator almost vanishes. The measured value [9] together with the SM prediction [10] give a first constraint on the coefficients of the two relevant operators,

$$\left.\begin{aligned}\Gamma_{meas} &= 2^{+0.47}_{-0.43}\,\mathrm{GeV}\\ \Gamma_{SM} &= 1.33\,\mathrm{GeV}\end{aligned}\right\}\frac{C_{\phi q}^{(3)}}{\Lambda^2} + 1.35\frac{C_{tW}}{\Lambda^2} = 4^{+2.8}_{-2.5}\mathrm{TeV}^{-2}. \tag{8}$$

3.3. The W helicities

The helicity of the W boson affects the angular distribution of top decay products defined by the helicity fractions,

$$\frac{1}{\Gamma}\frac{d\Gamma}{d\cos\theta} \equiv \frac{3}{8}(1+\cos\theta)^2F_R + \frac{3}{8}(1-\cos\theta)^2F_L + \frac{3}{4}\sin^2\theta F_0 \tag{9}$$

where θ is the angle between the top and the neutrino momenta in the W rest frame. The helicity fractions are only affected by $\mathcal{O}_{tW}$,

$$\begin{aligned}F_0 &= \frac{m_t^2}{m_t^2+2m_W^2} - \frac{4\sqrt{2}\mathrm{Re}C_{tW}v^2}{\Lambda^2 V_{tb}}\frac{m_t m_W(m_t^2-m_W^2)}{(m_t^2+2m_W^2)^2}\\ F_L &= \frac{2m_W^2}{m_t^2+2m_W^2} + \frac{4\sqrt{2}\mathrm{Re}C_{tW}v^2}{\Lambda^2 V_{tb}}\frac{m_t m_W(m_t^2-m_W^2)}{(m_t^2+2m_W^2)^2}\\ F_R &= 0.\end{aligned} \tag{10}$$

[2] As a matter of fact, only one operator interfere with the SM

F_R vanishes because the b quark is massless and the sum of the helicity fractions is one by definition. The SM prediction at NNLO [11] and the Atlas result [12] with the constraints $F_0 + F_L + F_R = 1$ and $F_R = 0$ strongly limit the coefficient of $\mathcal{O}_{tW}$

$$\left.\begin{aligned} F_0^{SM} &= 0.687 \pm 5 \\ F_0^{meas} &= 0.66 \pm 5 \end{aligned}\right\} \frac{C_{tW}}{\Lambda^2} = 0.44 \pm 0.9\text{TeV}^{-2}. \tag{11}$$

Together with the constraint from the total width (8), this result fixes the range allowed for the coefficient of $\mathcal{O}_{\phi q}^{(3)}$

$$\frac{C_{\phi q}^{(3)}}{\Lambda^2} = 3.4^{+4}_{-3.7}\text{TeV}^{-2}. \tag{12}$$

3.4. The leptons invariant mass

As shown in Eq. (6), the contribution of the four-fermion operator only differs from the SM by the power of the denominator. In fact, the remaining $(s - m_W^2)$ factor is due to the SM amplitude. Consequently, $\mathcal{O}_{ql}^{(3)}$ contribution on the leptons invariant mass changes of sign at the W mass as displayed on Fig. 2. Accidentally, the contributions from the two regions are

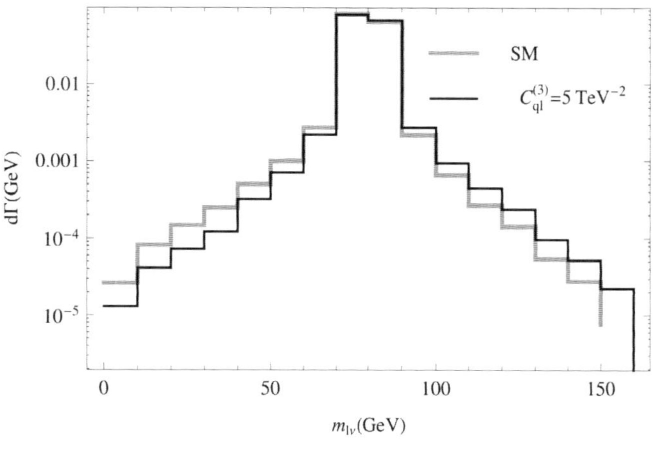

Figure 2. Invariant mass of the leptons pair in the top decay $t \to be\bar{\nu}_e$ for the SM and the SM and $\mathcal{O}_{ql}^{(3)}$ at the order Λ^{-2}, *i.e.* only the interference between the SM and the four-fermion operator is added to the SM distribution.

almost equal in modulus such that their sum nearly vanishes. The effect of the corresponding four-fermion operator for the light quarks is identical. The four-fermion operator cannot be constrained yet because the leptons invariant mass has not been measured so far.

3.5. An example of indirect constraints

A modification of the Wtb vertex affects the two points function of the W boson at the loop level as shown on the left diagram in Fig. 3. However, those corrections have infinities and are useless unless they can be absorbed using renormalization. Contrary to anomalous couplings, an EFT is renormalizable order by order in Λ. The contribution from $\mathcal{O}_{tW}$ to the $\widehat{S}$ parameter

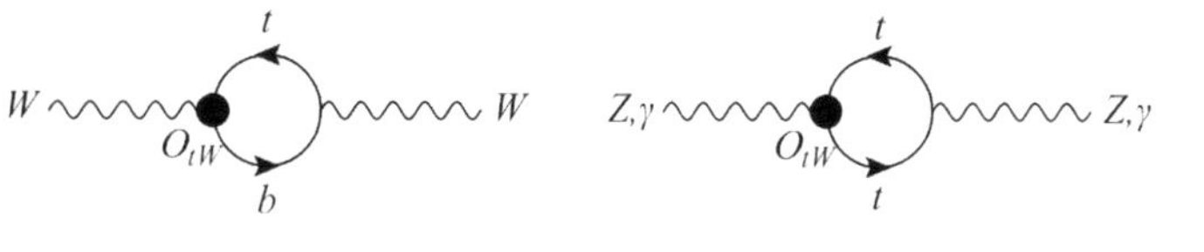

Figure 3. One loop corrections to the two points functions of the electroweak bosons due to the operator $\mathcal{O}_{tW}$.

is infinite and absorbed by the renormalization of the coefficient of a dimension-six operator affecting $\widehat{S}$ at the tree-level while the contribution to $\widehat{T}$ vanishes. However, the correction to the $\widehat{U}$ parameter is finite [13] and can be used to constrain the coefficient of this operator,

$$\left.\begin{array}{r}\widehat{U}=N_c\frac{gC_{tW}}{4\pi^2}\frac{\sqrt{2}vm_t}{4\Lambda^2}\\ \widehat{U}_{meas}=(-5.0\pm 8.4)\times 10^{-4}\end{array}\right\}\frac{C_{tW}}{\Lambda^2}=-0.7\pm 1.1\ \mathrm{TeV}^{-2}. \quad (13)$$

This constraint is comparable and in agreement with the one derived from the W helicities and presented in Eq. (11). Similar indirect constraints can be obtained from other processes as well. For example, the Wtb vertex also alters $b\to s\gamma$ [14].

4. Conclusion

If the new physics is heavy, it should affect the top width, the W helicity fractions and the invariant mass distribution of the leptons or light quarks according to the effective extension of the SM. All the four dimension-six operators relevant for the top decay are or can be constrained by measuring those quantities. However, the top width is poorly known and the invariant mass distribution has no been measured yet. Those operators can also be constrained using indirect measurements because effective field theories are renormalizable in the moderne sense. We have shown in addition that the effective field approach is more predictive and simpler, i.e. guarantee gauge invariance, than the alternative anomalous couplings approach. Finally, EFT can now be easily implemented into MadGraph thanks to the new FeynRules interface [15] and ALOHA [16]. In particular, the effective theory for top decay is available in MadGraph5 [17].

References

[1] Degrande C, Greiner N, Kilian W, Mattelaer O, Mebane H *et al.* 2012 (*Preprint* 1205.4231)
[2] Buchmuller W and Wyler D 1986 *Nucl.Phys.* **B268** 621
[3] Grzadkowski B, Iskrzynski M, Misiak M and Rosiek J 2010 *JHEP* **1010** 085 and ref. [10-13] therein (*Preprint* 1008.4884)
[4] Leung C N, Love S and Rao S 1986 *Z.Phys.* **C31** 433
[5] Aguilar-Saavedra J 2011 *Nucl.Phys.* **B843** 638–672 (*Preprint* 1008.3562)
[6] del Aguila F, Perez-Victoria M and Santiago J 2000 *Phys.Lett.* **B492** 98–106 (*Preprint* hep-ph/0007160)
[7] Zhang C and Willenbrock S 2010 *Nuovo Cim.* **C033N4** 285–291 (*Preprint* 1008.3155)
[8] Kane G L, Ladinsky G and Yuan C 1992 *Phys.Rev.* **D45** 124–141
[9] Abazov V M *et al.* (D0 Collaboration) 2012 *Phys.Rev.* **D85** 091104 (*Preprint* 1201.4156)
[10] Jezabek M and Kuhn J H 1989 *Nucl.Phys.* **B314** 1
[11] Czarnecki A, Korner J G and Piclum J H 2010 *Phys.Rev.* **D81** 111503 (*Preprint* 1005.2625)
[12] Aad G *et al.* (ATLAS Collaboration) 2012 *JHEP* **1206** 088 (*Preprint* 1205.2484)
[13] Greiner N, Willenbrock S and Zhang C 2011 *Phys.Lett.* **B704** 218–222 (*Preprint* 1104.3122)
[14] Martinez R, Perez M and Toscano J 1994 *Phys.Lett.* **B340** 91–95
[15] Degrande C, Duhr C, Fuks B, Grellscheid D, Mattelaer O *et al.* 2012 *Comput.Phys.Commun.* **183** 1201–1214 (*Preprint* 1108.2040)
[16] de Aquino P, Link W, Maltoni F, Mattelaer O and Stelzer T 2012 *Comput.Phys.Commun.* **183** 2254–2263 (*Preprint* 1108.2041)
[17] URL https://cp3.irmp.ucl.ac.be/projects/madgraph/wiki/Models/TopEffTh

Journal of Physics: Conference Series **452** (2013) 012010 doi:10.1088/1742-6596/452/1/012010

LHC searches for physics beyond the Standard Model with top quarks

Tobias Golling, on behalf of the ATLAS and CMS Collaborations
Department of Physics, Yale University, New Haven, CT, USA
E-mail: tobias.golling@yale.edu

Abstract. Searches are presented for physics beyond the Standard Model involving top-quark and related signatures. The results are based on proton-proton collision data corresponding to integrated luminosities between 1 fb^{-1} and 5 fb^{-1} collected at a center-of-mass energy of 7 TeV with the ATLAS and CMS detectors at the Large Hadron Collider in 2011. The data are found to be consistent with the Standard Model. The non-observation of a signal is converted to limits at the 95% confidence level on the production cross section times branching ratio and on the masses of the hypothesized new particles for appropriate benchmark models.

1. Introduction

The Standard Model of particle physics is believed to be an effective theory valid up to energies close to 1 TeV. However, no new physics beyond the Standard Model (SM) has been observed yet, and it is critical to explore a wide range of possible signatures. A promising avenue lies in final states that involve the heaviest of the particles presumed to be elementary, the top quark. This document summarizes the status of these searches using the ATLAS [1] and CMS [2] detectors at the CERN Large Hadron Collider (LHC). The results are based on proton-proton collision data corresponding to integrated luminosities between 1 fb^{-1} and 5 fb^{-1} collected at a center-of-mass energy of 7 TeV in 2011.

The fact that the top quark is the heaviest elementary particle might be a hint that it plays a special role in the theory of electroweak symmetry breaking. The so-called hierarchy problem refers to the fact that the large quantum contributions to the square of the Higgs-boson mass should make the Higgs mass many orders of magnitude larger than the electroweak scale. Either there is an incredible, unnatural fine-tuning cancellation or nature chose another mechanism to protect the Higgs mass to keep it at the observed low value [3, 4]. Many models introduce top partners to cancel these quantum corrections. Finally, there are tantalizing hints of new physics in the forward-backward asymmetry in $t\bar{t}$ events at the Tevatron [5, 6].

Benchmark signal models are used to define signature-based searches in final states involving one or two leptons (electrons and muons), jets, of which typically one or more are required to be identified as b-jets, and missing transverse momentum ($E_{\mathrm{T}}^{\mathrm{miss}}$). The benchmarks include models of fourth-generation and vector-like quarks [7, 8, 9, 10, 11, 12, 13, 14, 15, 16, 17, 18, 19, 20], top partners in little Higgs models [21, 22], as well as non-SM production of four top quarks [14], same-sign top-quark pair (tt) production [15], top+jet resonances in $t\bar{t}$+jet events [23, 24], $W' \to t\bar{b}$ resonances [25, 26], and flavor changing neutral currents (FCNC) in single top-quark production [27]. Searches for $t\bar{t}$ resonances, third generation supersymmetry and new physics in top-quark decays and properties are covered elsewhere in these proceedings [28].

2. Experimental techniques

The experimental approaches are similar between ATLAS and CMS. Searches are based on the l+jets or dilepton $t\bar{t}$ channels. Jets are reconstructed from three-dimensional calorimeter energy clusters using the anti-k_t jet clustering algorithm [29] with a radius parameter of 0.4 for ATLAS and 0.5 for CMS. Leptons are required to pass quality criteria and to be isolated [30, 31, 32, 33]. The transverse momenta of jets and leptons are typically required to be larger than 20 or 25 GeV or more. Multivariate tagging algorithms [34, 35] are used to identify b-jets.

Typically, the statistical tool BumpHunter [36] is used to check for deviations, an excess or deficit, from the background hypothesis. In the absence of a signal, cross-section and mass limits are derived for benchmark models using CLs [37, 38] or occasionally Bayesian methods [39]. All limits quoted in this document are obtained at the 95% confidence level. Unless mentioned otherwise the new particles are assumed to decay with 100% branching fraction to the corresponding final state under study.

The largest background typically originates from SM $t\bar{t}$ production and is estimated from Monte Carlo (MC) simulation using MC@NLO [40, 41] or POWHEG [42] at ATLAS, and MADGRAPH [43, 44] at CMS. Data-driven multi-jet background estimates are based on the matrix method and on binned likelihood fits to the $E_{\mathrm{T}}^{\mathrm{miss}}$ distribution. Data-driven W+jets estimates use the inherent W charge asymmetry in pp collisions. ALPGEN [45] (ATLAS) and MADGRAPH (CMS) are used to model W+jets in MC simulation. The composition of the flavor of the quarks produced in association with the W boson is measured in the low jet multiplicity bins and extrapolated using MC simulations. Small backgrounds, like the production of single top quarks, boson pairs, $t\bar{t}+W$, $t\bar{t}+Z$, and Z+jets are typically estimated from MC simulations.

The leading systematic uncertainties typically originate from the uncertainty in the jet energy scale, the b-tagging efficiency, and the $t\bar{t}$ MC modeling. In certain cases the impact of systematic uncertainties on the sensitivity of the search is reduced by using Gaussian constraints or other marginalization techniques, see e.g. Ref. [10].

3. Results

An ATLAS analysis [7] searches for the pair production of heavy non-SM quarks Q with decays according to $Q\bar{Q} \to W^+qW^-\bar{q}$ with $q = d, s, b$ for up-type Q or $q = u, c$ for down-type Q. The search is performed with 1.04 fb^{-1} of integrated luminosity. Dilepton final states are selected, requiring large $E_{\mathrm{T}}^{\mathrm{miss}}$ and at least two jets. No b-tagging is applied. Mass reconstruction of heavy quark candidates is performed by assuming that the W-boson decay products are nearly collinear. The resulting mass reconstruction is shown in Fig. 1. No deviation from the SM expectation is observed. Heavy non-SM quark masses below 350 GeV are excluded.

A similar CMS search [8] for pair production of heavy top-like quarks t' has been performed in the decay mode $t'\bar{t}' \to W^+bW^-\bar{b}$. The search uses 5.0 fb^{-1} of integrated luminosity. Again dilepton final states are selected, requiring large $E_{\mathrm{T}}^{\mathrm{miss}}$ and at least two jets, but this time exactly two of the jets have to be identified as b-jets. The minimum value of the four possible masses of the system defined by one of the two leptons and one of the two b-jets (M_{lb}^{min}) is found to be a good variable for distinguishing $t'\bar{t}'$ from $t\bar{t}$ events, as can be seen in Fig. 2. The observed number of events agrees with the expectation from SM processes. Heavy t' quarks with a mass less than 557 GeV are excluded.

Searches for heavy top-like quarks t' given the hypothesized decay mode $t'\bar{t}' \to W^+bW^-\bar{b}$ are also conducted in final states with a single charged lepton, $E_{\mathrm{T}}^{\mathrm{miss}}$ and, depending on the analysis, at least three or four jets, of which at least one must be identified as a b-jet [9, 10, 11].

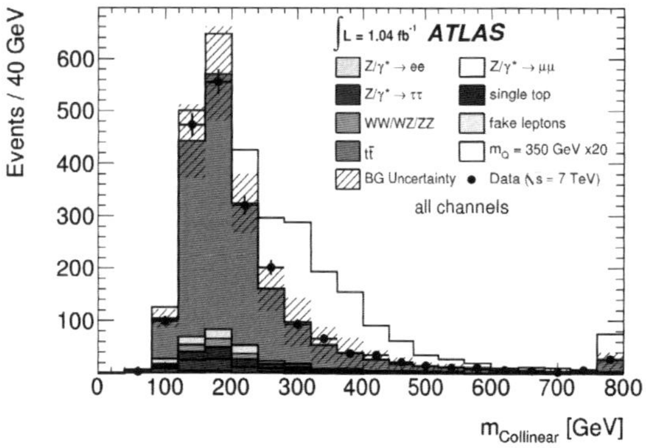

Figure 1. Expected and observed distributions of the collinear mass for the sum of ee, $\mu\mu$ and $e\mu$ channels [7]. The last bin contains overflow events. Samples are stacked. The signal has been amplified to 20 times the expected rate.

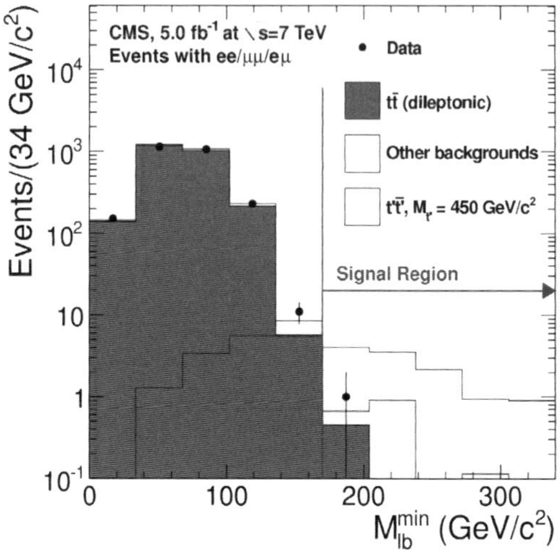

Figure 2. Comparison between the data and the simulated background for $M_{lb}^{\min}$ [8]. The expected distribution for a signal with $m_{t'} = 450$ GeV is also shown. The signal region is defined by $M_{lb}^{\min} > 170$ GeV.

CMS [9] uses the reconstructed t' mass $m_{\rm reco}$, obtained from a kinematic fit of the reconstructed four-momenta to the decay hypothesis $t'\bar{t}' \to W^+bW^-\bar{b} \to l\nu bq\bar{q}'\bar{b}$, as well as $H_{\rm T}$, defined as the scalar sum of the transverse momenta of the objects associated to the t' and $\bar{t}'$ decay products. The two-dimensional distributions of $H_{\rm T}$ versus $m_{\rm reco}$ are fitted with analytic functions for the signal S and the background B. All two-dimensional bins are then sorted in increasing order of the expected S/B ratio, using the functions. The resulting distribution of this so-called S/B rank is shown in Fig. 3 and is used to exclude t' masses below 570 GeV with 5.0 fb^{-1} of integrated luminosity.

ATLAS [11] uses the reconstructed mass $m_{\rm reco}$ of the candidate t' as the discriminant. A tight selection is applied targeting t' masses above 400 GeV. In this mass range the decay products have large momenta. The reconstruction of hadronically-decaying W bosons $W_{\rm had}$ takes advantage of this. They are either defined as a single jet with $p_T > 250$ GeV and jet mass in the range of 60-110 GeV or as a dijet system with $p_T > 150$ GeV, angular separation $\Delta R(j,j) < 0.8$, and mass within the range of 60-110 GeV. Additional kinematic selection criteria include $H_{\rm T}$ (defined as the scalar sum of the transverse momenta of the lepton, $E_{\rm T}^{\rm miss}$, and the jets from the hypothesized t' decays) to be larger than 750 GeV, $\Delta R(l,\nu) < 1.4$, $\Delta R(W_{\rm had}, b\text{-jet}) > 1.4$, and $\Delta R(l, b\text{-jet}) > 1.4$. With 4.7 fb^{-1} of integrated luminosity a t' quark with mass lower than 656 GeV is excluded.

In addition, in light of the recent discovery of a new boson of mass 126 GeV at the LHC, upper limits are derived, as shown in Fig. 4, for vector-like quarks of various masses in the two-dimensional plane of BR($t' \to Wb$) versus BR($t' \to Ht$), where H is the SM Higgs boson (BR($t' \to Zt$) = 1− BR($t' \to Wb$) − BR($t' \to Ht$).)

A search for pair-produced, heavy, vector-like charge-2/3 quarks is performed by CMS [12], assuming BR($t' \to Zt$)= 1. Events are selected by requiring two charged leptons from the Z-boson decay, as well as an additional isolated charged lepton. Using 1.14 fb^{-1} of integrated luminosity t' quarks with a mass less than 475 GeV are excluded.

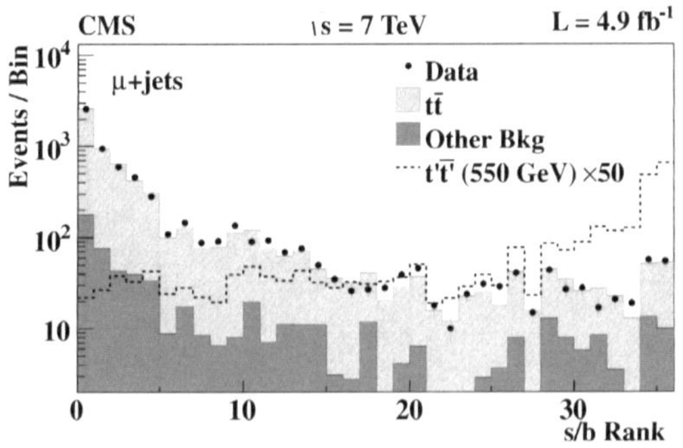

Figure 3. Number of events per bin in the two-dimensional H_T versus m_{reco} histogram, as a function of the S/B rank for the μ+jets channel [9].

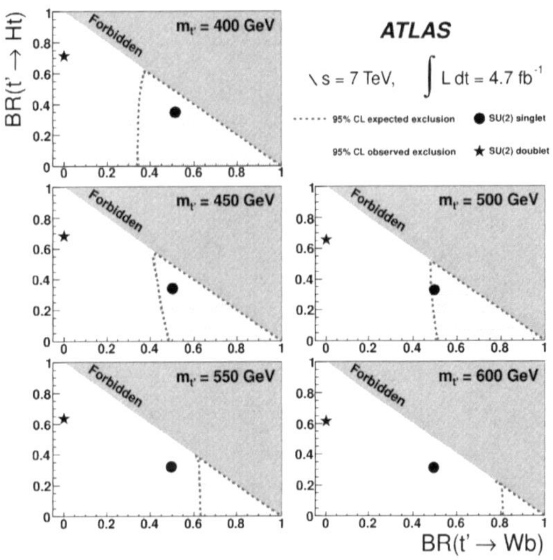

Figure 4. Observed (red filled area) and expected (red dashed line) exclusion limits in the plane of BR($t' \to Wb$) vs BR($t' \to Ht$) for different values of the vector-like t' quark mass [11]. The grey solid area corresponds to the unphysical region where the sum of the branching ratios exceeds unity.

Both ATLAS and CMS conduct searches for heavy pair-produced bottom-like quarks. These b' quarks are assumed to decay exclusively to Wt. The $W^+tW^-\bar{t}$ final state has the distinctive signature of three or more leptons or two leptons of same charge which is exploited.

CMS uses 4.9 fb^{-1} of integrated luminosity to select trilepton and same-sign-dilepton events with E_T^{miss} [13]. At least one jet must be identified as a b-jet. Events are rejected where any two leptons of the same flavor have an invariant mass consistent with the Z-boson mass. Furthermore, S_T, defined as the scalar sum of the transverse momenta of the leptons, E_T^{miss}, and the jets, is used to reject background, as shown in Fig. 5. b' quarks with mass below 611 GeV are excluded.

A similar analysis by ATLAS focuses on same-sign-dilepton events using 4.7 fb^{-1} of integrated luminosity [14]. The most important background is the contribution arising from fake leptons. Another significant background is from various SM processes where two real leptons are produced, but in which one of the leptons has a mis-identified charge. Additional signal hypotheses are considered for the interpretation of the results. Both single and pair production of new heavy quarks $T_{5/3}$, with charge 5/3, are considered with $T_{5/3} \to W^+t$. For the single production the assumed coupling constant λ of the $tWT_{5/3}$ vertex is varied as shown in Fig. 6. Assuming only $b'\bar{b}'$ or $T_{5/3}\bar{T}_{5/3}$ production quark masses below 670 GeV are excluded. In addition, first limits are set on non-SM production of four top quarks, yielding $\sigma_{4\mathrm{tops}} < 61$ fb. In an earlier version of the ATLAS same-sign-dilepton search [15] using 1.04 fb^{-1} of integrated luminosity the results are also interpreted for same-sign top-quark pair production. The results leave little room to explain the measurement of the forward-backward asymmetry in top-quark pair production at the Tevatron by a flavor-changing Z' boson.

CMS excludes $T_{5/3}$ masses below 645 GeV by analyzing very similar same-sign-dilepton final states, assuming $T_{5/3}\bar{T}_{5/3}$ production and using 5.0 fb^{-1} of integrated luminosity [16].

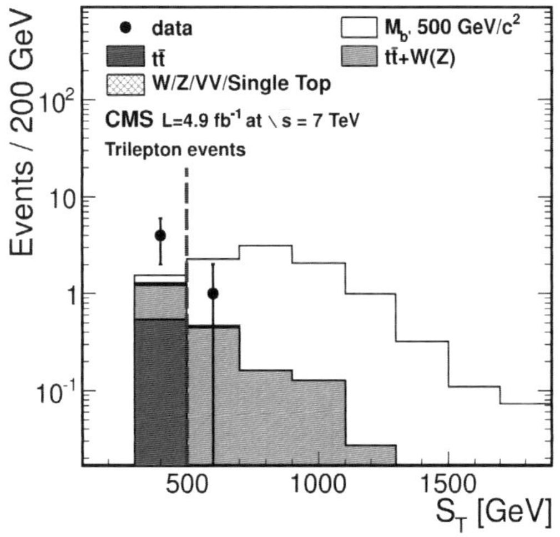

Figure 5. The S_{T} distributions for the trilepton channel [13]. The vertical dotted line indicates the lower S_{T} threshold used in the analysis.

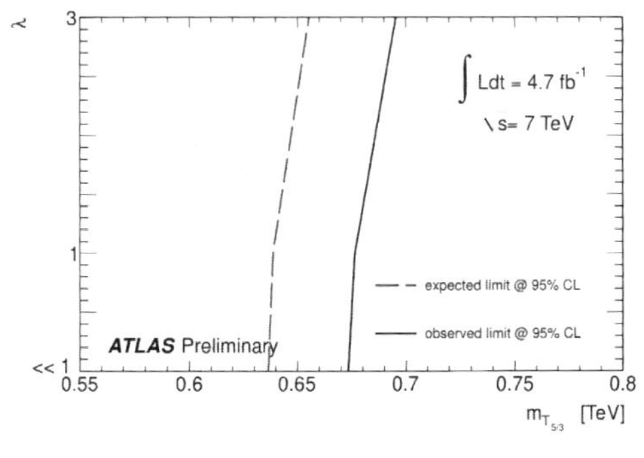

Figure 6. Expected and observed lower limits on the $T_{5/3}$ signal as a function of the $T_{5/3}$ mass and the coupling constant λ [14]. The shaded area is excluded.

Also the 1-lepton channel is used by ATLAS to set limits on $b'\bar{b}'$ production with $b' \to W^+t$ [17]. Similar to Ref. [11] hadronically decaying W bosons are reconstructed. The limits are not competitive since only 1.04 fb^{-1} of integrated luminosity was used for these results.

A combined search in the 1-lepton, same-sign-dilepton and trilepton final states is presented by CMS by assuming both single and pair production of both t' and b' [18]. By analyzing S_{T} and the invariant bW mass in bins of number of reconstructed hadronically decaying W bosons and b-jets, limits are presented as a function of the mass difference between t' and b' and as a function of the matrix element V_{tb}.

Both ATLAS [19] and CMS [20] set limits on $b'\bar{b}'$ production with at least one b' decaying to a Z boson and a bottom quark. Using 2.0 fb^{-1} of integrated luminosity ATLAS excludes vector-like singlet b' quarks mixing solely with the third SM generation with masses below 358 GeV. Assuming BR($b' \to Zb$)= 1 CMS excludes b' masses below 550 GeV using 4.9 fb^{-1}.

Searches for signatures of pair production of supersymmetric top partners also have sensitivity to spin-1/2 top partners in little Higgs models as shown by ATLAS analyses presented in Refs. [21, 22].

Searches for new heavy resonances, a color singlet W' or a color triplet ϕ, produced in association with a top quark are motivated by top-flavor violating processes designed to explain the $t\bar{t}$ forward-backward asymmetry observed at the Tevatron. Two-dimensional limits are set on the mass and the coupling of W' and ϕ by ATLAS [23] by analyzing the $\bar{t}$+jet and the t+jet invariant mass, respectively, in $t\bar{t}$+jet candidate events. Similar limits are set by CMS [24]. The limits leave little room for top-flavor violating processes to explain the $t\bar{t}$ forward-backward asymmetry observed at the Tevatron.

Both ATLAS [25] and CMS [26] search for resonances in the $t\bar{b}$ (and c.c.) spectrum and set lower limits on the mass of right handed $W' \to t\bar{b}$ of 1.85 TeV. ATLAS uses the invariant $t\bar{b}$ mass as the discriminant, while CMS makes use of boosted decision trees.

ATLAS uses a neural network analysis to search for FCNC single top-quark production [27]. Two-dimensional limits are set in the plane of BR($t \to ug$) and BR($t \to cg$).

4. Conclusions

Top quarks play an important role in ATLAS and CMS searches for physics beyond the SM. No hints of new phenomena could be established yet. Results from LHC *pp* collisions at a center-of-mass energy of 8 TeV and eventually close to 14 TeV are anticipated with great suspense.

Acknowledgments

The author is supported by grants from the Department of Energy Office of Science and by the Alfred P. Sloan Foundation.

References

[1] ATLAS Collaboration, JINST **3**, S08003 (2008).
[2] CMS Collaboration, JINST **3**, S08004 (2008).
[3] ATLAS Collaboration, Phys. Lett. B **716**, 1 (2012) [arXiv:1207.7214 [hep-ex]].
[4] CMS Collaboration, Phys. Lett. B **716**, 30 (2012) [arXiv:1207.7235 [hep-ex]].
[5] T. Aaltonen *et al.* [CDF Collaboration], Phys. Rev. D **83**, 112003 (2011) [arXiv:1101.0034 [hep-ex]].
[6] V. M. Abazov *et al.* [D0 Collaboration], Phys. Rev. D **84**, 112005 (2011) [arXiv:1107.4995 [hep-ex]].
[7] ATLAS Collaboration, Phys. Rev. D **86**, 012007 (2012) [arXiv:1202.3389 [hep-ex]].
[8] CMS Collaboration, Phys. Lett. B **716**, 103 (2012) [arXiv:1203.5410 [hep-ex]].
[9] CMS Collaboration, Phys. Lett. B **718**, 307 (2012) [arXiv:1209.0471 [hep-ex]].
[10] ATLAS Collaboration, Phys. Rev. Lett. **108**, 261802 (2012) [arXiv:1202.3076 [hep-ex]].
[11] ATLAS Collaboration, arXiv:1210.5468 [hep-ex] (2012).
[12] CMS Collaboration, Phys. Rev. Lett. **107**, 271802 (2011) [arXiv:1109.4985 [hep-ex]].
[13] CMS Collaboration, JHEP **1205**, 123 (2012) [arXiv:1204.1088 [hep-ex]].
[14] ATLAS Collaboration, ATLAS-CONF-2012-130, https://cds.cern.ch/record/1478217 (2012).
[15] ATLAS Collaboration, JHEP **1204**, 069 (2012) [arXiv:1202.5520 [hep-ex]].
[16] CMS Collaboration, CMS-PAS-B2G-12-003, https://cds.cern.ch/record/1478430 (2012).
[17] ATLAS Collaboration, Phys. Rev. Lett. **109**, 032001 (2012) [arXiv:1202.6540 [hep-ex]].
[18] CMS Collaboration, arXiv:1209.1062 [hep-ex] (2012).
[19] ATLAS Collaboration, Phys. Rev. Lett. **109**, 071801 (2012) [arXiv:1204.1265 [hep-ex]].
[20] CMS Collaboration, CMS-PAS-EXO-11-066, https://cds.cern.ch/record/1460386 (2011).
[21] ATLAS Collaboration, Phys. Rev. Lett. **108**, 041805 (2012) [arXiv:1109.4725 [hep-ex]].
[22] ATLAS Collaboration, JHEP **1211**, 094 (2012) [arXiv:1209.4186 [hep-ex]].
[23] ATLAS Collaboration, arXiv:1209.6593 [hep-ex] (2012).
[24] CMS Collaboration, Phys. Lett. B **717**, 351 (2012) [arXiv:1206.3921 [hep-ex]].
[25] ATLAS Collaboration, Phys. Rev. Lett. **109**, 081801 (2012) [arXiv:1205.1016 [hep-ex]].
[26] CMS Collaboration, arXiv:1208.0956 [hep-ex] (2012).
[27] ATLAS Collaboration, Phys. Lett. B **712**, 351 (2012) [arXiv:1203.0529 [hep-ex]].
[28] See contributions in these proceedings by J.-F. Arguin, S. Fleischmann and D. Pagano.
[29] M. Cacciari, G. P. Salam and G. Soyez, JHEP **0804**, 063 (2008) [arXiv:0802.1189 [hep-ph]].
[30] ATLAS Collaboration, Eur. Phys. J. C **72**, 1909 (2012) [arXiv:1110.3174 [hep-ex]].
[31] ATLAS Collaboration, ATLAS-CONF-2011-063, https://cds.cern.ch/record/1345743 (2011).
[32] CMS Collaboration, CMS-PAS-EGM-10-004, https://cds.cern.ch/record/1299116 (2010).
[33] CMS Collaboration, CMS-PAS-PFT-10-003, https://cds.cern.ch/record/1279347 (2010).
[34] ATLAS Collaboration, ATLAS-CONF-2012-043, https://cds.cern.ch/record/1435197 (2012).
[35] CMS Collaboration, arXiv:1211.4462 [hep-ex] (2012).
[36] G. Choudalakis, arXiv:1101.0390 (2011).
[37] T. Junk, Nucl. Instrum. Meth., vol. A434, pp. 435443 (1999).
[38] A. L. Read, vol. 28, no. 10, pp. 26932704 (2002).
[39] I. Bertram *et al.* [D0 Collaboration] FERMILAB-TM-2104 (2000).
[40] S. Frixione and B. R. Webber, JHEP **0206**, 029 (2002) [hep-ph/0204244].
[41] S. Frixione, P. Nason and B. R. Webber, JHEP **0308**, 007 (2003) [hep-ph/0305252].
[42] S. Frixione, P. Nason and C. Oleari, JHEP **0711**, 070 (2007) [arXiv:0709.2092 [hep-ph]].
[43] F. Maltoni and T. Stelzer, JHEP **0302**, 027 (2003) [hep-ph/0208156].
[44] J. Alwall, P. Demin, S. de Visscher, R. Frederix, M. Herquet, F. Maltoni, T. Plehn and D. L. Rainwater *et al.*, JHEP **0709**, 028 (2007) [arXiv:0706.2334 [hep-ph]].
[45] M. L. Mangano, M. Moretti, F. Piccinini, R. Pittau and A. D. Polosa, JHEP **0307**, 001 (2003) [hep-ph/0206293].

Measurements of new physics in top quark decay at LHC

Davide Pagano for the ATLAS and CMS collaborations
Université Catholique de Louvain
E-mail: davide.pagano@uclouvain.be

Abstract. A summary of ATLAS and CMS results on searches for new physics in top quark decay is presented. Three analysis are reported: the ATLAS and CMS searches for flavor changing neutral currents (FCNC) in top-quark decays and the CMS search for baryon number violation (BNV) in top-quark decays. ATLAS and CMS provided exclusion limits (at 95% CL) on the FCNC decay $t \to Zq$, using respectively 2.1 fb^{-1} and 5.0 fb^{-1} of pp collisions at $\sqrt{s}$ = 7 TeV collected in 2011. Upper limits on $B(t \to Zq)$ are respectively 0.73% and 0.24%. CMS has also set an exclusion limit (at 95% CL) on the BNV decay $t \to lbq$ using 5.0 fb^{-1} of data. Upper limit on $B(t \to lbq)$ is 0.67%.

1. Searches for Flavour Changing Neutral Currents in Top Quark Decays

In the Standard Model (SM), the top quark decays to a W boson and b-quark with a branching fraction of nearly 100%. Flavour changing neutral current (FCNC) decays, although allowed in the SM at the level of quantum correction, are highly suppressed by the GIM mechanism, with typical branching ratios (BR) of the order of 10^{-14}.

However, some SM extensions predict much higher values for FCNC decays involving the top-quark. Examples are supersymmetric models with R-parity violation, quark-singlet model and topcolor-assisted technicolor models, where the BR for FCNC decays can be of the order of 10^{-4} [1][2]. For this reason, direct searches for the flavor changing neutral currents in top quark decays have already been performed at Tevatron. CDF and D0 experiments set an upper limit on $B(t \to Zq)$ of 3.7% [3] and 3.2% [4] at the 95% confidence level (CL) respectively. The same search has been performed at LHC by ATLAS [5] and CMS [6] experiments, using respectively 2.1 fb^{-1} and 5.0 fb^{-1} of pp collision at $\sqrt{s} = 7$ TeV.

1.1. Search for the FCNC decay $t \to Zq$ with the ATLAS detector

ATLAS searched for $t \to Zq$ in $t\bar{t}$ events where the top (antitop) quark decayed to Zq, while the antitop (top) quark decayed to Wb [7]. Only the leptonic decays of the Z and W bosons were considered, resulting in a final state with three isolated charged leptons and at least two jets.

Two event selections were implemented, based on different requirements on the lepton candidates: in the so-called "3ID"analysis, all three lepton candidates were required to be reconstructed and identified using the full ATLAS detector ("ID"lepton candidates); in the so-called "2ID+TL"analysis, only two "ID"candidates were requested, while the third lepton candidate was required to be reconstructed using the ATLAS inner detector only ("TL"lepton candidate).

	3ID	2ID+TL
ZZ and WZ	9.5 ± 4.4	$1.0 \pm ^{0.5}_{0.6}$
$t\bar{t}W$ and $t\bar{t}Z$	0.51 ± 0.14	0.25 ± 0.05
$t\bar{t}$, WW	0.07 ± 0.02	
Z+jets	1.7 ± 0.7	7.6 ± 2.2
Single top	0.01 ± 0.01	
2+3 fake leptons	$0.0 \pm ^{0.2}_{0.0}$	
Expected background	11.8 ± 4.4	8.9 ± 2.3
Data	8	8
Signal efficiency	$(0.205 \pm 0.024)\%$	$(0.045 \pm 0.007)\%$

Table 1. ATLAS $t \to Zq$ search: Expected and observed yields and signal efficiencies for the 3ID and 2ID+TL selections [7].

In both selections, exactly three isolated lepton candidates with $p_T > 20$ GeV ($p_T > 25$ GeV for the leading lepton) and at least two jets with $p_T > 25$ GeV were requested. In the 2ID+TL selection a jet consistent with originating from a b-quark was also required. The transverse missing energy of the event was requested to be greater than 20 GeV and signal-like final states were selected through a χ^2 minimization with respect to jet and lepton assignments and the neutrino longitudinal momentum (p_z^ν). The χ^2 definition was the following:

$$\chi^2 = \frac{\left(m^{reco}_{j_a l_a l_b} - m_t\right)^2}{\sigma_t^2} + \frac{\left(m^{reco}_{j_b l_c \nu} - m_t\right)^2}{\sigma_t^2} + \frac{\left(m^{reco}_{l_c \nu} - m_W\right)^2}{\sigma_W^2} + \frac{\left(m^{reco}_{l_a l_b} - m_Z\right)^2}{\sigma_Z^2} \quad (1)$$

where $j_{a,b}$ were the jets and $l_{a,b,c}$ were the three lepton candidates. The following constraints were used on top-quark, W and Z bosons masses and resolutions: $m_t = 172.5$ GeV, $m_W = 80.4$ GeV, $m_Z = 91.2$ GeV, $\sigma_t = 14$ GeV, $\sigma_W = 10$ GeV, $\sigma_Z = 3$ GeV. From all combinations, the one with the smallest χ^2 was chosen. Events were retained if reconstructed top-quark, W and Z bosons masses were respectively within 40 GeV, 30 GeV and 15 GeV of their reference values.

For the 3ID selection the dominant background came from events with three real leptons, whose contribution was taken from MC, while for the 2ID+TL selection the dominant background was from events with at least one fake lepton and a data-driven method was used for its estimation.

Table 1 reports data and expected yields for the 3ID and 2ID+TL selections. Good agreement between data and expectations was found, with no evidence for the $t \to Zq$ decay mode. Upper limits at 95% CL on signal branching ratio $B(t \to Zq)$ were derived. A number of systematic uncertainties, which could affect the expected number of signal and background events, were taken into account. In the 3ID selection, the most important sources of systematic uncertainties were the jet energy scale and the shape for ZZ and WZ. For the 2ID+TL selection, where $\sim 90\%$ of background was estimated from data, the most important contribution was the uncertainty on the fake lepton prediction.

Observed (expected) 95% C.L. upper limits on $BR(t \to Zq)$ were found to be 0.81% (0.95%), 3.2% (3.31%) and 0.73% (0.61%) for the 3ID channel, the 2ID+TL channel and their combination respectively.

1.2. Search for the FCNC decay $t \to Zq$ with the CMS detector
Also CMS performed the search for the FCNC decay $t \to Zq$ using the final state with three charged leptons and two jets [8]. Exactly three leptons candidates (electrons or muons) were

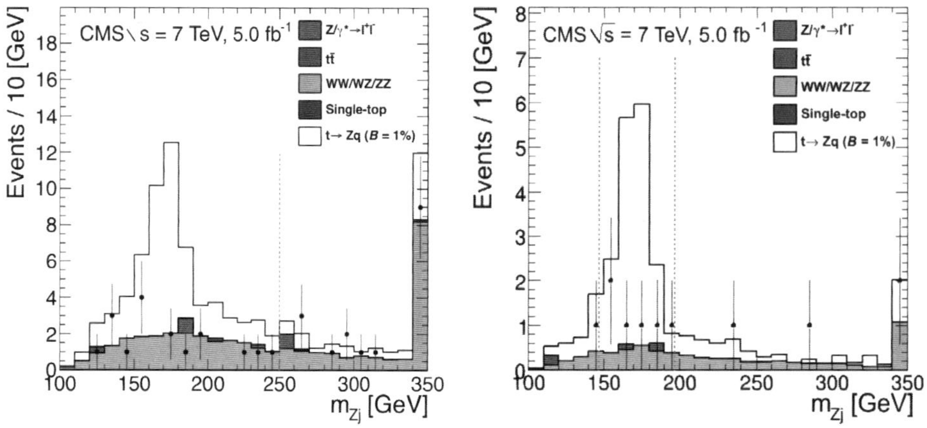

Figure 1. CMS $t \to Zq$ search: Comparison between data and simulated events of the m_{Zj} for the "ST"(left) and "b-tag"(right) selections before the requirements on m_{Zj} and m_{Wb} [8].

required to be isolated, with $p_T > 20$ GeV and coming from the same primary vertex. At least one opposite-sign same-flavour dilepton pair was required to have invariant mass between 60 GeV and 120 GeV. Finally events are requested to have at least two jets with $p_T > 30$ GeV and transverse missing energy $E_T^{miss} > 20$ GeV.

For both $t \to Zq \to llj$ and $t \to Wb \to l\nu$ decays, the full reconstruction of the top quark masses (m_{Zj} and m_{Wb} respectively) were possible. In the second case, the missing transverse energy was assumed as transverse component of the neutrino momentum and the invariant mass of the lepton and the neutrino ($m_{l\nu}$) was constrained to the W boson mass.

CMS performed the search for $t\bar{t} \to WbZq$ using two different selections. The first one required a minimum value of S_T[1] and loose requirements on m_{Zj} and m_{Wb}. The second selection had tighter requirements on m_{Zj} and m_{Wb} and also required that one of the jets were consistent with the hadronization of a b-quark. These two selections are referred as the "ST"and "b-tag"selections in the following.

In the "ST"selection, a candidate event was required to have S_T above 250 GeV, while m_{Zj} and m_{Wb} were requested to be between 100 GeV and 250 GeV. In the "b-tag"selection, one jet was required to be identified as b-quark initiated jet and the reconstructed top-quark masses m_{Zj} and m_{Wb} were requested to be within 25 GeV and 35 GeV of the assumed top mass (m_t = 172.5 GeV) respectively. Fig. 1 shows the comparison between data and simulated events of the m_{Zj} for the "ST"(left) and "b-tag"(right) selections, before the requirements on m_{Zj} and m_{Wb}. Dotted lines show the boundaries of the allowed mass region in the two selections.

Table 3 reports data and expected yields for the "ST"and "b-tag"selections. No excess beyond the SM expectations was observed and 95% CL upper limits on the signal branching ratio were derived. Many sources of systematic uncertainties were taken into account. The most important contrubutions came from the jet energy scale and missing energy resolution as well as the b-tagging efficiency for the "b-tag"selection.

Observed and expected 95% CL upper limits on the branching fraction $B(t \to Zq)$ are reported in Table 3. The expected limit for the "ST"selection is more sensitive and therefore the corresponding observed limit, 0.24%, was taken as the final result.

[1] S_T is defined as the sum of the p_T of the leptons, the p_T of the jets and the missing energy

Selection	S_T	b-tag
WZ background	$13.59 \pm 0.20 \pm 2.58$	$0.718 \pm 0.011 \pm 0.150$
ZZ background	$1.09 \pm 0.02 \pm 0.21$	$0.058 \pm 0.001 \pm 0.012$
Drell-Yan and $t\bar{t}$ background	$1.52 \pm 0.46 \pm 0.41$	$0.055 \pm 0.017 \pm 0.012$
Total background prediction	$16.20 \pm 0.51 \pm 2.62$	$0.83 \pm 0.02 \pm 0.15$
Observed events	11	0
Expected limit at the 95% CL	$\mathcal{B}(t \to Zq) < 0.36\%$	$\mathcal{B}(t \to Zq) < 0.42\%$
Observed limit at the 95% CL	$\mathcal{B}(t \to Zq) < 0.24\%$	$\mathcal{B}(t \to Zq) < 0.30\%$

Table 2. CMS $t \to Zq$ search: Data and expected yields, with statistical plus systematic uncertainties, for the "ST"and "b-tag"selections. Observed and expected limits at the 95% CL are also reported [8].

2. Search for baryon number violating top-quark decay with the CMS detector

Baryon number B is a conserved quantity in the SM though small violations can arise from non-perturbative effects. However, baryon number violation (BNV) naturally occurs in many scenarios beyond the SM, like supersymmetry, grand unified theories, and black hole physics. Baryon number violation in processes involving the top-quark production and decay have been recently investigated in [9], where it has been suggested the possibility of the BNV decays $t \to \bar{b}\bar{c}\mu^+$ $(\bar{t} \to bc\mu)$ and $t \to \bar{b}\bar{u}e^+$ $(\bar{t} \to bue)$, proceeding via an effective operator that makes it equivalent to a four fermion point interaction [9]. CMS searched for such decays in $t\bar{t}$ events where just one of the two top decayed to one lepton and two jets, while the other one decayed hadronically, resulting in final state with one isolated lepton, five jets and no neutrino [10]. This search was performed using 5.0 fb^{-1} of pp collision at $\sqrt{s} = 7$ TeV.

Top-quark BNV decays could appear not only in $t\bar{t}$ events, but also in events with tW production and single top production via s- and t-channels. However, the contribution to BNV from the two latter processes was proven to be negligible and therefore they were treated as non-top backgrounds.

A "basic"selection, where to normalize the top yield to data, and a signal region, with a "tight"selection, were defined. In order to reduce the impact of some systematic uncertainties, the expected top yield in the tight selection (N^T_{top}) was derived from quantities related to the basic selection. As results, expected total yield in the tight selection (N^T_{exp}) was expressed as:

$$N^T_{exp} = N^T_{top} - N^T_{bck} = \left(N^B_{obs} - N^B_{bck}\right)\left[\frac{N^B_{t\bar{t}}}{N^B_{t\bar{t}} + N^B_{tW}} \times \epsilon^{T|B}_{t\bar{t}} \frac{N^B_{tW}}{N^B_{tW} + N^B_{t\bar{t}}} \times \epsilon^{T|B}_{tW}\right] + N^T_{bck}, \quad (2)$$

where $N^B_{t\bar{t}}$ (N^B_{tW}) is the $t\bar{t}$ (tW) expected yield in the basic selection, N^B_{bck} (N^T_{bck}) is the non-top background yield in the basic (tight) selection and $\epsilon^{T|B}_{t\bar{t}}$ $(\epsilon^{T|B}_{tW})$ is the efficiency for $t\bar{t}$ (tW) events to pass the tight selection once the basic selection is passed. All terms in the square brackets of Eq. 2 are functions of the signal BR.

In the basic selection, events were requested to have exactly one isolated muon (electron), with $p_T > 35$ GeV and $|\eta| < 2.1$ (2.4), and at least five jets, with $p_T > 30$ GeV and $|\eta| < 2.4$. To reduce the contribution of non-top background, at least one jet consistent with originating from a b-quark was also required. In additon to the previous requirements, in the tight selection the transverse missing energy was requested to be less than 20 GeV and signal-like final states were selected by a χ^2 minimisation with respect to the jet assigment. The χ^2 was defined as:

Dataset	Cross section (pb)	Tight - Yield
$t\bar{t}$	157.5 ± 24.4	584 ± 81
W+jets	31310 ± 1560	76 ± 42
Z+jets	3048 ± 132	36 ± 20
WW	43.0 ± 1.5	0.97 ± 0.53
WZ	18.2 ± 0.7	0.92 ± 0.51
ZZ	5.9 ± 0.1	0.32 ± 0.18
tW	15.7 ± 0.8	12.8 ± 1.8
t-ch	64.6 ± 3.4	2.3 ± 1.3
s-ch	4.63 ± 0.19	0.26 ± 0.14
ttW	0.16 ± 0.02	2.0 ± 1.1
QCD	-	9.0 ± 9.0
Total Exp.	-	724 ± 39
Data	-	796 ± 28

Dataset	Cross section (pb)	Tight - Yield
$t\bar{t}$	157.5 ± 24.4	497 ± 72
W+jets	31314 ± 1558	88 ± 35
Z+jets	3048 ± 132	82 ± 33
WW	43.0 ± 1.5	0.80 ± 0.32
WZ	18.2 ± 0.7	1.10 ± 0.44
ZZ	5.9 ± 0.1	0.37 ± 0.15
tW	15.7 ± 0.8	14.6 ± 2.1
t-ch	64.6 ± 3.4	3.2 ± 1.3
s-ch	4.63 ± 0.19	0.30 ± 0.12
ttW	0.16 ± 0.02	1.77 ± 0.71
QCD	–	109 ± 54
Total Exp.	-	798 ± 66
Data	-	843 ± 29

Table 3. CMS $t \to lbq$ search: Adopted cross section values, expected and observed yields in the tight selection for the muon channel (left) and electron channel (right). Uncertainties are statistical plus systematic [10].

$$\chi^2 = \sum_i \frac{(x_i - \bar{x}_i)^2}{\sigma_i^2}, \tag{3}$$

where the x_i were the reconstructed invariant mass of the W boson from the hadronically decaying top-quark, the reconstructed invariant mass of the hadronically decaying top-quark and the reconstructed invariant mass of the BNV decaying top-quark. Events with minimised $\chi^2 > 20$ were rejected.

Table 4 left (right) reports the expected and the observed yields in the "tight"selection, after the tt and tW normalisation, for the muon (electron) channel. No significant excess beyond the SM expectations was observed and 95% CL upper limits on the signal branching ratio were derived in both channels as well as their combination. Many sources of systematic uncertainties were taken into account. The dominant contribution came from the jet energy scale and from the uncertainty on the ISR/FSR modelling in the MC samples.

The observed 95% CL limits on signal BR were derived to be 0.76% and 0.72% for the muon and electron channel, while the expected limits were 0.44% and 0.54% respectively. Results from the muon and electron channels were also combined by maximising the product of the two likelihood functions, assuming a common value of BR for the two channels. The resulting observed 95% CL upper limit on signal BR was 0.67% (expected limit was 0.41%) and it was taken as final result.

References

[1] J. A. Aguilar-Saavedra and B. M. Nobre, Phys. Lett. B **553** (2003) 251 [hep-ph/0210360].
[2] J. M. Yang, B. -L. Young and X. Zhang, Phys. Rev. D **58** (1998) 055001 [hep-ph/9705341].
[3] F. Abe *et al.* [CDF Collaboration], Phys. Rev. Lett. **80** (1998) 2525.
[4] V. M. Abazov *et al.* [D0 Collaboration], Phys. Lett. B **701** (2011) 313 arXiv:1103.4574 [hep-ex].
[5] G. Aad *et al.* [ATLAS Collaboration], JINST **3** (2008) S08003.
[6] S. Chatrchyan *et al.* [CMS Collaboration], JINST **3** (2008) S08004.
[7] G. Aad *et al.* [ATLAS Collaboration], JHEP **1209**, 139 (2012) arXiv:1206.0257 [hep-ex].
[8] S. Chatrchyan *et al.* [CMS Collaboration], arXiv:1208.0957 [hep-ex].
[9] Z. Dong, G. Durieux, J. M. Gerard, T. Han and F. Maltoni, Phys. Rev. D **85** (2012) 016006 arXiv:1107.3805.
[10] [CMS Collaboration], CMS-PAS-B2G-12-002.

Journal of Physics: Conference Series **452** (2013) 012012 doi:10.1088/1742-6596/452/1/012012

Top quarks objects definition and performance at ATLAS

V. Boisvert on behalf of the ATLAS Collaboration
Royal Holloway University of London, Egham, Surrey, TW20 0EX, UK
E-mail: veronique.boisvert@cern.ch

Abstract. The definition and performance of objects in events containing top quarks are presented. More specifically, the trigger, reconstruction and identification criteria for electrons, muons and taus are described. The selection and data calibration of jets and b-jets are also discussed. The effect of pile-up are described for each object.

1. Introduction

Top quark events contain almost the full list of objects that the ATLAS detector is capable of detecting. Since the top quark decays into a W boson and a b quark, the final state may contain electrons, muons or tau leptons from the W decay, associated with some missing transverse energy assigned to the corresponding neutrino. The W boson might also decay to a pair of quarks, which will hadronize into jets. Finally, top quark events also contain b-jets. In this note, each object definition and performance is described for the analyses that used the 7 TeV dataset of 2011 as well as the beginning of the 8 TeV dataset of 2012. The definition and performance of boosted objects in top quark events is described elsewhere in these proceedings [1].

As the instantaneous luminosity increased over time, the amount of pile-up also increased in the data, as illustrated in Fig. 1. There are two categories of pile-up: the in-time pile-up refers to multiple interactions within the same bunch crossing, while the out-of-time pile-up is associated with the fact that even though the bunch spacing in the LHC is 50 ns, some detectors collect signal for a longer time period, for example in the Liquid Argon calorimeters the drift time of the signal is between 250 and 600 ns. The pile-up present in events affects the reconstruction and performance of essentially every object in ATLAS. To mitigate the effect of pile-up, new variables have been used which preserve the object performance. For example, jet identification now requires that jets satisfy a certain amount of jet vertex fraction, defined as the fraction of track p_T associated to the jet coming from the hard scattering interaction. More details on the impact of pile-up is given in each following section.

2. Electrons

The trigger efficiency turn-on curve as a function of the electron E_T is shown in Fig. 2 [2]. The event filter electron E_T threshold varied between 20 GeV and 22 GeV over the 7 TeV dataset, while the analyses offline requirement is typically $E_T > 25$ GeV. In addition, to counter the effects of pile-up, the trigger also requests at Level 1 an upper threshold on hadronic leakage. The electron reconstruction algorithm first uses a sliding window algorithm with each electromagnetic

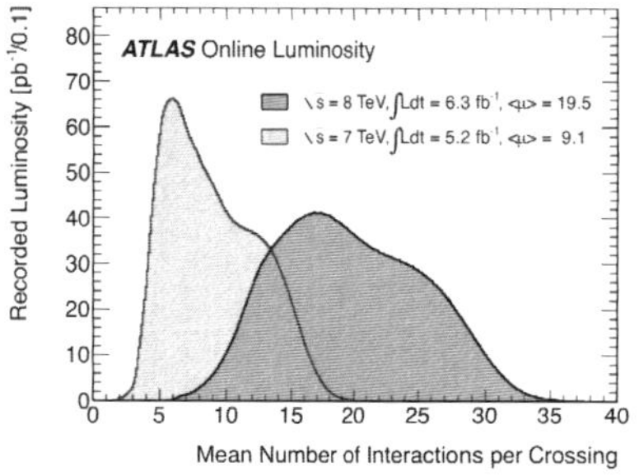

Figure 1. Mean number of interactions per crossings in the 7 TeV and 8 Tev datasets.

cluster, then it matches tracks to those clusters. Conversion photons are kept in the list to ensure high efficiency. The acceptance for electrons is defined by requesting that $|\eta_{cluster}| < 2.47$. The identification of electrons relies on a multi-variate analysis which contains various calorimeter variables. Examples of variables used for the track quality requirements include: the number of hits in the various detectors, the distance of closest approach d_0 to the primary vertex and the ratio of the number of transition radiation hits over the total number of hits on the track. At the identification stage a conversion veto is applied.

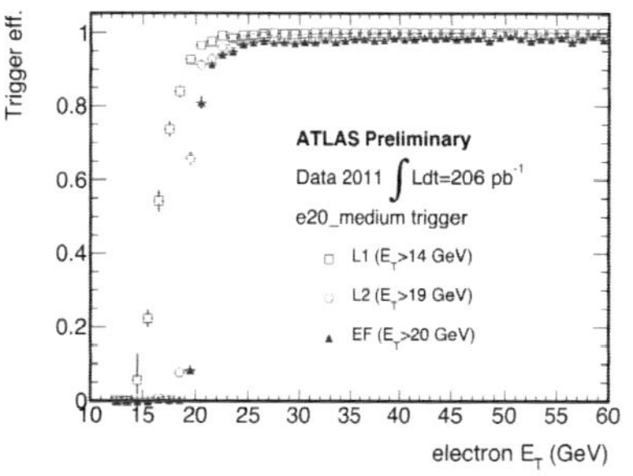

Figure 2. Trigger efficiency curve as a function of electron E_T in the 7 TeV dataset [2].

Since the electron comes from a W decay it is isolated from tracks or calorimeter clusters. The variables used to ensure that the electron is isolated are the sum of transverse energy (momentum) within a cone of $\Delta R^1 < 0.2$ (0.3) around the electron. Instead of having a fixed cut value, the cut values are different according to the electron p_T and η in order to keep the efficiency constant at 90%.

Improvements included in the reconstruction algorithms processing the 8 TeV dataset include taking into account Bremssthralung in the track pattern recognition and the use of a Gaussian Sum Filter to refit the track associated to electromagnetic clusters. The reoptimization of the identification criteria due to the additional amount of pile-up include relaxing variables that are sensitive to pile-up, like the fraction of energy leaking into the hadronic calorimeter, while tightening other variables that are more pile-up robust, like the lateral shower shape in the first sampling [3]. The efficiency improvement due to these changes are shown in Fig 3 comparing the 7 TeV and 8 TeV dataset efficiency as a function of the electron E_T and η [4].

The calibration of the electron efficiency as well as the energy scale and resolution is obtained from data samples, using the so-called Tag and Probe method, based on testing the algorithm on the decayed objects of standard candles for electrons like J/Ψ, Z and W events [5] [6]. The systematic uncertainty on the efficiency calibration is about 2.5% (of which about 1% is from

[1] $\Delta R = \sqrt{(\Delta\phi)^2 + (\Delta\eta)^2}$

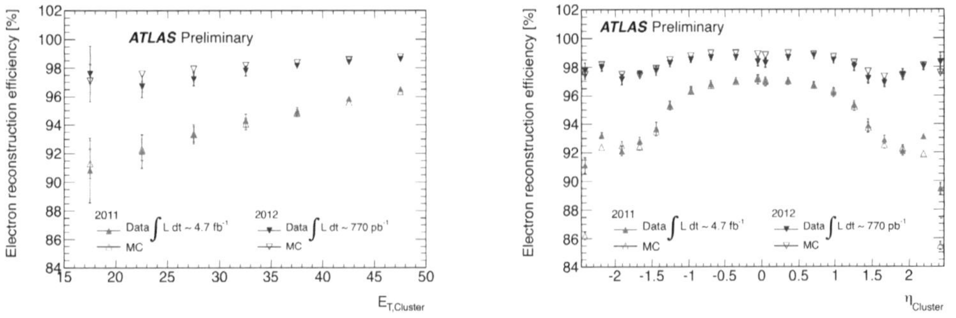

Figure 3. Electron reconstruction efficiency as a function of the electron E_T (left) and cluster η (right) showing the improvements between the 7 TeV and 8 TeV dataset [4].

the pile-up effect), while the systematic uncertainty on the energy scale and resolution is about 1.5%.

3. Muons

Shown in Fig. 4 is the muon trigger efficiency as a function of muon p_T for the 7 TeV dataset and 8 TeV dataset [7] [8]. The p_T threshold was kept at 18 GeV during the full period. Muon tracks are fit separately in the inner detector and muon spectrometer, and top quark analyses use combined tracks for muon objects. The acceptance for muons is $|\eta| < 2.5$.

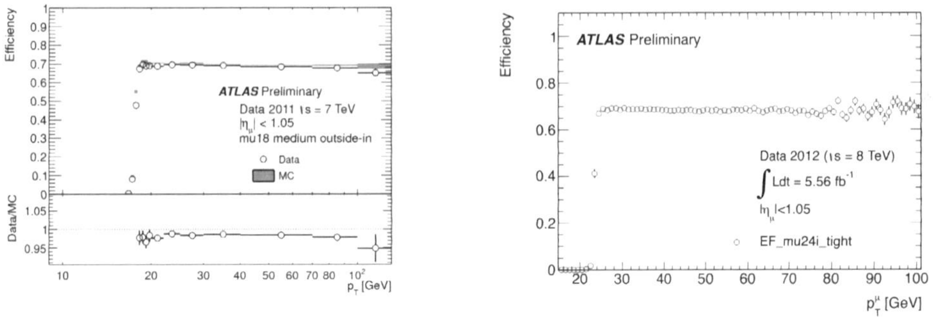

Figure 4. Muon trigger efficiency curves for the 7 TeV (left) [7] and 8 TeV (right) [8] dataset.

The isolation criteria for the muons were optimised for low dependence on pile-up and high multijet background rejection. They consist of requiring the E_T sum inside a cone of ΔR <0.2 around the muon direction to be less than 4 GeV and the p_T sum inside a cone of ΔR <0.3 to be less than 2.5 GeV. Since E_T is very affected by pile-up, it is best to require a small cone size and a large value, as opposed to p_T which is less affected by pile-up such that it is then best to require a larger cone size and a small cut value [9]. These improvements are shown in Fig. 5. As for the requirement that the muon must be a distance $\Delta R > 0.4$ from an identified jet, it is not affected by pile-up if the jet p_T >25 GeV and the jet vertex fraction is greater than 0.75. The calibration methods to determine the efficiency and resolution of muons are discussed in [10] and [11] respectively.

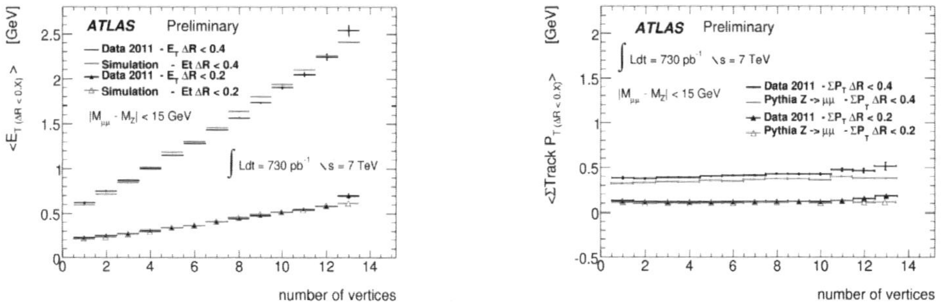

Figure 5. Comparison of different cone sizes for E_T (left) and p_T (right) showing the performance of these variables with pile-up [9].

4. Taus

The ATLAS measurement of the top quark production cross-section using taus is described in these proceedings [12]. The ATLAS reconstruction of hadronic taus is seeded by jets associated with one or greater or equal to three tracks. Various discriminants are used to remove jet and electron fakes. The jet rejection factor is about 300 for a 35% signal efficiency while the electron rejection factor is between 100 and 1000 for a 50% signal efficiency [13].

5. Jets

Reconstruction of jets relies on the anti-k_t algorithm with R=0.4 starting from topological clusters at the electromagnetic scale. A pile-up subtraction scheme at the electromagnetic scale is applied and those corrections depend on the number of primary vertices in the event, the number of primary interactions (μ) and η. Most top quark analyses require a jet vertex fraction for the jet to be greater than 0.75. The improvement provided by this requirement is shown in Fig. 6 [14].

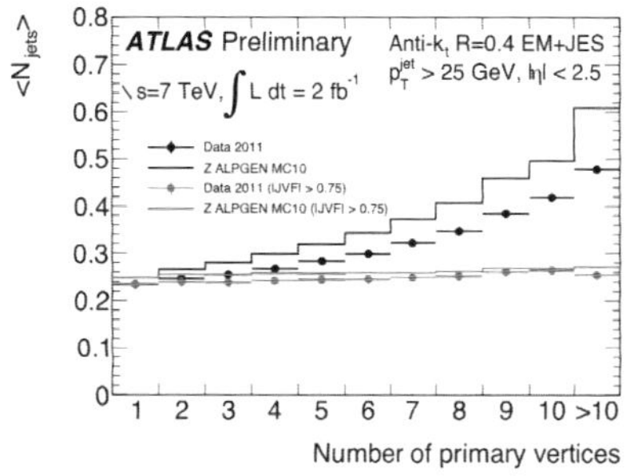

Figure 6. Average number of selected jets as a function of number of primary vertices showing the impact of requiring the jets to have a jet vertex fraction greater than 0.75 [14].

The calibration of the jet energy scale relies on Z+jets and γ+jets p_T balance for $|\eta| < 1.2$ while dijet p_T balance is used for $|\eta| > 1.2$. The jet energy scale correction factors are obtained from the full 2011 dataset and now allow for the various uncertainty sources to be nuisance parameters. Additional nuisance parameters take into account high p_T extrapolation, inter calibration for jets with high η and pile-up. The jet energy scales are expressed in such a way as to allow analyses to fit for them *in-situ*. Top quark analyses also apply uncertainties associated with: b-jets, light quark and gluon composition and close-by jet. The jet energy scale uncertainty using the Z+jets events is shown in Fig. 7 [15].

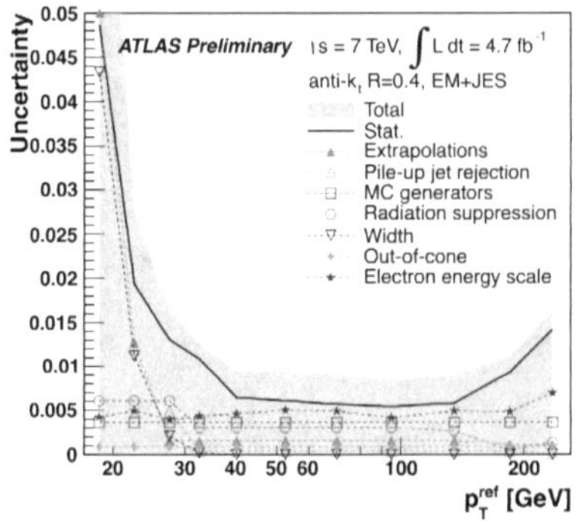

Figure 7. Jet energy scale uncertainty as a function of jet p_T using Z+jets events [15].

The jet energy resolution agrees between data and MC in both the dijet balance and bi-sector technique, such that no MC smearing is needed. There is however a systematic uncertainty on the jet energy resolution based on the uncertainty of the calibration [16].

6. b-jets

Since every top quark decays to a W and a b jets, b-jet tagging is crucial to top quark physics. The current most performant b-tagging algorithm (called MV1) is based on three tagging algorithms fed into a neural network parameterized in terms of jet p_T and η [17]. The first algorithm (JetFitterCombNN) exploits the topology of b and c hadron decays inside the jet using Kalman filtering, the likelihood uses vertex mass, momentum, flight-length significance and track multiplicity. The second algorithm (IP3D+SV1) uses the impact parameters significance as well as the z_0 significance as a reference pdf for b-jets and light-jets in the MC. Finally the third algorithm (SV0) searches for two-track vertices, and also uses the invariant mass of the tracks, the ratio of the track energy over the jet energy and the number of two-track vertices. The light jet rejection as a function of b-jet efficiency is shown in Fig. 8 for the various b-tagging algorithms [18]. The operating point used for most 7 TeV top quark analyses is at 70% efficiency.

The JetFitterCombNNc algorithm has been optimised for charm rejection to help with for example suppressing the W+c background in single top analyses. The performance of this algorithm is also shown in Fig. 8. There is also a Soft Muon Tagging algorithm which relies on a good muon inside a jet, also allowing for charge tagging of the jet.

The b-tagging efficiency scale factors between data and MC are obtained from various data-driven methods including: System 8 [19], p_T^{rel} [18] and D^* selection [20]. The methods produce compatible results, the combined calibration reduces the uncertainty, which is down below 10% and shown in Fig. 9 for the 70% operating point. Top quark events can also be used as a calibration method for b-tagging algorithms (not used in top quark analyses). The top anti-top quark calibration in the lepton+jets channel is also shown in Fig. 9. Dedicated c-jet efficiency scale factors are obtained from a dedicated D^* analysis, while for tau-jets the c-jets scale factors are used with increased uncertainty.

The probability of light jets being b-tagged is calibrated using two methods: the SV0-mass method relies on templates from MC of the invariant mass of the secondary vertex and is fit to the data, the second method is the negative tag rate method. The scale factors for efficiency and mistag rate are binned in p_T and η of the jet. Each individual systematic uncertainty is varied as well as bin-to-bin covariance matrices used. The scale factor uncertainties vary from 18% to 49% [22].

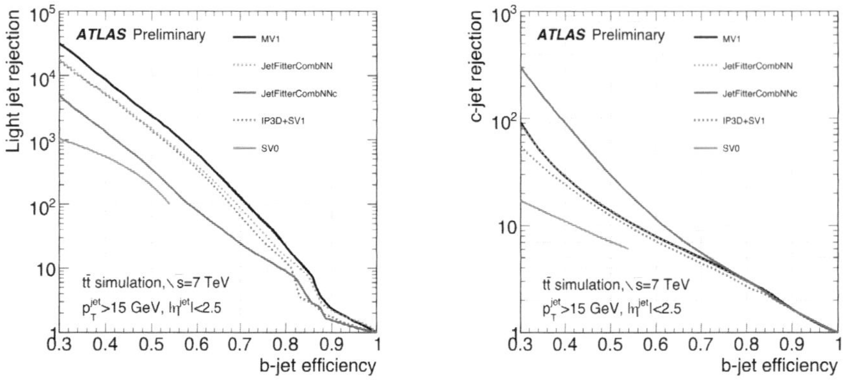

Figure 8. Comparison of the performance of various b-tagging algorithms, on the left is the light jet rejection as a function of b-jet tagging efficiency and on the right is the charm-jet rejection as a function of b-jet tagging efficiency. For most top quark analyses, the operating point is 70% b-jet tagging efficiency [18].

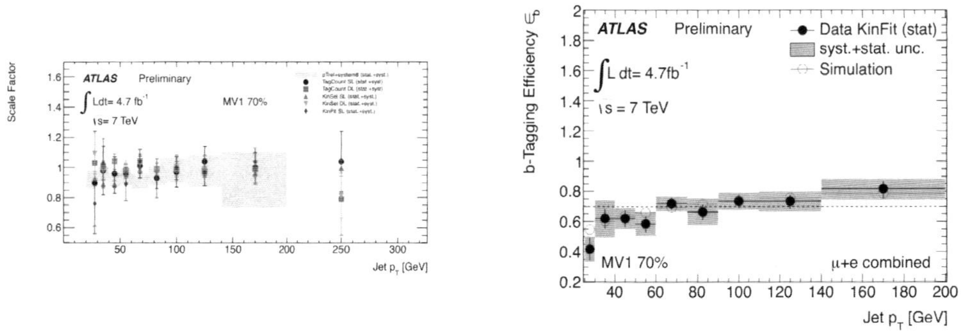

Figure 9. Left: b-jet tagging scale factors as a function of jet p_T for the different data calibration methods. Right: b-jet tagging efficiency as a function of jet p_T using the top anti-top quark events in the lepton+jets channel calibration method [21].

7. Missing Transverse Energy

The definition of missing transverse energy (E_T^{miss}), characteristic of for example the presence of a neutrino, is given by:

$$-E_{x,y}^{miss} = E_{x,y}^{RefElec} + E_{x,y}^{RefJet} + E_{x,y}^{RefSoftJet} + E_{x,y}^{RefMuon} + E_{x,y}^{CellOut}.$$

The E_T^{miss} used in top quark analyses uses top quark object definitions so that calorimeter cells are calibrated according to which objects they are associated with. Remaining energy cells are included as a CellOut term calibrated at the EM scale. Jets with $p_T > 20$ GeV are included at EM+JES scale while jets with 7 GeV $< p_T <$ 20 GeV at EM scale. The muon term uses the muon p_T of all isolated muons. Finally, the effect of pile-up will enter through the SoftJet and CellOut terms, they are a small fraction of the whole E_T^{miss} and 6.6% of the uncertainty on E_T^{miss} comes from the soft terms [23].

8. Conclusions

The ATLAS top quark object definitions and calibrations have evolved quite a lot over the 2011 7 TeV dataset. Top quark members are at the forefront of ATLAS on those topics, since top quark events involve the majority of ATLAS objects. A common thread is that a combination of methods typically offer the best performance. In some cases, the systematic uncertainty determination is quite sophisticated. Finally, the increasing amount of pile-up in the data has been a challenge but ATLAS has managed to rise to it.

References

[1] S. Fleischmann on behalf of the ATLAS and CMS collaborations, Boosted top quark techniques and searches for ttbar resonances at the LHC, *TOP 2012 Proceedings*
[2] ATLAS Collaboration, ATL-COM-DAQ-2011-032, *url: https://cds.cern.ch/record/1351827/*
[3] ATLAS Collaboration, ATL-COM-PHYS-2012-260, *url: https://cds.cern.ch/record/1430634/*
[4] ATLAS Collaboration, ATL-COM-PHYS-2012-783, *url: https://cds.cern.ch/record/1454809/*
[5] ATLAS Collaboration, ATL-COM-PHYS-2011-1637, *url: https://cds.cern.ch/record/1403075/*
[6] ATLAS Collaboration, ATL-COM-PHYS-2012-259, *url: https://cds.cern.ch/record/1430632/*
[7] ATLAS Collaboration, ATLAS-CONF-2012-099, *url: https://cds.cern.ch/record/1462601/*
[8] ATLAS Collaboration, *url: https://twiki.cern.ch/twiki/bin/view/AtlasPublic/MuonTriggerPublicResults#Muon_trigger_performances_in_201*
[9] ATLAS Collaboration, ATL-COM-PHYS-2011-1503, *url: https://cds.cern.ch/record/1395476/*
[10] ATLAS Collaboration, ATL-COM-PHYS-2011-1640, *url: https://cds.cern.ch/record/1403080/*
[11] ATLAS Collaboration, ATL-COM-PHYS-2011-1504, *url: https://cds.cern.ch/record/1395478/*
[12] G. Khoriauli on behalf of the ATLAS and CMS Collaborations, Top pair cross-section results at LHC in final states with taus or no leptons, *TOP 2012 Proceedings*
[13] ATLAS Collaboration, ATLAS-CONF-2011-152, *url: https://cds.cern.ch/record/1398195/*
[14] ATLAS Collaboration, *url: https://twiki.cern.ch/twiki/bin/view/AtlasPublic/ JetEtmissApproved2011PileupOffsetAndJVF*
[15] ATLAS Collaboration, ATLAS-CONF-2012-053, *url: https://cds.cern.ch/record/1452641/*
[16] ATLAS Collaboration, *url: https://twiki.cern.ch/twiki/bin/view/AtlasPublic/ JetEtmissApproved2011JetResolution*
[17] ATLAS Collaboration, ATLAS-CONF-2011-102, *url: https://cds.cern.ch/record/1369219/*
[18] ATLAS Collaboration, ATLAS-CONF-2012-043, *url: https://cds.cern.ch/record/1435197/*
[19] ATLAS Collaboration, ATLAS-CONF-2011-143, *url: https://cds.cern.ch/record/1386703/*
[20] ATLAS Collaboration, ATLAS-CONF-2012-039, *url: https://cds.cern.ch/record/1435193/*
[21] ATLAS Collaboration, ATLAS-CONF-2012-097, *url: https://cds.cern.ch/record/1460443/*
[22] ATLAS Collaboration, ATLAS-CONF-2012-040, *url: https://cds.cern.ch/record/1435194/*
[23] ATLAS Collaboration, ATLAS-CONF-2012-101, *url: https://cds.cern.ch/record/1463915/*

Journal of Physics: Conference Series **452** (2013) 012013 doi:10.1088/1742-6596/452/1/012013

Instrumental Backgrounds to $t\bar{t}$ and Single Top Production at Hadron Colliders

Joseph Boudreau, on behalf of ATLAS, CMS, CDF, and D0 Collaborations

Department of Physics and Astronomy, University of Pittsburgh, Pittsburgh, PA, 15260, USA

E-mail: boudreau@pitt.edu

Abstract. A number of backgrounds to top-quark production at the LHC and Tevatron experiments are characterized as instrumental: beam related-backgrounds, cosmic ray backgrounds, and backgrounds from non-prompt (or *fake*) leptons. A number of strategies have been deployed in the experiments to minimize these backgrounds and to estimate their rates. The strategies are applied at the level of the accelerator, the detector apparatus, the first few cuts which are common to many physics analyses, and more refined kinematic cuts. Remaining fake lepton background is estimated using sophisticated data-driven techniques. We present here a survey of the state-of-the-art in the reduction of instrumental backgrounds and also their estimation.

1. Introduction

Instrumental backgrounds are those coming from beam-related effects, detector noise, cosmic rays, and misidentified leptons. In this paper, we survey the effects leading to instrumental backgrounds, explain how they are minimized, and discuss how the remaining irreducible background of fake leptons is estimated.

Events with top quarks in the final state require the use of the entire detector system, since all types of physics objects are present in $t\bar{t}$ and single top signatures. Tracking systems and electromagnetic calorimetry are both important for electron identification, tracking and muon chambers for muon identification, calorimetry for jet-finding and energy estimation, tracking for their classification as b- or light-jets, and the whole detector for the estimate of the missing transverse energy ($\not\!\!E_{\mathrm{T}}$). This makes the analysis of $t\bar{t}$ and single top events susceptible to many effects, such as fake leptons from the accelerator beam, cosmic rays, pion and kaon decay-in-flight and punch-through.

2. Beam-related backgrounds

The first type of instrumental background we discuss is commonly called *beam halo*. Protons in the LHC (or Tevatron, as the case may be) strike limiting apertures, or beam gas, producing neutrons, γ-rays, pions and kaons which decay in flight along the beam direction to muons. These particles, having energies up to the beam energy, impinge on the detector from the side facing the LHC tunnel[1]. Events of this type reconstructed in the ATLAS and CMS detectors are shown in Fig. 1. Backgrounds not originating from the collision region include calorimeter noise bursts and cosmic rays, the latter overlapping approximately 1 event in 10^4.

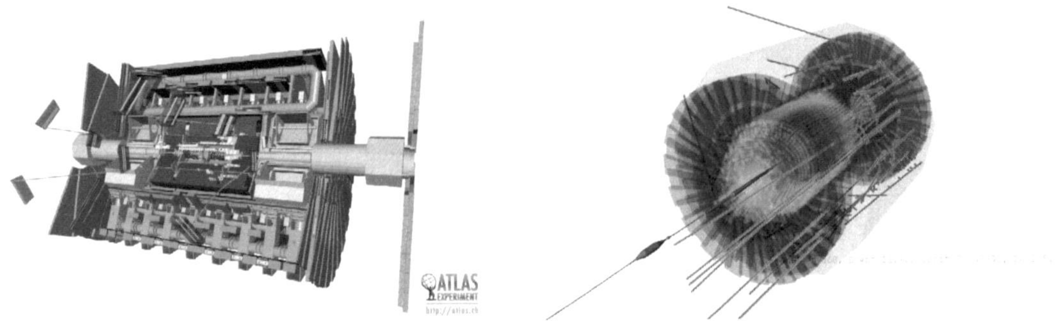

(a) (b)

Figure 1. Beam halo events from ATLAS (a) and CMS (b) showing muons tracks traversing the detectors quasi-parallel to the beam line and originating from a location outside the respective detector volumes.

These backgrounds were carefully considered during the construction of the LHC[2] and the CMS[3] and ATLAS[4] detectors. A beam cleansing system consisting of scrapers developed first at the Tevatron was incorporated into the design of the LHC. In addition, large steel shielding elements plug the opening of the LHC tunnel at the entrance to the caverns in both ATLAS and CMS. This suppresses hadrons and γ-rays, leaving muons as the dominant component of beam halo reaching the LHC detectors. The first few cuts applied to top-physics analysis remove the remaining beam-halo and cosmic ray background. CMS generally requires that the primary vertex be found within a cylinder of radius 2 cm and length 24 cm centered around the nominal interaction point[5]. ATLAS[6] requires a primary vertex with at least five tracks, and in order to reject cosmic rays, further discards events containing muon pairs with $\Delta\phi > 3.1$ and $|d_0| > 0.5$ mm where d_0 is the signed distance of closest approach to the primary vertex. In dilepton analyses, the d_0 of the two leptons are required to have opposite sign. Timing information from calorimeters and scintillators is no longer required to reject beam background, since the overlap rate is presently at the level of about 1 in 10^8.

3. Multijet production background

Multijet events with u, d, c, s or b quarks becomes a background to $t\bar{t}$ when they contain a reconstructed lepton from the weak decay of a hadron, a misidentified hadron, or a misidentified jet. We refer to all of these objects as *fake leptons*. Fake electrons can come from photon conversions in the detector material, tracks overlapping with photons, and jets with fluctuations to a larger neutral component (i.e. π^0's), implying low numbers of charged tracks and early showering in the calorimeter, and semileptonic decays of b- and c-quarks. Fake muons can arise from *sail-through* hadrons that pass through the calorimeter material without showering, *punch-through* hadrons which emerge from hadronic showers, semileptonic decays of b- and c-quarks, and the decay of pions and kaons within the tracking volume. This happens rarely, however the enormous multijet cross section, four orders of magnitude greater than that of $t\bar{t}$ production, makes them an important source of background. Their simulation is impractical because of the large sample sizes, and unreliable because the processes that lead to fake leptons are difficult to model precisely. Instead, data-driven techniques are used to estimate the multijet background to $t\bar{t}$ and single top production. Typically, these data-driven methods are variants of what is known as the *Matrix method*, the *Jet-electron method*, or the *Anti-electron method*.

3.1. The Matrix Method

The matrix method was developed and first applied by the D0 collaboration[7] to determine the number of fake leptons from multijet production (muons or electrons) in a selected event sample. Consider the case of a $t\bar{t}$ with one top-quark decaying semileptonically and the other hadronically (the lepton+jets channel). The standard selection is relaxed (e.g. by loosening an isolation cut) in order to obtain a loose selection. A simple model is now constructed to account for the observed number of events in each category, N^{loose} and N^{std}. In the loose sample, one has $N^{\text{loose}}_{\text{real}}$ events with real leptons and $N^{\text{loose}}_{\text{fake}}$ events with fake leptons. The probability for these events to survive the final selection is r for events with real leptons and f for events with fake leptons. The population of observed events is then predicted to be:

$$\begin{pmatrix} N^{\text{loose}} \\ N^{\text{std}} \end{pmatrix} = \begin{pmatrix} 1 & 1 \\ r & f \end{pmatrix} \cdot \begin{pmatrix} N^{\text{loose}}_{\text{real}} \\ N^{\text{loose}}_{\text{fake}} \end{pmatrix} \tag{1}$$

If the efficiencies r and f are known, the equation can be inverted to determine the composition of the loose sample; then the selection efficiencies can be used again to determine the composition of the final sample. In practice, r and f are commonly determined from pure or background-subtracted samples of real and fake leptons. This idea can be extended to more complicated situations. In events with two leptons (e.g. the top dilepton channel, or the single top Wt channel), more categories are required (std/std, loose/std, std/loose, loose/loose) and the data model is more complicated, but the basic idea is the same. Distributions in any kinematic variable for multijet background can also be determined, by assigning to each event in the loose sample a probability for being real or fake, according to whether the event passes the standard selection or only the loose selection. To obtain samples of real leptons, leptons from $Z \to e^+e^-$ or $Z \to \mu^+\mu^-$ are often used, and sometimes Monte Carlo. Fake leptons are taken from samples enriched in multijet events, with systematic uncertainties evaluated by varying the choice of the sample. Typically, the background estimation is conducted in bins of transverse momentum p_{T} or pseudorapidity η, after correcting the real and fake samples for contamination. Results from a CMS analysis of the dilepton channel [8] are shown in Fig. 2. In this example, where two leptons are involved, the level of fake lepton background from multijet production is relatively low. This is also a feature of the single top Wt channel.

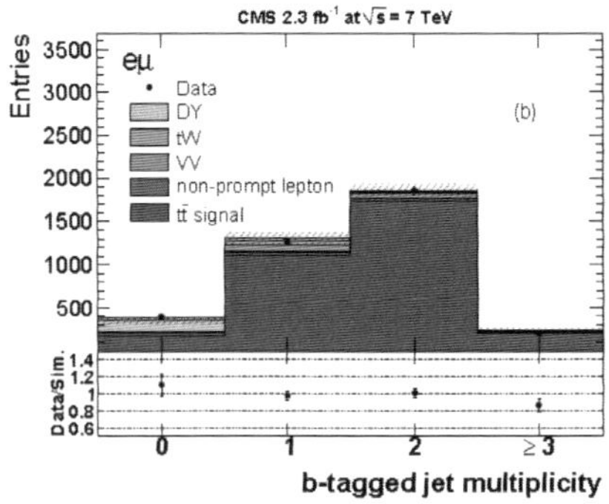

Fake leptons (CMS dilepton)

	e^+e^-	$\mu^+\mu^-$	$e^\pm\mu^\pm$
N_{W}	1.8± 4.8	9.8±5.6	42.4±14.6
N_{MJ}	0.6± 0.5	0.2±0.1	7.5±3.9

Figure 2. CMS dilepton analysis from Ref. [8]. Estimated backgrounds are shown in the table, where W (*W-like*) indicates one fake lepton and MJ (*Multijet-like*) indicates two fake leptons.

3.2. The Jet-electron Method

In the *Jet-electron* method, a sample of jets (the *jet-electron* sample) is selected whose properties resemble those of jets likely to fake electrons, i.e. those with few tracks which have showered early in the calorimeter. In an ATLAS analysis of single top t-channel production[9], for example, 80-95% of the energy must have been deposited in the electromagnetic section of the calorimeter; p_T and η selections are identical to those used for an electron selection, at least four tracks must be found in the jet, and no other electron or muon candidate may be present in the event. The shape of the multijet background is taken from the jet-electron sample, while the overall normalization is determined from a fit to the $\not{E}_T$ distribution before any cuts on $\not{E}_T$ are applied. While there is no *a priori* reason to believe that the jet-electron model should also describe the multijet background with fake muons, in practice the model adequately describes this background. The analysis from ATLAS[9] is illustrated in Fig. 3. A variant of the jet-electron method is the anti-electron method, which uses a sample of lepton candidates for which some but not all of the electron identification cuts have been satisfied. This method was used frequently in the CDF collaboration. Systematic uncertainties are estimated by varying the template shape. In CDF this was done by obtaining alternate templates from two-track (rather than one-track) lepton candidates. CMS has applied anti-lepton background estimates for muons, where non-isolated muons are used to derive muon template shapes, and for electrons, for which *marginal failures* (candidates failing two out of six electron identification cuts) were used. In the case of CMS, the normalization was determined from a sideband region $\not{E}_T < 20$ GeV and from a fit to the reconstructed vertex mass.

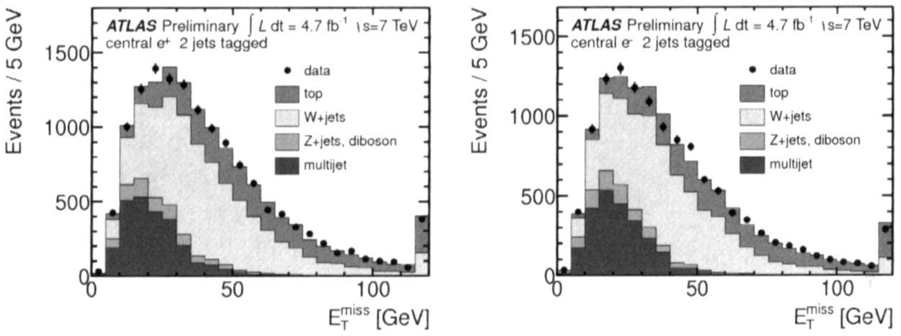

Figure 3. Results from ATLAS single top t-channel analysis from Ref. [9], which require a single lepton in addition to jets and $\not{E}_T$. The multijet background is relatively high. Here it has been estimated with the jet-electron method.

3.3. Other methods

Every analysis is unique, and occasionally one encounters background-estimate techniques which don't fall into any of the above categories. One can observe from the plots of Fig. 2 and Fig. 3 that the probability of fake leptons from multijet production is much lower when two leptons are required than when one lepton is required. The background to a dilepton analysis usually contains one real lepton from a W boson decay and one fake lepton. In Ref. [10], a CDF dilepton analysis, tracks with $p_T > 20$ GeV which point to electron-like or muon-like signatures in the calorimeters are called *fakeable objects*. Events in a high-p_T lepton sample accompanied by a fakeable object is called a *fakeable event*. Finally, the rate of fake dilepton events is computed by weighting each fakeable event by the fake rate for the fakeable objects in the event, the latter being measured in data in a sample of generic jets satisfying a trigger requirement of $E_T^{\mathrm{trig}} > 20$, 50, 70, and 100 GeV. The selection criteria are varied to obtain an estimate of the systematic uncertainty on the fake-lepton background, which is quoted to be 30%. This makes an important contribution to the overall uncertainty in the top dilepton cross section.

Table 1. Methods used in CMS, ATLAS, CDF, and D0 for a few example analyses. MM=Matrix Method. JE=Jet-electron Method. AE=Anti-electron method.

Exp.	Channel	Observable	$\int \mathcal{L}dt$ [fb^{-1}]	Ref.	Method
ATLAS	lepton+jets	$\sigma(t\bar{t})$	0.035	[11]	MM(μ), AE(e)
ATLAS	dilepton	$\sigma(t\bar{t})$	0.7	[12]	MM
ATLAS	single top lepton+jets	$\sigma(t)$, t-channel	1.04	[13]	JE
CMS	lepton+jets	$\sigma(t\bar{t})$	0.036	[14]	AE
CMS	dilepton	$\sigma(t\bar{t})$	2.3	[8]	MM
CMS	single top leptons+jets	$\sigma(t)$, t-channel	0.036	[15]	AE
CDF	lepton+jets	$\sigma(t\bar{t})$	2.7	[16]	AE
CDF	dilepton	$\sigma(t\bar{t})$	2.8	[10]	other
CDF	single top lepton+jets	$\sigma(t)$, s, Wt, t-channel	3.2	[17]	JE/AE hybrid
D0	lepton+jets	$\sigma(t\bar{t})$	0.9	[18]	MM
D0	dilepton	$\sigma(t\bar{t})$, m_t	1.0	[19]	AE
D0	single top	$\sigma(t)$, t-channel	2.3	[20]	MM

4. Conclusions

We have discussed a number of approaches to minimizing instrumental backgrounds and the methodology of estimating irreducible fake-lepton backgrounds. Good design of detector and accelerator systems have kept the former sources of background under control, while remaining fake-lepton backgrounds have been estimated in a variety of ways at CDF, D0, ATLAS, and CMS. Usually the methodology consists of either the matrix method, the jet-electron method, or the anti-electron method, but a creative physics community and different challenges leads to some variation in these techniques which occasionally defies the above classification. Table 1 gives a few examples analyses where one can find these techniques applied.

References

[1] Drozhdin A, Hutinin M and Mokhov N 1996 *Nucl. Instrum. Meth.* A**381** 531
[2] L. Evans and P. Bryant (editors) 2008 JINST **3** S08001
[3] CMS Collaboration 2008 JINST **3** S08004
[4] ATLAS Collaboration 2008 JINST **3** S08003
[5] CMS Collaboration 2011 *Eur. Phys. J.* C**71** 1721
[6] ATLAS Collaboration 2010 ATLAS-CONF-2010-038, https://cds.cern.ch/record/1277678
[7] D0 Collaboration 2000 *Phys. Rev.* D**61** 072001
[8] CMS Collaboration 2011 arXiv:1208.2671
[9] ATLAS Collaboration 2012 ATLAS-CONF-2012-056, https://cds.cern.ch/record/1453783
[10] CDF Collaboration 2010 *Phys. Rev.* D**82** 052002
[11] ATLAS Collaboration 2012 *Phys. Lett.* B**711** 244
[12] ATLAS Collaboration 2012 *JHEP* 1205 059
[13] ATLAS Collaboration 2012 *Phys. Lett.* B**717** 330
[14] CMS Collaboration 2011 *Phys. Rev.* D**84** 092004
[15] CMS Collaboration 2011 *Phys. Rev. Lett.* **107** 091802
[16] CDF Collaboration 2011 *Phys. Rev.* D**84** 031101
[17] CDF Collaboration 2010 *Phys. Rev.* D**82** 112005
[18] D0 Collaboration 2008 *Phys. Rev. Lett.* **100** 192004
[19] D0 Collaboration 2009 *Phys. Lett.* B**679** 177
[20] D0 Collaboration 2010 *Phys. Lett.* B**682** 363

Vector boson+jets as background to top-quark physics at Tevatron and LHC

Frédéric Déliot, on behalf of the ATLAS, CDF, CMS and D0 Collaborations
CEA-Saclay, DSM/Irfu/SPP, bat 141, 91191 Gif sur Yvette Cedex, France
E-mail: frederic.deliot@cea.fr

Abstract. I review here the latest measurements related to vector boson production in association with light or heavy flavor jets at the Tevatron and at the LHC. The methods to estimate the W+jets and Z+jets background in top-quark analyses are also presented.

1. Introduction

Vector boson production in association with jets (V+jets) is an important process. First, it allows to test perturbative quantum chromodynamics (QCD) in a stringent way. It is also used to check Monte Carlo (MC) simulation and stands as one of the favoured processes to tune this simulation. These processes are particularly interesting because they allow precision QCD tests with multiple renormalisation or factorisation scales set by the vector boson mass and by the associated light or heavy-flavor jet transverse momentum. Precise measurements of the V+jets properties are therefore of primary importance.

In top-quark physics, V+jets production is one of the main backgrounds in the lepton+jets (ℓ+jets) and dilepton ($\ell\ell$) channels. In both cases, vector boson production in association with heavy quarks (mainly b-quarks) leads to the same final state as the production of a top-antitop quark pair ($t\bar{t}$). From a theoretical point of view, predictions exist for $V+\leq 3$ jets at the Tevatron and $V+\leq 4$ jets at the LHC.

Different jet reconstruction algorithms are used at the Tevatron and the LHC. At the Tevatron, the midpoint cone algorithm is used with a cone radius of $\Delta R = \sqrt{\Delta\eta^2 + \Delta\Phi^2} = 0.4$ for CDF and $\Delta R = 0.5$ for D0. The anti-kt algorithm is used at the LHC with $\Delta R = 0.4$ for ATLAS [1] and $\Delta R = 0.5$ for CMS [2] which also exploits the technique of particle flow for jet reconstruction. The LHC results presented in this article are based on 7 TeV data. These reconstructed jets are calibrated for the electromagnetic and hadronic scales. For the MC/data distributions shown in the following, the measurements are corrected at the particle level within the acceptance cuts. This means that the results are valid for specific ranges of lepton and jets kinematics.

2. V+jets experimental measurements

2.1. W+jets experimental measurements

At the Tevatron the W+jets measurements rely on jets with $p_T > 20$ GeV and $|\eta| < 2$ for CDF or $|y| < 3.2$ for D0. The measured differential cross sections are normalized to the total inclusive W production cross section so that cancellations of some systematic uncertainties occur. In some cases, the normalization is rather performed using the next-to-next-leading order (NNLO) inclusive W prediction. The correction for reconstruction effects is completed by bin-by-bin unfolding in CDF or by regularized matrix inversion in D0. The relative uncertainty on the results range from around 4% $W + 1$ jet to 20% for $W + 3$ jets. In general, comparisons with normalized multileg generators or next-to-leading order (NLO) simulations show a good agreement. The predictions and measurements for the normalized differential cross section of $W + e\nu \geq n$ jets in CDF [3] and D0 [4] are shown in Figure 1.

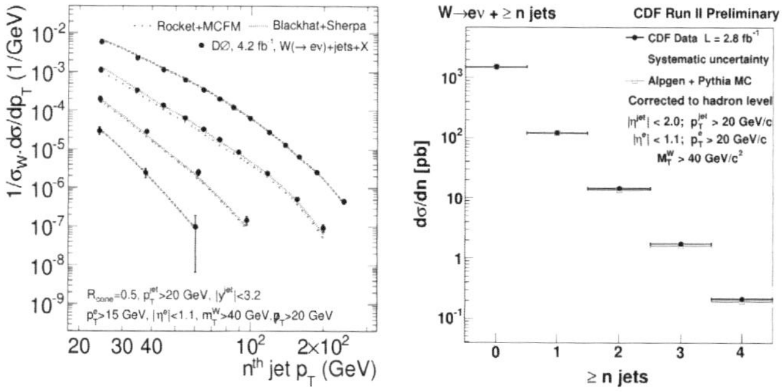

Figure 1. Measured W +n jet differential cross section at D0 for jet multiplicities $n = 1 - 4$ [4] (left) and jet multiplicity measurement normalized to the total inclusive $W \to e\nu$ cross section at CDF [3] (right).

At the LHC, jets with $p_T > 20$ or 30 GeV and $|y| < 4.4$ for ATLAS or $|\eta| < 2.4$ for CMS are selected. The normalization for the differential cross section measurements is performed in the same way as at the Tevatron. The correction for reconstruction effects is based on iterative Bayesian unfolding in ATLAS or singular value decomposition (SVD) in CMS. The uncertainties for the measurements of the W production vary from 8% for the $W+1$ jet production to 20% for $W + 3$ jets [5, 6]. At the LHC also, the comparison between multileg or NLO generators shows good agreement. Figure 2 shows the MC/data comparison for some differential cross section measurements in ATLAS and CMS.

At the Tevatron, D0 and CDF also measure the inclusive $W + c$ jets cross sections. In addition, CDF studies the $W + 1, 2b$ jets cross sections. $W + c$ jets production is interesting since it is sensitive to the s quark content of the proton and the process is calculable in multi-scale perturbative QCD. MC predicts that the process $W + c$ represents around 5% of the $W + 1$ jet production (when the jet has $p_T > 10$ GeV). Soft lepton tagging is used to identify the $W + c$ production. With 4.3 fb^{-1} of data, CDF measures $\sigma(W + c) \cdot Br(W \to \ell\nu) = 13.3^{+3.3}_{-2.9}(\text{stat} + \text{syst})$ pb for jets with $p_T > 20$ GeV and $|\eta| < 1.5$ in agreement with the NLO prediction of 11.3 ± 2.2 pb. With 1 fb^{-1} of data, D0 measures $\frac{\sigma(W+c)}{\sigma(W+jets)} = 7.4 \pm 1.9(\text{stat})^{+1.2}_{-1.4}(\text{syst})\%$ [7] in agreement with the prediction from Alpgen+Pythia simulation of $4.4 \pm 0.3\%$. With 1.9 fb^{-1} of data, CDF also measures the cross section with at least one tight b-tagged jet using a template fit of the vertex mass to get $\sigma(W + b/bb) \cdot Br(W \to \ell\nu) = 2.74 \pm 0.27(\text{stat}) \pm 0.42(\text{syst})$ pb [8] that appears to be higher by around 3σ than the NLO prediction of $1.22 \pm 0.14(\text{syst})$ pb. The LHC experiments also study the W+heavy flavor jet production. Using events with one b-tag and a template fit to the vertex mass distribution in 35 pb^{-1}, ATLAS measures $\sigma(W + b/bb) \cdot Br(W \to \ell\nu) = 10.2 \pm 1.9(\text{stat}) \pm 2.6(\text{syst})$ pb within the fiducial cuts [9] which is 1.8σ higher than the NLO prediction of 4.8 ± 1.0 pb. Using events with one tag and a template fit of a secondary vertex discriminant in 36 pb^{-1}, CMS measures $\frac{\sigma(W+c)}{\sigma(W+jets)} = 14.3 \pm 1.5(\text{stat}) \pm 2.4(\text{syst})\%$ using jets with $p_T > 20$ GeV and $|\eta| < 2.1$ [10] in

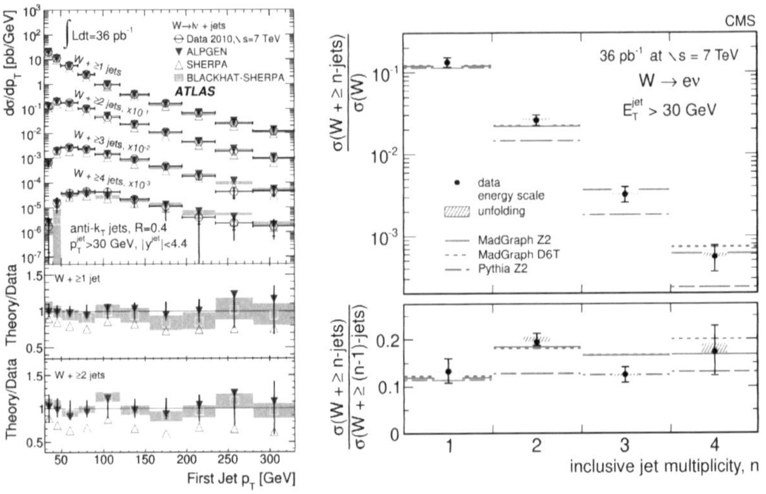

Figure 2. W+jets cross section as a function of the p_T of the first jet in the event in ATLAS [5] (left) and the ratio $\sigma(W + n\ jets)/\sigma(W)$ in the electron channel compared to the expectations in CMS [6] (right).

agreement with the NLO prediction of $13 \pm 2\%$.

2.2. Z+jets experimental measurements

Studies of Z+jets production at the Tevatron use jets with $p_T > 20$ or 30 GeV and $|\eta| < 2.1$ in CDF and $|y| < 2.5$ at D0. The uncertainties of the differential measurements typically range from 7 to 10% (for $Z + 2$ jets). Measurements [11] are compared to leading order (LO), multileg and NLO generators. A good agreement is reached with multileg and NLO generators. The measured differential cross sections as a function of jet multiplicity or jet momentum are generally more precise than MC predictions.

Measurements are also performed at the LHC using 36 pb^{-1} of data using jets with $p_T > 30$ GeV and $|y| < 4.4$ for ATLAS and $|\eta| < 2.4$ for CMS. For reconstruction corrections, ATLAS uses a bin-by-bin unfolding while CMS uses SVD. CMS also tests the so-called Berends-Giele scaling by measuring $\frac{\sigma(Z+n\ jets)}{\sigma(Z+n+1\ jets)} = \alpha + n\beta$ showing that β is compatible with 0. The uncertainties range between 13 to 20%. A good agreement is observed between data [12, 6] and multileg and NLO generators.

At the Tevatron, CDF measures the $Z + b$ jets cross section using jets reconstructed with the midpoint cone algorithm with $R = 0.7$ and $p_T > 20$ GeV and $|\eta| < 1.5$ while the leptons are selected using a neural network. Using 9.1 fb^{-1} of data, the cross section is extracted using a template fit of the secondary vertex mass leading to: $\frac{\sigma(Z+b\ jets)}{\sigma(Z+jets)} = 2.08 \pm 0.18(\text{stat}) \pm 0.27(\text{syst})\%$ in agreement with the predictions from MCFM and corresponding to a K-factor of 1.6 compared to the Alpgen prediction. At D0, jets with $p_T > 15$ or 20 GeV and $|\eta| < 2.5$ are used. A template fit of a discriminant variable built with the secondary vertex mass and the probability from tracks to originate from the primary vertex is used. With 4.2 fb^{-1} of data, D0 measures $\frac{\sigma(Z+b\ jets)}{\sigma(Z+jets)} = 1.93 \pm 0.22(\text{stat}) \pm 0.15(\text{syst})\%$ [13] in agreement with the prediction from MCFM of $1.92 \pm 0.22\%$.

At the LHC, the Z+b jets cross section is measured with jets with $p_T > 25$ GeV and $|y| < 2.1$ with a template fit of the secondary vertex mass yielding $3.55^{+0.82}_{-0.74}(\text{stat})^{+0.73}_{-0.55}(\text{syst}) \pm$

0.12(lumi) pb [14] in agreement with the predictions from MCFM of 3.88±0.58 pb. With 2.2 fb^{-1} of data, CMS measures independently the $Z + 1b$ and $Z + 2b$ jets cross sections using jets with $p_T > 25$ GeV and $|y| < 2.1$ and a template fit to the secondary vertex mass of the leading jet. The results give $\sigma(Z + b) = 5.84 \pm 0.08(\text{stat}) \pm 0.72(\text{syst})^{+0.25}_{-0.55}(\text{theory})$ pb [15] compatible with the prediction from MCFM, and $\sigma(Z + bb) = 0.37 \pm 0.02(\text{stat}) \pm 0.07(\text{syst}) \pm 0.02(\text{theory})$ pb [16] in agreement with the predictions from Madgraph.

2.3. Summary of the V+jets measurements

Even if a lot of V+jets measurements have been performed at the Tevatron and the LHC, only a few of them have been listed above. The V+light jets measurements agree with predictions from NLO and multileg generators (scaled to the NNLO inclusive prediction). The measurements are already enabling to constrain the MC predictions since their precisions are around 20% for $W + 3$ jets and 15% for $Z + 2$ jets. In general, measurements of V+heavy flavor jets agree with predictions from NLO and scaled multileg generators. However, some discrepancy exists in the measurement of the $W + b$ jets cross section at CDF while the LHC measurements so far agree with the predictions for the same process. The measurement uncertainties are generally larger than the ones on MC predictions. Typically, the precisions obtained in data are around 25% for $W + c$ jets, 20-30% for $W + b$ jets and around 15-30% for $Z + b$ jets. Further measurements with higher statistics are expected, both, from Tevatron and LHC in the near future. LHC measurements with $\sqrt{s} = 8$ TeV are also expected.

3. V+jets background determination in top-quark analyses

Depending on the experiments, on the colliders and on the analyses, different techniques are used to evaluate the background from V+jets in top-quark measurements. Only some of them will be discussed below which are based on the methods used for $t\bar{t}$ and single top-quark cross section measurements. The general rule (which suffers from exception) is that the V+jet background shapes in top-quark analyses are usually taken from MC (either from Alpgen or Madgraph) while the V+jets normalization is usually measured directly on data. The ratio of V+heavy flavor (hf) jets over V+light jets is rescaled using NLO predictions and/or checked or constrained using data itself. When both light and hf jet MC samples are produced separately, there is some overlap of hf jets in the light flavor jet samples because hf jets can originate from matrix element (ME) or parton shower (PS). A procedure to remove the overlap needs to be established. A possible removal can be based on jet angles, i.e. for a given heavy flavor jet, the jet from PS is kept if there is a jet j with $\Delta R(j, hf) < 0.4$ otherwise the jet from ME is kept.

3.1. W+jets background determination

The W+jets background is of particular importance in the lepton+jets channel.

At CDF, Alpgen+Pythia is used to model the shapes of the W+jets background. In the $t\bar{t}$ lepton+jets cross section measurements, the W+jets background is normalized to data before b-tagging and after subtraction of the other backgrounds. The $t\bar{t}$ cross section was measured using a fit to the distribution of a neural network based flavor separator in 9 samples ($1- \geq 5$ jets, 1 or 2 b-tag) [17]. The background is constrained with the 1 or 2 jets sample with floating normalization for the $b\bar{b}$, $c\bar{c}$ and single c MC samples. Fitting the normalization allows to reduce the W-jets related-systematic uncertainties. The measured K-factor related to Alpgen+Pythia is $K_{b\bar{b}} = 1.39^{+0.28}_{-0.22}$ in agreement with the theoretical prediction. In CDF single top cross section measurements, the W+n jets pretag data sample is used to scale the prediction from Alpgen. The flavor separator is fitted in the 1 jet sample to get $K_{hf} = 1.4 \pm 0.4$. A 30% uncertainty is assessed for the extrapolation to the W+2,3 jets samples.

D0 is also using Alpgen+Pythia for modelling the W+jets background. In the $t\bar{t}$ lepton+jets cross section measurement, the normalization is fitted after subtraction of the other background together with the $t\bar{t}$ cross section in bins of number of jets. After checking on data, the NLO predictions for the ratio $R_{hf} = N(W + bb/cc)/N(W + light) = 1.47 \pm 0.22$ and $R_c = N(W+c)/N(W+light) = 1.27 \pm 0.15$ are used which translates into an uncertainty of 0.2 pb on the measured $t\bar{t}$ cross section [18]. R_{hf} can also be fitted together with the $t\bar{t}$ cross section using samples with 1 or 2 jets, with or without b-tagging leading to $R_{hf} = 1.55 \pm 0.09(\text{stat})^{+0.17}_{-0.19}(\text{syst})$. The uncertainty on R_{hf} translates into an uncertainty of 0.02 pb in the fitted $t\bar{t}$ cross section. In the single top quark cross section analysis, the W+jets background is normalized with the multijet background before b-tagging using the lepton p_T, the missing transverse energy ($\not\!\!E_T$) and the W transverse mass distributions. The W+hf scale factor is taken from the NLO predictions and cross-checked with the 0-tag 2 jet sample. A 12% uncertainty is assigned to this normalization. Figure 3 shows the CDF and D0 distributions on control samples in the $t\bar{t}$ cross section analyses.

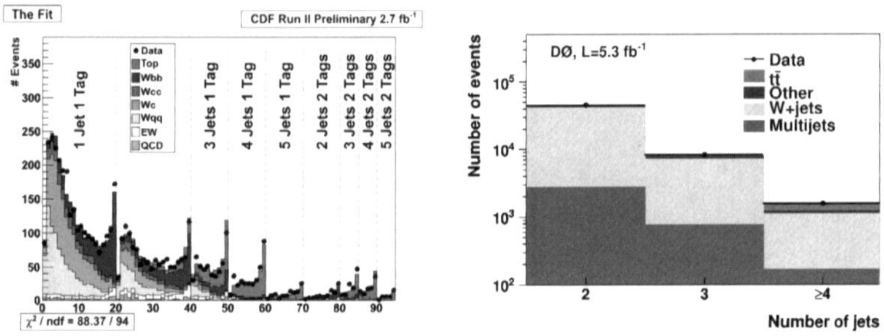

Figure 3. Output of the flavor separator after the fit split by the number of jets and tags in CDF [17] (left), jet multiplicity distribution for events with 0 b-tagged jets at D0 [18] (right).

In ATLAS, Alpgen+Herwig is used to generate the W+jets MC samples. In the $t\bar{t}$ lepton+jets cross section analysis, the W+jets normalization is fitted together with the $t\bar{t}$ cross section. Alternatively, the ratio of the number of produced W^+ and W^- is used. At the LHC, due to the quark content of the protons, more W^+ are produced than W^- in W+jets processes. The ratio $r_{MC} = W^+/W^-$ is well known theoretically. Hence the total number of W+jets before tagging (in the 4 jet inclusive bin) can be extracted from the value of r_{MC} in the MC with the formula: $N_{\geq 4pretag} = \frac{r_{MC}+1}{r_{MC}-1}(D^+ - D^-)$ where D^+, D^- is the number of events with W^+ or W^- in data after selection. The hf scale factor is determined in the 1, 2 jet sample without and with b-tagging. In single top cross section measurements, the W+jets sample is scaled by a factor K=1.2 which corresponds to the inclusive NNLO predictions from FEWZ. The W+hf normalization is fitted together with the signal.

CMS is using Madgraph+Pythia to generate the W+jets MC samples. In their untagged $t\bar{t}$ cross section analysis, a template fit of discriminating variables (like the distribution of $\not\!\!E_T$) allows to extract the W+jets normalization and the $t\bar{t}$ cross section. In the tagged analysis, a template fit of the secondary vertex mass is performed to get the W+jets normalization and the hf scale factor. The extracted hf scale K_{hf} is found to be slightly higher than the NNLO predictions: $K_{W+b\ jets} = 1.9^{+0.6}_{-0.5}$ and $K_{W+c\ jets} = 1.4 \pm 0.2$. For the single top quark cross section measurements, a template fit of the non-tagged jet pseudo-rapidities is exploited in sidebands using the 2 jets, 1 b-tag sample. Figure 4 shows typical distributions for the W+jets, $t\bar{t}$ combined fit at the LHC.

3.2. Z+jets background determination

The Z+jets background is mostly relevant for the analyses in the dilepton channel. As for the W+jets samples, the Tevatron experiments are using Alpgen+Pythia to generate the Z+jets MC

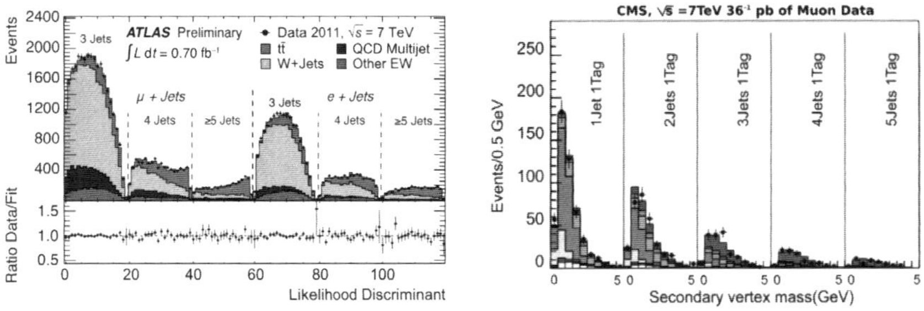

Figure 4. Result of combined fit to data in the 3 to 5-jet bin in ATLAS [19] (left), Results of the combined fit in the muon channel for exactly 1 b-tag at CMS [20] (right).

background while ATLAS is using Alpgen+Herwig and CMS Madgraph+Pythia. Depending on the collaborations, the number of Z+light jets or Z+hf jets is taken directly from the predictions of the multileg generators scaled to the NNLO computations, or it is evaluated using a data-driven method. The data-driven method scales the data in the Z mass window after $\not{E}_T$ selection. The MC is used afterwards to predict the ratio of events inside and outside the Z mass region. The data driven extraction agrees with the MC predictions but usually leads to smaller uncertainties (1-3% uncertainties on the $t\bar{t}$ cross section).

4. Conclusion

A lot of V+jets measurements have been performed by the CDF, D0, ATLAS and CMS experiments at the Tevatron and the LHC. The measurements agree with the NLO MC predictions. Multilegs generators are suitable to reproduce most of the V+jets kinematics. Still, more results and checks are expected to come. Various methods are used to measure the V+jets background in top-quark analyses. Depending on the needed precision, multileg generators are scaled to NLO predictions or data driven normalization are used or fit together with the top-quark signal. Thanks to the advanced background estimation technique, the achieved uncertainty on the V+jets background is usually found to be small.

References

[1] ATLAS Collaboration, JINST **3**, S08003 (2008).
[2] CMS Collaboration, JINST **3**, S08004 (2008).
[3] T. Aaltonen *et al.* [CDF Collaboration], CDF Note 10423.
[4] V. M. Abazov *et al.* [D0 Collaboration], Phys. Lett. B **705** (2011) 200 [arXiv:1106.1457 [hep-ex]].
[5] ATLAS Collaboration, Phys. Rev. D **85**, 092002 (2012) [arXiv:1201.1276 [hep-ex]].
[6] CMS Collaboration, JHEP **1201**, 010 (2012) [arXiv:1110.3226 [hep-ex]].
[7] V. M. Abazov *et al.* [D0 Collaboration], Phys. Lett. B **666**, 23 (2008) [arXiv:0803.2259 [hep-ex]].
[8] T. Aaltonen *et al.* [CDF Collaboration], Phys. Rev. Lett. **104**, 131801 (2010) [arXiv:0909.1505 [hep-ex]].
[9] ATLAS Collaboration, Phys. Lett. B **707**, 418 (2012) [arXiv:1109.1470 [hep-ex]].
[10] CMS Collaboration, CMS-PAS-EWK-11-013, https://cdsweb.cern.ch/record/1369558.
[11] V. M. Abazov *et al.* [D0 Collaboration], Phys. Lett. B **678**, 45 (2009) [arXiv:0903.1748 [hep-ex]].
[12] ATLAS Collaboration, Phys. Rev. D **85**, 032009 (2012) [arXiv:1111.2690 [hep-ex]].
[13] V. M. Abazov *et al.* [D0 Collaboration], Phys. Rev. D **83**, 031105 (2011) [arXiv:1010.6203 [hep-ex]].
[14] ATLAS Collaboration, Phys. Lett. B **706**, 295 (2012) [arXiv:1109.1403 [hep-ex]].
[15] CMS Collaboration, JHEP **1206**, 126 (2012) [arXiv:1204.1643 [hep-ex]].
[16] CMS Collaboration, CMS-PAS-SMP-12-003, https://cdsweb.cern.ch/record/1428117.
[17] T. Aaltonen *et al.* [CDF Collaboration], Phys. Rev. D **84**, 031101 (2011) [arXiv:1103.4821 [hep-ex]].
[18] V. M. Abazov *et al.* [D0 Collaboration], Phys. Rev. D **84**, 012008 (2011) [arXiv:1101.0124 [hep-ex]].
[19] ATLAS Collaboration, ATLAS-CONF-2011-121, http://cds.cern.ch/record/1376413.
[20] CMS Collaboration, Phys. Rev. D **84**, 092004 (2011) [arXiv:1108.3773 [hep-ex]].

Object definition and performance at CMS

Carmen Diez Pardos for the CMS Collaboration
Deutsches Elektronen-Synchrotron (DESY), Notkestrasse 85, 22607 Hamburg, Germany
E-mail: carmen.diez@desy.de

Abstract. The performance of the different object reconstruction and identification algorithms in CMS has been studied on data collected in pp collisions at $\sqrt{s} = 7$ TeV and at $\sqrt{s} = 8$ TeV at the LHC. We present measurements of the reconstruction efficiencies, fake rates, and momentum scale and resolution, which are relevant for top physics analyses.

1. Introduction

Top quark physics requires a good understanding of all objects as the products of top decays involve jets, high momentum isolated leptons (in leptonic decay modes), soft leptons from b-jets, E_T^{miss} from leptonic W decays etc. Within the CMS [1] Collaboration, the Particle Flow (PF) approach [2] is used for the event reconstruction and widely used in top analyses. It combines information from all detector components, prior to jet clustering and missing energy calculation, to reconstruct particles (hadrons, photons, muons and electrons). From these particles composite objects, such as jets, taus or E_T^{miss}, are reconstructed. This yields to higher response and a big improvement in energy resolution and tau identification.

2. Leptons

2.1. Muon and electron

Muons are reconstructed using information from the silicon tracking and muon systems. Different algorithms are combined to reach a robust and efficient identification [3]. Electron reconstruction includes information from the tracker system and the calorimeters. Shower shapes, ratio of the energy deposited in the hadronic and electronic calorimeters, and conversion vetoes for electrons [4] are used for its identification. Lepton identification and isolation efficiencies are measured using the tag and probe approach with Z resonances. Fig. 1 shows an example of the reconstruction and identification efficiency for muons (as a function of p_T) and electrons in the central region of the detector (as a function of η). In both cases, there is a remarkable agreement between data and Monte Carlo and the reconstruction efficiencies are very high.

The momentum scale and resolution of muons are studied using $Z \rightarrow \mu\mu$ resonances and cosmic ray data, in the range $20 < p_T < 100$ GeV, where the momentum measurement is provided by the tracker. The average bias in the muon momentum scale was measured with a precision better than 0.2% and was found to be consistent with zero. The relative p_T resolution is between 1.3% to 2.0% for muons in the barrel and better than 6% in the endcaps. The resolution measured from di-electron resonances is around 1–2% in the barrel region and better than 4% in the endcaps.

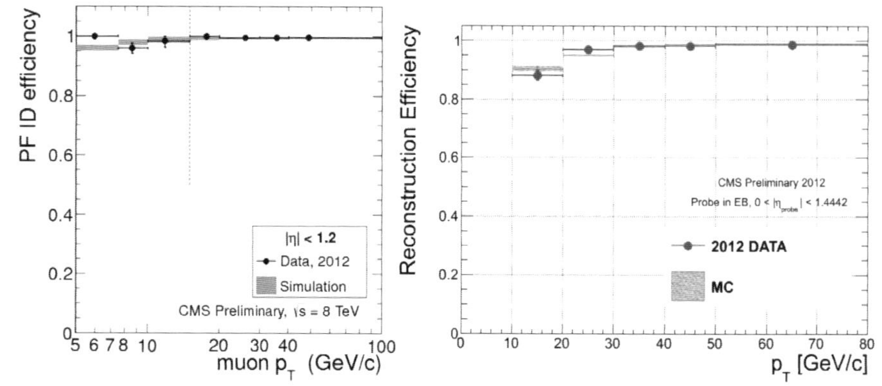

Figure 1. Muon (left) and electron (right) reconstruction efficiency in the central region of the detector obtained with a sample of $Z \to ll$ at $\sqrt{s}$=8 TeV.

2.2. *Taus*

τ-leptons decay predominantly hadronically, being their products one or three charged mesons (mostly π^+ and π^-), up to three neutral pions, and τ neutrino. The main algorithm used at CMS to reconstruct taus is the Hadron plus Strips (HPS) algorithm [5]. The HPS combines PF electromagnetic particles in strips, to take into account possible broadening of calorimeter depositions from photon conversions. The neutral objects are then combined with existing charged hadrons to reconstruct the hadronic tau decay products.

To measure the efficiency of its reconstruction and identification in data, a sample of $Z \to \tau\tau \to \mu\tau_{\rm had}$ events enriched in real taus was used with relatively small contribution from other processes. The identification efficiency results approximately 50%, keeping the misidentification rate for jets at the level of ~1%, see Fig. 2 left. The tau energy scale measured with the same method was found to be close to unity with the measured energy scale uncertainty better than 3%, see Fig. 2 right.

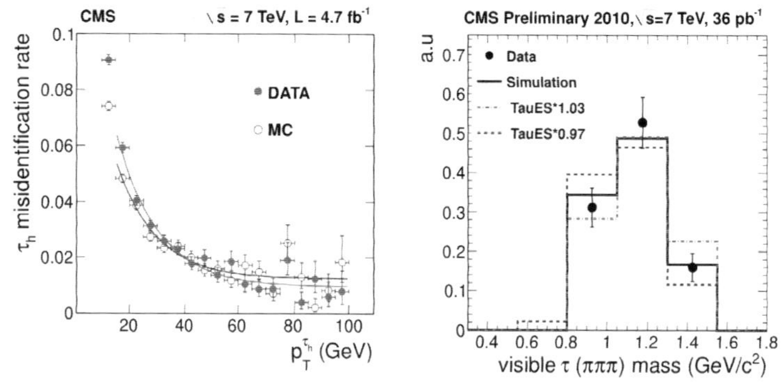

Figure 2. Left: fake-rate measured in data (7 TeV) compared to MC prediction. Right: the reconstructed invariant mass of taus decaying into three charged pions compared to predictions of the simulation.

Isolated electrons passing the identification/isolation criteria of the algorithm, misidentified as pions originating from $\tau_{\rm h}$ are also an important source of background in many analyses with

τ_h in the final state. The misidentification rate, found by applying the tag and probe method with a $Z \rightarrow ee$ sample, is smaller than 2%. For muons, the misidentification rate is below 0.2%.

3. Jets

Jets in CMS are reconstructed using PF candidates with the anti-kT algorithm [6] with a spatial separation parameter of R = 0.5. The CMS Collaboration uses a factorised approach for jet calibration to combine the advantages of Monte-Carlo simulation based studies with robust data-driven methods [8]. First, an offset correction is applied, to remove the additional energy inside jets due to pile-up events and electronic noise. Second, a MC calibration factor is used to correct the non-uniformity and non-linearity of energy response in pseudorapidity and jet transverse momentum, respectively. This factor is estimated from MC simulation of multijet events by requiring the energy of the generator jets to match that of the reconstructed jets. Fig. 3 shows the jet energy response before and after applying PU and MC truth corrections.

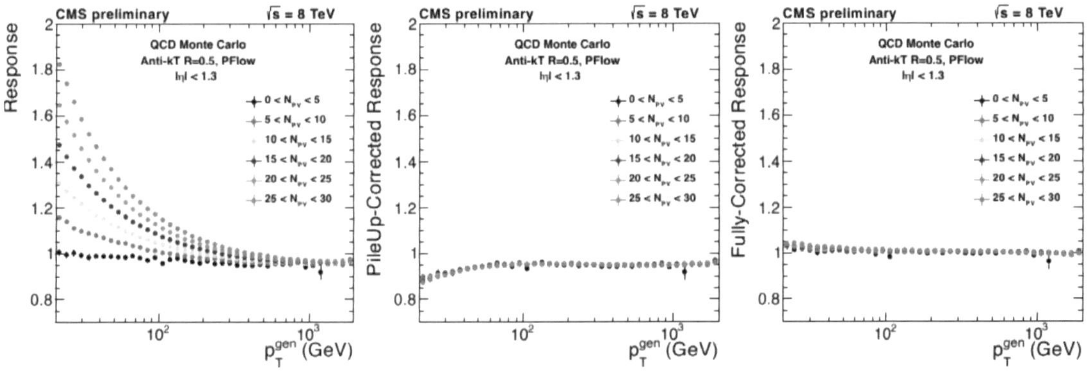

Figure 3. Jet energy response before applying any correction (left), after the pile-up correction (middle) and the corrections to the generator level (right), which show closure over the whole p_T range.

Finally, the residual corrections account for the small differences between data and simulation. The dijet p_T balance technique [8] is used to measure the energy of a jet at any $|\eta|$ relative to the other jet identified in the central region with $|\eta| < 1.3$. The relative jet energy response measured in data is compared to one obtained from the simulation and the correction is derived to correct for the observed small differences. The absolute jet energy response is measured in the reference region of $|\eta| < 1.3$ using γ^*/Z+jet events in data and MC simulation where a jet energy scale is derived by exploiting the balance in transverse momentum between a boson and a jet in the final state.

The uncertainties on the transverse momentum of a corrected jet strongly depend on its p_T and η, see Fig. 4. The jet energy scale is determined with a precision of about 1% in the central part of the detector and better than 3% in the endcaps.

4. B-tagging

The ability to identify the jets from the hadronisation of heavy flavour quarks plays a central role in precision measurements of top quark properties. Flavour tagging algorithms are thus used to identify these jets and reduce the otherwise overwhelming background from processes involving jets from gluon and light quarks fragmentation and from c-quark hadronisation. Analyses also need an accurate estimate of the performance (efficiency and misidentification rate) of the chosen

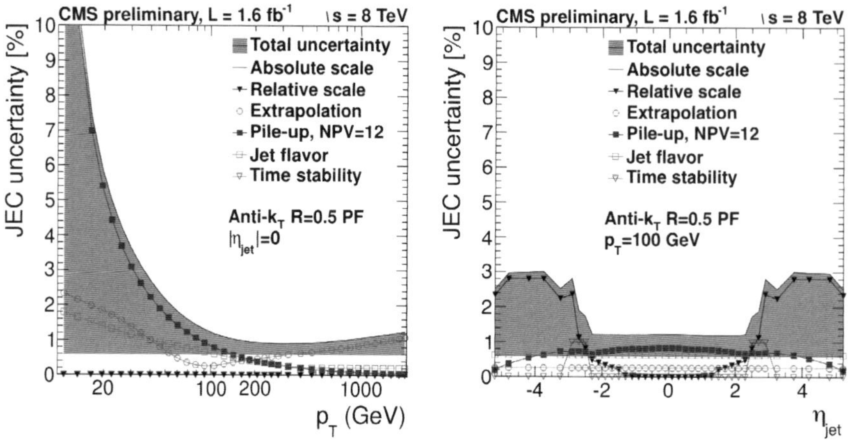

Figure 4. Uncertainty of the final jet energy corrections as a function of p_T (left) and η (right) split up into the considered sources of systematic uncertainties [7].

algorithm. The main features to identify b-quarks are their large lifetime ($\sim$1.5 ps) and their decay length ($\sim$1.8 mm).

Different algorithms are used, based on the track impact parameter calculated in three dimensions or/and the presence of a secondary decay vertex and the kinematic variables associated to it. These variables are for instance the flight distance and direction or properties of the system of associated tracks (multiplicity, energy, mass) [10]. Fig. 5 presents the performance of various identification algorithms obtained from MC simulation.

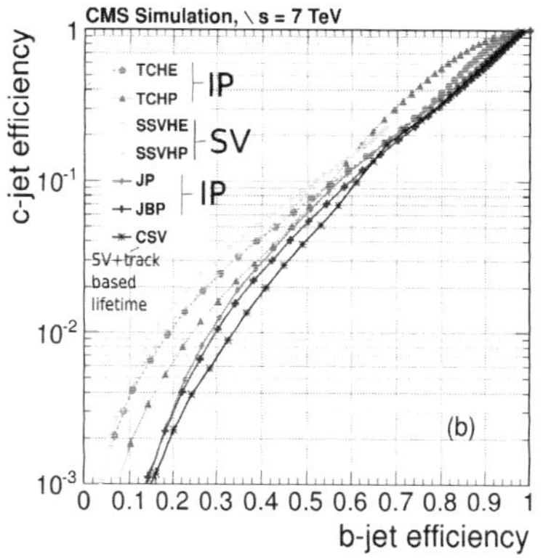

Figure 5. Curves of c-efficiencies as a function of the b-efficiency obtained from simulation for the different algorithms developed by the CMS Collaboration.

The performance of these taggers is studied using multijet events and $t\bar{t}$ events with different methods. Due to a large b-quark mass, the muon momentum component transverse to the jet axis ($p_{\rm rel}$) or impact parameter (IP) of the muon track is larger for muons from b hadron decays than that for muons in light-flavour jets. Modelling of $p_{\rm rel}$ and IP in MC simulation to represent distributions expected from b-jets is used as a template and that is compared to the distribution observed in multijet data events. The distribution of $p_{\rm rel}$ is used to measure the tagging efficiency for low/average-p_T jets, while that of IP allows to measure efficiency for high-p_T jets. The measurements from different methods are combined based on the weighted mean taking into account the correlation between uncertainties. Individual results are combined to provide the optimal measurement of data/MC SF for $30 < p_T < 670$ GeV. The results are shown in Fig. 6, left. Various methods using $t\bar{t}$ events [9] in the lepton+jets and dilepton channels provide inclusive results suited for measurements other than the cross section, see Fig. 6 right.

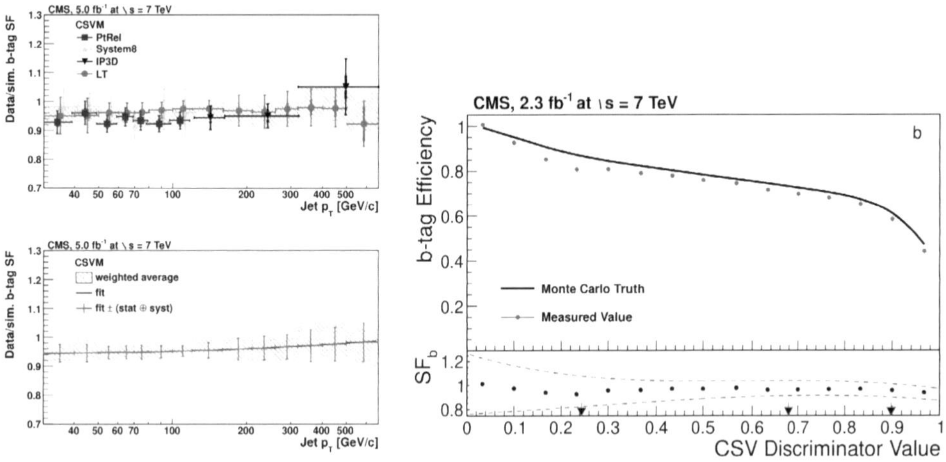

Figure 6. Left: individual (top) and combined (bottom) measurements of the ratio of efficiencies in data and simulation for the CSV method. Thick lines indicate the statistical error, while thin lines correspond to the combined statistical and systematic uncertainties. Top analyses apply those scale factors to correct MC yields. Right: measured b-tagging efficiency as a function of the discriminator threshold for the CSV algorithm from data and predicted from simulation are shown in the upper histogram. The scale factors are shown in the lower histogram. The total uncertainties are represented by a blue band for the scale factor. The arrows indicate the standard operating points.

5. Missing transverse energy

Events are reconstructed with the Particle Flow technique: E_T^{miss} is computed as the negative vectorial sum of all particle candidates, including the JES correction. It is a challenging variable as it is easy to obtain fake E_T^{miss} due to large shower fluctuations, non-linear calorimeter response, instrumental noise or poorly instrumented areas. The missing transverse energy resolution is studied in $Z \rightarrow \mu\mu$ events. In a well-measured event, the transverse momentum of a boson and hadronic activity should be balanced: $q_T^Z + u_T + E_T^{miss}=0$ where q_T^Z is the transverse momentum of the boson and u_T is the hadronic recoil, defined as the vector sum of the transverse momenta of all particles except the vector boson. The projection of the latter into the axis parallel to the Z boson momentum, $u_{\|}$, is used to obtain the energy correction to the E_T^{miss} in data. The E_T^{miss} resolution is assessed as the standard deviation of the perpendicular projection of the hadronic recoil to the Z transverse momentum ($\sigma(u_\perp)$) [11, 12].

The E_T^{miss} resolution as a function of the number of primary vertices is shown in Fig. 7, left. Compared to previous data taking periods, an improved resolution for a fixed number of primary vertices was achieved in 2012 by the use of dedicated pile-up corrections. Furthermore, improved calorimeter reconstruction techniques have reduced the impact of pile-up on the E_T^{miss} measurements. The resolution degradation per additional primary vertex only amounts to 3 GeV. The correction applied to jets are propagated to E_T^{miss} to restore the consistency of

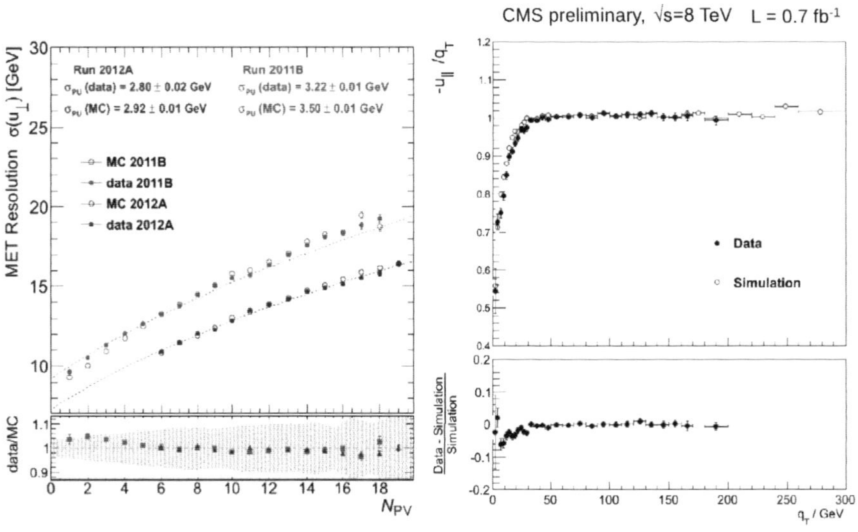

Figure 7. E_T^{miss} resolution, $\sigma(u_\perp)$, as a function of the number of primary vertices (left) and corrected response, $u_{||}/q_T$, as a function of transverse momentum of the Z boson q_T (right).

the event. This corrected E_T^{miss} is presented in Fig. 7 right. The response is very close to unity over a wide p_T range and the results in data agree with Monte Carlo simulation.

6. Conclusions

The CMS Collaboration has developed a number of sophisticated algorithms to identify jets, b-jets, leptons, etc. These algorithms are studied with data and simulation, showing a great performance of the object identification both at 7 and 8 TeV and a remarkable agreement between data and simulation. Furthermore, CMS is performing well in the LHC environment, adapting to the increasing luminosity (applying PU dependent corrections, developing customised isolation), which allows for precision measurements of top quark properties.

References

[1] CMS Collaboration, The CMS experiment at the CERN LHC, JINST 03 (2008) S08004
[2] CMS Collaboration, Commissioning of the Particle Flow Event Reconstruction with the First Collisions Recorded in the CMS Detector, CMS-PAS-PFT-10-001 (2010).
[3] CMS Collaboration, Performance of CMS muon reconstruction in pp collision events at sqrt(s) = 7 TeV, JINST 7 (2012) P10002
[4] CMS Collaboration, CMS-PAS-EGM-10-004, CMS-PAS-EGM-10-001
[5] CMS Collaboration, Performance of tau reconstruction algorithms in 2010 data collected with CMS, CMS-PAS-TAU-11-001.
[6] M. Cacciari, G. P. Salam, and G. Soyez, JHEP 04, 063, (2008).
[7] CMS Collaboration, Jet Energy Corrections and Uncertainties, CMS DP-12-012.
[8] CMS Collaboration, Determination of jet energy calibration and transverse momentum resolution in CMS, 2011 JINST 6 P11002.
[9] CMS Collaboration, Measurement of the b-tagging efficiency using $t\bar{t}$ events, CMS PAS-BTV-11-003.
[10] CMS Collaboration, b-Jet Identification in the CMS Experiment, PAS-BTV-11-004.
[11] CMS Collaboration, E_T^{miss} Performance in 2011 CMS Data, CMS DP-12-003.
[12] CMS Collaboration, E_T^{miss} Performance in 2012 RunA CMS Data, CMS DP-12-013.

Journal of Physics: Conference Series **452** (2013) 012016 doi:10.1088/1742-6596/452/1/012016

SM single-top production at hadron colliders

Pietro Falgari

Institute for Theoretical Physics and Spinoza Institute,
Utrecht University, 3508 TD Utrecht, The Netherlands

E-mail: p.falgari@uu.nl

Abstract. We give an overview of the status of theoretical predictions for single-top production in the Standard Model. We focus in particular on recent developments, including calculations of off-shell effects at next-to-leading order beyond the narrow-width approximation, all-order resummation of soft corrections and matching of next-to-leading order parton-level results to Monte Carlo parton showers.

Preprint numbers: ITP-UU-13/01, SPIN-13/01.

1. Introduction

The hadroproduction of a single top quark, first observed at the Tevatron [1, 2], is a process of significant phenomenological relevance, providing informations complementary to those that can be obtained from top-quark pair production. Being mediated by electroweak-boson exchange, single-top production represents an unique window into the charged-current interactions of the top quark, enabling tests of the $V-A$ structure of the Wtb vertex and a direct extraction of the CKM matrix element V_{tb}, which is at the moment only indirectly constrained. Single-top production is also very sensitive to new-physics effects and anomalous couplings, whose strength can be assessed by a precise measurement of the production cross sections. Furthermore, single-top production probes the bottom-quark parton distribution inside the proton, which at present is weakly constrained by other experimental data. For these reasons, the study of the production of a single top quark will be an important part of the physics programme of the Large Hadron Collider (LHC) at Cern, where several results are already available [3, 4, 5, 6].

In the Standard Model (SM) single-top production is customary divided into three production channels: t-channel production, which involves the exchange of a space-like W boson, s-channel production, where the intermediate W boson is time-like, and associated production of a top quark and a real W boson. The tree-level Feynman diagrams contributing to the three different channels are shown in Figure 1. Both at Tevatron and the LHC the t-channel process is the dominant one, accounting for about 63% and 71% of the cross section, respectively [7] [1]. At the Tevatron s-channel production is the second most important channel ($\sim 30\%$) but it is negligible at the LHC, where Wt production has the second largest cross section (about 25%). It has to be mentioned that the classification into three separate channels is delicate and somewhat artificial, since, starting at next-to-leading order (NLO) in α_s t-channel and s-channel production mix with each other. Furthermore, at NLO Wt production interferes with $t\bar{t}$ production, which makes

[1] The relative size of the three channels changes slightly depending on the theoretical prediction used.

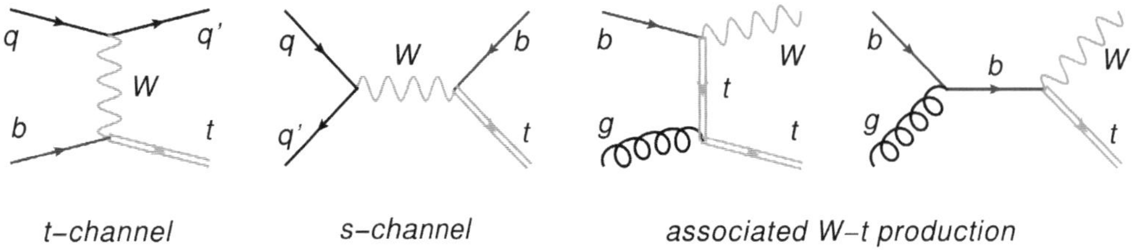

Figure 1. Tree-level single-top production in the Standard Model.

the theoretical definition of two separate processes difficult and the extraction of the Wt signal from the much larger $t\bar{t}$ background experimentally challenging. The prospects of a meaningful theoretical and experimental definition of the Wt process were thoroughly investigated in [8, 9].

NLO QCD predictions for single-top production in the approximation of a stable top have been available for more than ten years, both at the inclusive [10, 11, 12, 13, 14] and differential level [15, 16]. Electroweak and SUSY-QCD corrections have been also computed and found to be small, typically below 5% of the tree-level result [17, 18]. More recently the relation between the five-flavour (5F) and four-flavour (4F) scheme at NLO has been investigated [19, 20]. The two approaches differ in the treatment of the initial bottom quark in t-channel single-top production, which originates from a non-vanishing bottom PDF in the 5F scheme, and is generated by splitting of an initial gluon in the 4F scheme. In [19, 20] the two schemes were found to be in good agreement at NLO, except for distributions related to the spectator b-jet, which in the 5F scheme are effectively LO observables. Beyond the stable-top approximation, NLO predictions for single-top production and decay have been computed in the Narrow Width Approximation (NWA) [21, 22, 23, 24, 25, 26], and were implemented in the numerical code MCFM [27]. In this approach the top quark is produced on shell and let decay, with full spin correlations, to its final products.

Recent developments beyond a fixed-level NLO calculation for production and decay of an on-shell top include the calculation of off-shell and non-factorizable corrections to t-channel and s-channel single top production, soft-logarithm resummation of the partonic cross sections and the matching of NLO parton-level calculations to Monte Carlo parton showers. We will cover these topics in the following sections. New-physics effects in single-top production in extensions of the SM have also been widely investigated, but they are beyond the scope of this review.

2. Off-shell and non-factorizable corrections to single-to production

In the framework of the NWA NLO contributions are given by factorizable corrections to the on-shell production and decay of the top quark, while non-factorizable contributions connecting initial- and final-state light partons, as well as off-shell and finite-width effects, are neglected. These terms are small for the total cross section, of order of the top width-to-mass ratio $\Gamma_t/m_t \sim 1\%$, due to large cancellations between virtual and real corrections. However non-factorizable corrections could be a priori large for exclusive observables, like arbitrary kinematical distributions. This was investigated in [28, 29], where off-shell and non-factorizable corrections to t-channel and s-channel production were computed.

The calculation of [28, 29] is based on an effective-field theory (EFT) description of the single-top production process [30], built upon the hierarchy $\Gamma_t \ll m_t$. In this approach contributions to the amplitude are divided into hard corrections, encoding physics at the large momentum scale $q \sim m_t$, and soft contributions, describing the long-distance physics associated with the low scale $q \sim \Gamma_t$. In the effective theory only soft modes are described by dynamical fields, while hard contributions are encoded into the effective couplings (matching coefficients) of the

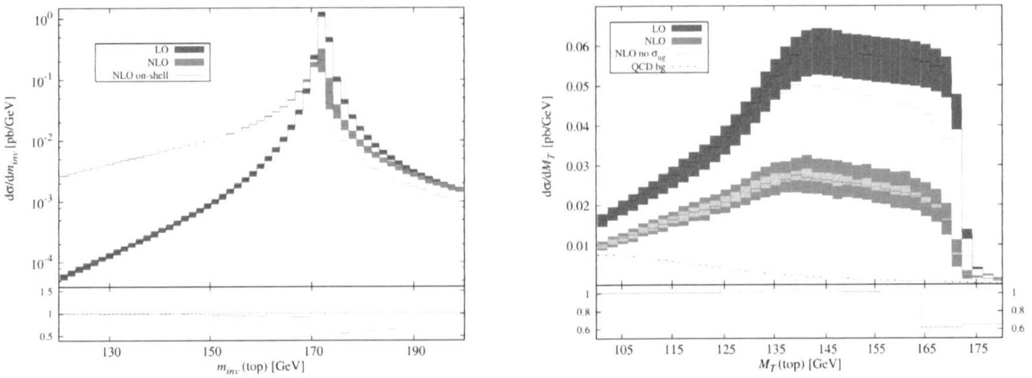

Figure 2. Top-quark invariant-mass (left) and transverse-mass (right) distributions for t-channel single top production at the LHC with $\sqrt{s} = 7\,\mathrm{TeV}$ [29]. See the text for explanation.

EFT Lagrangian. In this framework an NLO calculation requires $\mathcal{O}(\alpha_s)$ corrections to the hard matching coefficients and one-loop soft contributions to the EFT matrix elements. These can be defined in terms of an expansion of full QCD diagrams in hard and soft momenta using the method of regions [31].

Two examples of the effect of off-shell and non-factorizable corrections are shown in Figure 2, where the top invariant-mass and transverse-mass distributions for t-channel single-top production at a 7 TeV LHC are given for $m_t = 172\,\mathrm{GeV}$ [29]. The blue and red curves represent the LO and NLO off-shell result, respectively, the bands the corresponding uncertainty obtained from scale variation. The green curve in the lower inset is the ratio of the NLO on-shell prediction in the NWA to the off-shell result. One can seen that off-shell and non-factorizable effects are large around the peak of the invariant-mass distribution, with deviation of up to 50% from the on-shell result. However these effects change sign around the peak, which explains the small corrections in the transverse-mass distribution for $M_T < m_t$, where large cancellations occur due to averaging over different values of the invariant mass for a fixed M_T. These cancellations are less effective close the the distribution edge $M_T \sim m_t$, due to phase-space restrictions. In this region off-shell effects are large, about 40% of the NLO on-shell result, and are necessary for a reliable theoretical prediction of the distribution.

3. Resummation of soft logarithms

Single-top production observables contain logarithmic contributions $\alpha_s^m[\ln^k s_4/s_4]_+$ (with $k = 0, ..., 2m-1$) related to suppression of soft-gluon emission near kinematical thresholds, with s_4 the (observable dependent) kinematical variable that vanishes at threshold. Such terms can give a sizeable contribution to the cross sections, and all-order resummation of these threshold terms can be used as a mean to improve fixed-order predictions. For single-top production, this has been studied separately by two different groups, using a formalism based on resummation in Mellin-moment space [32, 33] and a framework involving soft-collinear effective theory (SCET) and renormalization group evolution equations [34, 35].

Both methods are based on the factorization of the cross section at threshold into a hard function H describing the short-distance scattering process, a soft function S encoding soft radiation and additional terms accounting for collinear emission from initial- and final-state particles. The all-order resummation of soft contributions in S is controlled by matrix-valued soft anomalous dimensions Γ_s, which have been computed up to two loops [36, 37, 38, 39].

Explicit results for the NNLL resummation of the single-top production cross section are

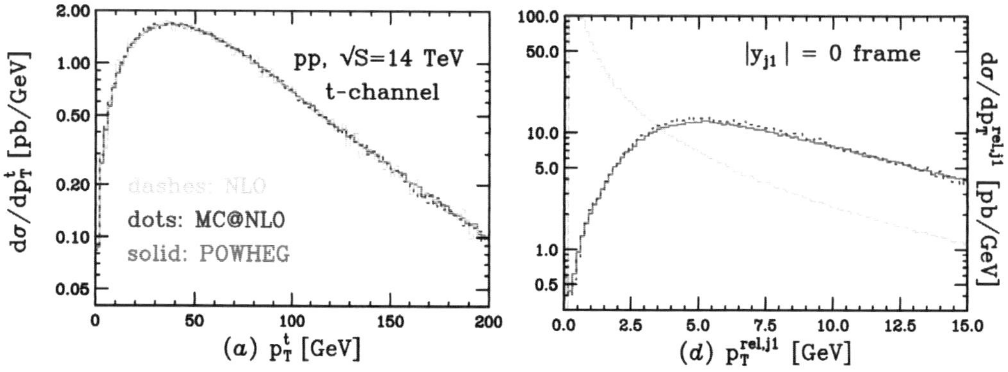

Figure 3. Comparison of fixed-order NLO results for t-channel single-top production at the LHC with MCPS results obtained with POWHEG and MC@NLO (plots taken from [44]). See the text for explanation.

available for the three separate channels in the Mellin formalism [40], and for t-channel [35] and s-channel production [34] in the SCET approach. Both formalisms find small resummation effects, of order of few percents, in t-channel single-top production. On the other hand, in s-channel production the results of the two groups show bigger discrepancies, with far larger soft effects (13%−15%) found in the Mellin approach [36]. This discrepancy has not been investigated yet. However it should be noted that in the two approaches different kinematical variables s_4 are chosen, so that sets of different logarithms are resummed. While the two calculations should give formally equivalent results at NNLL accuracy for the total cross section, power-suppressed contributions, which are not controlled by resummation, could be numerically important and be responsible for the observed differences.

4. Matching of NLO results to Monte Carlo parton showers

While a fixed-order NLO calculation gives an accurate description of wide-angle radiation, in the low p_T region multiple collinear emission becomes important and has to be taken into account. Furthermore, to realistically describe physical final states, a parton-level calculation has to be interfaced to some kind of hadronization model describing the evolution of energetic coloured partons into colour-singlet hadronic states. This can be done in the framework of Monte Carlo parton showers (MCPS), which gives a probabilistic description of multiple collinear splitting and can be easily interfaced to one's preferred hadronization model. Recently a lot of effort has been put into the matching of NLO fixed-order calculations to Monte Carlo parton showers, in an attempt to obtain a framework in which both large-angle radiation at NLO and all-order collinear emission are correctly described. This requires addressing double-counting issues in the collinear region, which has been consistently done in two different frameworks, MC@NLO [41] and POWHEG [42]. The production of a single top have been implemented in both approaches [43, 44, 45].

Figure 3 shows the comparison of the two NLO Monte Carlo showers with a fixed-order NLO result for the specific case of t-channel single-top production at the LHC with $\sqrt{s} = 14\,\text{TeV}$. The two plots (taken from [44]) represent the top-quark transverse-momentum distribution (left) and the distribution of the relative momentum of all the partons clustered inside the hardest jet, p_T^{rel,j_1}, in the reference frame in which the rapidity of the jet is zero. In both cases the good agreement between the results of the two NLO parton showers is evident. For the top transverse-momentum distribution one can also note the good agreement between the fixed-

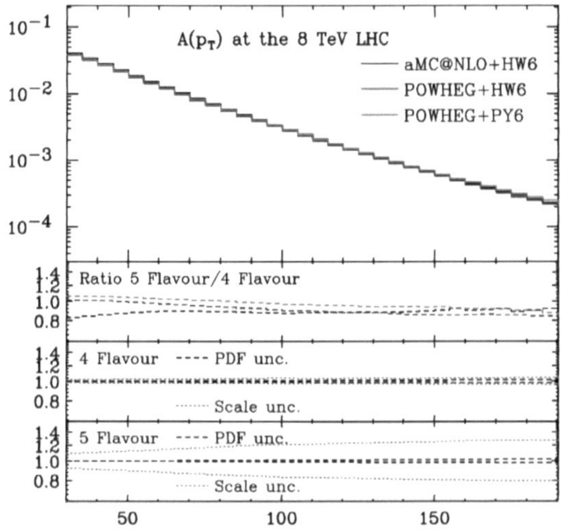

Figure 4. Acceptance as a function of the second-hardest b-jet at the 8 TeV LHC in the 4F scheme. Main plot: results for MC@NLO and POWHEG interfaced to Herwig and Pythia. Lower insets: ratio of the results in the 5F and 4F scheme and scale and PDF dependence of the 4F and 5F prediction computed with MC@NLO. See [46] for details.

order NLO prediction and the showered results. This is not the case for p_T^{rel,j_1}, where the parton-shower and fixed-order results show stark differences. This is expected, since the total relative momentum gives a measure of the spreading of the hardest jet, whose correct description requires the inclusion of multiple collinear emission when $p_T^{\mathrm{rel},j_1} \to 0$. Comparisons for the s-channel process and associated tW production [45] have shown a similar good agreement between MC@NLO and POWHEG, and confirm the importance of a NLO parton-shower treatment to correctly describe observables in both the small and large transverse-momentum regions.

Recently a comparison of the two NLO MCPS frameworks has also been performed in the four-flavour scheme [46]. Good agreement was again found between POWHEG and MC@NLO. As already observed for the fixed-order NLO cross section [19, 20], a comparison of results in the 4F and 5F schemes at the parton-shower level shows that the two descriptions are consistent at NLO, though the former gives more accurate predictions for observables which are related to the spectator b-jet. This can be seen for example in Figure 4, where the acceptance $A(p_T) = 1/\sigma \int_{p_T}^{\infty} dp_T^{(j_b,2)} \partial\sigma/\partial p_T^{(j_b,2)}$ is shown as a function of the transverse momentum of the second-hardest b-jet. From the insets is clear that the 4F scheme result has a much smaller theoretical uncertainty, though 4F and 5F predictions are consistent within their respective errors.

5. Conclusions

In view of its phenomenological relevance, in the last few years a lot of effort has been put into providing an accurate theoretical description of single-top production at hadron colliders. State-of-the-art results are represented by fixed-order NLO predictions matched to Monte Carlo parton showers, which provide an exact NLO description of the first emission and resum all-order radiation in the collinear region. In this framework the production and decay of the top quark is treated in the on-shell approximation. Studies of off-shell and non-factorizable effects have been performed for t-channel and s-channel production, and they have shown that these effects, though most of the time small, can be enhanced close to kinematical thresholds, and should be taken into account for a correct description of the shape of distributions close to kinematical edges. Resummation of Sudakov logarithms related to soft-gluon emission has also been studied by two different groups. In both cases, small resummation effects were found for t-channel production, but somewhat contrasting results were presented for s-channel production, which would motivate a further investigation of these effects.

References

[1] T. Aaltonen *et al.* [CDF Collaboration], Phys. Rev. Lett. **103** (2009) 092002 [arXiv:0903.0885 [hep-ex]].
[2] V. M. Abazov *et al.* [D0 Collaboration], Phys. Rev. Lett. **103** (2009) 092001 [arXiv:0903.0850 [hep-ex]].
[3] G. Aad *et al.* [ATLAS Collaboration], Physics Letters B **717** (2012) 330 [arXiv:1205.3130 [hep-ex]].
[4] G. Aad *et al.* [ATLAS Collaboration], Phys. Lett. B **716** (2012) 142 [arXiv:1205.5764 [hep-ex]].
[5] S. Chatrchyan *et al.* [CMS Collaboration], JHEP **1212** (2012) 035 [arXiv:1209.4533 [hep-ex]].
[6] S. Chatrchyan *et al.* [CMS Collaboration], [arXiv:1209.3489 [hep-ex]].
[7] W. Bernreuther, J. Phys. G **35** (2008) 083001 [arXiv:0805.1333 [hep-ph]].
[8] S. Frixione, E. Laenen, P. Motylinski, B. R. Webber and C. D. White, JHEP **0807** (2008) 029 [arXiv:0805.3067 [hep-ph]].
[9] C. D. White, S. Frixione, E. Laenen and F. Maltoni, JHEP **0911** (2009) 074 [arXiv:0908.0631 [hep-ph]].
[10] G. Bordes and B. van Eijk, Nucl. Phys. B **435** (1995) 23.
[11] T. Stelzer, Z. Sullivan and S. Willenbrock, Phys. Rev. D **56** (1997) 5919 [arXiv:hep-ph/9705398].
[12] M. C. Smith and S. Willenbrock, Phys. Rev. D **54** (1996) 6696 [arXiv:hep-ph/9604223].
[13] W. T. Giele, S. Keller and E. Laenen, Phys. Lett. B **372** (1996) 141 [arXiv:hep-ph/9511449].
[14] S. Zhu, Phys. Lett. B **524** (2002) 283 [Erratum-ibid. B **537** (2002) 351].
[15] B. W. Harris, E. Laenen, L. Phaf, Z. Sullivan and S. Weinzierl, Phys. Rev. D **66** (2002) 054024 [arXiv:hep-ph/0207055].
[16] Z. Sullivan, Phys. Rev. D **70** (2004) 114012 [arXiv:hep-ph/0408049].
[17] M. Beccaria *et al.*, Phys. Rev. D **77** (2008) 113018 [arXiv:0802.1994 [hep-ph]].
[18] G. Macorini, S. Moretti and L. Panizzi, arXiv:1006.1501 [hep-ph].
[19] J. M. Campbell, R. Frederix, F. Maltoni and F. Tramontano, Phys. Rev. Lett. **102** (2009) 182003 [arXiv:0903.0005 [hep-ph]].
[20] J. M. Campbell, R. Frederix, F. Maltoni and F. Tramontano, JHEP **0910** (2009) 042 [arXiv:0907.3933 [hep-ph]].
[21] J. M. Campbell, R. K. Ellis and F. Tramontano, Phys. Rev. D **70** (2004) 094012 [arXiv:hep-ph/0408158].
[22] Q. H. Cao and C. P. Yuan, Phys. Rev. D **71** (2005) 054022 [arXiv:hep-ph/0408180].
[23] Q. H. Cao, R. Schwienhorst, J. A. Benitez, R. Brock and C. P. Yuan, Phys. Rev. D **72** (2005) 094027 [arXiv:hep-ph/0504230].
[24] J. M. Campbell and F. Tramontano, Nucl. Phys. B **726** (2005) 109 [arXiv:hep-ph/0506289].
[25] S. Heim, Q. H. Cao, R. Schwienhorst and C. P. Yuan, Phys. Rev. D **81** (2010) 034005 [arXiv:0911.0620 [hep-ph]].
[26] R. Schwienhorst, C. P. Yuan, C. Mueller and Q. H. Cao, arXiv:1012.5132 [hep-ph].
[27] J. M. Campbell and R. K. Ellis, arXiv:1204.1513 [hep-ph].
[28] P. Falgari, P. Mellor and A. Signer, Phys. Rev. D **82** (2010) 054028 [arXiv:1007.0893 [hep-ph]].
[29] P. Falgari, F. Giannuzzi, P. Mellor and A. Signer, Phys. Rev. D **83** (2011) 094013 [arXiv:1102.5267 [hep-ph]].
[30] M. Beneke, A. P. Chapovsky, A. Signer and G. Zanderighi, Nucl. Phys. B **686** (2004) 205 [hep-ph/0401002].
[31] M. Beneke and V. A. Smirnov, Nucl. Phys. B **522** (1998) 321 [hep-ph/9711391].
[32] N. Kidonakis, Phys. Rev. D **74** (2006) 114012 [hep-ph/0609287].
[33] N. Kidonakis, Phys. Rev. D **75** (2007) 071501 [arXiv:hep-ph/0701080].
[34] H. X. Zhu, C. S. Li, J. Wang and J. J. Zhang, arXiv:1006.0681 [hep-ph].
[35] J. Wang, C. S. Li, H. X. Zhu and J. J. Zhang, arXiv:1010.4509 [hep-ph].
[36] N. Kidonakis, Phys. Rev. D **81** (2010) 054028 [arXiv:1001.5034 [hep-ph]].
[37] N. Kidonakis, Phys. Rev. D **82** (2010) 054018 [arXiv:1005.4451 [hep-ph]].
[38] N. Kidonakis, Phys. Rev. D **83** (2011) 091503 [arXiv:1103.2792 [hep-ph]].
[39] T. Becher and M. Neubert, Phys. Rev. D **79** (2009) 125004 [Erratum-ibid. D **80** (2009) 109901] [arXiv:0904.1021 [hep-ph]].
[40] N. Kidonakis, arXiv:1210.7813 [hep-ph].
[41] S. Frixione and B. R. Webber, JHEP **0206** (2002) 029 [hep-ph/0204244].
[42] S. Frixione, P. Nason and C. Oleari, JHEP **0711** (2007) 070 [arXiv:0709.2092 [hep-ph]].
[43] S. Frixione, E. Laenen, P. Motylinski and B. R. Webber, JHEP **0603** (2006) 092 [arXiv:hep-ph/0512250].
[44] S. Alioli, P. Nason, C. Oleari and E. Re, JHEP **0909** (2009) 111 [Erratum-ibid. **1002** (2010) 011] [arXiv:0907.4076 [hep-ph]].
[45] E. Re, Eur. Phys. J. C **71** (2011) 1547 [arXiv:1009.2450 [hep-ph]].
[46] R. Frederix, E. Re and P. Torrielli, JHEP **1209** (2012) 130 [arXiv:1207.5391 [hep-ph]].

Journal of Physics: Conference Series **452** (2013) 012017 doi:10.1088/1742-6596/452/1/012017

Single Top Quark Measurements at the Tevatron

J Joshi (for the DØ and CDF collaborations)
University of California, Riverside, CA 92521-0413, USA
E-mail: jyoti@fnal.gov

Abstract. This paper reports the measurement of electroweak production of single top quarks in lepton plus jets final state by the DØ and CDF collaborations. At Fermilab's 1.96 TeV proton-antiproton collider, events are selected from several inverse femtobarns of data in the W+jets topology consistent with the signature of an isolated charged lepton (electron or muon), large missing transverse energy ($\not\!\!E_T$) from the W boson decay and two or more jets, at least one of them is required to be identified as originating from a bottom quark. Sophisticated multivariate analysis techniques are employed to separate the small single top quark signal from background. Both experiments measure the single top production cross section in s-channel, t-channel and $s + t$-channel. The data is also used to extract limits on the CKM matrix element $|V_{tb}|$.

1. Introduction

The focus of this workshop, the top quark, is by far the heaviest and most interesting known particle of the standard model (SM). Top quarks are mostly produced in particle-antiparticle pairs via the strong interaction from a very high energy virtual gluon [1, 2]. The cross section at the Tevatron 1.96 TeV proton-antiproton collider is about 7.5 pb [3]. They can also be produced singly from a highly energetic virtual W boson via the electroweak interaction [4, 5]. There are three modes of single top quark production at a hadron collider [6]: t-channel production ($tqb = tq\bar{b}+\bar{t}qb$) [7], where a W boson and a b quark fuse to produce the top quark, and there are a spectator light quark and a bottom antiquark; s-channel production ($tb = t\bar{b} + \bar{t}b$) [8], where a W boson decays to a top quark and bottom antiquark. A third tW process, in which the top quark is produced together with a W boson, has a small cross section at the Tevatron [9] and is therefore not considered in this analysis. Figure 1 shows the lowest level Feynman diagrams for s- and t-channel production. Single top quark production is very distinct from $t\bar{t}$ pair production since it comes from an electroweak Wtb vertex instead of a strong gtt one. Within the SM, the single top signal allows a direct measurement of the Cabibbo-Kobayashi-Maskawa (CKM) matrix element $|V_{tb}|$ without assuming unitarity or three generations [10, 11]. Furthermore, since the top quark decays before hadronization, its polarization can be directly observed in the angular correlations of its decay products [12]. Single top processes are expected to be sensitive to several kinds of new physics. We present an improved measurement of the production rate of tb+tqb from DØ and CDF collaborations. We also present a measurement of the production rates of the individual tb and tqb processes, without any theoretical assumptions on either the tb or tqb production rate or the ratio of their yields. Finally, we present a new direct measurement of $|V_{tb}|$ extracted from the measured tb+tqb cross section. These results have been described in more details in [13, 14].

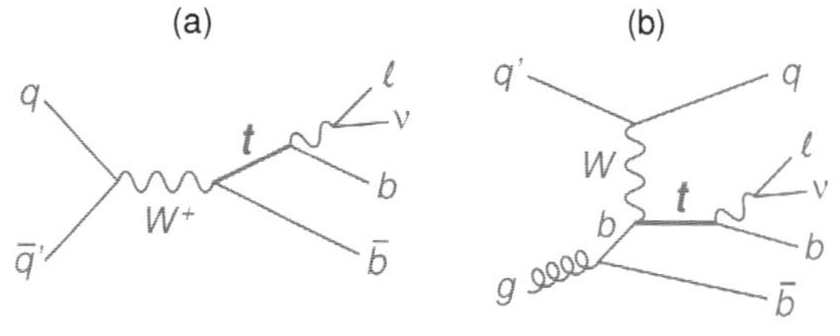

Figure 1. Lowest level Feynman diagrams for (a) tb and (b) tqb single top quark production.

2. Selecting Single Top Quark Events

The result presented in this document is based on 5.4 fb^{-1} of integrated luminosity recorded between 2002 and 2010 with the D0 detector [15], and 7.5 fb^{-1} of integrated luminosity recorded up to March 2011 with the CDF detector [16]. The data were collected with a logical OR of many trigger conditions at D0, that results in a fully efficient trigger selection for the single top signal and at CDF from the high-p_T electron/muon data stream and high-$\not\!\!E_T$ data stream. According to the SM, the top quark is expected to decay almost exclusively into a b-quark and a W boson, as other decays are suppressed by the small values of $|V_{ts}|$ and $|V_{td}|$ matrix elements. Thus, the final state consists of an isolated electron or muon, missing transverse energy, at least one b quark jet from the decay of the top quark and a second b quark jet in the s-channel, or a light quark jet and a spectator b quark jet in the t-channel. In both cases, gluon radiation can give rise to additional jets. The data are divided into different mutually exclusive subsamples based on the jet multiplicity (2 or 3 jets), and the number of jets identified as originating from b quarks (1 or 2 b-tags) in order to take advantage of the different signal:background ratios and dominant sources of backgrounds. D0 also included events with a fourth jet because there could have been an initial-state- or final-state-radiated quark or gluon. CDF required the electron or muon to have $p_T > 20$ GeV and pseudorapidity $|\eta| < 1.6$. D0 required $p_T > 15$ (20) GeV and $|\eta| < 1.1$ for electrons events with 2 (3 or 4) jets, and $p_T > 15$ GeV and $|\eta| < 2.0$ for muons. CDF set the threshold for $\not\!\!E_T > 25$ GeV and $\not\!\!E_T$ is corrected for the presence of muons and jets. D0 required $\not\!\!E_T$ not to be aligned with the direction of the lepton or the leading jet to limit the number of events originating from multijet production entering the candidate samples and also $\not\!\!E_T$ is required to be in the range of $(20, 200)$ GeV for events with 2 jets and $(25, 200)$ GeV for events with 3 or 4 jets. Events are selected that contain one jet with transverse momentum $p_T > 25$ (20) GeV and at least a second jet with $p_T > 15$ $(12-20)$ GeV, both within pseudorapidity $|\eta| < 3.4$ (2.8) in case of D0 (CDF).

The background processes that these cuts selected are mainly from W+jets events at low multiplicity and $t\bar{t}$ pairs at high jet multiplicity, with small contributions from Z+jets, dibosons, and multijet events where one of the jets is misidentified as an electron, or a b jet decayed to produce a muon that is misleadingly reconstructed as isolated from the jet. All the background processes (except multijets) and the signal events are simulated using Monte Carlo (MC) models. Single top signal events are modeled for a top-quark mass $m_t = 172.5$ GeV using the COMPHEP-based next-to-leading order (NLO) MC event generator SINGLETOP [17] (D0) and POWHEG [18] (CDF). PYTHIA [19] is used to model the hadronization of any generated partons. The $t\bar{t}$, W+jets, and Z+jets backgrounds are simulated using the ALPGEN leading-log MC event generator [20], with PYTHIA used to model hadronization. The multijet background is modeled using data. Figure 2 shows the comparisons between data and simulation for some important distributions after applying b-tagging.

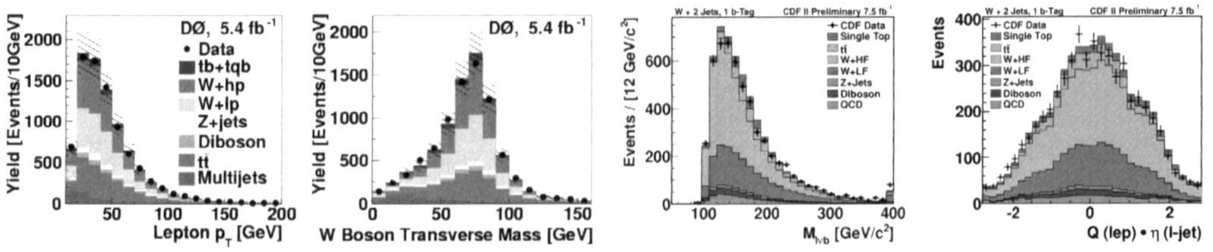

Figure 2. Comparison between the data and background model: lepton p_T and W boson transverse mass distributions from D0 analysis (first and second from left); reconstructed top quark mass and light quark jet pseudorapidity multiplied by lepton charge from CDF analysis (third and fourth from left).

3. Cross Section Measurement

The measurement of single top quark cross section present substantial experimental challenges comparing with $t\bar{t}$ production as it suffers from lower SM production rate and a large kinematically similar background. Simple counting experiments will not yield a precise measurement of the single top production cross section. Since the expected single top contribution is smaller than the uncertainty on the background, different multivariate analysis (MVA) methods are used to extract the signal. Following subsections describes the cross section measurement by D0 and CDF analyses.

3.1. DØ measurement

Three different MVA techniques are applied in this analysis: (i) boosted decision trees (BDT) [21], (ii) bayesian neural networks (BNN) [22], and (iii) neuroevolution of augmented topologies (NEAT) [23]. All three methods use the same data and model for background and consider the same sources of systematic uncertainties. Each MVA method is trained separately for the two single top quark production channels: for the tb (tqb) discriminants, with tb (tqb) considered signal and tqb (tb) treated as a part of the background. Using ensembles of datasets containing contributions from SM signal and background, correlation among the outputs of the individual MVA methods is $\approx 70\%$. An increase in sensitivity can therefore be obtained by combining these methods to form a new combined discriminant that takes as inputs the three discriminant output of BDT, BNN and NEAT. Figure 3 shows the combined outputs of the tb, tqb and tb+tqb discriminants, where good agreement is observed over the entire range. In these plots, the bins are sorted and merged ("ranked") as a function of the expected signal-to-background ratio (S:B) such that S:B increases monotonically within the range of the discriminant.

The single top quark production cross section is measured using a Bayesian inference approach [24, 25]. Figure 4 shows the resulting expected and observed posterior density distributions for tb, tqb and $tb + tqb$ for the combined discriminants. All of the results are consistent with SM predictions for a top quark mass of 172.5 GeV.

3.2. CDF measurement

CDF uses artificial neural networks NeuroBayes® package to separate signal from background events [14]. The output distributions of s- and t-channel events are combined into one signal distribution, where the ratio between them is as predicted by the SM. The background processes whose output distributions look very similar and are hence difficult to distinguish. On this account, some of the processes are merged into one template with a ratio given by background estimation and this results in total six background templates: $t\bar{t}$, $W\!+\!HeavyFlavor$, $W + LightFlavor$, Dibosons, Z+jets, and multijets. NN is trained separately for s-channel (tb) and t-channel (tqb): (i) tb discriminant is trained on 2jet 2b-tag subsample with tqb as a part of

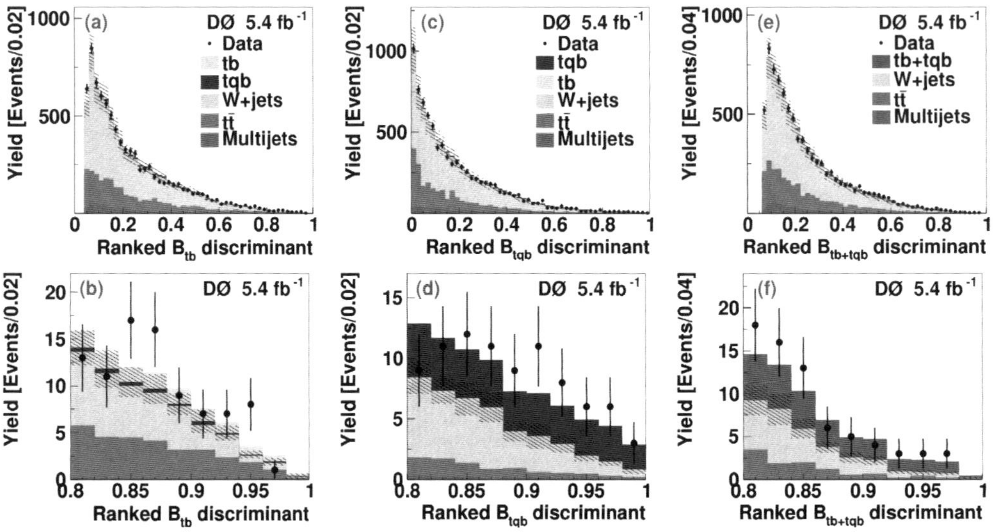

Figure 3. Distributions of the (a-b) B_{tb}, (c-d) B_{tqb} and (e-f) B_{tb+tqb} discriminant for the entire range ([0-1]) of the output, and for the signal region ([0.8-1]). The bins have been "ranked" by their expected signal-to-background ratio. The hatched bands show the 1σ uncertainty on the background prediction.

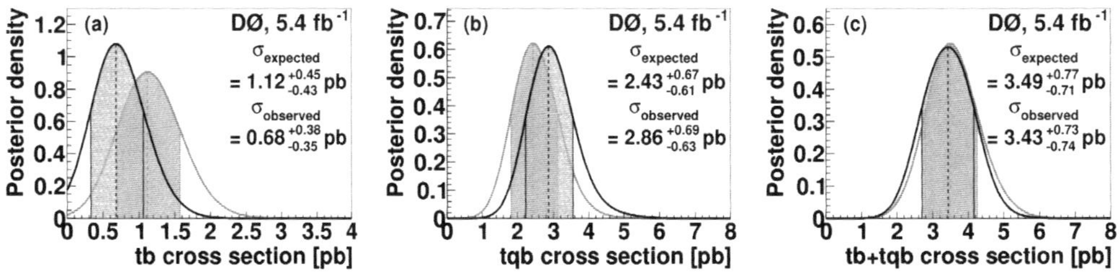

Figure 4. The expected (grey) and observed (black) posterior probability densities for (a) *tb*, (b) *tqb*, and (c) *tb*+*tqb* production. The shaded bands indicate the 68% C.L.s from the peak values.

background, and (ii) *tqb* discriminant is trained over rest of the subsample with *tb* treating as a part of background. Figure 5 (left) shows the NN discriminant distribution in the 2 and 3 jet signal region.

Using a Bayesian statistical approach, the total cross section of single top quark production assuming the SM ratio among *s*-channel and *t*-channel production is measured. The resulting posterior distribution for $s + t$-channel is shown in Figure 5 (right).

3.3. Independent measurement of s- and t-channel cross section

To measure the individual *tb* and *tqb* production cross sections without any assumption on their values or their ratio, a two-dimensional (2D) posterior probability density is constructed as a function of the *tqb* and *tb* production rates assuming a flat prior for each signal [26]. The resulting 2D density is shown in Figure 6, together with the points for the measured cross sections and the value expected in the SM value. From these 2D posteriors, we form a *t*-channel (*s*-channel) 1D posterior probability density function by integrating over the *tb* (*tqb*) axis, and thereby avoiding any dependency on the *tb* (*tqb*) cross section. The best-fit cross section is the one for which the posterior is maximized, and corresponds to $\sigma_s = 1.05^{+0.48}_{-0.45}$ pb (D0) and $1.81^{+0.63}_{-0.58}$ pb (CDF)

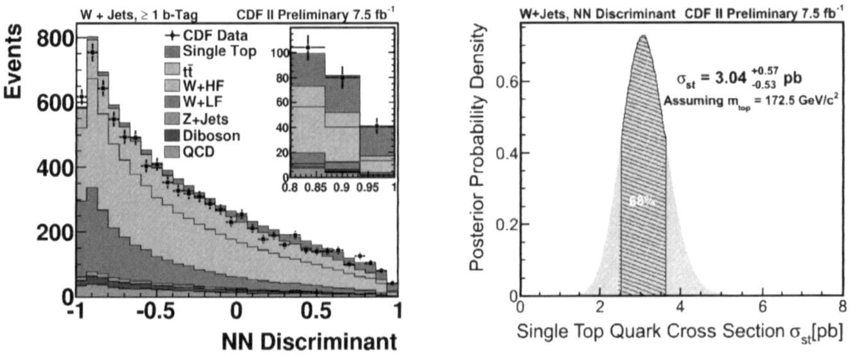

Figure 5. The resulting NN discriminant distribution (left) and the posterior probability density (right) for tb+tqb production.

and $\sigma_t = 2.32^{+0.68}_{-0.63}$ pb (D0) and $1.49^{+0.47}_{-0.42}$ pb (CDF), which are in good agreement with SM predictions [9] for the assumed $m_t = 172.5$ GeV.

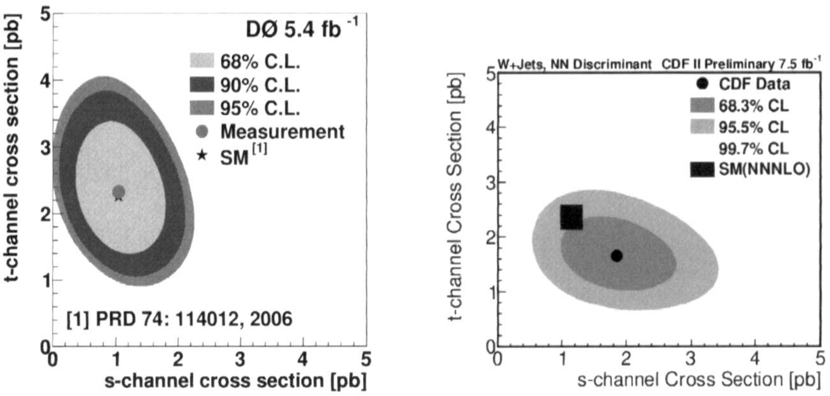

Figure 6. The posterior probability density for tqb vs. tb cross sections for D0 (left) and for CDF (right). The points corresponding to measured cross section and the SM expectation are shown.

4. $|V_{tb}|$ Measurement

The single top quark production cross section is directly proportional to the square of the CKM matrix element, $|V_{tb}|^2$, enabling us to measure $|V_{tb}|$ directly without any assumption on the number of quark families or the unitarity of the CKM matrix. Assuming the top quark decays exclusively to Wb and that the Wtb interaction is CP-conserving, we measured the strength of $V - A$ coupling, which can be > 1. Using the measured $tb + tqb$ cross section and restricting the prior to the SM region [0,1], we extract the limit of $|V_{tb}| > 0.79$ (D0) and $|V_{tb}| > 0.78$ (CDF) at the 95% C.L. The result is also shown in Figure 7.

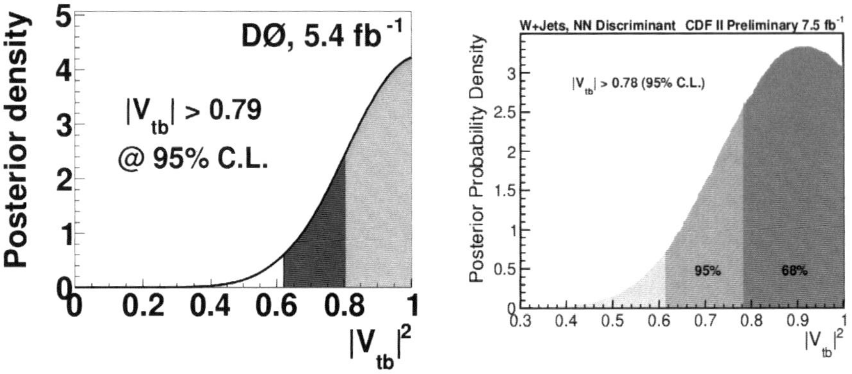

Figure 7. The posterior probability for $|V_{tb}|^2$, D0 (left) and CDF (right).

5. Summary

In summary, we have measured the single top quark production cross section using 5.4 fb^{-1} of data collected by the DØ experiment and 7.5 fb^{-1} of data collected by the CDF experiment at the Fermilab Tevatron Collider. We have also measured the individual production rates of tb and tqb processes and provided the new direct limit on the CKM matrix element $|V_{tb}|$ at the 95% C.L.

References

[1] F. Abe *et al.* (CDF Collaboration), Phys. Rev. Lett. **74**, 2626 (1995).
[2] S. Abachi *et al.* (D0 Collaboration), Phys. Rev. Lett. **74**, 2632 (1995).
[3] S. Moch and P. Uwer, Phys. Rev. D **78**, 034003 (2008).
[4] V. M. Abazov *et al.* (D0 Collaboration), Phys. Rev. Lett. **103**, 092001 (2009).
[5] T. Aaltonen *et al.* (CDF Collaboration), Phys. Rev. Lett. **103**, 092002 (2009).
[6] A. P. Heinson *et al.*, Phys. Rev. D **56**, 155 (1997).
[7] S. S. D. Willenbrock and D. A. Dicus, Phys. Rev. D **34**, 155 (1986); C.-P. Yuan, Phys. Rev. D **41**, 42 (1990).
[8] S. Cortese and R. Petronzio, Phys. Lett. B **253**, 494 (1991).
[9] N. Kidonakis, Phys. Rev. D **74**, 114012 (2006).
[10] N. Cabibbo, Phys. Rev. Lett. **10**, 531 (1963); M. Kobayashi and T. Maskawa, Prog. Theor. Phys. **49**, 652 (1973).
[11] G. V. Jikia and S. R. Slabospitsky, Phys. Lett. B **295**, 136 (1992).
[12] G. Mahlon and S. J. Parke, Phys. Rev. D **55**, 7249 (1997); G. Mahlon and S. J. Parke, Phys. Lett. B **476**, 323 (2000).
[13] V. M. Abazov *et al.* (D0 Collaboration), Phys. Rev. D **84**, 112001 (2011).
[14] `http://www-cdf.fnal.gov/physics/new/top/confNotes/cdf10793_SingleTop_7.5_public.pdf`
[15] V. M. Abazov *et al.* (D0 Collaboration), Nucl. Instrum. Methods Phys. Res., Sect. A **565**, 463 (2006).
[16] D. Acosta *et al.* (CDF Collaboration), Phys. Rev. D **71**, 032001 (2005).
[17] E. E. Boos *et al.*, Phys. Atom. Nucl. **69**, 1317 (2006). We used SINGLETOP version 4.2p1.
[18] S. Alioli *et al.*, J. High Energy Phys. **0909**, 111 (2009).
[19] T. Sjöstrand, S. Mrenna, and P. Skands, J. High Energy Phys. **05**, 026 (2006).
[20] M. L. Mangano *et al.*, J. High Energy Phys. **07**, 001 (2003).
[21] L. Breiman *et al.*, *Classification and Regression Trees* (Wadsworth, Stamford, 1984).
[22] R.M. Neal, *Bayesian Learning for Neural Networks* (Springer-Verlag, New York, 1996).
[23] K. O. Stanley & R. Miikkulainen, *Evolving Neural Networks through Augmenting Topologies.* Evolutionary Computation 10(2): 99-127
[24] V. M. Abazov *et al.* (DØ Collaboration), Phys. Rev. D **84**, 112001 (2011).
[25] V. M. Abazov *et al.* (DØ Collaboration), Phys. Lett. B **705**, 313 (2011).
[26] V.M. Abazov *et al.* (D0 Collaboration), Phys. Lett. B **682**, 363 (2010).

Measurement of the single top quark t-channel cross section with a template fit analysis

Mario Merola for the CMS Collaboration
INFN Napoli, Italy
E-mail: mario.merola@na.infn.it

Abstract. We present a measurement of the inclusive single top t-channel production cross section in proton-proton collisions at the LHC, using data collected with the CMS experiment during 2011 and 2012. The analyzed data correspond to an integrated luminosity of 1.17/1.56 fb^{-1} for muon/electron channel respectively at the centre-of-mass energy of 7 TeV, and to 5.0 fb^{-1} at 8 TeV. The analysis exploits the pseudorapidity distribution of the recoil jet and reconstructed top-quark mass using background estimates determined from control samples in data. The measurement is used to determine the CKM matrix element $|V_{\rm tb}|$.

1. Introduction

The top quark, the last of the six quarks discovered at the Tevatron proton anti-proton collider in 1995 [1, 2], can be produced via strong interaction in association with the top anti-quark and via electroweak interaction as single top. Among the three different single top production modes, the t-channel, the s-channel and the tW production, the t-channel is the one with higher production cross section at the LHC, $\sigma^{\rm th}_{t-{\rm ch.}} = 64.6^{+2.1}_{-0.7}{}^{+1.5}_{-1.7}$ pb [3], at a centre-of-mass energy of 7 TeV and for a top-quark mass of $m_{\rm t} = 172.5$ GeV/c^2. For this reason it was the first single top process observed at the LHC and up to now the one which permits the most precise determination of the CKM matrix element $|V_{\rm tb}|$.

In Figure 1 the leading order Feynman diagrams for the t-channel production are shown. The typical signature expected for such events comprises one forward jet recoiling against the heavy top quark and one central b jet coming from top decay. As in the past analysis [4], only the leptonic decays of the W boson are considered, so one isolated charged lepton and missing energy are present as well in the final state. In addition, a second b jet can be present, coming from gluon splitting and characterized by softer transverse momentum spectrum with respect to the central b jet.

In this contribution the cross section measurements using data collected with the CMS detector [5] at a centre-of-mass energy $\sqrt{s}$ = 7 TeV and 8 TeV are described: the first and more sophisticated analysis represents the most precise measurement of t-channel cross section and of $|V_{tb}|$ up to now [6], while the second constitutes the first measurement of single top production cross section at 8 TeV [7].

2. Event Selection and Analysis strategy

The most important backgrounds, processes whose final states fake our signal, are the top pair production $t\bar{t}$ in which one top decays hadronically and the other leptonically, W+jets, in

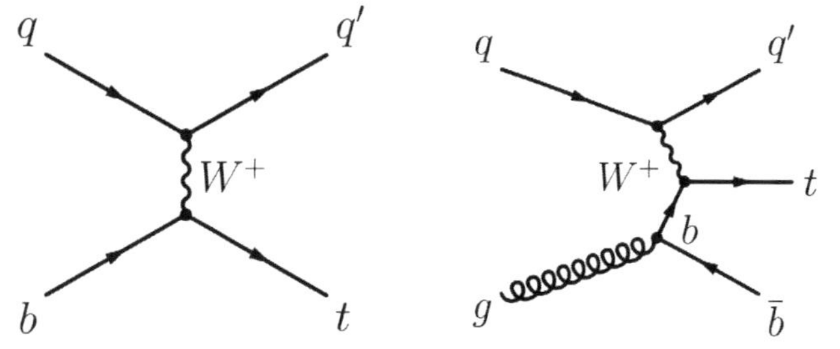

Figure 1. Leading order Feynman diagrams for t-channel single top production in the t-channel: $2 \to 2$ (left) and $2 \to 3$ (right).

particular W+heavy flavoured jets, Z+jets, QCD multijets and minor ones. The event selection is therefore optimized in order to obtain a sub-sample of data enriched in t-channel events, and can be summarized as follows:

- **lepton selection**: exactly one muon or one electron is required to pass tight quality and isolation criteria, and to have a transverse momentum greater than 20 GeV/c (26 GeV/c at 8 TeV) / 30 GeV/c (muon/electron). This step provides the rejection of most of QCD background;
- **jet selection**: exactly two jets are required to have $p_T > 30$ GeV/c (> 60 GeV/c at 8 TeV) for $|\eta| < 4.5$. This allows a tight rejection of W+jets and QCD backgrounds together with $t\bar{t}$ which is characterized by high jet multiplicity;
- ***b*-tagging**: one and only one jet is required to come from b-quark fragmentation, that is it should pass a tight threshold on the Track Counting High-Purity (TCHP) b-tagging algorithm [8]. This requirement helps reducing W+jets and especially the $t\bar{t}$ background whose final states contain in almost 100% of the cases at least two b jets;

In order to further reduce the QCD multijet background a cut on the W transverse mass $m_T > 40$ GeV/c^2 or 50 GeV/c^2 at 8 TeV (muons), or $\not{E}_T > 35$ GeV (electrons) is applied. Finally a cut is applied on the reconstructed top-quark mass defining a Signal Region (SR) for $130 < m_{\ell\nu b} < 220$ GeV/c^2 and a W+jets and $t\bar{t}$ enriched Sideband Region (SB) outside the mass window.

3. Backgrounds estimation and signal extraction procedure

The cross section measurement is performed with a template fit to the characteristic pseudorapidity of the forward jet, $|\eta_{j'}|$, in the Signal Region. Four are the event yield parameters fitted: signal, top category ($t\bar{t}$ + single top s and tW channels), EWK category (W/Z+jets and diboson production) and QCD. The templates used as models for signal and backgrounds are 1-dimensional histograms either from simulation or from data-driven estimations.

It has been observed, in agreement with other measurements [9, 10], that the Wb+X and Wc+X production is mismodeled in Monte Carlo simulations and a slight excess of such events is present in data. To not rely on simulation, **W+jets $|\eta_{j'}|$ template** and yield are taken from data: in the Sideband Region we subtract from data all the processes except for W/Z+jets, and then take what remains as W/Z+jets data-driven estimation. It has been assumed that Z+jets processes scale as W+jets and have similar $|\eta_{j'}|$ shape. Furthermore it has been proven that the shape of $|\eta_{j'}|$ for this background processes does not change significantly from Sideband to Signal Region.

The **QCD template** is taken from a control region obtained inverting the isolation cut on the selected lepton (for electrons failing identification criteria is also requested). Its yield in the Signal Region is obtained for muons (electrons) performing a fit to the W transverse mass ($\not{E}_T$)

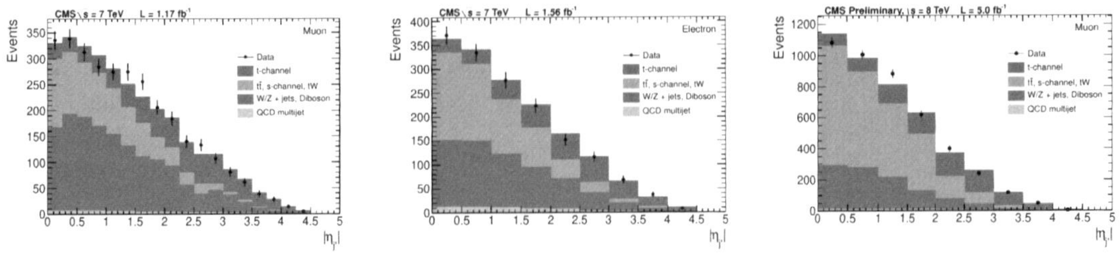

Figure 2. Results of the fit to $|\eta_{j'}|$ for muons and electrons at 7 TeV (left and centre) and for muons only at 8 TeV (right).

distribution and then counting the number of QCD events above the thresholds established in the selection. The QCD yield is considered fixed in the fit to $|\eta_{j'}|$.

Lastly, the $t\bar{t}$ **template** is taken from simulation at 7 TeV and from a control sample on data at 8 TeV (sample obtained selecting 3 jets with two of them passing b-tagging requirements, henceforth '3jets 2tags'). All the other processes are taken from simulation. Figure 2 shows the fit results at 7 and 8 TeV.

4. Systematic uncertainties

The systematic uncertainties of the measurement are determined by use of pseudo-experiments which take into account the effect of the systematic sources on the $|\eta_{j'}|$ distribution and on the event yield of the different processes. For each systematic source the pseudo-experiments are generated with templates varied by $\pm 1\sigma$ of the corresponding uncertainty, and the fit to $|\eta_{j'}|$ is repeated. The mean shift with respect to the nominal measurement is taken as the corresponding systematic error.

The uncertainty related to W+jets and $t\bar{t}$ data-driven estimates has been also evaluated dicing pseudo-experiments in the Sideband Region and in the '3jets 2tags' sample, repeating the extraction procedure and repeating the fit to $|\eta_{j'}|$. The corresponding uncertainty is taken as the root mean square of the distribution of the fit results. It is worth noticing that the procedure adopted for data-driven background estimation, although sensitive to the available statistics in the control regions, nonetheless makes the measurement less dependent on the theoretical uncertainties involving W+jets and $t\bar{t}$ processes.

For the 7 TeV analysis the uncertainties with higher impact on the cross section measurement are: W+jets rate (5.9%), jet energy scale (4.0%), $t\bar{t}$ rate (3.3%) and b-tagging (3.1%). For the 8 TeV analysis they are: jet energy scale (6.8%), signal model (5.5%), muon trigger and reconstruction (5.1%), b-tagging (4.6%).

5. Results and Conclusions

In conclusion the analyses reported the following cross section measurements, analyzing 1.17/1.56 fb^{-1} for muon/electron channel respectively at 7 TeV, and 5.0 fb^{-1} at 8 TeV:

$$\sigma_{t-\text{ch.}} = 70.0 \pm 6.0\ (\text{stat.}) \pm 7.4\ (\text{syst.}) \pm 1.5\ (\text{lum.})\ \text{pb} \quad (7\,\text{TeV}),$$
$$\sigma_{t-\text{ch.}} = 80.1 \pm 5.7\ (\text{stat.}) \pm 11.0\ (\text{syst.}) \pm 4.0\ (\text{lum.})\ \text{pb} \quad (8\,\text{TeV}).$$

when the Standard Model expectation is 64.6 pb at 7 TeV and 87.1 pb at 8 TeV [3, 11]. The main differences between the 7 TeV and the 8 TeV analyses are due to the different pileup conditions (increased number of multiple interactions in the same event) and to the increased contribution

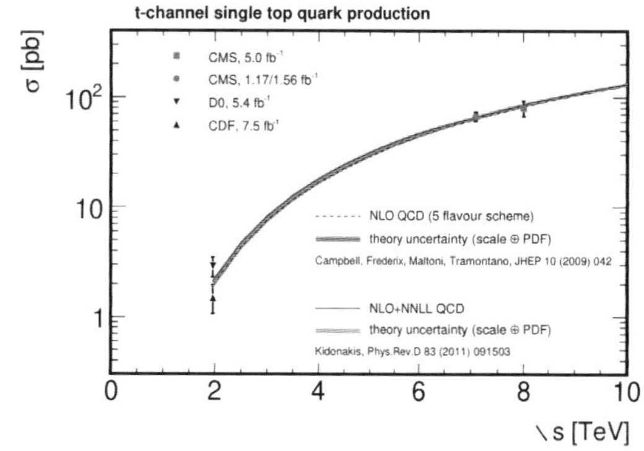

Figure 3. Single top t-channel cross section as function of the centre-of-mass energy. The error band (width of the curve) is obtained by varying the top mass within its current uncertainty [13], estimating the PDF uncertainty according to the HEPDATA recommendations [14], and varying the factorization and renormalization scales coherently by a factor two up and down.

of the $t\bar{t}$ background with respect to W+jets. Moreover, in the 7 TeV analysis both muons and electrons are considered in the final states instead of muons only at 8 TeV.

In addition, the measurement at 7 TeV has been combined using BLUE method [12] with two multivariate analyses (Neural Network and Boosted Decision Trees) measuring the single top t-channel production cross section [6]. The result obtained is:

$$\sigma_{t\text{-ch.}} = 67.2 \pm 6.1 \text{ pb} = 67.2 \pm 3.7 \text{ (stat.)} \pm 4.6 \text{ (syst.)} \pm 1.5 \text{ (lum.) pb}$$

All the results are summarized in Figure 3 where the measurements are compared with the theoretical prediction.

From the cross section a measurement of the CKM element $|V_{\text{tb}}|$ can be obtained. Assuming $|V_{\text{td}}|$ and $|V_{\text{ts}}| << |V_{\text{tb}}|$, and allowing for the presence of a possible anomalous form factor f_{L_V} [15] we have $|f_{L_V} V_{\text{tb}}| = \sqrt{\sigma_{t\text{-ch.}}/\sigma^{\text{th}}_{t\text{-ch.}}}$ and measure

$$\begin{aligned} |f_{L_V} V_{\text{tb}}| &= 1.020 \pm 0.046 \text{ (meas.)} \pm 0.017 \text{ (theor.)} \quad (7\,\text{TeV}), \\ &= 0.96 \pm 0.08 \text{ (meas.)} \pm 0.02 \text{ (theor.)} \quad (8\,\text{TeV}), \end{aligned}$$

where $\sigma^{\text{th}}_{t\text{-ch.}}$ is the SM prediction calculated assuming $|V_{\text{tb}}| = 1$. The 7 TeV result represents the most precise measurement of $|V_{\text{tb}}|$ currently available.

In Figure 4, for the 8 TeV analysis, the verification of two important features characterizing the t-channel production is shown: the first is the charge asymmetry between top quark and anti-quark production (the first is produced with a cross section double compared to the second), the other is the top quark polarisation due to the V-A nature of the couplings. It can be accessed via $\cos\theta^*$ distribution, where θ^* is the angle, in the top quark rest frame, between the charged lepton and the non-b-tagged jet.

References

[1] Abe F *et al.* (CDF Collaboration) 1995 *Phys. Rev. Lett.* **74**(14) 2626–2631
[2] Abachi S *et al.* (D0 Collaboration) 1995 *Phys. Rev. Lett.* **74**(14) 2632–2637
[3] Kidonakis N 2011 *Phys. Rev.* **D 83** 091503 (*Preprint* `1103.2792 [hep-ph]`)
[4] Chatrchyan S *et al.* (CMS Collaboration) 2011 *Phys. Rev. Lett.* **107** 091802 (*Preprint* `1106.3052 [hep-ex]`)
[5] Chatrchyan S *et al.* (CMS Collaboration) 2008 *JINST* **03** S08004
[6] Chatrchyan S *et al.* (CMS Collaboration) 2012 *JHEP* **2012**(12)
[7] CMS Collaboration 2012 Measurement of the single-top-quark t-channel cross section in pp collisions at $\sqrt{s}$ = 8 TeV CMS Physics Analysis Summary CMS-PAS-TOP-12-011 URL `https://cds.cern.ch/record/1478935`
[8] CMS Collaboration 2011 b-jet identification in the CMS experiment CMS Physics Analysis Summary CMS-PAS-BTV-11-004 URL `http://cdsweb.cern.ch/record/1427247`

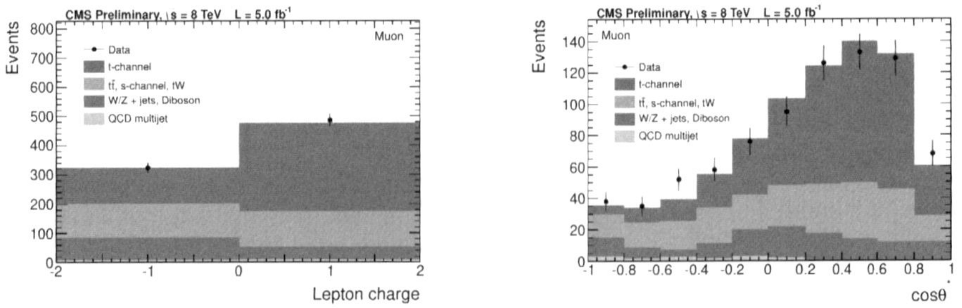

Figure 4. Two characteristic features of the single top t-channel production: charge asymmetry (left) and top quark polarisation (right). Signal and background processes are normalised to the fit results and the cut $|\eta_{j'}| > 2.0$ is applied to highlight the features.

[9] Aad G *et al.* (ATLAS Collaboration) 2012 *Phys.Lett.* **B707** 418–437 (*Preprint* 1109.1470 [hep-ex])
[10] CMS Collaboration 2011 Measurement of associated charm production in W final states at $\sqrt{s}$ = 7 TeV CMS Physics Analysis Summary CMS-PAS-EWK-11-013 URL http://cdsweb.cern.ch/record/1369558
[11] Kidonakis N 2012 (*Preprint* 1205.3453 [hep-ph])
[12] Lyons L, Gibaut D and Clifford P 1988 *Nucl. Instr. and Meth. A* **270** 110
[13] Group T E W (CDF and D0 Collaborations) 2009 (*Preprint* 0908.2171 [hep-ex])
[14] Campbell J M, Huston J W and Stirling W J 2007 *Rept. Prog. Phys.* **70** 89 (*Preprint* hep-ph/0611148)
[15] Aguilar-Saavedra J A 2009 *Nucl. Phys. B* **812** 181 (*Preprint* 0811.3842 [hep-ph])

Journal of Physics: Conference Series **452** (2013) 012019 doi:10.1088/1742-6596/452/1/012019

Single top-quarks at the LHC

D B Ta on behalf of the ATLAS and CMS collaborations
Nikhef, Science Park 105, 1098 XG Amsterdam, The Netherlands

Abstract. This article summarises the status of measurements of single top-quark production in all channels at the LHC for a centre-of-mass energy of 7 TeV with the full 2011 data set and for a centre-of-mass of 8 TeV for the t-channel cross-section. Results are reported from the ATLAS and CMS experiments. Both experiments also report evidence for the Wt-channel production which has never been observed before. ATLAS also sets a limit on the s-channel cross-section. So far, the measured cross sections are in good agreement with the predictions by the Standard Model.

1. Introduction

At the LHC, top quarks are expected to be produced singly via weak, charged-current interactions in the Wtb vertex. There are three relevant subprocesses that are distinguished by the virtuality of the exchanged W boson. The dominant process is the t-channel exchange of a virtual W (t-channel). As opposed to the case at the Tevatron, the subprocess with the second highest cross section is the associated production of an on-shell W boson and a top quark (Wt-channel). The Drell-Yan type production of a single top-quark (s-channel) has the smallest cross section. For all three channels the Feynman diagrams are depicted in Fig. 1. The approximate NNLO calculations [1, 2, 3] for all the single top-quark production processes at both 7 TeV and 8 TeV are given in Table 1 for the sum of single t and $\bar{t}$ production. The ratio of t and $\bar{t}$ in the t- and s-channel is roughly 2, which corresponds to the ratio of up and down valence-quarks in the colliding protons. The total cross section for single top-quark production amounts to about half of that of $t\bar{t}$ production [4].

Table 1. Theoretical cross-section for the different single top-quark channels.

centre-of-mass energy	t-channel [1]	Wt-channel [2]	s-channel [3]
7 TeV	$64.6^{+2.7}_{-2.0}$	15.7 ± 1.1	4.6 ± 0.2
8 TeV	$87.8^{+3.4}_{-1.9}$	22.4 ± 1.5	5.6 ± 0.2

Measuring the cross section and properties of single top-quarks tests the structure of the Wtb vertex, which in the Standard Model (SM), is governed by the weak interaction. The CKM matrix element $|V_{tb}|$ can be directly extracted from the cross-section measurement without the three fermion generations assumption, allowing to test for a 4th or higher generation hypothesis. The cross sections are also sensitive to the parton distribution functions (PDFs). The total t-channel cross-section is sensitive to the b-quark PDF, while the ratio of t and $\bar{t}$ production cross-sections is sensitive to the up- and down-quark PDF ratio.

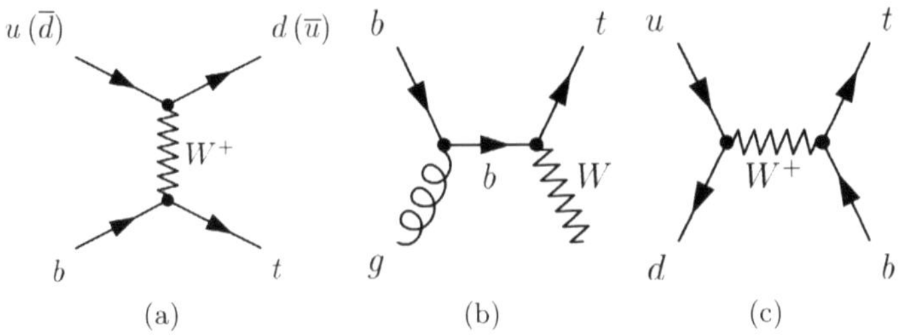

Figure 1. Leading-order Feynman diagrams for the three single-top channels: (a) t-channel, (b) Wt-channel and (c) s-channel.

Single top-quarks final states occur in many models of physics beyond the SM [5], for example in the production of a heavy W'-boson that decays preferentially into quarks (see ATLAS and CMS measurements in Refs. [6, 7]), in the production of a charged Higgs boson, in the decays of fourth generation or excited quarks (see Ref. [8]) or in the final state for flavour-changing neutral-current interactions (see Ref. [9]). Cross-section measurements in each individual channel are important as the different channels are sensitive to different types of physics models beyond the SM.

2. General physics object reconstruction and selection

The general object reconstruction and object definitions details for ATLAS [10] and CMS [11] can be found in Refs. [12, 13, 14, 15, 16] and Refs. [17, 18, 19], respectively.

In general, events are selected at ATLAS with high transverse momentum single lepton triggers (isolation from calorimeter energy depositions at trigger level is required for the 8 TeV analyses). Leptons are required to have transverse momentum, $p_T > 25$ GeV, pseudorapidity, $|\eta| < 2.5$ and have to be isolated from energy depositions in the calorimeter and from reconstructed tracks. Jets are reconstructed with the anti-k$_t$ algorithm [20] with radius parameter $R = 0.4$, and are required to have $p_T > 25$ GeV (30 GeV at 8 TeV). Central jets reach a pseudorapidity $|\eta| < 2.5$ and forward jets, used in the t-channel analyses, reach $|\eta| < 4.5$. Information about jets containing b quarks is also used (b-tagging). A neural network combines lifetime-related information reconstructed from the tracks associated with each jet. The analyses use working points with $50 - 60\%$ efficiency with a misidentification probability of $1/500$. The missing transverse energy, E_T^{miss}, is reconstructed from the negative vector sum of the p_T of the physics objects and the transverse energy of unassociated calorimeter cells.

In CMS analyses, events are selected using single lepton triggers with isolation requirements at the trigger level, but also combined triggers, such as electron+b-jet trigger or dilepton triggers, are used. The physics objects are reconstructed from particle flow (PF) objects, that identify if tracks and energy depositions belong to electrons, muons, photons, charged or neutral hadrons and applies different calibrations accordingly. Leptons are required to be central ($|\eta| < 2.5$) and to have $p_T > 20 - 30$ GeV and isolation from other PF objects is also required. For the veto on additional leptons the p_T-requirement is usually lower than the p_T of the selected leptons. Jets are also built from PF objects with the anti-k$_t$ algorithm with radius parameter $R = 0.5$, and they must have $p_T > 30$ GeV and can reach $|\eta| < 4.5$. The identification of jets from b quarks is using a threshold on a discriminator that is based on the secondary vertex reconstruction. The working point of the discriminator has $\sim 60\%$ efficiency and a misidentification probability of $1/1000$. The E_T^{miss} is reconstructed from the vector sum of the PF objects.

A commonly used variable is the transverse mass of the lepton-E_T^{miss} system, $m_T = \sqrt{2p_T E_T^{miss}[1 - cos(\Delta\phi(\vec{l}, \vec{E_T^{miss}}))]}$, with the p_T of the lepton, and the azimuthal angle $\Delta\phi$

between the $E_{\mathrm{T}}^{\mathrm{miss}}$ vector and the lepton vector, $\vec{l}$.

Top quarks decay virtually 100% to b quarks and a W boson. From the decay mode of the W boson (in case of a dilepton analysis, there are two W bosons) the events are classified into hadronic or leptonic modes. Most single-top analyses use the single-lepton mode using the lepton to trigger the event, while in the Wt-channel analysis the dilepton mode is preferred due to the lower background fraction.

3. Background estimations

The major background process for single top-quark analyses is top-pair ($t\bar{t}$) production due to its large cross section and similarity of the decay chain. This background's shape is obtained from Monte-Carlo (MC) simulations while its normalisation is obtained from $t\bar{t}$ dominated regions and varied within the theoretical cross-section uncertainty.

The next important background is W-boson production in association with additional jets (W+jets, light flavour and heavy flavour jets). The ATLAS analyses described here either simultaneously fit the normalisation of this background with the signal in the $E_{\mathrm{T}}^{\mathrm{miss}}$ distribution or estimate it with the tag-counting method. In this method the overall normalisation is taken from W+jets dominated data samples and the fraction of W+jets events that contain jets from heavy-flavour decays are then estimated by comparing data events with and without the b-tag requirement.

CMS analyses described here use the shape and the normalisation of data events from a sideband of the analysis (see Section 4) that is dominated by W+jets events. After the subtraction of non-W+jets events using MC samples, these events are used to model the W+jets background in the signal region.

Z-boson production with associated jets is an important background for analyses in the dilepton final state. ATLAS analyses scale the MC events by comparing data to MC yields in sidebands of a 2D distribution in $E_{\mathrm{T}}^{\mathrm{miss}}$ and the dilepton invariant mass (see Section 5). Some corrections are applied to account for non-Z-boson background contributions. The CMS analyses correct the $E_{\mathrm{T}}^{\mathrm{miss}}$ distribution, so that it agrees with data in the peak region.

The multijet background, in which a jet is misidentified as a lepton, is modelled in the ATLAS single-top analysis with two methods. The jet-electron model obtains shape templates for both the electron and the muon channel from events in which jets with electron-like energy depositions in the calorimeter are treated as electrons. The normalisation is taken from a fit to the $E_{\mathrm{T}}^{\mathrm{miss}}$ data distribution after the preselection. The second method, the matrix method, uses equations that connect the number of selected data events to their sources of events with real and misidentified leptons for two definitions of the leptons. The efficiencies for real and misidentified leptons to pass the looser or both lepton definitions criteria that enter the method were determined in separate control regions dominated by the respective lepton types.

In the CMS analyses, the multijet background is modelled with template distributions from events using a looser lepton definition and that are dominated by multijet events. The normalisation is then determined from a fit to either the $E_{\mathrm{T}}^{\mathrm{miss}}$ distribution (electron channel) or the m_{T} distribution (muon channel) before the cut on the respective variable in the analysis.

4. t-channel cross-section measurement and single t- and $\bar{t}$-quark production cross-section ratio

The cross section of the single-top t-channel has been measured at both centre-of-mass energies, 7 TeV and 8 TeV. Since t-channel events contain typically forwards jets from the scattered initial parton, jets with $|\eta| < 4.5$ are used.

The ATLAS analysis at $\sqrt{s} = 7$ TeV uses 1.04 fb^{-1} of data [12]. It employs a multivariate analysis (MVA) that uses a neural network (NN) to separate the signal from background events in events with 2 and 3 jets. The multijet background is obtained from the jet-electron model, while the W-boson and $t\bar{t}$ processes are normalized during the signal extraction. To suppress the multijet background, events are required to have $m_{\mathrm{T}} + E_{\mathrm{T}}^{\mathrm{miss}} > 60$ GeV. The NN has been trained on the 12 (18) highest ranked variables in events with 2 (3) jets and exactly one b-tagged jet. The NN output distribution can be seen in Fig. 2(a). Among the highest ranked variables is the invariant mass of the lepton-b-jet-neutrino system as depicted in Fig. 2(b). The cross section is determined from a maximum likelihood fit on the full output of the NN. The result is summarised in Table 2. This analysis has been cross checked with a cut-and-count analysis and also with another multivariate analysis using a boosted decision tree (BDT). The largest systematic uncertainties for this analysis are due to the modelling of the initial- and final-state radiation (ISR/FSR) and of the b-tagging.

With the full 7 TeV dataset of 4.7 fb^{-1} integrated luminosity and a similar analysis strategy, the ratio of single t- and $\bar{t}$-quark cross sections has been measured [14]. NNs have been trained separately for events with a positive or negative charged lepton. The total cross section is compatible with the previous result and the ratio of the cross section is $R = 1.81 \pm 0.10\,(\text{stat.}) \pm 0.20\,(\text{syst.})$. This result is shown in Fig. 2(c) alongside predictions using different PDF sets. The largest systematic uncertainties come from the background normalisation, ISR/FSR, multijet-background modelling and the jet energy scale.

For the analysis at $\sqrt{s} = 8$ TeV with 5.8 fb^{-1} of data a similar strategy has been used [13]. Due to the more challenging pile-up conditions present in this dataset, tighter requirements on the jets and event variables are imposed. Jets are only selected if they have $p_{\mathrm{T}} > 30$ GeV and events need to satisfy $E_{\mathrm{T}}^{\mathrm{miss}} > 30$ GeV and $m_{\mathrm{T}} > 50$ GeV. The background is modelled with the same methods as in the previous analysis. NNs were trained with 11 variables on events with 2 and 3 jets, separately. The largest systematic uncertainties for this analysis remain the modelling of the ISR/FSR radiation and of the b-tagging. The result is summarised in Table 2.

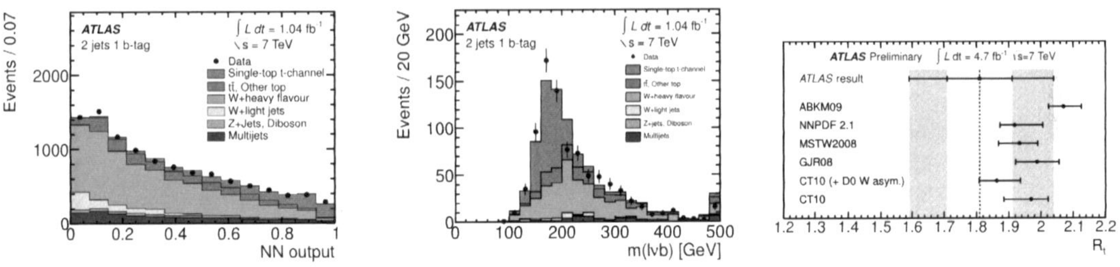

(a) (b) (c)

Figure 2. Selected ATLAS 7 TeV t-channel cross-section measurement figures showing (a) the NN output for events with 2 jets [12], (b) one of the most sensitive variables for the neural network, the invariant mass of the lepton-b-jet-neutrino system for events with 2 jets [12] and (c) the measured ratio of single t- and $\bar{t}$-quark cross-section in comparison with predictions using different PDF sets [14].

The t-channel cross-section measurement performed by CMS at $\sqrt{s} = 7$ TeV uses 1.17 fb^{-1} (muon channel) and 1.56 fb^{-1} (electron channel) data [17]. Events are required to have $E_{\mathrm{T}}^{\mathrm{miss}} > 35$ GeV (electron channel) or $m_{\mathrm{T}} > 40$ GeV (muon channel) and to have exactly one b-tagged jet. The measurement combines the result from three analyses with the BLUE method [21]. Selected output distributions can be found in Fig. 3. Common to all analyses is the multijet-background estimation using a template from a multijet enriched region. The first analysis performs a maximum likelihood fit on the full range of the $|\eta_{j'}|$ distribution where j' is the highest p_{T} non-tagged jet. A signal region with the requirement 130 GeV $< m_{l\nu v} <$ 220 GeV is defined to further enhance the signal. W+jets events are modelled from the sidebands. In the likelihood fit, the W+jets, Z+jets and diboson samples are combined into a single background template. The other templates are from the signal, $t\bar{t}$ events and from the multijet template. The NN analysis uses 6 events regions that have 2, 3 and 4 jets with 1 or 2 b-tagged jets, respectively. For the electron (muon) channel 38 (37) variables are used. A BDT analysis uses the same event regions, but it is only trained on events with 2 and 3 jets and 1 b-tagged jet with 11 observables. Both MVAs determine the cross section with a Bayesian method that uses a flat prior for the signal strength and Gaussian priors to include the systematic uncertainties as nuisance parameters. The theoretical uncertainties are not included as nuisance parameters and treated separately. The largest uncertainties after the combination are the statistical uncertainties, followed by the uncertainty on the W+jets normalisation and the generator uncertainties.

For the analysis at $\sqrt{s} = 8$ TeV using 5.0 fb^{-1} only the single muon topology and the fit on the $|\eta_{j'}|$ distribution is used [18]. Similarly to the ATLAS analysis, the harsher pile-up conditions required adjustments to the object selection: for example the trigger threshold has been raised, the lepton is required to have $p_T > 26$ GeV and $|\eta| < 2.1$ and tighter isolation requirements. Jets in the event are required to have p_T larger than 40 GeV and events must have $m_T > 50$ GeV. The background determination strategy is more elaborate and it fits the multijet template separately in the signal and sideband region. The $t\bar{t}$ shape is taken from events with 3 jets of which 2 jets have to be b-tagged. The largest uncertainty arise from statistical limitations of the selected data sample, the jet energy scale and the uncertainty in the t-channel MC generator. All results are summarised in Table 2.

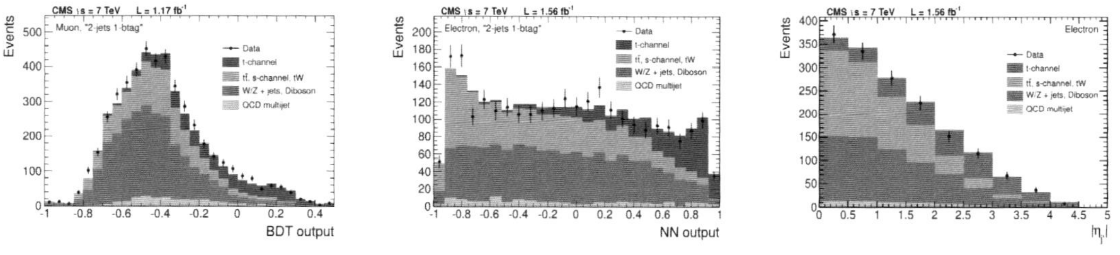

(a) (b) (c)

Figure 3. Selected output distributions of the analyses that were combined for the 7 TeV t-channel analysis with CMS [17], (a) the BDT output for events with 2 jets and a muon, (b) the NN output for electron events with 2 jets and (c) the $|\eta_{j'}|$ distribution of the hardest non-tagged jet.

Table 2. Summary of the single top-quark cross-section measurements and $|V_{tb}|$ determinations for both experiments, ATLAS and CMS.

	t-channel [pb]	$\lvert V_{tb}\rvert$, $\lvert V_{tb}\rvert$ @95CL for $\lvert V_{tb}\rvert$ in [0, 1]
ATLAS 7 TeV [12]	83 ± 4(stat.) ± 20(syst.)	1.13 ± 0.14, > 0.75
ATLAS 8 TeV [13]	95 ± 2(stat.) ± 18(syst.)	1.04 ± 0.11, > 0.80
CMS 7 TeV [17]	67 ± 4(stat.) ± 3(syst.) ± 4(theo.) ± 2(lumi.)	$1.02 \pm 0.05 \pm 0.02$(theo.), > 0.92
CMS 8 TeV [18]	80 ± 6(stat.) ± 11(syst.) ± 4(lumi.)	$0.96 \pm 0.08 \pm 0.02$(theo.), > 0.81

5. Wt-channel cross-section measurement

The ATLAS Wt-channel cross-section analysis uses 2.05 fb^{-1} of data at $\sqrt{s} = 7$ TeV [15]. The events are selected to have at least one central jet and no b-tagging requirement is imposed on the events. To reject background events, $E_T^{miss} > 50$ GeV and a cut on the angle between the leptons and the E_T^{miss} vector is required. In addition for the ee and $\mu\mu$ events, the invariant dilepton mass has to be outside of the Z-boson mass window. The Z-boson background normalisation in the signal region is determined as described in Section 3. The cross section is obtained from a profile likelihood method on the output of a BDT that can be seen in Fig. 4. The most sensitive variable here is the p_T of the system, which is the scalar sum of the p_T of all physics objects and E_T^{miss}. The BDTs are trained separately for events with 1, 2 and ≥ 3 jets using 22 observables. The largest systematic uncertainties are from the jet energy scale uncertainty, the parton shower modelling and the MC generator modelling. The result of 17 ± 6 pb presents an evidence of the single-top Wt-channel production at a level of 3.3σ significance (3.4σ expected significance).

The CMS analysis uses 2.04 fb^{-1} of data at $\sqrt{s} = 7$ TeV and requires in the selection at least one b-tagged jet [19]. The E_T^{miss} requirements are $E_T^{miss} > 30$ GeV and $E_T^{miss}{}_{\text{tracker}} > 30$ GeV, which is the E_T^{miss} using tracks only. Again, for the ee and $\mu\mu$ events the invariant dilepton mass has to be outside of the Z-boson mass window. The $t\bar{t}$ background is constrained in enriched sample regions with 2 jets and 1 and 2 b-tagged jets, respectively. The E_T^{miss} in the Z boson MC sample is corrected as described in Section 3. A binned likelihood fit to a BDT output, which is shown in Fig. 4, is performed in all signal and control regions to extract the cross section. The largest uncertainties on the signal modelling comes from statistical uncertainties, uncertainties on the jet energy scale and on the matrix-element/parton-shower matching threshold. The result of 16 ± 5 pb corresponds to an evidence of the single-top Wt-channel production at CMS at a level of 4.0σ significance (3.6σ expected significance). All results are summarised in Table 3.

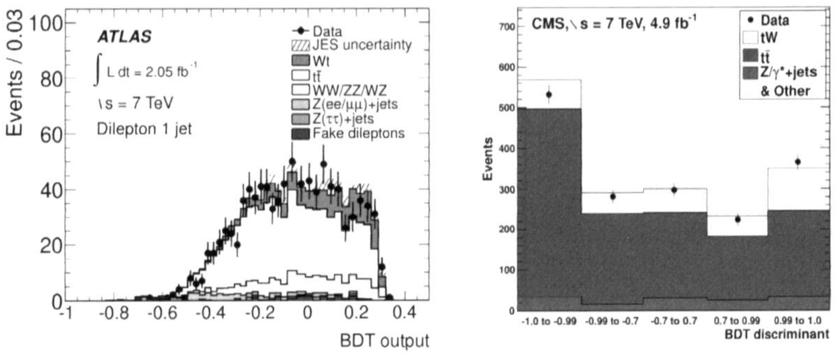

(a) (b)

Figure 4. Selected BDT output distributions for the Wt-channel cross-section measurement, (a) the output for the ATLAS analysis for events with 1 jet [15] and (b) the output distribution for the CMS analysis [19].

6. s-channel cross-section measurement

An attempt to measure the s-channel cross-section was only performed by ATLAS [16]. The analysis uses 0.70 fb^{-1} of data at $\sqrt{s} = 7$ TeV, employing a cut-and-count method. Events are required to have exactly two central jets with exactly one b-tagged jets. These stringent jet requirements are necessary to suppress the largest background processes. To reduce the multijet background, $E_T^{miss} > 25$ GeV and $m_T + E_T^{miss} > 60$ GeV are also required. The W+jets background is estimated with the tag-counting method and the multijet background is estimated with the jet-electron model. The cut-based analysis uses a sequence of cuts optimized for best statistical significance. The cuts include several mass constraints on the mass of the reconstructed top and jets, the p_T of the jets and angular distances between the b-jet/jet-system and the lepton. A likelihood ratio method on the number of expected events is used to extract a cross-section limit. So far no evidence of single-top s-channel production have been found at ATLAS. The major uncertainties are from the limited data and MC statistics and the MC generator uncertainties.

Table 3. Summary of the single top-quark cross-section measurements for the Wt- and s-channel and $|V_{\mathrm{tb}}|$ determinations for both exeriments, ATLAS and CMS.

	Wt-channel	s-channel
ATLAS 7 TeV [15, 16]	$17 \pm 3(\text{stat.}) \pm 5(\text{syst.})$ pb @ 3.3σ	< 26.5(expected 20.5) pb
$\|V_{\mathrm{tb}}\|$	$1.03^{+0.16}_{-0.19}$	
CMS 7 TeV [19]	16^{+5}_{-4} pb @ 4.0σ	
$\|V_{\mathrm{tb}}\|$, $\|V_{\mathrm{tb}}\|$ @95CL for $\|V_{\mathrm{tb}}\|$ in $[0, 1]$	$1.01^{+0.16}_{-0.13} \pm 0.04(\text{theo.}), > 0.79$	

7. Summary

The status of measurements of single top-quark production at the LHC has been presented here. The analyses at $\sqrt{s} = 7$ TeV use up to the full 2011 data set and the analysis at $\sqrt{s} = 8$ TeV use approximately a quarter of the available data in 2012. The t-channel cross-section has been measured at ATLAS and CMS for both centre-of-mass energies of 7 TeV and 8 TeV. These measurements are already limited by systematic uncertainties and precision measurements of the cross section will need to constrain the systematic uncertainties even more. With this clean selection of t-channel single-top events the measurements of the properties of single top-quark production can begin. Both experiments have found first evidence for Wt-channel production, which had not been observed before, at $\sqrt{s} = 7$ TeV at 3.3σ and 4.0σ level of significance for ATLAS and CMS, respectively. For this channel, characterised by an almost irreducible $t\bar{t}$ background, the statistical and systematical uncertainties are at the same level. With more data becoming available at $\sqrt{s} = 8$ TeV a discovery seems to be possible in the near future. All measurements also determine the CKM-matrix element $|V_{\mathrm{tb}}|$ and it seems to be compatible with the expectation from the SM. For the subleading process, the s-channel, a cross-section limit has been set by ATLAS and this analysis is clearly dominated by statistical uncertainties. So far, the measured cross sections show good agreement with the predictions by the SM.

References

[1] Kidonakis N 2011 *Phys. Rev. D* **83** 091503 (*Preprint* `arXiv:1103.2792[hep-ph]`)
[2] Kidonakis N 2010 *Phys. Rev. D* **82** 054018 (*Preprint* `arXiv:1005.4451[hep-ph]`)
[3] Kidonakis N 2010 *Phys. Rev. D* **81** 054028 (*Preprint* `arXiv:1001.5034[hep-ph]`)
[4] Aliev M, Lacker H, Langenfeld U, Moch S, Uwer P *et al.* 2011 *Comput. Phys. Commun.* **182** 1034–1046 hATHOR version 1.2 is used. (*Preprint* `1007.1327`)
[5] Tait T M and Yuan C P 2000 *Phys. Rev. D* **63** 014018 (*Preprint* `hep-ph/0007298`)
[6] ATLAS Collaboration 2012 *Phys. Rev. Lett.* **109** 081801 (*Preprint* `1205.1016`)
[7] CMS Collaboration 2012 (*Preprint* `1208.0956`)
[8] ATLAS Collaboration 2013 (*Preprint* `1301.1583`)
[9] ATLAS Collaboration 2012 *Phys. Lett. B* **712** 351–369 (*Preprint* `1203.0529`)
[10] ATLAS Collaboration 2008 *J. Inst.* **3** S08003
[11] CMS Collaboration 2008 *JINST* **3** S08004
[12] ATLAS Collaboration 2012 *Physics Letters B* **717** 330–350 (*Preprint* `1205.3130`)
[13] ATLAS Collaboration ATLAS–CONF–2012–132 URL `https://cds.cern.ch/record/1478371`
[14] ATLAS Collaboration ATLAS–CONF–2012–056 URL `http://cds.cern.ch/record/1453783`
[15] ATLAS Collaboration 2012 *Phys. Lett. B* **716** 142 (*Preprint* `1205.5764`)
[16] ATLAS Collaboration ATLAS–CONF–2011–118 URL `http://cds.cern.ch/record/1376410`
[17] CMS Collaboration 2012 Submitted to JHEP (*Preprint* `1209.4533`)
[18] CMS Collaboration CMS–PAS–TOP–12–011 URL `http://cds.cern.ch/record/1478935`

[19] CMS Collaboration 2012 *Phys. Rev. Lett.* (*Preprint* 1209.3489)
[20] M Cacciari, G P Salam and G Soyez 2008 *JHEP* **0804** 063 (*Preprint* 0802.1189)
[21] Lyons L, Gibaut D and Clifford P 1988 *Nucl.Instrum.Meth.* **A270** 110

Top quark mass results at Tevatron

Pavol Bartoš on behalf of CDF and DØ collaborations

Faculty of Mathematics, Physics and Inforamtics, Comenius University, Mlynska Dolina F1, Bratislava 842 48, Slovakia

E-mail: pavol.bartos@cern.ch

Abstract. The importance of the precise measurement of the top-quark mass (and W-boson mass) origin in possibility to predict the mass of the Higgs boson. The top-quark mass has been measured by the CDF and DØ experiments in many channels using different methods. The most precise measurement is still the Tevatron combination using the collected data with integrated luminosity of up to $5.8\,\text{fb}^{-1}$. The determined mass value is $173.18 \pm 0.56(\text{stat}) \pm 0.75(\text{syst})$ GeV or 173.18 ± 0.94 GeV, what gives a precision of $\pm 0.54\%$. However this article also reports the results using the full datasets, which are not included in the combination. In addition, the measurement of the mass difference of top and anti-top quarks is present showing the difference of 1.95 ± 1.26 GeV.

1. Introduction

Since the discovery of the top quark [1, 2] many properties has been studied. Top quark due to its big mass plays important role in the QCD corrections. The precise measurement of the top-quark and W-boson masses leads to better constrain of the Higgs-boson-mass window. The fact that the top-quark mass is close to the scale of electro-weak symmetry breaking evoke the question if the top quark have more fundamental role in this mechanism. The top-quark sector is expected to be sensitive to a new physics.

At the Tevatron $p\bar{p}$ collider top quarks are produced mainly in pairs through strong force quark-antiquark annihilation ($\sim 85\%$) and gluon-gluon fusion ($\sim 15\%$) processes. Due to the very short life time, it decays before hadronization into W boson and bottom quark in almost 100% of the time. We distinguish three different topologies of $t\bar{t}$ events, which can be defined by decays of W bosons - *dilepton channel*, *all-hadronic channel* and *lepton+jets channel*. In case of dilepton channel both W bosons decay leptonically - into electron or muon and corresponding antineutrinos. This channel is also called the cleanest one as it has the smallest amount of background what is caused by the presence of two leptons. However dilepton channel has smallest branching ratio. The lepton+jets channel covers the events with one W boson decaying leptonically and the other one hadronically. The lepton+jet channel is also called "golden channel" as the branching ratio is relatively large and the amount of background is manageable. The largest branching ratio has the all-hadronic channel, in which both W bosons decay hadronically. However in this case the amount of QCD multijets background is big.

2. Methodology

Event though there are several methods used to measure top-quark mass, we describe only two of them as they are used in the measurements mentioned in this article:

Template method uses variables sensitive to the top quark mass, which is reconstructed from its decay products. After generating the Monte Carlo (MC) samples with different top-quark masses, the templates are obtained by reconstructing the variables. The result of the measured top-quark mass is then obtained by comparing the template of the sensitive variables obtained from the data with the ones extracted from different MC samples. This method is fast, but the statistical uncertainty is worse then the ones from the other methods.

Matrix element method is based on calculating the event probability density P_E, which express the probability that the event with the certain input top-quark mass (m_t) leads to the measured variables which are seen in the detector:

$$P_E(x; m_t, f_{t\bar{t}}) = f_{t\bar{t}} P_{t\bar{t}}(x; m_t) + (1 - f_{t\bar{t}}) P_b \tag{1}$$

where x stands for measured kinematic variables, $f_{t\bar{t}}$ is fraction of signal events, and $P_{t\bar{t}}$ (P_b) is the probability that the event represent signal $t\bar{t}$ (background) event. The probability $P_{t\bar{t}}$ can be expressed by parton distribution functions of the colliding proton and anti-proton, the detector response, and the differential partonic cross section, which is defined by matrix element. This method uses the full event information and take into account also event-by-event differences. The final result of measured top-quark mass is extracted from the sample likelihood defined as $L = \prod P_E$.

3. Measurements

3.1. CDF measurement using data sample of $8.7\,fb^{-1}$ *in lepton+jets channel*

The measurement is based on the template method, where templates of three sensitive variables are obtained from MC samples generated with 76 different m_t assuming 29 possible shifts in jet energy scale (JES).

The events are required to contain one isolated electron (muon) with the transverse energy E_T (momentum p_T) greater than 20 GeV (GeV/c) and at least four jets with $E_T > 20$ GeV and $|\eta| < 2.4$. The jets can be tagged as b-jets using SECVTX b-tagging algorithm [3]. Due to the presence of neutrino the amount of the missing transverse energy is required ($\not{E}_T > 20$ GeV). For the further suppression of the background in the events with zero and one b-tagged jets, we require the H_T (sum of the $\not{E}_T$ and the E_T of lepton and jets) to be greater than 20 GeV and minimum ϕ angle between $\not{E}_T$ and jets to be greater than 0.5. To increase the subsamples of events with one and two b-tagged jets we allow the fourth jet to have $E_T > 12$ GeV and $|\eta| < 2.4$.

To obtain the background predictions we use MC simulations or in case of QCD multijets data-driven method. The individual sources are then combined using their relative cross-section and acceptances. The backgrounds are assumed to have no dependence on top-quark mass, but all MC-based background are allowed to have Δ_{JES} dependence.

To reconstruct the top-quark mass m_t^{reco}, we use the kinematic fitter [4] which assigns the four highest-p_T jets to the four final-state quarks from $t\bar{t}$ decay and associate the leptonically decaying W boson to the b jet originating in the same top-quark decay. It minimizes the χ^2 variable defined as:

$$\chi^2 = \sum_{i=\ell,4jets} \frac{(\hat{p}_T^i - p_T^i)^2}{\sigma_i^2} + \sum_{j=x,y} \frac{(\hat{p}_j^{UE} - p_j^{UE})^2}{\sigma_j^2} + \frac{(M_{jj} - M_W)^2}{\sigma_W^2} + \frac{(M_{\ell\nu} - M_W)^2}{\sigma_W^2} + \frac{(M_{bjj} - m_t^{reco})^2}{\sigma_t^2} + \frac{(M_{b\ell\nu} - m_t^{reco})^2}{\sigma_t^2} \tag{2}$$

The first term constrain the best-fit value $\hat{p}_T$ of the jets and lepton, within their uncertainties (σ_i), to remain close to their measured values p_T. Similarly the second term constrain the value of the unclustered energy, which represents the energy in the calorimeter towers not associated with the jets nor lepton. The last four terms represent the differences between the W-boson mass M_W (top-quark mass) and the invariant mass of its decay products.

As we mentioned before the measurement define three sensitive variables. Two of them are defined as the reconstructed top-quark masses m_t^{reco} and $m_t^{reco(2)}$ obtained from the minimum and 2nd minimum χ^2 combination, respectively. The third sensitive variable m_{jj} is defined as dijet invariant mass obtained from non-b-tagged jets, which is closest to M_W.

To obtain the result of the measurement we define the likelihood by number of events, signal and background probability density functions, expected number of background, and finally by two free parameters: number of signal or background events. The signal and background probability density functions are derived by Kernel Density Estimate (KDE) method [5] as they are three dimensional (due to the three sensitive variables).

The measured top-quark mass is $172.85 \pm 0.71(\text{stat.}) \pm 0.84(\text{syst.})\,\text{GeV}/c^2$. The main source of the systematic uncertainty comes from residual JES and MC generator. [6]

3.2. DØ measurement using data sample of $3.6\,fb^{-1}$ *in lepton+jets channel*

This measurement is based on matrix element method. It uses $2.6\,\text{fb}^{-1}$ of the collected data and at the end combine the result with the one obtained previously, what leads to most precise measurement of DØ collaboration.

The events are required to contain exactly one isolated electron (muon) with $E_T > 20\,\text{GeV}$ and $|\eta| < 1.1$ ($p_T > 20\,\text{GeV}/c$ and $|\eta| < 2$), and exactly four jets with $|\eta| < 2.5$ with the highest p_T jet having $p_T > 40\,\text{GeV}/c$ and the other three jets $p_T > 20\,\text{GeV}/c$. The missing transverse energy is required to be greater than 20 GeV (25 GeV) in case of electon+jets (μ+jets) events. The QCD multijets background is suppressed by requiring a minimal azimutal separation between the lepton and missing transverse energy $\Delta\phi(e, \not{E}_T) > 0.7\pi - 0.045.\not{E}_T$ or $\Delta\phi(\mu, \not{E}_T) > 2.1 - 0.035.\not{E}_T$ in electron+jets or μ+jets events, respectively. We further require at least one jet to be b-tagged by NN b-tagging algorithm [7].

The method used in this measurement is described in Sec. 2. The event probability density (used to express the overall likelihood) is defined by the probability of representing the signal $t\bar{t}$ ($P_{t\bar{t}}$) and the probability of representing background (P_b). The $P_{t\bar{t}}$ is dependent on top-quark mass (m_t) and jet energy scale factor (k_{JES}), while the P_b is independent on top-quark mass and depend on jet energy scale factor.

The measured value of the top-quark mass done on the dataset of $2.6\,\text{fb}^{-1}$ is $176.01 \pm 1.01(\text{stat.}) \pm 0.79(\text{JES}) \pm 1.02(\text{syst.})\,\text{GeV}/c^2$. This result is combined with the previous measurement done on $1\,\text{fb}^{-1}$ of data using the BLUE method [8, 9]. The combine result leads to $m_t = 174.94 \pm 0.83(\text{stat.}) \pm 0.78(\text{JES}) \pm 0.96(\text{syst.})\,\text{GeV}/c^2$ [10]. The main systematic uncertainty (except of JES) comes from modeling of hadronization and underlying events.

3.3. CDF measurement using data sample of $8.7\,fb^{-1}$ *in MET+jets channel*

Similarly as in case of previously mentioned CDF measurement, this template based measurement uses MC samples generated with 76 different m_t assuming 29 possible shifts in jet energy scale (JES).

The events are selected by multijet trigger and are required to have at least four and at most six jets with $E_T > 15\,\text{GeV}$ and $|\eta| < 2$. In addition to that no lepton is required and the significance of $\not{E}_T$ ($\not{E}_T/\sqrt{\sum E_T}$) has to be less than $3\,\text{GeV}^{1/2}$. For further suppression of the background at least one jet has to be b-tagged using SECVTX b-tagging algorithm. The Neural Network (NN, described in Ref. [11]) is trained to better separation of the background from signal events. Only events with 1 (2) b-tagged jet(s) and NN output above 0.9 (0.8) are used.

The background is dominated by QCD multijets production and so is estimated by data-driven method. A per-jet b-tag rate matrix is build from a $t\bar{t}$-signal-negligible data sample, which consists of events with exactly three jets. The matrix is applied to events with higher jet-multiplicity to estimate the background rate. The background estimation is then validated at different jet-multiplicities at the background dominant region.

As in case of the measurement described in Sec. 3.1 the top mass kinematic fitter is used to reconstruction top-quark mass. However the χ^2 is defined by modified equation:

$$\chi^2 = \sum_{i=\ell,4jets} \frac{(\hat{p}_T^i - p_T^i)^2}{\sigma_i^2} + \sum_{j=x,y} \frac{(\hat{p}_j^{UE} - p_j^{UE})^2}{\sigma_j^2} + \frac{(M_{jj} - M_W)^2}{\sigma_W^2} + \frac{(M_{bjj} - m_t^{reco})^2}{\sigma_t^2} + \frac{(M_{b\not{E}_T} - m_t^{reco})^2}{\sigma_t^2} \quad (3)$$

where the first four terms are the same as in Eq. 2 and the last term represent the difference of top-quark mass and the invariant mass of b-jet and $\not{E}_T$ assuming that $\not{E}_T$ information include not only the neutrino, but also the undetected lepton information.

The final results is obtain by the same method as is described in Sec. 3.1 using three top-quark-mass sensitive variables m_t^{reco}, $m_t^{reco(2)}$ and m_{jj}. The measured top-quark mass is 173.9±1.6(stat.+JES)±0.9(syst.) GeV/c^2. The main source of the systematic uncertainty comes from the residual JES and MC generator. [11]

3.4. DØ measurement using data sample of $5.3\,fb^{-1}$ *in dilepton channel*

This measurement is based on neutrino weighting technique (νWT, kind of matrix element method) [12]. It uses 4.3 fb^{-1} of the collected data and at the end combines the result with the one obtained previously.

The events with two isolated leptons (ee, $e\mu$, $\mu\mu$) with $p_T > 15$ GeV/c are further required to have at least two jets with $E_T > 20$ GeV and $|\eta| < 2.5$. The electrons (muons) are selected if satisfy the requirement of $|\eta| < 2.5$ (< 2). In addition to that $\mu\mu$ events are required to have $\not{E}_T > 40$ GeV, sum of the leading-lepton p_T and jets' p_T has to be higher than 120 GeV in case of $e\mu$ events, and in case of ee and $\mu\mu$ events, the $\not{E}_T$ has to be significantly different than typical values found in the distribution from Z boson events. Detailed description of the selection criteria can be found in Ref. [13].

The background coming from $Z/\gamma^\star$ and dibosons are simulated by MC, while background from instrumental effects resulting in lepton misidentification are modeled by data.

To extract the top-quark mass, we employ the νWT technique. We integrate over the η distributions of both neutrinos and solve the event kinematics to allow a calculation of $\not{E}_T$ from the neutrino momentum solutions. By comparing the calculated $\not{E}_T$ to the measured one for each event, we calculate a weight for a given m_t. For each neutrino rapidity sampling, we sum the weight values calculated from all combinations of neutrino momentum solutions and jet assignments. We therefore arrive at a distribution of relative weight for a range of m_t for each event (MC samples are generated for m_t range from 140 to 200 GeV). Requiring the integral of this distribution to be nonzero excludes events with a measured $\not{E}_T$ that is incompatible with coming from neutrinos from top quark decay. We generate distributions of $t\bar{t}$ signal as a function of m_t and the mean and RMS values of the event weight distributions. Finally we perform a binned maximum likelihood fit of an analyzed event sample to these probability distributions according to the total signal and background yields expected in our data. [14]

Applying this method to the data sample of 4.3 fb^{-1} we obtain the value of the top mass 173.7 ± 2.8(stat.)±1.5(syst.) GeV/c^2. This result is combined with the previous measurement done on 1 fb^{-1} of data using the νWT and matrix weighting methods [15]. The combine measurements yields $m_t = 174.0 \pm 2.4$(stat.)±1.4(syst.) GeV/c^2 [14]. The systematic uncertainty coming from jet energy calibration is reduced by using a correction obtained from DØ lepton+jet measurement described in Sec. 3.2.

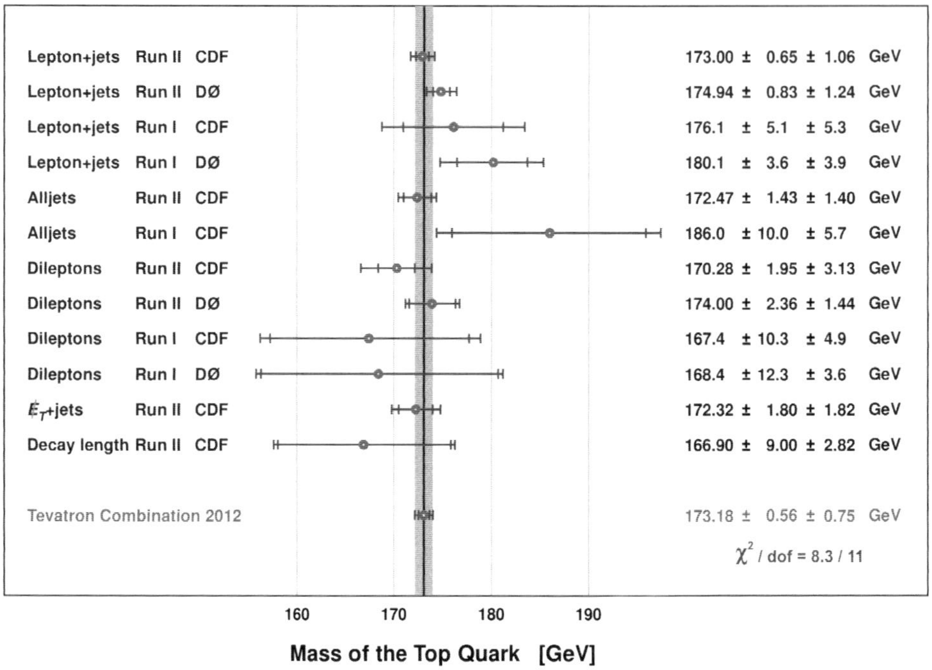

Figure 1. The 12 input measurements of m_t from the Tevatron collider experiments along with the resulting combined value of m_t^{comb}. The gray region corresponds to uncertainty of $\pm 0.94\,\mathrm{GeV}$.

4. Tevatron combination

The CDF and DØ collaborations measure the top-quark mass in the different final states using the different techniques. There are twelve individual measurements which are combined to result in a more precise value of the mass, than any individual decay channel can provide. In Fig. 1 we show the results of the individual measurements using the data sample with integrated luminosity of up to $5.8\,\mathrm{fb}^{-1}$. The determined combined top-quark mass value is $m_t^{comb} = 173.18 \pm 0.56(\mathrm{stat.}) \pm 0.75(\mathrm{syst.})\,\mathrm{GeV}/c^2$ or $173.18 \pm 0.94\,\mathrm{GeV}/c^2$, which has a precision of $\pm 0.54\%$. It makes the combined result the most precise one. [16].

The precision can be improved by including the measurements done on the full statistics of available collected data samples.

5. Top anti-top mass difference

This measurement is done on the lepton+jets data sample selected by the same criteria as are described in Sec. 3.1. To reconstruct the mass difference of top and anti-top quark, the kinematic fitter is used. The χ^2 value calculated by the fitter can be expressed by doing simple substitution of m_t^{reco} by term $(\overline{M}_t + dm_{reco}/2)$ in Eq. 2. The value of $\overline{M}_t$ is set to $172.5\,\mathrm{GeV}/c^2$.

The measurement uses the template method and define two sensitive variables Δm^{reco} and $\Delta m^{reco(2)}$ by lepton charge $(Q(l))$ and dm_{reco} obtained from the minimum and 2nd minimum χ^2 combination as follows:

$$\Delta m^{reco} = -Q(l).dm_{reco}^{min} \tag{4}$$

$$\Delta m^{reco(2)} = -Q(l).dm_{reco}^{2^{nd}min} \tag{5}$$

There are 20 MC samples generated with different top-quark-mass difference.

The final result is obtained by unbinned maximum likelihood fit, where the likelihood is defined by two-dimensional probability density functions obtained by KDE method. We measure top anti-top mass difference of $\Delta M_{top} = -1.95 \pm 1.11(\text{stat.}) \pm 0.59(\text{syst.})\,\text{GeV}/c^2$ or $\Delta M_{top} = -1.95 \pm 1.26\,\text{GeV}/c^2$. This result is consistent with CPT-symmetry expectation ($\Delta M_{top} = 0\,\text{GeV}/c^2$). [17]

6. Conclusions

We present several new CDF and DØ measurements. CDF Collaboration comes with the results obtained using full statistics of collected data. The most precise measurement is still the Tevatron combination, which does not include the CDF full statistics measurements, yet. The combination using the collected data with integrated luminosity of up to 5.8 fb1 and determines the mass value of 173.18±0.56(stat)±0.75(syst) GeV or 173.18±0.94 GeV, what gives a precision of 0.54% and makes this result the most precise measurement. In addition, the measurement of the mass difference of top and anti-top quarks is present showing the mass difference of 1.95 ± 1.26 GeV.

7. Acknowledgments

It is a pleasure to thank the CDF and DØ collaborators for their well done work, the top-group conveners for their help and the organizers of the TOP 2012 for a very interesting workshop. My travel was supported by Ministry of Education, Science, Research and Sport of the Slovak Republic.

References

[1] Abe F *et al.* (CDF Collaboration) 1995 *Phys. Rev. Lett.* **74** 2626
[2] Abachi S *et al.* (D0 Collaboration) 1995 *Phys. Rev. Lett.* **74** 2632
[3] Acosta D *et al.* (CDF Collaboration) 2005 *Phys. Rev.* D **71** 052003
[4] Abulencia A *et al.* (CDF Collaboration) 2006 *Phys. Rev.* D **73** 032003
[5] Cranmer K S *Comput. Phys. Commun.* 2001 **136** 198
[6] Aaltonen T *et al.* (CDF Collaboration) 2012 *Phys. Rev. Lett.* **109** 152003
[7] Abazov V M *et al.* (D0 Collaboration) 2010 *Nucl. Instrum. Meth. Res. Sec.* A **620** 490
[8] Lyons L, Gilbaut B and Clifford P 1988 *Nucl. Instrum. Meth. Phys. Res. Sec.* A **270** 110
[9] Valassi A 2003 *Nucl. Instrum. Meth. Phys. Res. Sec.* A **500** 391
[10] Abazov V M *et al.* (D0 Collaboration) 2011 *Phys. Rev.* D **84** 032004
[11] Aaltonen T *et al.* (CDF Collaboration) 2012 *Conference note* **10810**
[12] Abachi S *et al.* (D0 Collaboration) 1998 *Phys. Rev. Lett.* **80** 2063
[13] Abazov V M it et al. (D0 Collaboration) 2011 *Phys. Lett.* B **704** 403
[14] Abazov V M *et al.* (D0 Collaboration) 2012 *Phys. Rev.* D **86** 051103(R)
[15] Abazov V M *et al.* (D0 Collaboration) 2009 *Phys. Rev.* D **80** 092006
[16] Aaltonen T *et al.* (CDF and D0 Collaborations) 2012 *Phys. Rev.* D **86** 092003
[17] Aaltonen T *et al.* (CDF Collaboration) 2012 *Conference note* **10777**

Journal of Physics: Conference Series **452** (2013) 012021 doi:10.1088/1742-6596/452/1/012021

Experimental and theoretical uncertainties in top-quark mass measurements at the Tevatron and the LHC

Giorgio Cortiana,
for the ATLAS, CDF, CMS and DØ Collaborations
Max-Planck-Institut für Physik (Werner-Heisenberg-Institut),
Föhringer Ring 6, D-80805 München, Germany

E-mail: cortiana@mpp.mpg.de

Abstract.
Since the discovery of the top-quark at the Tevatron collider in 1995, enormous efforts have been devoted to the precise determination of its properties and in particular of its mass, m_{top}. Together with precision electroweak measurements, m_{top} can be used to check the Higgs boson mass (m_{H}) dependencies on m_{top} and the W boson mass (m_{W}) within the Standard Model, and to constrain the allowed mass range for heavy particles predicted by Standard Model extensions. In this paper a review of the current experimental and theoretical uncertainties affecting the top-quark mass measurements at the Tevatron and the LHC is provided: strategies and techniques adopted to evaluate the main systematic uncertainties within the various experiments are summarised and compared, highlighting, when appropriate, complementarities and differences which are important ingredients to set the basis of a future world m_{top} combination.

1. Introduction

The top-quark mass is a fundamental parameter of the Standard Model of particle physics (SM). Together with precision electroweak measurements, m_{top} can be used to check the Higgs boson mass (m_{H}) dependencies on m_{top} and the W boson mass (m_{W}), and to constrain the allowed mass range for heavy particles predicted by extensions of the SM. A large effort has been undertaken to the precise m_{top} determination both at the Tevatron and the LHC colliders. A summary of the most recent results can be found in [1]. As overview, the time evolution of the Tevatron and the LHC measurements is displayed in Figures 1 and 2, respectively. In the figures, the horizontal bands represent the current m_{top} combination values: $m_{\mathrm{top}}^{\mathrm{Tevatron}} = 173.2 \pm 0.6\,(\mathrm{stat}) \pm 0.8\,(\mathrm{syst})$ GeV [2], $m_{\mathrm{top}}^{\mathrm{LHC}} = 173.3 \pm 0.5\,(\mathrm{stat}) \pm 1.3\,(\mathrm{syst})$ GeV [3][1]. Except for the very early Tevatron measurements systematic uncertainties are dominating; these can be grouped into three main categories: theoretical, experimental and background related uncertainties. They will be briefly described in the following sections. The discussion will be based on information from direct m_{top} measurements included either in the present Tevatron and LHC combinations, or presented for the first time at the TOP2012 conference [1, 2, 3].

[1] The LHC combined value is based on a partial set of measurements from the 2011 datasets, and it is expected to be updated in the near future by including the latest results. Measurements included in the present combination are highlighted by the horizontal grey band in Figure 2.

Journal of Physics: Conference Series **452** (2013) 012021 doi:10.1088/1742-6596/452/1/012021

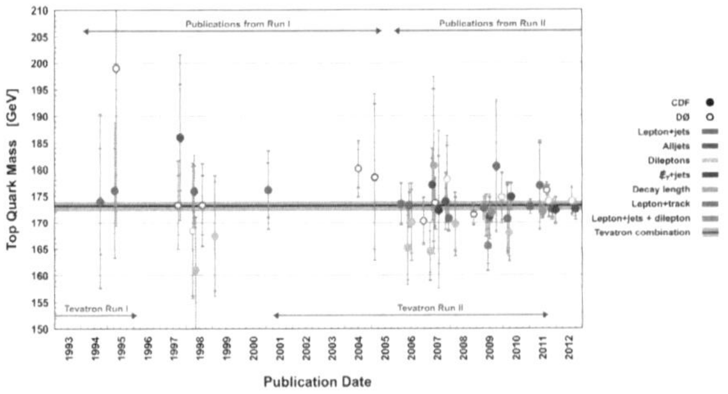

Figure 1. Time evolution of Tevatron $m_{\rm top}$ publications, from [2].

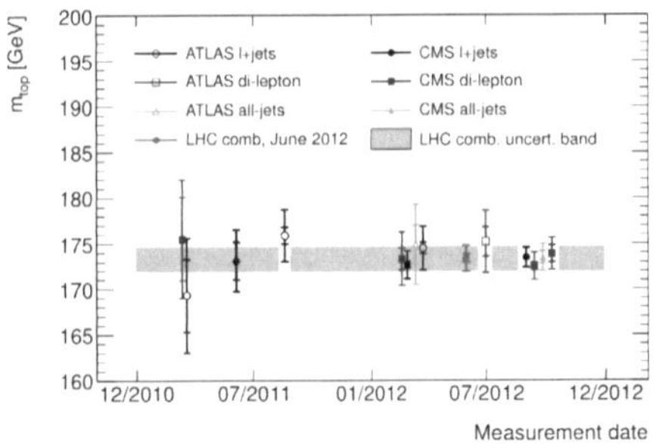

Figure 2. Time evolution of LHC $m_{\rm top}$ measurements.

Table 1. Summary of theoretical, experimental and background related uncertainties on $m_{\rm top}$ in different experiments at the Tevatron and the LHC colliders [1, 2, 3]. Typical uncertainty values and ranges are given in GeV. See text for details. $^{(*)}$ denotes cases in which the hadronisation systematics is included in the JES uncertainty, while $^{(**)}$ denotes cases in which the UE systematics is included in the hadronisation systematics.

Uncertainty source	CDF	DØ	ATLAS	CMS
Parton distribution functions (PDF)	0.1	0.2	0.1 – 0.6	0.1 – 0.5
MC generator choice	0.1 – 0.7	0.3 – 0.6	0.3 – 1.3	< 0.1 – 0.4
Hadronisation model	0.2 – 0.3	0.6	0.2 – 0.9	(*)
Underlying event (UE)	(**)	(**)	0.2 – 0.6	0.2 – 1.4
Colour reconnection (CR)	0.2 – 0.5	0.3 – 0.5	0.6 – 1.2	0.1 – 0.5
QCD radiation (ISR/FSR)	0.1 – 2.7	0.3 – 0.4	0.5 – 2.5	0.3 – 1.2
Pileup	0.1 – 0.2	0.1 – 1.2	0.05 – 0.7	0.1 – 1.0
b-tagging and b-jet modelling	0.2 – 2.5	0.1 – 0.2	0.3 – 0.5	0.1 – 0.5
Jet resolution and reconstruction	–	0.2 – 0.3	0.1 – 0.9	0.1 – 0.5
$E_{\rm T}^{\rm miss}$modelling	–	–	0.1	0.1 – 0.4
in-situ jet energy scale stat. component	0.5 – 1.5	0.5 – 0.6	0.4	0.4
Light-jet energy scale (JES)	0.5 – 4.4	0.6 – 3.4	0.7 – 2.1	0.2 – 2.1
b-jet energy scale (bJES)	0.1 – 0.8	0.3 – 0.7	1.4 – 2.5	0.5 – 1.1
Background from MC	< 0.01 – 1.7	< 0.01 – 1.0	0.1 – 1.8	0.1 – 0.2
Background from data	0.1 – 0.6	0.2	0.6 – 1.9	0.2 – 0.4
Statistics	0.5 – 10.3	0.8 – 12.3	0.6 – 4.0	0.3 – 4.6
Total systematics	1.0 – 5.7	1.2 – 3.9	2.3 – 4.6	1.0 – 4.6
Total	1.1 – 11.5	1.5 – 12.8	2.4 – 6.3	1.1 – 6.5

The typical $m_{\rm top}$ uncertainty values and ranges, across the various analyses and for each of the sources listed below are reported in Table 1.

2. Theoretical uncertainties

Theory based uncertainties are related to the simulation of $t\bar{t}$ signal events, to the event modelling and the description of the hard scattering environment. Choices to be made in the signal simulation are the (anti) proton distribution functions (PDF), the Monte Carlo generator (MC) and the hadronisation model. On the event modelling side, important ingredients are related to the description of the underlying event (UE), via MC tunes, and the settings adopted for the modelling of colour reconnection (CR), extra QCD radiation and the description of additional interactions accompanying the hard scatter (pileup). A detailed description of each of these aspects can be found in [4].

2.1. Parton distribution functions, PDF

Uncertainties on the partonic composition of the (anti) proton impact the top-pair production cross section and the $t\bar{t}$ event kinematics. These effects are evaluated by varying PDF eigenvectors within their uncertainty and comparing the m_{top} results obtained from different PDF sets [5]. At the Tevatron, the PDF uncertainties are evaluated by comparing m_{top} results obtained using MRST98L and CTEQ6M sets (CTEQ6M only for DØ). At LHC, CMS investigates m_{top} changes using MSTW08, CTEQ6.6 and NNPDF2.0; m_{top} variations changing CTEQ6.6 eigenvectors are considered in ATLAS. In all cases, the generated events are reweighted to reflect the change with respect to the default PDF used in the simulation. Uncertainties on m_{top} in the most recent analyses are sub-dominant, nevertheless, in the near future, PDF fits including additional LHC data will further help in reducing this uncertainty contribution.

2.2. MC generator choice

Various simulation programs [6] are available for the generation of $t\bar{t}$ (and background) events. They differ both in technical implementation aspects and in the perturbation order at which calculations are performed. Typical choices are the leading order (LO) PYTHIA and HERWIG programs, the LO multi-leg programs ALPGEN and MADGRAPH, for the simulation of the $t\bar{t}$ event in association with extra jets, and the next-to-leading order (NLO) programs MC@NLO and POWHEG. Typically, only a subset of the available MC programs are considered per experiment. At the Tevatron, differences between PYTHIA and MC@NLO, ALPGEN and MC@NLO are quoted by CDF and DØ, respectively. At the LHC, ATLAS quotes differences between MC@NLO and POWHEG, while CMS investigates m_{top} changes within MADGRAPH, POWHEG, and ALPGEN (documented for di-lepton analyses only).

2.3. Hadronisation models

Hadronisation models (cluster or string fragmentation as implemented in HERWIG and PYTHIA, respectively) describe the transition from final state partons to colourless hadrons. The change in m_{top} obtained by exchanging cluster and string models in a fixed MC setup is generally quoted as hadronisation uncertainty. This source of uncertainty is typically also considered as one of the components of the Jet Energy Scale uncertainty (JES) and as a contribution to the uncertainty stemming from the MC modelling. Sizeable double counting effects between these two sources are anticipated, but no quantitative information is available to date. For the time being, experiments choose either to quote (this is the case for CDF, DØ, and ATLAS) or to exclude (CMS) the hadronisation systematics related to the $t\bar{t}$ modelling in the MC. Given the relative large size of the uncertainty (Table 1), a harmonisation of the treatment of this systematics and an in-depth investigation of the level of double counting effects involved when considering both components is important for the next generation of measurements and m_{top} combinations.

2.4. Underlying event, colour reconnection and MC tunes

Additional low p_T processes can contribute to the total observed activity in hard hadron collisions. This is modelled using different settings for the description of interactions between proton remnants (underlying event) [7]. The cross-talk between the hard interaction and the beam remnants in different colour systems (colour reconnection) can change the event kinematics and introduce biases in the determination of m_{top}. These effects are in general evaluated by changing PYTHIA tunes, and using tunes with and without colour reconnection effects included [8]. With the large $t\bar{t}$ samples collected especially at the LHC, in the near future MC tunes will be improved and CR effects and their modelling can be further constrained using data.

2.5. Radiation systematics

Increased or decreased levels of initial- and final-state QCD radiation (ISR/FSR) accompanying the $t\bar{t}$ system modify the event kinematics and topology, and alter the performances of the $t\bar{t}$ reconstruction algorithms. Radiation levels can be constrained using $t\bar{t}$ events directly, for example, by studying the jet multiplicity in the l+jets, or di-lepton channels. However, in general, these measurements suffer from large systematic uncertainties due to detector effects and from the uncertainty in the luminosity determination. As a consequence, alternative methods to constrain the parameters controlling ISR/FSR have been derived. At the Tevatron, Drell-Yan events ($q\bar{q} \to \gamma^*/Z \to l^+l^-$) are used. These processes have the same initial state as most of $t\bar{t}$ events ($q\bar{q} \to t\bar{t}$), but no final-state radiation. The mean p_T of the produced di-lepton pairs is sensitive to the initial-state radiation (ISR) settings. The values of Λ_{QCD} and Q^2 in PYTHIA that bracket the data define the ISR variations [9], and the same parameter ranges are used to vary the final-state radiation settings when evaluating the m_{top} systematics. At the LHC [10], as an alternative to the study of the jet multiplicity in $t\bar{t}$ events, jet-veto analyses are exploited. In the $t\bar{t}$ di-lepton channel, the fraction of events which do not contain an extra jet[2] above a certain p_T threshold is compared for data and MC to constrain the modelling of quark and gluon radiation. The veto criterion can be extended to probe jet activity beyond the leading additional jet, adding sensitivity to all hard emissions accompanying the system. Variation of ISR/FSR parameters within PYTHIA, constrained by data, are used to evaluate m_{top} systematics due to the radiation effects at the Tevatron and by the ATLAS experiment. In CMS, where $t\bar{t}$ events are generated using multi-leg MC, samples with varied Q^2 and parton-jet matching scales are used to address this systematics. Investigations from [3] and [10] indicate that the two approaches describe to a large extent the same physics effect. Larger datasets are anticipated to further constrain ISR/FSR effects and help reducing their impact on the determination of m_{top}.

2.6. Pileup

Depending on the instantaneous luminosity, additional interactions (pileup) accompanying the hard process occur during the same bunch crossing. On average about 2, 9, and 20 interactions per bunch crossing are registered at the Tevatron, and at the LHC during 2011 and 2012 running conditions, respectively. To simulate the impact of this, additional low-p_T events from MC (or data for the DØ experiment) are overlaid to the $t\bar{t}$ process. These events contribute to building up the total amount of energy and cause colour exchanges between the remnants, thereby increasing the number of particles produced in the hadronisation. They affect the JES determination and its related uncertainty, the m_{top} resolution and the event kinematics. In all cases, a reweighting procedure in the MC simulation is applied to match the luminosity profile observed in data. The uncertainty due to the pileup modelling is obtained from the uncertainties in the reweighting procedure. Except for the early LHC measurements, the corresponding m_{top} uncertainty is sub-dominant. A separate JES uncertainty contribution is associated to jet-calibration methods implementing pileup energy suppression techniques.

3. Experimental uncertainties

Experimental uncertainties stem from the modelling of the physics objects used in the analysis for the $t\bar{t}$ reconstruction and from the description of the detector response. These are related to

[2] An additional jet besides the two b-jets from the top-quark decay.

the identification, reconstruction and calibration of leptons, jets, and missing transverse energy, $E_{\mathrm{T}}^{\mathrm{miss}}$, in the events.

3.1. b-tagging, jet energy resolution and reconstruction efficiency

The identification of b-jets from the top-quark decays (b-tagging) is an important handle for the physics background reduction in m_{top} analyses. Algorithms, based on B-hadron properties, are used and calibrated to account for detector effects and possible mis-modelling of the jet, tracking and event variables in the MC with respect to the data. Uncertainties in these calibrations are propagated to the analyses in form of p_T and η dependent scale factors (the statistical component of the scale factor uncertainty, in some cases dominating, will decrease as new data is collected and analysed). Moreover, differences in the jet energy resolution between data and MC, which translate into broader or narrower invariant mass distributions, are accounted for in the analyses by applying smearing procedures to the simulated MC events. Finally, the jet reconstruction efficiency measured in the data, relative to track-based jets, is propagated in the simulation and thus to the analyses.

3.2. $E_{\mathrm{T}}^{\mathrm{miss}}$ modelling, and lepton energy and momentum scales

The $E_{\mathrm{T}}^{\mathrm{miss}}$ absorbs the scale and resolution uncertainty of all physics objects like e, γ, μ, τ and jets. In addition, there are soft-terms including un-clustered energy or energy not associated to physics objects. The uncertainty in these components, typically of O(10%), is propagated to the analyses. Uncertainties affecting the lepton momentum and energy scales constitute in general a sub-dominant contribution to the m_{top} uncertainty, and vary between 0.1 GeV and $0.2 - 0.3$ GeV for $t\bar{t} \to l$+jets and di-lepton analyses, respectively.

3.3. Light-quark jet energy scale

The Jet Energy Scale (JES) accounts for the transition from particle-level jets (or partons in the CDF case) to reconstructed jets. Its determination is highly non-trivial and involves both MC and in-situ techniques (*i.e.* γ/Z+jet, di-jet balance). The corresponding uncertainty is derived from in-situ measurements along with systematic variations of the MC simulation, and typically it is at the level of a few percent, varying with jet properties (p_T, η). Uncertainties on the JES can easily introduce O(few GeV) uncertainties on m_{top}. Most of the recent m_{top} measurements in the l+jets, $E_{\mathrm{T}}^{\mathrm{miss}}$+jets, and all-hadronic channels have developed techniques to constrain in-situ the light-quark jet energy scale using information from the $W \to jj$ decays from top-quarks, exploiting a simultaneous fit to the data to determined m_{top} along with a global jet energy scale factor. This procedure allows for a large reduction of the JES related systematics on the m_{top} determination, at the price of a relatively small additional contribution to the statistical uncertainty from the increased dimensionality of the fit to the data. Although the method generally is very effective, at the level of precision characteristic of the latest measurements (and JES determinations) it can become less powerful due to its potential sensitivity to additional relative differences in the light- and b-jet energy response introduced by variations of the event modelling setting like UE, and CR (see CMS all-hadronic analysis in [1]).

3.4. b-jet energy scale

The detector response to heavy-flavour jets can differ to that of light-quark or gluon originated jets due to b/c semileptonic decays, and to differences in the shower and particle content of b-jets with respect to the inclusive jets. Complementary strategies are in place for different experiments to reduce and evaluate the impact of the b-jet energy scale uncertainty (bJES) on the measurements. CDF evaluates the b-to-light-quark energy scale difference by applying single particle responses measured in data or predicted by MC to samples of jets in $t\bar{t}$ events.

In the latest measurements, DØ applies flavour dependent jet energy corrections to the MC to reduce the data/MC response difference and thus minimise the uncertainty. At the LHC, in ATLAS the bJES uncertainty receives contributions from uncertainties in the calorimeter single particle response, the b-jet fragmentation, and the limited knowledge of the detector material. In CMS, the maximum PYTHIA-HERWIG jet energy response differences, for pure quark flavour at low p_T and for pure gluon flavour at high p_T, are used as a flavour uncertainty applicable to any jet flavour mixture. Detector effects are considered to be negligible due to the extensive use of tracking to reconstruct individual particles in the jets (Particle-Flow algorithm) [3]. The bJES uncertainty is among the largest uncertainty sources in current LHC m_{top} measurements. Improved analysis techniques aimed at constraining in-situ the bJES, or exploiting flavour dependent jet energy corrections are being investigated.

4. Background related uncertainties

Uncertainties on the background normalisation and shape introduce an uncertainty on m_{top}. The effects are analysis, kinematical selection, and decay channel dependent. The related uncertainties are classified into two main categories, based on their origin: MC-based or data-driven. The latter category apply for example to the data-driven estimate of the contribution of the fake leptons to the signal and to the normalisation of the W+jets background component.

5. Conclusions

The m_{top} uncertainty is currently dominated by the systematic contributions stemming from the modelling of signal events, and the knowledge of the jet energy scale for light- and b-quark originated jets. Experiments at the Tevatron and at the LHC have been and are doing a great job in reducing the m_{top} uncertainty related to the physics objects and the detector description. As an example, the development of in-situ techniques to constrain the light-jets energy scale allowed the mitigation of the traditional dominant JES uncertainty on m_{top}. Further improvements are expected for the next generation of measurements investigating techniques for the reduction of the flavour dependent JES uncertainty. As outline for the near future, and to reach and consolidate the sub-GeV m_{top} precision, experimental efforts need to be complemented by a consistent reduction of the signal modelling uncertainties, in close collaboration with the theory community, and by extensive studies aimed at reducing source- by-source possible double counting effects.

References

[1] (Tevatron) P. Bartos and (LHC) M. Siedel, these proceedings; and the ATLAS Collaboration, ATLAS-CONF-2012-082; the CMS Collaboration: arXiv:1209.2319; Eur. Phys. J. C72 (2012) 2202; CMS-PAS-TOP-11-017.
[2] The CDF and DØ Collaborations, Phys. Rev. D 86, 092003 (2012).
[3] The ATLAS and CMS Collaborations, ATLAS-CONF-2012-095 or CMS-PAS-TOP-12-001.
[4] (ATLAS) I. Fleck, (CMS) M. Gosselin, and (Tevatron) V. Shary, these proceedings.
[5] (MRST98L) A.D. Martin et al, Eur. Phys. J C4, 463 (1998); (CTEQ6M) J. Pumplin, et al, J. High Energy Phys. 07, 012 (2002); (CTEQ6.6) P. M. Nadolsky et al., Phys. Rev. D78 (2008) 013004; (MSTW2008) A. D. Martin et al, Eur. Phys. J. C63 (2009) 189; (NNPDF2.0) R. D. Ball et al, Nucl. Phys. B838, 136 (2010).
[6] (PYTHIA) T. Sjostrand, S. Mrenna, and P. Z. Skands, JHEP 05 (2006) 026; (HERWIG) G. Corcella et al., JHEP 0101 (2001) 010; (ALPGEN) M.L. Mangano et al.,JHEP, 307 (2003) 1; (MADGRAPH) J. Alwall, M. Herquet, F. Maltoni et al., JHEP 06 (2011) 128. (MC@NLO) S. Frixione, P. Nason, B.R. Webber, JHEP 0308 (2003) 007; (POWHEG) S. Frixione, P. Nason and C. Oleari, JHEP 0711 (2007) 070.
[7] R. Field, arXiv:1010.3558 ; the ATLAS Collaboration, ATL-PHYS-PUB-2011-009.
[8] P.Z. Skands, Phys. Rev. D82 (2010) 074018.
[9] The CDF Collaboration, Phys. Rev. D73, 032003 (2006).
[10] The ATLAS Collaboration, Eur.Phys.J. C72 (2012) 2043; the CMS Collaboration, CMS-PAS-TOP-12-023.

Journal of Physics: Conference Series **452** (2013) 012022 doi:10.1088/1742-6596/452/1/012022

Measurement of the top-quark mass with all-jets final states in pp collisions at 7 TeV

Eike Schlieckau for the CMS Collaboration

Phd Student, Universität Hamburg, Luruper Chaussee 149, 22761 Hamburg, Germany

E-mail: eike.schlieckau@desy.de

Abstract. A measurement is presented of the top-quark mass in the all-jets channel using a sample of $t\bar{t}$ candidate events requiring at least six jets in the final state. The data have been collected by CMS in pp collisions at 7 TeV and correspond to an integrated luminosity of 3.54 fb^{-1}. In this channel, a very tight event selection is needed to reduce the huge amount of multi-jet events. In addition, the signal fraction and mass resolution is improved by the use of a kinematic fit. An ideogram method is used to measure the top-quark mass with the background estimated from data through an event mixing technique. The top-quark mass is measured to be $m_t = 173.5 \pm 0.7$ (stat.) $\pm$ 1.3 (syst.) GeV.

1. Introduction

A first top-quark mass (m_t) measurement in all-jets final states is presented with the CMS detector [1]. This final state yields the largest signal, however, it is dominated by multijet background. The detailed descriptions of this analysis can be found in Ref. [2].

The event selection is very similar to the one from the CMS $t\bar{t}$ cross section measurement in the same final state [3]. The analysis employs a kinematic fit of the $t\bar{t}$ final state and likelihood functions for each event ("ideograms") that depend on m_t and jet energy scale (JES).

2. Data Samples and Event Selection

The analysis uses two multijet triggers which collected 3.54 fb^{-1} of the 2011 data sample. The method has been developed and evaluated with simulated events, while the background from multijet events is estimated from data employing an event mixing method.

Jets are clustered with the anti-k_t algorithm ($R = 0.5$) [4, 5] from particles reconstructed by a particle flow algorithm [6]. Events are selected with at least four (five, six) central jets with a transverse momentum of $p_T > 60$ $(50, 40)$ GeV. Additional central jets are considered with $p_T > 30$ GeV. At least two jets originating from bottom quarks are required, being tagged with an algorithm that combines reconstructed secondary vertices and track-based lifetime information.

For the final selection, a kinematic least-squares fit [7] is applied, exploiting the characteristic topology: two W bosons reconstructed from untagged jets with an invariant mass of 80.4 GeV [8] and two top quarks of equal mass reconstructed from the W bosons and b-tagged jets.

The fit procedure is repeated for every experimentally distinguishable jet permutation using all jets that pass the selection criteria, to find the correct jet combination. Per event, the permutation with the smallest χ^2 is chosen and accepted with a goodness-of-fit probability $P_{\text{gof}} = P\left(\chi^2, n = 3\right) > 0.09$.

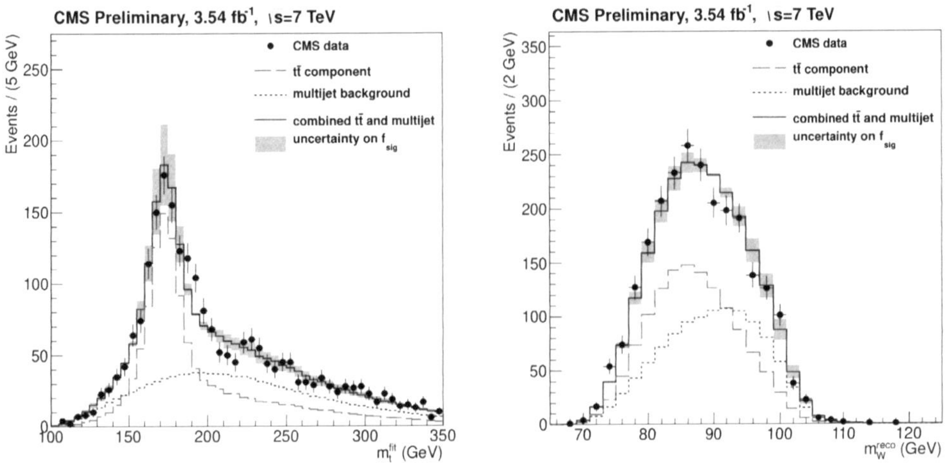

Figure 1. Reconstructed top-quark mass from the kinematic fit (left) and average reconstructed W-boson mass before being constrained by the kinematic fit (right). The simulated $t\bar{t}$ signal ($m_{t,gen}$ = 172.5 GeV) and background from event mixing are normalized to data with an uncertainty band from the signal fraction f_{sig}.

To further reduce the multijet background from bb production, an additional criterion on the distance of the two bottom quark candidates, $\Delta R_{b\bar{b}} = \sqrt{\Delta\phi^2_{b\bar{b}} + \Delta\eta^2_{b\bar{b}}} > 1.5$, is imposed.

The $t\bar{t}$ events are classified into three categories based on the jet-parton associations in simulation: correct permutations (f_{cp} = 27.9%), wrong permutations (f_{wp} = 22.6%) where at least one jet is not associated to the correct parton, and unmatched permutations (f_{un} = 49.4%), with at least one quark from the $t\bar{t}$ decay not matched unambiguously to a jet.

3. Background Modeling

The multijet background is estimated by an event mixing technique. All events after the b-tagging selection are taken as input, mixing the jets between the different events such that in every newly generated event all jets are originating from different events, keeping at least two b-tagged jets in the new event.

The simulated $t\bar{t}$ sample and modeled background are normalized to data with an expected signal fraction (f_{sig}) from simulation, which depends on the cross section [9, 10, 11] and the selection efficiency for $t\bar{t}$ events.

4. Ideogram Method

A likelihood function is constructed that allows the simultaneous determination of JES and m_t. Based on this likelihood function, two different estimates of m_t are performed: with a fixed JES ("1-D analysis") or simultaneously with the JES ("2-D analysis").

The observable used for measuring m_t is the top-quark mass from the kinematic fit (m_t^{fit}). The average reconstructed W-boson mass before they are constrained by the kinematic fit (m_W^{reco}) is taken as an estimator for measuring a global JES. Figure 1 compares data, expectation from simulation, and modeled background for m_t^{fit} and m_W^{reco}.

A likelihood ($\mathcal{L}$) to estimate m_t and JES given the observed data sample can be defined as:

$$\mathcal{L}(m_t, \text{JES}|\text{sample}) \;\propto\; \mathcal{L}(\text{sample}|m_t, \text{JES}) = \prod_{\text{events}} P\left(m_t^{fit}, m_W^{reco}|m_t, \text{JES}\right)^{P_{gof}}.$$

There is no correlation between m_t^{fit} and m_W^{reco}, hence, the probability $P\left(m_{t,i}^{fit}, m_W^{reco}|m_t, \mathrm{JES}\right)$ factorizes into

$$P\left(m_t^{fit}, m_W^{reco}|m_t, \mathrm{JES}\right) = f_{sig} \cdot \sum_j f_j P_j\left(m_t^{fit}|m_t, \mathrm{JES}\right) \cdot P_j\left(m_W^{reco}|m_t, \mathrm{JES}\right) + (1 - f_{sig}) \cdot P_{bkg}\left(m_t^{fit}\right) \cdot P_{bkg}\left(m_W^{reco}\right),$$

where f_j and P_j with $j \in \{cp, wp, un\}$ are the relative fraction and the probability density functions for signal of the three different permutation cases, which are determined from simulated $\mathrm{t\bar{t}}$ events with different generated top-quark masses ($m_{t,gen}$) and different JES. The m_t^{fit} distributions are fitted with a Breit-Wigner function convoluted with Gaussian resolution for the *cp* case and with the sum of a Landau function and a Gaussian function with common means for the *wp* and *un* cases for different $m_{t,gen}$ and JES. The m_W^{reco} distributions are fitted with asymmetric Gaussian functions. The parameters of all fitted signal functions are parametrized linearly in terms of $m_{t,gen}$, JES, and the product of the two. As the background is modeled from data, its probability density distributions do not depend on m_t nor JES.

In the 1-D analysis, m_t is estimated from $-2\ln\{\mathcal{L}(m_t, \mathrm{JES} = 1|\mathrm{sample})\}$. In the 2-D analysis, the most likely m_t and JES are obtained by minimizing $-2\ln\{\mathcal{L}(m_t, \mathrm{JES}|\mathrm{sample})\}$.

For each combination of m_t and JES 10 000 pseudo-experiments are conducted using simulated $\mathrm{t\bar{t}}$ events and modeled background events from event mixing to calibrate the measurement and its statistical uncertainty.

5. Systematic Uncertainties

An overview of the different sources of systematic uncertainties is shown in Table 1.

As expected, the main systematic uncertainty in the 1-D measurement stems from the uncertainty in JES and the 2-D measurement reduces this uncertainty to a small p_T- and η-dependent JES uncertainty. However, the 2-D approach leads to increased uncertainties for color reconnection effects, underlying event, and the modeling of the non-$\mathrm{t\bar{t}}$ background.

Overall, the 1-D measurement offers a better precision because of the strong impact of the underlying event on the top-quark mass measurement in the 2-D approach.

Table 1. Overview of systematic uncertainties. The total is defined by adding in quadrature the contributions from all sources.

	1-D analysis	2-D analysis	
	δ_{m_t} (GeV)	δ_{m_t} (GeV)	δ_{JES}
Fit calibration	0.13	0.14	0.001
Jet energy scale	0.97	0.10	0.002
b-JES	0.49	0.52	0.001
Jet energy resolution	0.15	0.13	0.003
b tagging	0.06	0.10	0.001
Trigger	0.24	0.26	0.006
Pileup	0.06	0.10	0.001
Parton distribution functions	0.06	0.10	0.001
Q^2 scale	0.22	0.34	0.005
ME-PS matching threshold	0.24	0.34	0.003
Underlying event	0.32	0.88	0.007
Color reconnection effects	0.15	0.58	0.006
Non-$\mathrm{t\bar{t}}$ background	0.20	0.62	0.008
Total	1.25	1.46	0.015

6. Results

Out of 3.54 fb^{-1} of 2011 data, 2418 events are selected and with a fixed JES=1 result in:

$$m_t = 173.49 \pm 0.69 \text{ (stat.)} \pm 1.25 \text{ (syst.) GeV}$$

The overall uncertainty of the presented 1-D analysis is 1.43 GeV.

A simultaneous fit of m_t and JES to the same data yields:

$$m_t = 174.28 \pm 1.00 \text{ (stat.+JES)} \pm 1.46 \text{ (syst.) GeV}$$
$$\text{JES} = 0.991 \pm 0.008 \text{ (stat.)} \pm 0.015 \text{ (syst.)}$$

The measured JES confirms the JES in data measured in events with Z bosons and photons [12]. The overall uncertainty in the top-quark mass of the presented 2-D analysis is 1.77 GeV.

The measured top-quark masses in both analyses are in agreement, with the 1-D analysis yielding a higher precision than the 2-D analysis.

7. Summary

A measurement of the top-quark mass is presented using events with all-jets final states, collected by CMS in pp collisions at $\sqrt{s} = 7$ TeV in 2011. A constrained fit reconstructs the complete kinematics of each event. For each selected event a likelihood is calculated as a function of assumed top-quark mass. Using a data sample with an integrated luminosity of 3.54 fb^{-1}, 2418 candidate events are observed and the mass of the top-quark is measured to be $m_t = 173.5 \pm 0.7$ (stat.) $\pm$ 1.3 (syst.) GeV. To date, this measurement constitutes the most precise determination of the top-quark mass in all-jets final states.

References

[1] Chatrchyan S *et al.* (CMS Collaboration) 2008 *JINST* **3** S08004
[2] CMS Collaboration 2012 *CMS Physics Analysis Summary* **CMS-PAS-TOP-11-017**
[3] CMS Collaboration 2011 *CMS Physics Analysis Summary* **CMS-PAS-TOP-11-007**
[4] Cacciari M and Salam G P 2006 *Phys. Lett. B* **641** 57
[5] Cacciari M, Salam G P and Soyez G 2008 *JHEP* **04** 063
[6] CMS Collaboration 2010 *CMS Physics Analysis Summary* **CMS-PAS-PFT-10-002**
[7] D'Hondt J *et al.* 2006 Fitting of event topologies with external kinematic constraints in CMS Tech. Rep. CMS-NOTE-2006-023 CERN Geneva
[8] Beringer J *et al.* (Particle Data Group) 2012 *Phys. Rev. D* **86** 010001
[9] Kidonakis N 2010 *Phys. Rev. D* **82** 114030
[10] Aliev M, Lacker H, Langenfeld U, Moch S, Uwer P *et al.* 2011 *CPC* **182** 1034–1046
[11] Ahrens V *et al.* 2010 *JHEP* **09** 097
[12] Chatrchyan S *et al.* (CMS Collaboration) 2011 *JINST* **6** P11002

Journal of Physics: Conference Series **452** (2013) 012023 doi:10.1088/1742-6596/452/1/012023

Top quark mass combination techniques and treatment of uncertainties at the Tevatron and LHC

Elizaveta Shabalina for the CDF, D0, ATLAS and CMS collaborations

Georg-August-Universität Göttingen, Friedrich-Hund-Platz 1, D-37077 Göttingen, Germany

E-mail: `Elizaveta.Shabalina@cern.ch`

Abstract. We present an overview of the statistical method used to combine top quark mass (m_t) measurements at the Tevatron and the LHC. We discuss the breakdown of uncertainties and their correlations. Finally we present the results of the latest m_t combinations from the Tevatron and the LHC.

1. Introduction

The top quark plays a special role in the standard model (SM) because it is by far the most massive elementary particle. Top quark mass represents one of the fundamental parameters of the SM, and its precise knowledge provides an important consistency check of the SM. Through the SM radiative corrections to the mass of the W boson, top quark mass imposes the constraint on the mass of the Higgs boson, and together with other electroweak observables included in the global fit it constrains the contributions from physics beyond the SM.

Since the discovery of the top quark in 1995 [1] a lot of efforts went into the measurement of its mass with high precision. The measurements were performed in various decay channels of the $t\bar{t}$ pair and using different techniques. Depending on the channel and technique the measurements have different sensitivity to the effects of jet energy scale calibration, signal and background modelling. Given the limited statistics of the $t\bar{t}$ sample at the Tevatron majority of m_t measurements were statistically limited. Thus performing the combination of these measurements was important to improve the precision and to check the consistency of the different measurements.

2. BLUE method

The first combination of the direct m_t measurements at the Tevatron was performed in 2004 followed by the updates each year as new measurements become available [2]. The statistical method used for combination calculates a linear weighted sum of individual results with weights determined such that they minimize the total uncertainty on the combined result [3] and is known as Best Linear Unbiased Estimate (BLUE). It takes into account statistical and systematic uncertainties of each measurement and their correlations and allows combination of correlated measurements of one or more parameters. The output provides χ^2 value of the fit to evaluate the consistency of the input measurements.

Important features of the method can be understood by considering a combination of two measurements $x_1 \pm \sigma_1$ and $x_2 \pm \sigma_2$ with correlation $\rho \leq 1$, one of which, x_1, is more precise than the other, i.e., $\sigma_2 > \sigma_1$. BLUE method will return a combined value of $x \pm \sigma_x$, where $x = \alpha x_1 + \beta x_2$ with $\alpha + \beta = 1$. The weight of the second measurement can be calculated as $\beta = (1 - \rho z)/(1 - 2\rho z + z^2)$, where z is the ratio of the uncertainties of the two measurements $z = \sigma_2/\sigma_1 > 1$. Then relative improvement of the uncertainty on the combined result with respect to the most precise one is $\sigma_x/\sigma_1 = \sqrt{z^2(1-\rho^2)/(1-2\rho z + z^2)}$. Figure 1 shows the relative improvement as a function of correlation ρ between the measurements for different values of z. If two uncorrelated ($\rho = 0$) measurements of the same precision ($z = 1$, blue curve) are combined the combination will have 30% better uncertainty than each of the measurements. However, the combination of two 50% correlated measurements one of which is twice more precise than the other (turquoise curve in figure 1) will have the same uncertainty as the most precise of the two. In general, figure 1 demonstrates that the relative improvement of the combination and the weights of the input measurements depend only on their precisions and correlations and not on the actual measured values x_1 and x_2. Furthermore, figure 2 shows that depending on the precision of the individual measurements and their correlation the less precise measurement can acquire negative weight in the combination. This happens if $\rho > 1/z$. In such case the combined value will be outside the interval between the two input values. The measurement with a negative weight still contributes to the improvement of the uncertainty of the combined result.

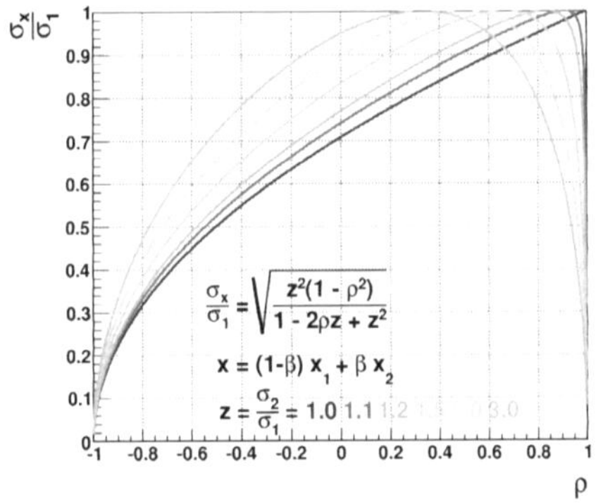

Figure 1. The relative improvement of the uncertainty on the combined measurement as a function of correlation for different values of z [4].

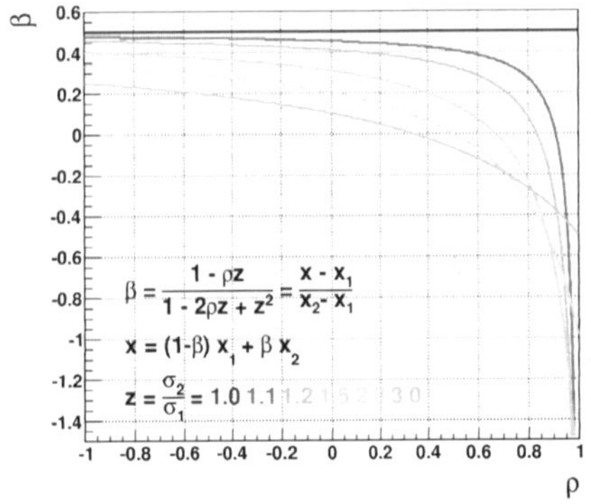

Figure 2. The weight of a less precise measurement in the combination as a function of correlation for different values of z [4].

3. Treatment of uncertainties

The most important part of the combination work is to understand correlations between the uncertainties of the individual measurements. The latter are combined in categories based on the physics origin of an uncertainty and the correlation pattern between analysis channels, data taking periods and experiments.

In the Tevatron m_t combination [2] the following correlation patterns are defined:

- **Uncertainties correlated between all measurements**: This class includes uncertainties related to the $t\bar{t}$ signal model and b-jets model based on Monte Carlo (MC) simulation.
- **Uncertainties correlated between all measurements of the same experiment**: This class includes uncertainties coming from calibration procedures used within the experiment.

- **Uncertainties correlated between measurements of the same experiment within the same data taking period**: This class includes uncertainties coming from calibration procedures used within the experiment which changed between data taking periods or use data from a specific period.
- **Uncertainties correlated between all measurements in the same channel**: These include uncertainties on the background determination based on simulation.
- **Uncorrelated uncertainties**: This class includes uncertainties of the statistical origin or the ones specific to a method.

The first LHC m_{t} combination [5] followed the breakdown of uncertainties introduced in the Tevatron combination as much as possible and reasonable. There are five categories of uncertainties which are common for both: statistical, method, background estimate by data driven techniques, background estimate from simulation and lepton model. Full breakdown consists of 14 categories for the Tevatron and 17 categories for LHC, 7 of which are related to jet energy calibration. The necessity of several categories for the jet energy calibration uncertainties is motivated by the fact that these were the dominating uncertainties on the m_{t} measurements and by the complex calibration procedures which differ between experiments and change between run periods. The calibration is based on both data and MC driven techniques which require different correlation patterns.

With the recently dramatically increased sample of top quarks the coherent approach to evaluation of various signal modelling uncertainties and understanding of their correlations became critical. At the Tevatron $t\bar{t}$ signal modelling uncertainties include hadronization model, higher order effects (since leading order MC is used as a default), uncertainty on the amount of initial and final state radiation, on modelling of color reconnection and the choice of PDF. All of them are combined in one class and taken as fully correlated between all measurements. Similar uncertainties are evaluated by the CMS and ATLAS collaborations. However, unlike at the Tevatron, the methods used to derive them differ significantly between ATLAS and CMS. For this reason LHC m_{t} combination keeps different modelling uncertainties as separate sources. Variation of the assumed correlation between two sources, the choice of the signal MC generator and the amount of radiation, turned out to have a non-negligible effect on the uncertainty on the combined m_{t}. A study showed that the uncertainty is the largest when correlation of these sources between the CMS and ATLAS measurements are assumed to be 50%. This value has been chosen to ensure a conservative uncertainty of the combined result.

4. Tevatron top quark mass combination results

The latest Tevatron top quark mass combination [2] is based on 12 measurements: 8 by the CDF and 4 by the D0 experiment, 5 performed in Run I of the Tevatron and 7 from Run II, 5 results from the l+jets, 4 from the dilepton and 2 from the all hadronic channel and one requiring $\not{E}_T$+jets. Table 1 summarizes the input measurements with the uncertainties and provides a breakdown of the uncertainty on the combined result which yields: $m_t = 173.18 \pm 0.56(\mathrm{stat}) \pm 0.75(\mathrm{syst})$ GeV with $\chi^2/NDF = 8.3/11$ corresponding to χ^2 probability of 69%. Figure 3 presents a summary of the input measurements and the combined m_{t} at the Tevatron. The combined measurement is 25% more precise than the most precise CDF measurement in the l+jets channel. Dominating uncertainties come from the limited statistics of the top quarks, expected to go down for the measurements using full Tevatron data set, and from the $t\bar{t}$ signal model.

5. LHC top quark mass combination results

The first LHC top quark mass combination [5] is based on 7 measurements: 3 by the ATLAS and 4 by the CMS collaboration, 3 performed using 2010 and 4 2011 data, 4 results from

Table 1. The uncertainty in GeV from each component for the twelve measurements of m_t and the resulting Tevatron combination [2]. The total uncertainties are obtained by adding the components in quadrature. The entries "n/a" stand for "not applicable" and "n/e" for "not evaluated." The non-evaluated uncertainties were not considered as significant sources of uncertainty for Run I measurements.

Channel	Exp.-Run	Light-jet response (1)	Light-jet response (2)	Out-of-cone correction	Offset	Model for b jets	Response to $b/q/g$ jets	*In-situ* light-jet calibration	Jet modeling	Lepton modeling	Signal modeling	Multiple interactions model	Background from theory	Background based on data	Calibration method	**Statistical uncertainty**	**Total JES uncertainty**	**Other systematic uncertainty**	**Total uncertainty**
		Jet energy scale systematics							Other systematics										
Lepton+jets	CDF-II	0.41	0.01	0.27	n/a	0.23	0.13	0.58	0.00	0.14	0.56	0.10	0.27	0.06	0.10	0.65	0.80	0.67	1.23
Lepton+jets	D0-II	n/a	0.63	n/a	n/a	0.07	0.26	0.46	0.36	0.18	0.77	0.05	0.19	0.23	0.16	0.83	0.83	0.94	1.50
Lepton+jets	CDF-I	3.4	0.7	2.7	n/a	0.6	n/e	n/a	n/e	n/e	2.7	n/e	1.3	n/e	0.0	5.1	4.4	2.8	7.3
Lepton+jets	D0-I	n/a	2.5	2.0	1.3	0.7	n/e	n/a	n/e	n/e	1.3	n/e	1.0	n/e	0.6	3.6	3.5	1.6	5.3
Alljets	CDF-II	0.38	0.04	0.24	n/a	0.15	0.03	0.95	0.00	n/a	0.64	0.08	0.00	0.56	0.38	1.43	1.06	0.91	2.00
Alljets	CDF-I	4.0	0.3	3.0	n/a	0.6	n/e	n/a	n/e	n/a	2.1	n/e	1.7	n/e	0.6	10.0	5.0	2.6	11.5
Dileptons	CDF-II	2.01	0.58	2.13	n/a	0.33	0.14	n/a	0.00	0.27	0.80	0.23	0.24	0.14	0.12	1.95	3.01	0.88	3.69
Dileptons	D0-II	n/a	0.56	n/a	n/a	0.20	0.40	0.55	0.50	0.35	0.86	0.00	0.00	0.20	0.51	2.36	0.90	1.11	2.76
Dileptons	CDF-I	2.7	0.6	2.6	n/a	0.8	n/e	n/a	n/e	n/e	3.0	n/e	0.3	n/e	0.7	10.3	3.9	3.0	11.4
Dileptons	D0-I	n/a	1.1	2.0	1.3	0.7	n/e	n/a	n/e	n/e	1.9	n/e	1.1	n/e	1.1	12.3	2.7	2.3	12.8
$\not{E}_T$+jets	CDF-II	0.45	0.05	0.20	n/a	0.00	0.12	1.54	0.00	n/a	0.78	0.16	0.00	0.12	0.14	1.80	1.64	0.78	2.56
Decay length	CDF-II	0.24	0.06	n/a	n/a	0.15	n/e	n/a	0.00	n/a	0.90	0.00	0.80	0.20	2.50	9.00	0.25	2.80	9.43
Tevatron Combination		0.12	0.19	0.04	0.00	0.15	0.12	0.39	0.11	0.10	0.51	0.00	0.14	0.11	0.09	0.56	0.49	0.57	0.94

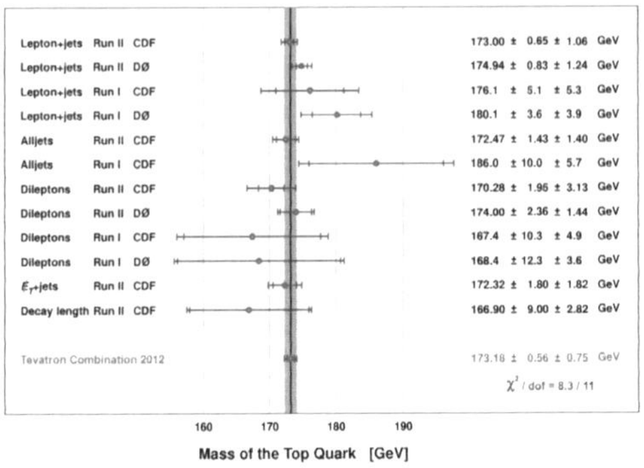

Figure 3. Summary of the Tevatron input m_t measurements and the resulting combined top quark mass [2].

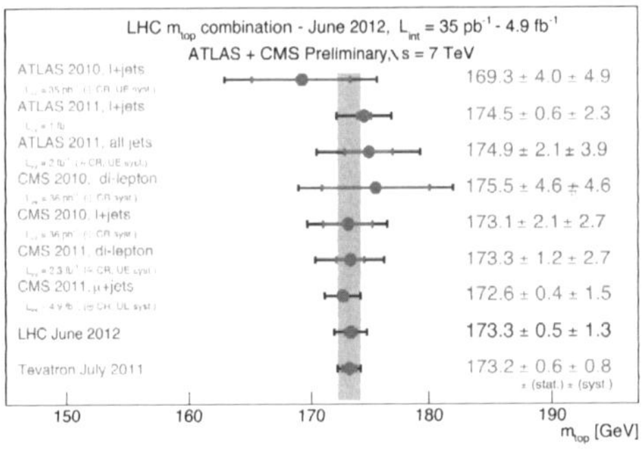

Figure 4. Summary of input measurements and result of LHC combination compared to the Tevatron combined m_t.

the l+jets, 2 from the dilepton and 1 from the all hadronic channel. Table 2 summarizes the input measurements with the uncertainties and provides a breakdown of the uncertainty on the combined result which yields: $m_t = 173.3 \pm 0.5(\text{stat}) \pm 1.3(\text{syst})$ GeV with $\chi^2/NDF = 2.5/6$ corresponding to χ^2 probability of 87%. The categories of uncertainties and their naming follow the breakdown used for the latest preliminary Tevatron m_t combination [6]. Figure 4 presents a summary of the input measurements and the combined m_t at the LHC compared to the Tevatron result. The combined measurement is 9% more precise than the most precise CMS measurement in the μ+jets channel. Dominating uncertainties come from the signal model, mainly from radiation and color reconnection models, from the b-jets and underlying event models.

Table 2. Inputs to the LHC combination and the results obtained using the categories grouping and correlation as defined in [5]. The entries "n/a" stand for "not applicable" and "n/e" for "not evaluated."

	ATLAS			CMS				LHC
	2010	2011		2010		2011		
	l+jets	l+jets	all jets	di-l	l+jets	di-l	μ+jets	comb.
[GeV]								
Measured m_t	169.3	174.5	174.9	175.5	173.1	173.3	172.6	173.34
Stat	4.0	0.6	2.1	4.6	2.1	1.2	0.4	0.47
iJES	n/a	0.4	n/a	n/a	n/a	n/a	0.4	0.38
aJES	n/a	n/a	n/a	n/a	n/a	n/a	n/a	n/a
bJES	2.5	1.6	1.4	0.9	0.9	1.1	0.7	0.68
cJES	n/a	n/a	n/a	n/a	n/a	n/a	n/a	n/a
dJES	2.1	0.7	2.1	2.1	2.1	2.0	0.2	0.07
rJES	n/a	n/a	n/a	3.3	n/a	n/a	n/a	0.06
Lept	n/e	n/e	n/e	0.3	n/e	0.2	n/e	0.01
MC	1.0	0.4	0.5	0.4	n/e	0.1	n/e	0.04
Rad	2.5	1.0	1.7	0.9	1.2	0.8	0.8	0.69
CR	0.6	0.6	0.6	0.5	0.5	0.5	0.5	0.55
PDF	0.5	0.1	0.6	0.5	0.1	0.4	0.1	0.01
DTMO	1.2	0.3	0.5	0.6	0.4	0.7	0.3	0.19
UE	0.6	0.6	0.6	1.4	0.2	0.6	0.6	0.47
BGMC	1.8	0.1	n/a	0.1	0.2	n/a	0.1	0.01
BGDT	0.6	0.5	1.9	n/a	0.4	0.4	n/a	0.16
Meth	0.4	0.1	1.0	0.3	0.1	0.4	0.2	0.13
MHI	0.7	< 0.05	n/e	1.0	0.1	0.2	0.4	0.25
[GeV]								
Total Syst. Unc	4.9	2.3	3.9	4.6	2.7	2.7	1.5	1.33
Total Unc.	6.3	2.4	4.4	6.5	3.4	3.0	1.5	1.40

6. Conclusions and outlook

The technique to perform top quark mass combination is well established and was used at the Tevatron and LHC. New m_t measurements using full Tevatron data set and large LHC data sets are expected to have significantly reduced statistical uncertainties and uncertainties related to the detector model that depend on the amount of data available for the calibration of objects. The uncertainties on the various aspects of the $t\bar{t}$ signal model will be by far the dominating ones. Thus for the future LHC top quark mass combinations and for the first combination of m_t measurements at the Tevatron and LHC uniform approach to evaluating signal model systematic uncertainties becomes critical. Given a regime of high correlations in the future combinations a careful breakdown of all important uncertainties into model-related and detector-related is required to correctly take into account correlations between input measurements.

References

[1] F. Abe *et al.* (CDF Collaboration), *Phys. Rev. Lett.* **74**, 2626 (1995);
S. Abachi *et al.* (D0 Collaboration), *Phys. Rev. Lett.* **74**, 2632 (1995).
[2] CDF and D0 collaborations, *Phys. Rev.* D **86**, 092003 (2012).
[3] L. Lyons, D. Gibaut, and P. Clifford, *Nucl. Instrum. Methods in Phys. Res. Sect.* A **270**, 110 (1988);
A. Valassi, *Nucl. Instrum. Methods in Phys. Res. Sect.* A **500**, 391 (2003).
[4] R. Nisius, private communication.
[5] ATLAS and CMS collaborations, ATLAS-CONF-2012-095 (2012), CMS PAS TOP-12-001 (2012).
[6] The Tevatron Electroweak Working Group for the CDF and D0 collaborations, arXiv:1107.5255 (2011).

5th International Workshop on Top Quark Physics (TOP2012) IOP Publishing
Journal of Physics: Conference Series **452** (2013) 012024 doi:10.1088/1742-6596/452/1/012024

Top-quark mass results at the LHC

Markus Seidel

Universität Hamburg, on behalf of the ATLAS and CMS collaborations

E-mail: markus.seidel@uni-hamburg.de

Abstract. The mass of the top quark has been measured by the ATLAS and CMS collaborations at the Large Hadron Collider (LHC) in pp collisions at $\sqrt{s} = 7$ TeV. Both experiments use large samples of 2011 data in all decay channels of top-quark pairs to obtain very precise results. An overview of the recent measurements is given in this report.

1. Introduction

The top-quark mass, m_t, is an important parameter of the standard model of particle physics and has been measured with great precision by the CDF and D0 collaborations at the Fermilab Tevatron [1]. At the CERN LHC both multi-purpose experiments, ATLAS [2] and CMS [3], study the production and properties of top quarks. Besides being significant by themselves, these measurements also constitute benchmarks for detector performance.

A new challenge for the reconstruction of top-quark events at the LHC is due to the large amount of high-energetic QCD radiation at $\sqrt{s} = 7$ TeV. This complicates the correct assignment of reconstructed jets to the quarks from the decay $t\bar{t} \rightarrow bW^+\bar{b}W^-$ where the subsequent decays of the W bosons result in either two charged leptons and the corresponding neutrinos ("dilepton channel"), one charged lepton, the corresponding neutrino and two light flavor jets ("lepton+jets") or four light-flavor jets ("all-hadronic"). The higher top-quark production cross-section and instantaneous luminosity compared to the Tevatron results in nearly 1 million top-quark pairs produced at each experiment in 2011, where an integrated luminosity of up to 5 fb^{-1} was recorded. Efficient trigger systems and algorithms for identifying b-jets allow the selection of top-quarks in unprecedented quantity. The uncertainties on the jet energy corrections reach a level of 1% [4, 5], giving an excellent opportunity for studying top-quarks with high precision.

2. Direct measurements using lepton+jets final states

Due to its large branching ratio and decent background contamination, the lepton+jets channel is considered to be the "golden channel" for direct top-quark mass measurements, where electrons and muons are considered as well-identificable leptons. The background from W+jets production can be controlled using b-tagging and the constrained kinematics allow for a complete reconstruction of the event.

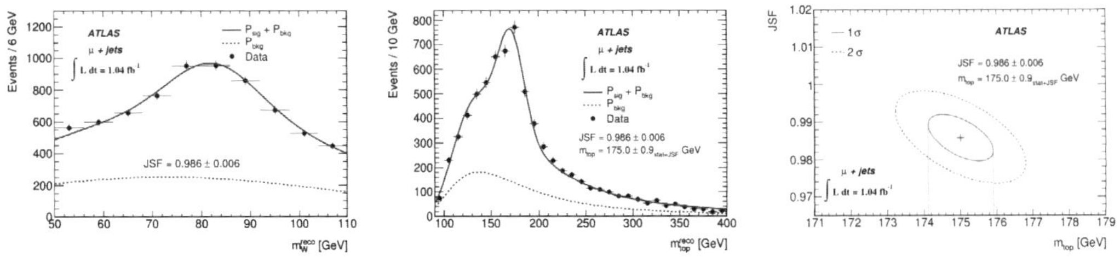

(a) (b) (c)

Figure 1: The reconstructed masses of (a) W boson candidates and (b) top-quark candidates in μ+jets events selected by the ATLAS experiment with 2D template fits. (c) Extracted top-quark mass and jet scale factor in the μ+jets channel using data corresponding to 1 fb^{-1}. [6]

2.1. ATLAS: Measurements using template methods [6]

In the event selection exactly one isolated muon/electron with transverse momentum (p_T) above 20/25 GeV within$|\eta| < 2.5$ and at least four jets with $p_T > 25$ GeV, $|\eta| < 2.5$ are required, where at least one jet has to be b-tagged. A significant amount of missing transverse energy ($E_{\mathrm{T}}^{\mathrm{miss}}{}_{e/\mu} > 20/35$ GeV, $E_{\mathrm{T}}^{\mathrm{miss}} + m_{\mathrm{W,lep}}^{\mathrm{T}} > 60$ GeV) and an untagged jet pair with 50 GeV $< m_{jj} <$ 110 GeV as a W boson candidate are required. The top-quark candidate is then formed by choosing the jjb triplet with maximum p_T and constraining m_{jj} to the W boson mass. In total, 11 781 events are selected in 1 fb^{-1} data with an estimated purity of 72% $t\bar{t}$ events.

Two different approaches are used for extracting the top-quark mass from the selected sample. In the 2D template method, templates for $m_{\mathrm{W}}^{\mathrm{reco}}$ as a function of an assumed global jet energy scale factor (JSF) and for $m_{\mathrm{top}}^{\mathrm{reco}}$ as a function of the input top-quark mass and JSF are constructed. An unbinned likelihood fit is performed to extract the top-quark mass, JSF and the number of background events simultaneously as shown in Fig. 1 and yields a result of $m_t = 174.5 \pm 0.6$ (stat) $\pm$ 2.3 (syst) GeV. The systematic uncertainty is dominated by the uncertainty on the amount of additional radiation (ISR/FSR) and the b-jet energy scale.

The alternative approach uses improved event reconstruction through a likelihood fit, selecting a subset of 2875 events in 1 fb^{-1} data. 1D templates are created for $R_{32} \equiv m_{\mathrm{top}}^{\mathrm{reco}}/m_{\mathrm{W}}^{\mathrm{reco}}$ as a function of m_t in an attempt to minimize the impact of the jet energy scale uncertainty. A binned likelihood fit is employed to extract $m_t = 174.4 \pm 0.9$ (stat) $\pm$ 2.5 (syst) GeV, where the systematic uncertainty is dominated by radiation and the jet energy scale.

2.2. CMS: Measurement using the ideogram method [7]

At CMS exactly one isolated lepton with $p_T > 30$ GeV, $|\eta| < 2.1$, at least four jets with $p_T > 30$ GeV, $|\eta| < 2.4$, and exactly two b-tags are required. For the event reconstruction b-tagged jets are assigned to b-quarks and untagged jets to light quarks. A constrained kinematic fit checks the compatibility of each possible permutation with the $t\bar{t}$ hypothesis, i.e. $m_W = 80.4$ GeV, $m_t = m_{\bar{t}}$. The permutations are weighted by their goodness-of-fit probability $P_{\mathrm{gof}}\left(\chi^2\right) = \exp\left(-\frac{1}{2}\chi^2\right)$ and are required to fulfill $P_{\mathrm{gof}}\left(\chi^2\right) > 0.2$. The selected sample contains 5194 events in 5 fb^{-1} data, with an estimated purity of 96% $t\bar{t}$ events and 44% correct permutations (Fig. 2).

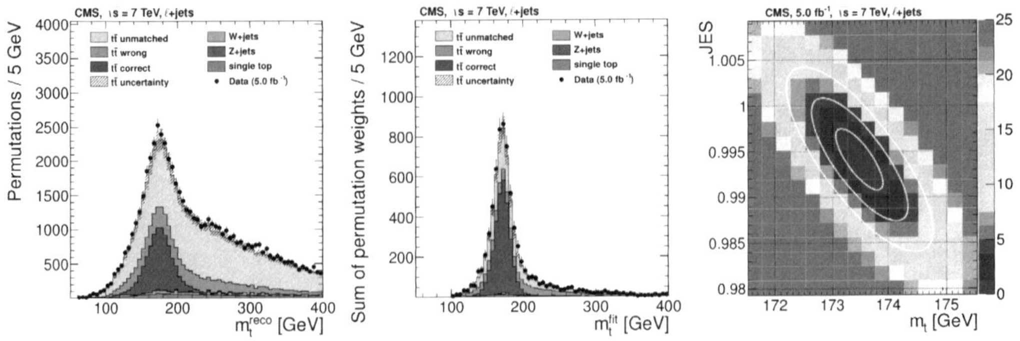

(a) (b) (c)

Figure 2: (a) The reconstructed top-quark mass after preselection for all possible permutations and (b) the top-quark mass after the kinematic fit and requiring $P_{\mathrm{gof}}(\chi^2) > 0.2$. (c) The 2D likelihood $(-2 \cdot \Delta \log(\mathcal{L}))$ measured for the ℓ+jets final states. [7]

For each permutation case j (representing correct, wrong and unmatched permutations), probability densities P_j are constructed for m_t and a global jet energy scale (JES) factor, so that the likelihood for observing a single event with n permutations is given by

$$\mathcal{L}(\mathrm{event}|m_t, \mathrm{JES}) = \sum_{i=0}^{n} P_{gof}(i)\, P\left(m_{t,i}^{fit}, m_{W,i}^{reco}|m_t, \mathrm{JES}\right),$$

$$P\left(m_{t,i}^{fit}, m_{W,i}^{reco}|m_t, \mathrm{JES}\right) = \sum_{j} f_j P_j\left(m_{t,i}^{fit}|m_t, \mathrm{JES}\right) \cdot P_j\left(m_{W,i}^{reco}|m_t, \mathrm{JES}\right).$$

The most likely m_t and JES given the data sample are then extracted by maximizing a weighted product over all event likelihoods.

Using data corresponding to 5 fb^{-1}, the top mass is measured very precisely to be $m_t = 173.49 \pm 0.43$ (stat+JES) ± 0.98 (syst) GeV. The dominating systematic effects are the uncertainties on b-JES and color reconnection, where m_t^{fit} and m_W^{reco} are shifted in opposite direction.

2.3. CMS: Measurement of the top mass difference using the ideogram method [8]

A check of CPT invariance is performed by measuring the mass difference between top- and antitop-quarks. Requiring no b-tagged jets the data is split in two samples by the lepton charge ($\ell^+ \to \bar{t}_{\mathrm{had}}, \ell^- \to t_{\mathrm{had}}$). The top-quark mass is measured in each sample with the ideogram method using 5 fb^{-1} of data, yielding $\Delta m_t = m_{t,\mathrm{had}} - m_{\bar{t},\mathrm{had}} = -0.44 \pm 0.46$ (stat) ± 0.27 (syst) GeV.

3. Direct measurements using dilepton final states

The dilepton final states are characterized by their low branching ratio but very clean signature. The underconstrained kinematics require special techniques for mass reconstruction.

3.1. ATLAS: Measurement using the $\overline{m}_{T2}$ method [9]

Events with exactly one isolated muon and one isolated electron with opposite charge and $p_{T,e/\mu} > 20/25$ GeV, $|\eta| < 2.5$ are selected. At least two b-tagged jets

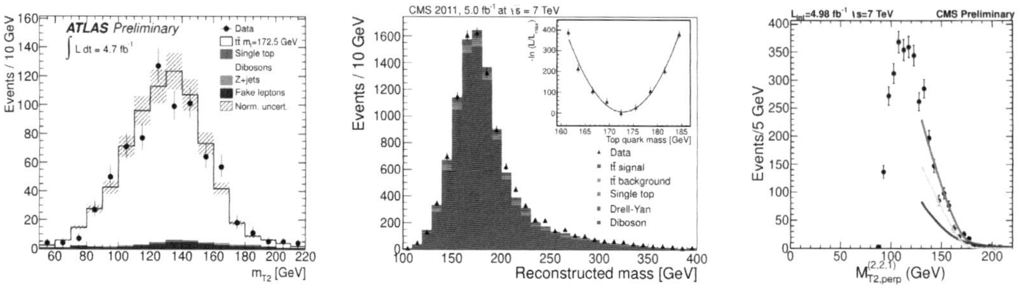

(a) (b) (c)

Figure 3: Measurements using dilepton final states: (a) m_{T2} distribution (ATLAS [9]), (b) reconstructed mass and likelihood (CMS AMWT [10]) and (c) fit to the $m^{221}_{T2\perp}$ endpoint (CMS [11]).

($p_T > 25$ GeV, $|\eta| < 2.5, \Delta\phi\,(\text{b-jet}_1, \text{b-jet}_2) > 2.2$), $E_{\mathrm{T}}^{\mathrm{miss}} > 20$ GeV and $H_T > 130$ GeV are required, where H_T is defined as the scalar sum of transverse momenta of the two selected leptons and all selected jets in the event.

Under the $t\bar{t}$ hypothesis, m_{T2} represents the lower bound of the parent particle mass:

$$m_{T2}\,(m_{\text{invis}}) \quad = \quad \min_{\vec{p}_T^{(1)},\vec{p}_T^{(2)}} \left\{ \max \left[m_T \left(m_{\text{invis}}, \vec{p}_T^{(1)} \right), m_T \left(m_{\text{invis}}, \vec{p}_T^{(2)} \right) \right] \right\}.$$

Requiring $m_{T2} < 220$ GeV and taking the permutation with smallest m_{T2}, 731 events are selected in 4.7 fb^{-1} data, with a purity of 95% $t\bar{t}$ (Fig. 3a). As m_{T2} is proportional to the simulated input mass, the mean value $\overline{m}_{T2}$ serves as an estimator for m_t. The calibration is cross-checked using simulated same-flavor events and the measurements yields $m_t = 175.2 \pm 1.6$ (stat)$^{+3.1}_{-2.8}$ (syst) GeV, where the JES, b-JES and color reconnection are the leading systematic uncertainties.

3.2. CMS: Measurement using the AMWT method [10]

Two isolated, oppositely charged leptons (e or μ, $p_T > 20$ GeV, $|\eta| < 2.4$, $m_{\ell\ell} > 20$ GeV) and at least two jets ($p_T > 30$ GeV, $|\eta| < 2.4$, 1 b-tag) are required. Same-flavor events have to fulfill additionally $E_{\mathrm{T}}^{\mathrm{miss}} > 40$ GeV and $m_{\ell\ell} < 76$ GeV and $m_{\ell\ell} > 106$ GeV. In 5 fb^{-1} data 11 627 events are selected, with an estimated purity of 89% $t\bar{t}$ signal events (Fig. 3b).

For the analytical reconstruction of the unknown neutrino momenta, a scan of top mass hypotheses in the range 100 – 300 GeV is performed and up to 2×4 solutions per event and top mass hypothesis are retrieved. Each solution is assigned a weight based on the LO matrix element. For each event the calculation is repeated 1000 times with the jets smeared within detector resolutions and the m_{top} hypothesis with the highest averaged weight is taken as m_{AMWT}. A likelihood fit of the m_{AMWT} distribution is performed and yields a result of 172.5 ± 0.4 (stat) ± 1.5 (syst) GeV, where the JES is the largest contribution to the systematic uncertainty.

3.3. CMS: Measurement using the endpoint method [11]

A new method fits the endpoints of $m^{221}_{T2\perp} = m_t^{min}$ (using only momentum components transverse to $p_T^{t\bar{t}}$), and of $m_{b\ell}^{\max}$ directly in data to retrieve a result with minimal dependence

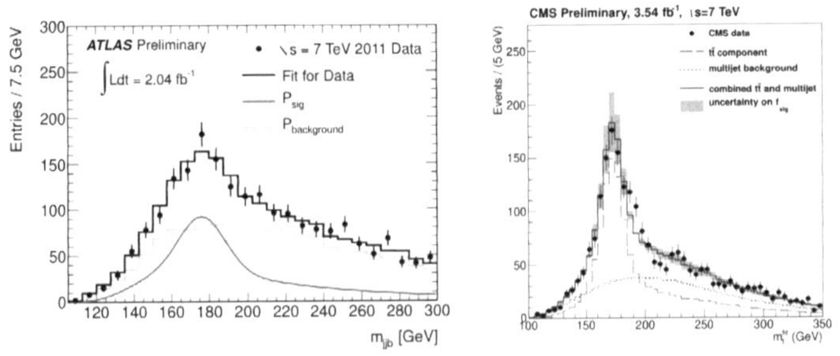

(a) (b)

Figure 4: Measurements using all-hadronic final states: Reconstructed top-quark mass of (a) the ATLAS [12] and (b) the CMS [13] measurement.

on simulation and thus complementary systematic uncertainties. Two leptons and two b-tagged jets are required so that 8678 events are selected in 5 fb^{-1} of data (Fig. 3c). The result $m_t = 173.9 \pm 0.9\ (\text{stat})^{+1.2}_{-1.8}\ (\text{syst})$ GeV is limited by the JES uncertainty.

4. Direct measurements using all-hadronic final states

All-hadronic final states are produced with the highest branching ratio and their kinematics are fully constrained. Measurements are challenging due to a large fraction of multijet background.

4.1. ATLAS: Measurement using the template method [12]

Events are selected with at least five jets ($p_T > 55$ GeV) and a sixth jet ($p_T > 30$ GeV), where two b-tagged jets with $p_T > 55$ GeV, $|\eta| < 2.5$ have a separation in $\eta - \phi$ space of at least $\Delta R_{bb} > 1.2$, and $E_{\mathrm{T}}^{\mathrm{miss}}/\sqrt{H_T} < 3$. The permutation that minimizes a χ^2 based on kinematic constraints ($m_W = 80.4$ GeV, $m_t = m_{\bar{t}}$) and fulfills $50 < m_{jj} < 110$ GeV, $\chi^2 < 8$ is chosen. The light-jet energies are rescaled with m_W/m_{jj} and the two values of m_{jjb} are taken as mass estimator from each event (Fig. 4a). The background is modeled using an event mixing technique, where jets from $\geq$ 6-jet events are added to events with exactly 5 jets. The template fit to the m_{jjb} distribution yields $m_t = 174.9 \pm 2.1$ (stat)± 3.8 (syst) GeV and the total uncertainty is dominated by statistics, background modeling, ISR/FSR, and JES.

4.2. CMS: Measurement using the ideogram method [13]

The preselection requires events with at least four jets with $p_T > 60$ GeV, $|\eta| < 2.4$, at least five jets with $p_T > 50$ GeV, at least six jets with $p_T > 40$ GeV and at least two jets with b-tag and $p_T > 30$ GeV. A kinematic fit with the constraints $m_W = 80.4$ GeV and $m_t = m_{\bar{t}}$ is performed and the permutation with minimum χ^2 and $P_{\text{gof}} > 9\%, \Delta R_{bb} > 1.5$ is chosen. In 3.5 fb^{-1} of data 2418 events are selected with a composition of 51% $t\bar{t}$ and 49% multijet (Fig. 4b).

For modeling the multijet background new events are constructed from the preselected sample (minor $t\bar{t}$ contribution of 17%), where each jet is taken from a random event in the order of the jet p_T ranking. Only events containing at least two b-tagged jets are used.

The signal contribution contains 28% correct $t\bar{t}$ permutations. Templates are created for each permutation case and the ideogram method is employed to extract the top-quark

Table 1: Results of the top-quark mass extraction from the top pair production cross-section. The results are given for different cross-section calculations and mass schemes.

	ATLAS [14]	CMS [15]	
Approx. NNLO × MSTW08NNLO	m_t^{pole} / GeV	m_t^{pole} / GeV	$m_t^{\overline{\text{MS}}}$ / GeV
Langenfeld et al. [16]	$166.4^{+7.8}_{-7.3}$	$170.3^{+7.3}_{-6.7}$	$163.1^{+6.8}_{-6.1}$
Kidonakis [17]	$166.2^{+7.8}_{-7.4}$	$170.0^{+7.6}_{-7.1}$	–
Ahrens et al. [18]	$162.2^{+8.0}_{-7.6}$	$167.6^{+7.6}_{-7.1}$	$159.8^{+7.3}_{-6.8}$

mass result of $m_t = 173.49 \pm 0.69$ (stat) ± 1.25 (syst) GeV. The result is robust against radiation uncertainties and the leading uncertainty stems from the JES.

5. Extraction from the top pair production cross-section

The extraction of the top-quark mass from the production cross-section exhibits the usage of a well-defined mass definition and different systematic uncertainties. Cross-section measurements in all final states can be used as input.

The ATLAS measurement [14] is based on the $t\bar{t}$ cross-section measured in the lepton+jets channel with 35 pb^{-1}, while the CMS measurement [15] uses the $t\bar{t}$ cross-section measured in the dilepton channel with 1.1 fb^{-1}. The measured and predicted cross-sections are parametrized in dependency of m_t and the top-quark mass is extracted using a joint-likelihood approach. The evaluated uncertainties are: Experimental, PDF, scale variation and $\alpha_S(M_Z)$ (CMS). The results are summarized in Table 1.

6. Summary

Measurements of the top-quark mass at the LHC have reached a great precision, profiting from formidable detector performance and large samples of $t\bar{t}$ events that can be selected with low background contamination. In the future an even larger integrated luminosity will allow for tighter constraints on systematic uncertainties and will enable new measurement methods with reduced or complementary uncertainties.

References

[1] CDF and D0 Collaborations 2012 *Phys.Rev.* **D86** 092003 (*Preprint* 1207.1069)
[2] ATLAS Collaboration 2008 *JINST* **3** S08003
[3] CMS Collaboration 2008 *JINST* **3** S08004
[4] ATLAS Collaboration 2011 (*Preprint* 1112.6426)
[5] CMS Collaboration 2011 *JINST* **6** P11002 (*Preprint* 1107.4277)
[6] ATLAS Collaboration 2012 *Eur.Phys.J.* **C72** 2046 (*Preprint* 1203.5755)
[7] CMS Collaboration 2012 *JHEP* **1212** 105 (*Preprint* 1209.2319)
[8] CMS Collaboration 2012 *JHEP* **1206** 109 (*Preprint* 1204.2807)
[9] ATLAS Collaboration 2012 (ATLAS-CONF-2012-082)
[10] CMS Collaboration 2012 *Eur.Phys.J.* **C72** 2202 (*Preprint* 1209.2393)
[11] CMS Collaboration 2012 (CMS-PAS-TOP-11-027)
[12] ATLAS Collaboration 2012 (ATLAS-CONF-2012-030)
[13] CMS Collaboration 2012 (CMS-PAS-TOP-11-017)
[14] ATLAS Collaboration 2011 (ATLAS-CONF-2011-054)
[15] CMS Collaboration 2011 (CMS-PAS-TOP-11-008)
[16] Langenfeld et al 2009 *Phys.Rev.* **D80** 054009 (*Preprint* 0906.5273)
[17] Kidonakis N 2010 *Phys.Rev.* **D82** 114030 (*Preprint* 1009.4935)
[18] Ahrens et al 2010 *JHEP* **1009** 097 (*Preprint* 1003.5827)

Journal of Physics: Conference Series **452** (2013) 012025 doi:10.1088/1742-6596/452/1/012025

Precise top quark cross-section results at LHC

J. Andrea

23 rue du Loess, BP28, 67037 Strasbourg cedex 2

E-mail: jandrea@cern.ch

Abstract.
The measurements of the inclusive $t\bar{t}$ cross section is presented for proton-proton collisions recorded at $\sqrt{s}$ =7 and 8 TeV by the ATLAS and the CMS experiments using the lepton+jets and the dilepton events. The most precise single measurement at 7 TeV is $\sigma_{t\bar{t}} = 161.9 \pm 2.5(\text{stat.})^{+5.1}_{-5.0}(\text{syst.})\pm 3.6(\text{lumi.})$ pb, using a luminosity of 2.3 fb^{-1}. The measurements from the two experiments are also combined to reach higher precision. With a luminosity of about 1 fb^{-1}, the combined cross-section is $\sigma_{t\bar{t}} = 173.3 \pm 2.3(\text{stat.})\pm 9.8(\text{syst.})$ pb. At $\sqrt{s}$ =8 TeV, $\sigma_{t\bar{t}}$ is measured in the lepton+jets and dilepton channels with the CMS detector and the corresponding combined measurement is $\sigma_{t\bar{t}} = 227 \pm 3(\text{stat.})\pm 11(\text{syst.})\pm 10(\text{lumi.})$ pb.

1. Introduction

With the large $t\bar{t}$ cross sections $\sigma_{t\bar{t}}$ at the LHC and the high integrated luminosity, top physics enters into a new area. Despite the large backgrounds from multi-jets, typical of hadronic colliders, the proton-proton collisions data recorded at the LHC, together with highly performing detectors, such as the ATLAS and the CMS detectors, allows to measure the $t\bar{t}$ cross section to a high level of precision. Precise measurement of $\sigma_{t\bar{t}}$ is a very good test of the perturbative QCD, which leads to an approximate NNLO $t\bar{t}$ cross sections of 167^{+17}_{-18} pb and 252^{+27}_{-29} pb [1] at 8 TeV. Low uncertainty on the $t\bar{t}$ normalization is also important for the search for new physics, as possible deviations from the Standard Model (SM) predictions, or to reach a good control of the $t\bar{t}$ background in the exclusive searches for Beyond Standard Model signatures. The measurement of $\sigma_{t\bar{t}}$ with a very high level of precision at the LHC becomes possible because of large the statistic available, both in signal and control regions. It is used to estimate the various backgrounds first, but also to have a better understanding of systematic effects. One can reach even better precision by combining the different cross-section measurements. Top quarks are decaying into a bottom quark and a W boson in almost 100% of the cases. As the subsequent W can either decay into a lepton (anti-lepton) and its associated anti-neutrino (neutrino), or into a pair of quarks, one can observe three main decay channels for $t\bar{t}$ events : the dilepton channel (in about 10% of the cases), the lepton+jets channel (in about 45% of the cases) and all-hadronic channel (in about 45% of the cases). The most precise measurements are usually perform using $t\bar{t}$ events with at least one W boson decaying leptonically, with the charged lepton being either a muon or an electron, leptons from τ decays are normally included. Those channels are less contaminated by QCD multi-jets backgrounds, and allows to reach good trigger efficiencies. In addition, the missing transverse energy ($\not{E}_{\text{T}}$), can be used to further reject QCD multi-jets and electroweak backgrounds. Leptonic channels are also less sensitive to the description of jets, and thus conduct to lower uncertainties. The lepton+jets channel benefits from a higher statistic of signal events, but is affected by the QCD multi-jet and the W+jets backgrounds. The dileptonic channels produce less signal events, but it is well compensated by the high luminosity reached at the LHC, and by a lower background contamination.

2. Top pair cross section at $\sqrt{s}$=7 TeV in the lepton+jets channel

In this section, the ATLAS and CMS $t\bar{t}$ cross section measurements in the lepton+jets channel are described. They are performed at $\sqrt{s}$=7 TeV, using different integrated luminosities.

First, the ATLAS measurement [2] is performed using 0.70 fb^{-1} of proton-proton collisions recorder by the ATLAS experiment. It is using the electron and muon+jets channels. The event selection asks for the events to pass a single-lepton trigger and to contain exactly one reconstructed, isolated and identified lepton with $E_T > 25$ GeV for electrons ($p_T >$20 GeV for muons) and $|\eta| < 2.5$. In addition, events are required to contain at least 3 reconstructed jets with $p_T > 25$ GeV and $|\eta| < 2.5$, and to satisfy $\not{E}_T >$ 35 GeV for the electron+jets channel (25 GeV for the muon+jets), $m_T(W) > 25$ GeV and $\not{E}_T + m_T(W) > 60$ GeV, with $m_T(W)$ the transverse mass of W boson, determined using the lepton 4-momenta and the $\not{E}_T$. Lepton selection efficiencies in the simulation is corrected by comparing data and simulation for $Z \to l^+l^-$ events. The contamination of W+jets events is estimated from data using the charge asymmetry of the W events, and the QCD multi-jets contamination is determined using a Matrix Method. Other background events, like DY, single top-quark or dibosons events, are determined using simulations. $\sigma_{t\bar{t}}$ is determined using a maximum likelihood (profile) fit on a discriminating distribution, constructed from the pseudo-rapidity $|\eta|$ of the lepton, the p_T of the leading jet, the aplanarity (related to the "sphericity" of top events) and the $H_{T,3p}$ variable (related to the "transversity" of top events). The Likelihood used in the analysis is written as : $\mathcal{L}(\vec{\beta}, \vec{\delta}) = \prod_{k=1}^{120} \mathcal{P}(\mu_k, n_k) \times \prod_j G(\beta_j, \Delta_j) \times \prod_i G(\delta_i, 1)$, with the first term being a Poisson probability density to observed n_k events when μ_k is expected, the second term drives the normalization of signal and backgrounds events using Gaussian distributions and the third term introduces systematic uncertainties as nuisance parameters. In some other measurements presented in this note, similar techniques are used. Various systematic sources are considered in the measurement, the dominant ones being related to the uncertainty on the integrated luminosity, to the $t\bar{t}$ modeling and to the selections of jets and leptons. After the fit, the likelihood discriminant distribution can be seen on figure 1.

The measured cross section is found to be $\sigma_{t\bar{t}} = 179.0 \pm 3.9$(stat.)$\pm$9.0(syst.)$\pm$6.6(lumi.) pb. The dependence of the cross section on the top-quark mass is $\sigma_{t\bar{t}} = 411.9 - 1.35 \times m_{top}(GeV)$ pb. Another ATLAS measurement [3] uses a luminosity of 4.66 fb^{-1} and, in addition to the previously described event selection, uses the semi-muonic decay of b-hadrons in order to identified jets originating from b quarks. A detailed study of such soft-muon tagger performances is done in order to have a good control of the related systematic uncertainties. $\sigma_{t\bar{t}}$ is measured from a counting experiment, and found to be $\sigma_{t\bar{t}} = 165 \pm 2$(stat.)$\pm$17(syst.)$\pm$3(lumi.) pb.

Within the CMS experiment, $\sigma_{t\bar{t}}$ in the lepton+jets channel is measured using integrated luminosities of 0.8 to 1.1 fb^{-1} [4], depending on the channel. The events passing single lepton triggers are required to contain exactly one isolated and identified electrons with $p_T > 45$ GeV (35 GeV) and $|\eta| <2.5$($|\eta| <2.1$) for electrons (respectively for muons), at least one jet with $p_T >$30 GeV and $|\eta| < 2.4$, possibly identified as a b-jet, and $\not{E}_T >$30 GeV for the electron+jets channel ($\not{E}_T >$20 GeV for the muon+jets). Events with another loosely isolated lepton are rejected. The trigger and lepton selection efficiencies are determined using Z events from data, and used to correct the simulation. The cross section measurement is based on a profile likelihood fit of the multidimensional distribution of the jet multiplicity N_{jet}, the b-tagged jet multiplicity N_{b-tag} and the possibly reconstructed secondary vertices mass M_{SV}. The secondary vertices reconstructed within jets allow to discriminated light-flavor quarks and

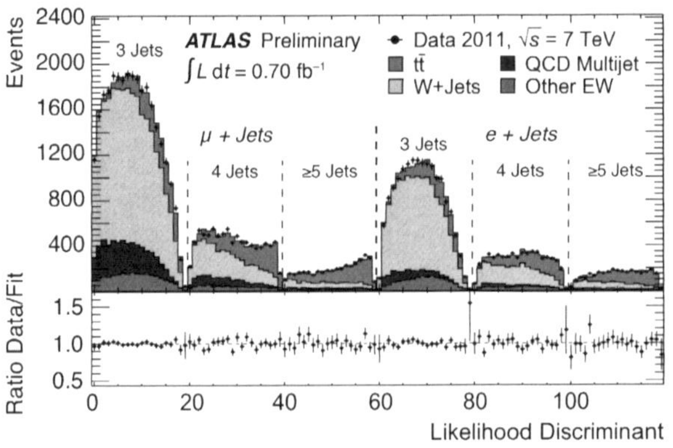

Figure 1. Result of combined fit to data in the exclusive three-jet bin, the exclusive four-jet bin and the inclusive five-jet bin of the electron+jets and muon+jets channels. The lower plot shows the ratio of data to the sum of fitted signal and background contributions.

gluon jets against b-quark jets. The likelihood fit treat the main systematic sources as nuisance parameters, and the normalization of almost all the background processes are free parameters. The QCD multi-jets is however determined using a control sample enriched in multi-jets events. The main sources of systematic uncertainties are the luminosity, the jet energy scale and resolution (JES/JER), the lepton selection and the description of Parton Density Functions (PDF). After combining the electron+jets and muon+jets channel, the $t\bar{t}$ cross section is found to be $\sigma_{t\bar{t}} = 164.4 \pm 2.8(\text{stat.}) \pm 11.9(\text{syst.}) \pm 7.4(\text{lumi.})$ pb.

3. Top pair cross section at $\sqrt{s}$=7 TeV in the dilepton channel

In this section, the ATLAS and CMS $t\bar{t}$ cross-section measurements in the dilepton channel are described. They are performed at $\sqrt{s}$=7 TeV using different integrated luminosities.

For the ATLAS measurement [5], the integrated luminosity used is 0.7 fb^{-1}. After selecting a dataset based on single lepton trigger, the event selection asks for 2 reconstructed, isolated and identified electrons ($p_T > 25$ GeV, $|\eta| < 2.5$) and muons ($p_T > 20$ GeV, $|\eta| < 2.5$) with opposite charges. Another selection requires the presence of a selected lepton and an isolated tracks with opposite charges. In addition, the selected events are required to contain at least 2 jets with p_T >25 GeV and $|\eta| < 2.5$, possibly b-tagged. For the di-electron and the di-muon channels, the dilepton invariant mass m_{ll} is required to be > 15 GeV, and $\not\!\!E_{\rm T}$> 60 GeV (40 GeV in case of b-tagged jets). To further remove events containing a Z boson a cut on the dilepton invariant mass is also applied, such as $|m_{ll}$-$m_Z| > 10$ GeV, with m_Z the mass of the Z boson. For the electron-muon channel, the sum of p_T of the selected jets is require to be above 130 GeV. The trigger and lepton selections efficiencies are determined using Z events from data. The DY background is estimated using a control sample enriched in Z mass. The $t\bar{t}$ (lepton+jets) and W+jets events are estimated from data using an extended Matrix Method. The contamination of the other sub-dominant backgrounds, such as single top-quark, Z→ $\tau\tau$ and dibosons events, are determined from simulations. The main systematics sources are coming from the luminosity, the jet and $\not\!\!E_{\rm T}$ selections and the modeling of signal events. After combining the dilepton channels, the $t\bar{t}$ cross section is measured to be $\sigma_{t\bar{t}} = 176 \pm 5(\text{stat.})^{+14}_{-11}(\text{syst.}) \pm 8(\text{lumi.})$ pb.

Within CMS, $\sigma_{t\bar{t}}$ is measured using an integrated luminosity of 2.3 fb^{-1} [6]. The dataset used is based on dilepton triggers. The event selection requires the presence of 2 reconstructed, isolated and identified leptons with opposite charges, with p_T >20 GeV and $|\eta|$ <2.5 for electrons

($|\eta|$ <2.1 for muons). The following requirements are applied in order to remove DY events : m_{ll} >20 GeV, $|m_{ll}$-$m_Z|$ > 15 GeV and $\not{E}_T$>40 GeV. The last two cuts are applied only on the dielectron and dimuon channels. In addition, only events with at least 2 jets (possibly b-tagged), with p_T >30 GeV and $|\eta|$ <2.5, are used to performed the measurement. Dilepton trigger efficiencies are estimated using an independent sample triggered by the $\not{E}_T$. The lepton selection efficiencies are estimated from Z$\rightarrow$ ll events and the $\not{E}_T$ selection efficiency is estimated from $t\bar{t}$ signal events in the electron-muon channel. $\sigma_{t\bar{t}}$ is measured using a profile likelihood ratio fit on the 2-dimensional distribution of the jet and b-tagged jet multiplicities, with the systematic sources treated as nuisance parameters. A counting analysis is also performed as a cross-check. The Z$\rightarrow$ ll background contamination in the dielectron and dimuon channels is estimated using the expected ratio of events outside and inside the Z mass cut. The Z$\rightarrow \tau\tau \rightarrow e\mu$ background contamination in the electron-muon channel is estimated from a template fit of the m_{ll} distribution. The number of events with at least one non-prompt lepton (not coming from γ^*/Z and W decays) is estimated using an extended Matrix Method. The contamination of other backgrounds is estimated from simulations. Data-to-simulation comparisons of the b-tagged jet multiplicities are shown in figure 2.

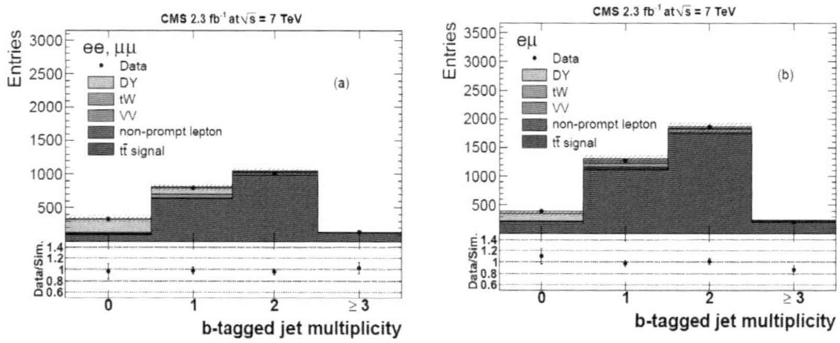

Figure 2. The multiplicity of b-tagged jets in events passing full event selections for (a) the summed ee and $\mu - \mu$ channels, and (b) the $e\mu$ channels.

The main systematic sources are coming from the uncertainty on JES and on the leptonic branching ratio of W bosons. After combining the dileptonic channels, the $t\bar{t}$ cross section is measured to be $\sigma_{t\bar{t}} = 161.9 \pm 2.5(\text{stat.})^{+5.1}_{-5.0}(\text{syst.}) \pm 3.6(\text{lumi.})$ pb. This corresponds to the most precise $t\bar{t}$ cross section measurement at the LHC. The dependence of $\sigma_{t\bar{t}}$ on the mass of the top-quark is found to be $\sigma_{t\bar{t}}/\sigma_{t\bar{t}}(m_t = 172.5) = 1.00 - 0.005 \times (m_t - 172.5) - 0.000137 \times (m_t - 172.5)^2$. This precise measurement of the $t\bar{t}$ cross sections allows to estimate other quantities, like the strong coupling constant α_S [7].

4. Combinations of top pair cross sections at $\sqrt{s}$=7 TeV

A way to further increase the precision on $\sigma_{t\bar{t}}$ is to combined statistically uncorrelated measurements. The combination of the different $\sigma_{t\bar{t}}$ measurements within the same experiment are already performed within the ATLAS [8] and the CMS [9] collaborations. A first combination of the LHC results [10],[11] is provided by the TOP-LHC working group with 0.7 to 1.1 fb^{-1}, thus without including the last ATLAS and CMS measurements. For sake of simplicity, the individual ATLAS and CMS combinations are used as input to a Best Linear Unbiased Estimator (BLUE [12]) method. The BLUE method was found to give similar results than more complex methods based on likelihood fits. To be able to perform the combination, the different systematic sources and their correlations are compared among the two experiments. The systematic sources coming from the detector modeling, Jet Energy Scale and background determination from data are assumed to be uncorrelated. Those coming from the signal and backgrounds modeling, taken from simulations, are assumed to be fully correlated. Special care is ported on the systematics from the luminosity, as the uncertainty from the bunch charges is considered as fully correlated, while the uncertainty related the detectors (which are different) is taken as uncorrelated. The combination of ATLAS and CMS cross-section measurements is found to be $\sigma_{t\bar{t}} = 173.3 \pm 2.3(\text{stat.}) \pm 9.8(\text{syst.})$ pb. After combination, the uncertainty on $\sigma_{t\bar{t}}$ becomes 5.8%, with an observed gain of 7% with respect to the most precise individual measurement.

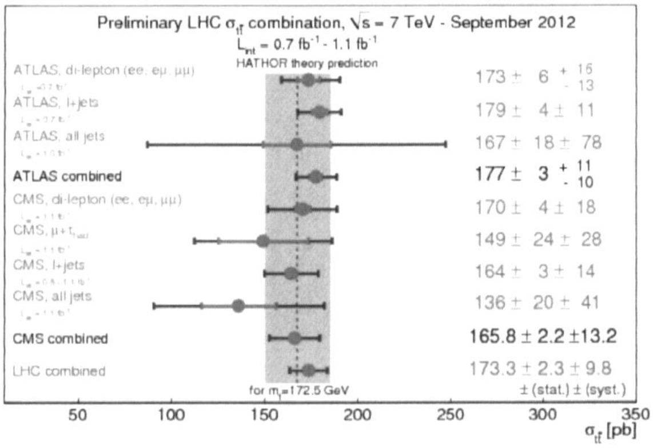

Figure 3. Input $\sigma_{t\bar{t}}$ measurements by the ATLAS and CMS collaborations and the result of the LHC combination. The band corresponds to the approximate NNLO in QCD calculation.

5. Top pair cross sections and their combinations at $\sqrt{s}$=8 TeV with the CMS detector

The $t\bar{t}$ cross section is also measured at $\sqrt{s} = 8$ TeV with the CMS detector in the dilepton [13] and the lepton+jets [14] channels, using luminosities between 2.4 to 2.8 fb^{-1}. The ratio of $\sigma_{t\bar{t}}$ at 7 and at 8 TeV is also calculated in the dilepton channel. $\sigma_{t\bar{t}}$ in the lepton+jets channel is measured after an event selection similar to the one presented in section 2. Instead of performing a profile likelihood fit of the multidimensional distribution of the N_{jet}, N_{b-tag} and M_{SV}, the $t\bar{t}$ cross section is determined by requiring the selected events to contain at least one b-tagged jet and by performing a binned likelihood fit of the invariant mass of the lepton and the closest b-tagged jet M_{lb}. A cross-check analysis is performed using the M3 variable (invariant mass of the combination of 3 jets which has the highest sum p_T). The measured cross section is found to be $\sigma_{t\bar{t}} = 228.4 \pm 9.0(\text{stat.})^{+29.0}_{-26.0}(\text{syst.}) \pm 10.0(\text{lumi.})$ pb. In the dilepton channel, $\sigma_{t\bar{t}}$ is measured with a similar event selection than the one presented in section 3. The measurement is based on a counting experiment using the number of observed event after requiring at least one b-tagged jet in the events. The measured cross section is found to be $\sigma_{t\bar{t}} = 227 \pm 3(\text{stat.}) \pm 11(\text{syst.}) \pm 10(\text{lumi.})$ pb. The combination of these cross sections using a BLUE method is found to be highly dominated by the dilepton channel, which has a very good precision, and leads to the same cross-section value. The ratio of the 8 TeV (combination) and the 7 TeV cross sections is measured to be 1.41±0.10.

6. Summary

The high luminosities provided by the LHC allows to perform high precision measurement of the $t\bar{t}$ cross section. All measurements are agreeing and are dominated by systematics uncertainties. For the same luminosity (about 1 fb^{-1}) ATLAS and CMS measurements have similar precisions, with total uncertainties of 11.4 pb (6.4%) for ATLAS and 14.4 pb (8.7%) for CMS. The combination of these measurements allows to be to an uncertainty as low as 10 pb. With higher luminosity, very high precision can be reached : 7 pb (< 5%) in the dilepton channel measured with the CMS detector. Similar precisions can be reached with the 8 TeV data, despite the higher multiplicities of pileup events. Finally, the measurements of the $t\bar{t}$ cross section start to be more accurate than the approximate NNLO calculations.

References

[1] M. Aliev et al., "HATHOR: HAdronic Top and Heavy quarks cross section calculatoR.", Comput. Phys. Commun. **182** (2011) 1034, doi:10.1016/j.cpc.2010.12.040, arXiv:1007.1327.
[2] ATLAS Collaboration, "Measurement of the $t\bar{t}$ production cross-section in pp collisions at $\sqrt{s}$=7 TeV using kinematic information of lepton+jets events.", ATLAS-CONF-2011-121.
[3] ATLAS Collaboration, "Measurement of the top quark pair production cross section with ATLAS in pp collisions at $\sqrt{s}$=7 TeV in the single-lepton channel using semi-leptonic b decays", ATLAS-CONF-2012-131.
[4] The CMS Collaboration,"Measurement of the $t\bar{t}$ production cross section in pp collisions at $\sqrt{s}$=7 TeV with lepton + jets final states", arXiv:1212.6682.
[5] The ATLAS Collaboration, "Measurement of the cross section for top-quark pair production in pp collisions at $\sqrt{s}$=7 TeV with the ATLAS detector using final states with two high-pT leptons", *JHEP* **07** (2011) 059.
[6] S. Chatrchyan et al. "Measurement of the $t\bar{t}$ production cross section in the dilepton channel in pp collisions at $\sqrt{s}$=7 TeV, *JHEP* **11** (2012) 067.
[7] The CMS Collaboration, "First Determination of the Strong Coupling Constant from the ttbar Cross Section", CMS PAS TOP-12-022.
[8] The ATLAS Collaboration, "Statistical combination of top quark pair production cross-section measurements using dilepton, single-lepton, and all-hadronic final states at $\sqrt{s}$=7 TeV with the ATLAS detector", ATLAS-CONF-2012-024.
[9] The CMS Collaboration, "Combination of top pair production cross section measurements", CMS PAS TOP-11-024.
[10] The ATLAS Collaboration, "Combination of ATLAS and CMS top-quark pair cross section measurements using up to 1.1 fb^{-1} of data at 7 TeV", ATLAS-CONF-2012-134.
[11] The CMS Collaboration, "Combination of ATLAS and CMS top-quark pair cross section measurements using proton-proton collisions at $\sqrt{s}$=7 TeV", CMS PAS TOP-12-003.
[12] L. Lyons, D. Gibaut, and P. Clifford, "How to combine correlated estimates of a single physical quantity", *Nucl. Instrum. Meth. A* **270** (1988) 110, doi:10.1016/0168-9002(88)90018-6.
[13] The CMS Collaboration, "Top pair cross section in e/mu+jets at 8 TeV", CMS PAS TOP-12-006.
[14] The CMS Collaboration, "Measurement of the $t\bar{t}$ production cross section in the dilepton channel in pp collisions at $\sqrt{s}$=8 TeV", CMS PAS TOP-12-007.

Precision top pair production at hadron colliders

Michał Czakon[1]
Institut für Theoretische Teilchenphysik und Kosmologie, RWTH Aachen University, D-52056 Aachen, Germany
E-mail: mczakon@physik.rwth-aachen.de

Alexander Mitov
Theory Division, CERN, CH-1211 Geneva 23, Switzerland
E-mail: alexander.mitov@cern.ch

Abstract. We demonstrate the impact of recent NNLO calculations on precision top quark phenomenology.

1. Introduction

This writeup was inspired by three recent developments in collider physics. First, a Higgs-like boson was discovered at CERN [1, 2], yet its precise identity remains unclear at present. Second, the otherwise widely appreciated connection between the top quark and Standard Model's Higgs boson was further strengthened in Refs. [3, 4]. Finally, new results in precision top pair production appeared [5, 6, 7], driving the precision in top physics to a quantitatively new level. In the following we give two examples that illustrate the power of such precision results.

2. Tevatron: a comparison between alternative resummation approaches

Circa early 2012, the best theoretical precision in top pair production was achieved through either NNLL threshold resummation [8, 9, 10] on top of the NLO results or through approximate NNLO [11] calculations based on truncation of the resummed NNLL results. A number of phenomenological studies appeared [10, 12, 13, 14, 15, 16, 17, 18, 19] with numerical predictions that showed a significant spread.

The main reason behind such a spread was identified in Ref. [19] as due to missing NNLO corrections that are not enhanced at threshold. Having the exact NNLO results for all partonic channels but the gg-initiated one, one can try to validate this assertion. As a test case we take the results of Refs. [19] and [18], both including NNLL soft-gluon resummation matched through approximate NNLO. Both results resum directly the total inclusive cross-section. [2] There are two main differences between these two calculations: different resummation approaches (in N-

[1] speaker

[2] It is possible [9] to arrive at the resummed total inclusive cross-section by integration over the resummed differential distribution, as was done in [10, 14, 15, 16] . Different power suppressed terms are generated in this case.

and x-spaces, respectively) as well as different choices for the unknown two loop hard matching constants.

The completion of the exact NNLO calculation for the Tevatron [5] allowed the extraction of the relevant two-loop hard matching constant and the fixing of all dominant two-loop power suppressed terms. Upon updating the results of Refs. [19] and [18] to promote them to full NNLO+NNLL accuracy (at the Tevatron) it was established that the spread of the predictions decreased dramatically, see table 1. Such a comparison can be easily performed with the help

	N-space	x-space
approximate NNLO + NNLL	$6.72^{+0.24}_{-0.41}$ [pb]	$7.22^{+0.29}_{-0.46}$ [pb]
exact NNLO + NNLL	$7.07^{+0.14}_{-0.23}$ [pb]	$7.15^{+0.21}_{-0.20}$ [pb]

Table 1. Comparison between resummed predictions in N- and x-space matched to approximate or exact NNLO.

of the programs `Top++` [20] and `TOPIXS` [21], respectively, and it demonstrates that indeed the spread in the numerical predictions is not (mainly) due to the different resummation formalisms. A similar comparison has been performed in Ref. [21]. It will be very interesting to perform such a comparison also at the LHC, once the complete gg-initiated NNLO correction becomes available.

3. Examining the quality of an approximation based on the high-energy limit of the cross-section

The variety of NNLO approximations for the total inclusive top pair cross-section was extended in Ref. [22] to include its high-energy limit, where the partonic energy s is much larger than the mass of the top quark, i.e. $\rho \equiv 4m^2/s \approx 0$. It is well understood [23, 24, 25, 26, 27, 28] that in this limit the cross-section exhibits logarithmic divergence:

$$\sigma_{\rm tot} \approx c_1 \ln(\rho) + c_0 + \mathcal{O}(\rho)\,. \tag{1}$$

The analytical result for the constant c_1 in all partonic channels can be found in Ref. [29].

In Ref. [22] elegant arguments were given that allowed the authors to predict approximate numerical values for the constant c_0 in all partonic channels. It was noted there that the value of the constant is important for stabilizing the behavior of the high-energy approximation. Subsequently, the constants for the $q\bar{q}$ and qg channels were computed in Refs. [6, 7] and found to agree with the prediction of Ref. [22], within the numerical uncertainties. Two different applications of the high-energy approximation have emerged in the literature since then; we discuss them in the following.

3.1. Application 1: Devising an approximation to the exact result

The authors of Ref. [22] argue that the high-energy limit of the cross-section, combined with its known threshold behavior, can be used to provide an approximation that is valid in the full kinematical region. In particular, they also argue that it can be used for improved phenomenological predictions for top pair production at hadron colliders.

It was first pointed out in Ref. [30] that theoretical predictions including (or not) the approximation of Ref. [22] differ about 7% with respect to each other. The source of this difference was later investigated in Ref. [7] where the approximate result of Ref. [22] was compared with the exact NNLO result in the qg initiated reaction. The shift in the hadronic cross-section at the LHC at 7 and 8 TeV due to the qg reaction alone was found to be about 5%.

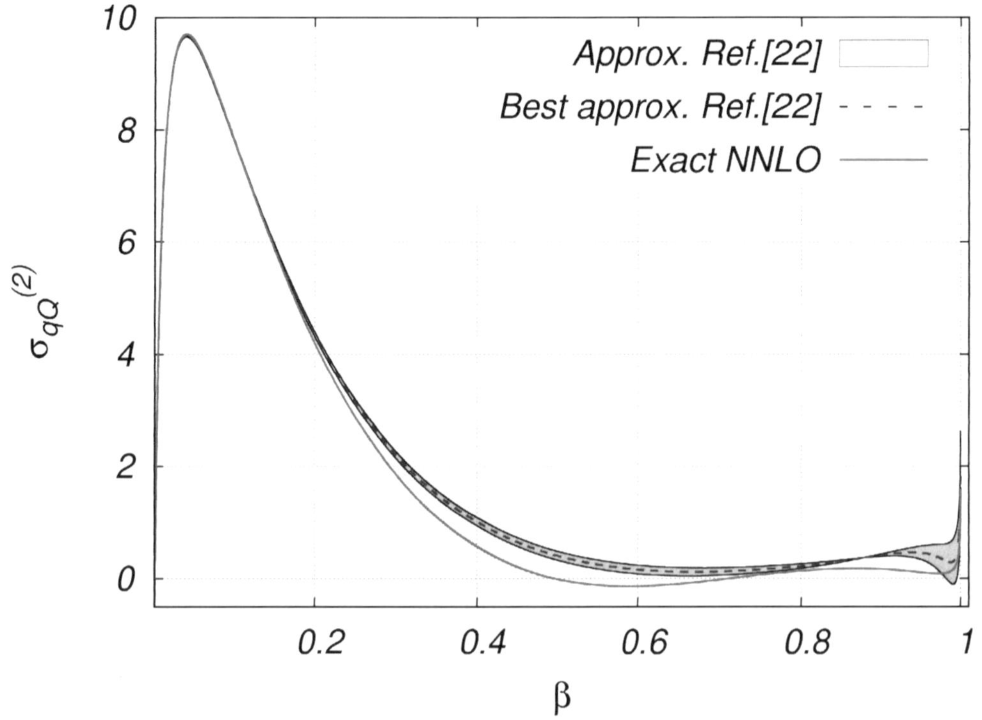

Figure 1. Various NNLO corrections to the partonic cross-section: the exact NNLO result (red) and the approximation of Ref. [22] (black band; best prediction is in blue).

Such a difference is numerically very significant since the overall contribution of the qg reaction at NNLO is only around 1%.

The conclusion in Ref. [7] was that the shape of the approximation [22] in the mid-region is qualitatively different from the exact result. This region contributes significantly to the total cross-section for typical LHC c.m. energies due to the shape of the parton fluxes. Further arguments why such an approximation may not describe adequately the mid-range region can be deduced from the exactly known NLO result: it was shown in Ref. [31] that power expansions of the cross-section, around either the threshold or the high-energy limit are not well convergent and cannot describe the mid-region even after the inclusion of a number of powers. In the following we investigate the quality of the approximation of Ref. [22] also in the $q\bar{q}$ reaction, which is now also known at NNLO [5, 6].

Similarly to the qg-initiated reaction [7], from fig. 1 we conclude that in the mid-region the approximation of the partonic cross-section deviates noticeably with respect to the exact result. The significance of this difference becomes even more pronounced once the partonic cross-sections are multiplied by the respective partonic flux, see fig. 2.

It is also easy to check that the contribution to the total cross-section at the Tevatron due to the the difference between the exact NNLO and the best prediction of Ref. [22] leads to a shift of around 0.14 [pb] (a relative shift in the total cross-section of about 2%), which is similar in size to the current scale variation at this collider.

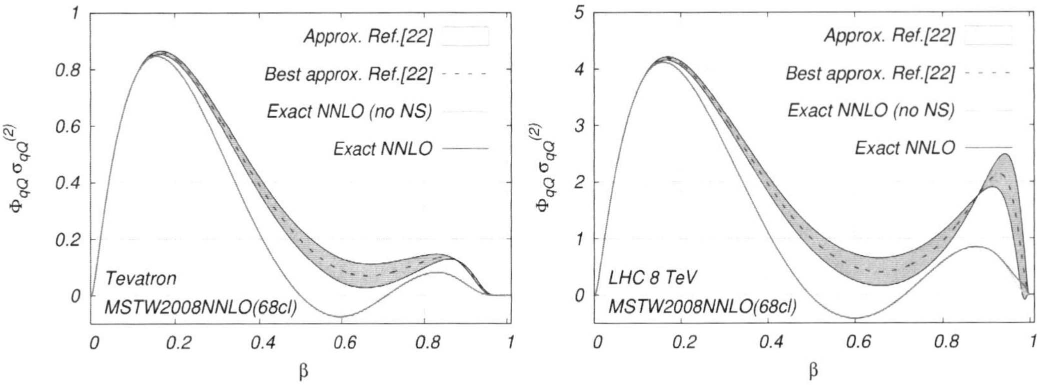

Figure 2. Various NNLO corrections to the partonic cross-section times $q\bar{q}$ partonic flux at Tevatron (left) and LHC 8 TeV (right): exact NNLO result (red), exact NNLO excluding the non-singlet correction (NS) [6] that contributes all the high-energy rise (green), and the approximation of Ref. [22] (black band; best prediction is in blue).

3.2. Application 2: Improving the limiting behavior of the full (numeric) result

The leading high-energy behavior Eq. (1) is utilized in Refs. [7, 6] in the following way. The calculation of the NNLO partonic cross-sections [5, 6, 7] is numeric, on a grid of finite number of points (typically around 80). This implies that one cannot calculate the partonic cross-section at, or close to, the two endpoints (threshold and high-energy limits) since both points are singular and therefore any numerical calculation will inevitably break there. It is in these regions, between the two kinematical endpoints and the lowest (highest) numerically computed point, where the knowledge of the asymptotic behavior proves to be very useful: one requires the fit of the numerical result to agree in both limits with the analytically known expansions, while the rest is covered by the numerical calculation which is reliable away from the two endpoints.

As explained in Ref. [7] the error that one produces in this inherently numerical approach is very small and phenomenologically irrelevant for productions of top quarks at any near-future hadron collider. Applications at the LHC for very light quarks - like bottom - are likely also possible, but the errors have to be re-assessed based on the criterion given in Ref. [7].

4. Conclusions

The level of precision in the description of top pair production at hadron colliders has increased dramatically in the last one year. The NNLO results for all partonic channels but the gg-initiated one have been calculated, and implemented in precision phenomenological studies. Comparing these exact result with various approximations we conclude that the mid-region behavior of the partonic cross-sections is very important and cannot be deduced from any available partial result or approximation.

Acknowledgments

We would like to thank the organizers for creating such a wonderful scientific atmosphere. We would also like to thank Sven Moch for very interesting discussion that triggered some of the studies presented in this proceedings.

References

[1] G. Aad *et al.* [ATLAS Collaboration], Phys. Lett. B [arXiv:1207.7214 [hep-ex]].
[2] S. Chatrchyan *et al.* [CMS Collaboration], Phys. Lett. B [arXiv:1207.7235 [hep-ex]].

[3] G. Degrassi, S. Di Vita, J. Elias-Miro, J. R. Espinosa, G. F. Giudice, G. Isidori and A. Strumia, JHEP **1208**, 098 (2012) [arXiv:1205.6497 [hep-ph]].
[4] M. L. Mangano and J. Rojo, JHEP **1208**, 010 (2012) [arXiv:1206.3557 [hep-ph]].
[5] P. Baernreuther, M. Czakon and A. Mitov, Phys. Rev. Lett. **109**, 132001 (2012) [arXiv:1204.5201 [hep-ph]].
[6] M. Czakon and A. Mitov, arXiv:1207.0236 [hep-ph].
[7] M. Czakon and A. Mitov, arXiv:1210.6832 [hep-ph].
[8] M. Beneke, P. Falgari and C. Schwinn, Nucl. Phys. B **828**, 69 (2010) [arXiv:0907.1443 [hep-ph]].
[9] M. Czakon, A. Mitov and G. F. Sterman, Phys. Rev. D **80**, 074017 (2009) [arXiv:0907.1790 [hep-ph]].
[10] V. Ahrens, A. Ferroglia, M. Neubert, B. D. Pecjak and L. L. Yang, JHEP **1009**, 097 (2010) [arXiv:1003.5827 [hep-ph]].
[11] M. Beneke, M. Czakon, P. Falgari, A. Mitov and C. Schwinn, Phys. Lett. B **690**, 483 (2010) [arXiv:0911.5166 [hep-ph]].
[12] U. Langenfeld, S. Moch and P. Uwer, Phys. Rev. D **80**, 054009 (2009) [arXiv:0906.5273].
[13] M. Beneke, P. Falgari, S. Klein and C. Schwinn, Nucl. Phys. Proc. Suppl. **205-206** (2010) 20 [arXiv:1009.4011 [hep-ph]].
[14] N. Kidonakis, Phys. Rev. D **82** (2010) 114030 [arXiv:1009.4935 [hep-ph]].
[15] V. Ahrens, A. Ferroglia, M. Neubert, B. D. Pecjak and L. -L. Yang, JHEP **1109**, 070 (2011) [arXiv:1103.0550 [hep-ph]].
[16] V. Ahrens, A. Ferroglia, M. Neubert, B. D. Pecjak and L. L. Yang, Phys. Lett. B **703**, 135 (2011) [arXiv:1105.5824 [hep-ph]].
[17] N. Kidonakis and B. D. Pecjak, Eur. Phys. J. C **72**, 2084 (2012) [arXiv:1108.6063 [hep-ph]].
[18] M. Beneke, P. Falgari, S. Klein and C. Schwinn, Nucl. Phys. B **855**, 695 (2012) [arXiv:1109.1536 [hep-ph]].
[19] M. Cacciari, M. Czakon, M. Mangano, A. Mitov and P. Nason, Phys. Lett. B **710**, 612 (2012) [arXiv:1111.5869 [hep-ph]].
[20] M. Czakon and A. Mitov, arXiv:1112.5675 [hep-ph].
[21] M. Beneke, P. Falgari, S. Klein, J. Piclum, C. Schwinn, M. Ubiali and F. Yan, JHEP **1207**, 194 (2012) [arXiv:1206.2454 [hep-ph]].
[22] S. Moch, P. Uwer and A. Vogt, Phys. Lett. B **714**, 48 (2012) [arXiv:1203.6282 [hep-ph]].
[23] P. Nason, S. Dawson and R. K. Ellis, Nucl. Phys. B **303**, 607 (1988).
[24] S. Catani, M. Ciafaloni and F. Hautmann, Phys. Lett. B **242**, 97 (1990).
[25] J. C. Collins and R. K. Ellis, Nucl. Phys. B **360**, 3 (1991).
[26] S. Catani, M. Ciafaloni and F. Hautmann, Nucl. Phys. B **366**, 135 (1991).
[27] S. Catani, M. Ciafaloni and F. Hautmann, Phys. Lett. B **307** (1993) 147.
[28] S. Catani and F. Hautmann, Nucl. Phys. B **427**, 475 (1994) [hep-ph/9405388].
[29] R. D. Ball and R. K. Ellis, JHEP **0105**, 053 (2001) [hep-ph/0101199].
[30] [CMS collaboration] CMS note CMS-PAS-TOP-12-022.
[31] M. Czakon and A. Mitov, Nucl. Phys. B **824**, 111 (2010) [arXiv:0811.4119 [hep-ph]].

Journal of Physics: Conference Series **452** (2013) 012027 doi:10.1088/1742-6596/452/1/012027

Measurement of Differential $t\bar{t}$ Cross Sections at 7 TeV

Martin Görner for the CMS Collaboration
Phd Student, Universität Hamburg, Luruper Chaussee 149, 22761 Hamburg, Germany
E-mail: martin.goerner@desy.de

Abstract. A measurement is presented of the normalized differential cross sections of top-quark pair production in pp collisions at a center of mass energy of 7 TeV, using 1.1 fb^{-1} of data collected in 2011 by the CMS experiment. The measurement is performed using five different final states with one or two muons or electrons in the final state. The event selections yield high purity $t\bar{t}$ samples. Kinematic reconstructions are performed to obtain full information of the top quarks, the $t\bar{t}$ system and the final state leptons. The results are compared to several QCD predictions up to next-to-leading order.

1. Introduction

The production cross section of top-quark pairs ($t\bar{t}$) is measured as a function of several top quark related quantities, utilizing final states with leptons (ee, $\mu\mu$, eμ, e+jets, μ+jets). Such measurements are a precision test of perturbative quantum chromodynamics (pQCD) and an important check for searches beyond the standard model (SM) where $t\bar{t}$ is a major background. Furthermore, they are sensitive to phenomena beyond the SM such as the production of a massive Z-like vector boson that could manifest itself as a resonance in the $t\bar{t}$ mass spectrum. This measurement is performed using 1.1 fb^{-1} data recorded by the CMS experiment [1] in 2011. All details are discussed in Ref. [2].

2. Event Selection and Background Estimation

The event selection is optimized to obtain a high-purity $t\bar{t}$ sample. To identify events of interest, single lepton or dilepton triggers are required for the corresponding decay channel.
Exploiting the event topology, events are required to contain isolated leptons in the final state. Two opposite sign leptons ($p_T > 20$ GeV, $|\eta| < 2.4$) are required in the dileptonic channels and exactly one lepton ($p_T > 30$ GeV, $|\eta| < 2.1$) in the e/μ+jets channels. Events with additional loosely isolated leptons are vetoed for the e/μ+jets channels while events with invariant dilepton masses smaller than 12 GeV are vetoed for the dileptonic channels to suppress QCD multijet background.
Furthermore, the events are required to contain at least two jets for the dileptonic channels and at least four jets for the e/μ+jets channels. Jets are clustered from particles reconstructed by a particle flow algorithm [3] using the anti-k_t algorithm with R=0.5 [4] and are selected if they have a transverse momentum $p_T > 30$ GeV and pseudo-rapidity $|\eta| < 2.4$. Additionally, at least one jet in the dilepton channels and at least two jets in the e/μ+jets channels are required to originate from bottom quarks. To identify them, an algorithm based on the reconstruction of

secondary vertices is used [5].
To further reduce background stemming from Drell-Yan processes, events in the ee and $\mu\mu$ channels are required to contain a minimal missing transverse energy of 30 GeV and the invariant mass of the two leptons is required to be outside a ± 15 GeV window around the Z-boson mass. For the eμ and e/μ+jets channels, the selected samples are expected to contain at least 90% $\mathrm{t\bar{t}}$ events and events with single top quarks are the largest background.
In the ee and $\mu\mu$ channels, the fraction of $\mathrm{t\bar{t}}$ events is expected to be above 80%. Leptonically decaying Z bosons accompanied by additional jets in the final state constitute the main background and are estimated from data by extrapolating the measured number of events within the veto region around the Z-boson mass into the signal region.

3. Kinematic Event Reconstruction

In order to fully reconstruct the event topology, the final-state objects (jets, leptons, and neutrinos) have to be correctly associated to their parent top quark. The measured missing transverse energy is assumed to reflect the transverse component of the neutrino momentum. Kinematic constraints, e.g. the reconstructed W-boson masses and equal reconstructed top-quark masses, are used in order to compensate for the unmeasured z component of the neutrino momentum and to build an event hypothesis.
In the e/μ+jets channels a kinematic fit is performed by varying the four-momenta of the final state objects to fulfill the constraints [6]. The five leading jets are considered to construct a hypothesis, using the b-tag information to assign jets as b jets from the t$\rightarrow$bW decay. In case of several hypotheses, the one with the lowest variation concerning the object resolution (minimum χ^2) is chosen.
Due to the underconstrained system in the dileptonic final state with two neutrinos, additional constraints are needed [7]. A set of hypotheses is created by fixing the top-quark mass and performing a scan from 100 to 300 GeV in 1 GeV steps. Hypotheses containing b-tagged jets are preferred. Finally, the hypothesis with the most probable reconstructed neutrino energy in comparison to a reference spectrum from simulation is chosen.

4. Normalized Differential Cross Sections

In order to quantify migration effects, the two quantities purity (p) and stability (s) are used. Purity (stability) is the number of particles generated and correctly reconstructed in a certain bin i divided by the total number of reconstructed (generated) particles in the same bin:

$$\mathrm{p} = \frac{\mathrm{N^i_{gen\&rec}}}{\mathrm{N^i_{rec}}} \quad ; \quad \mathrm{s} = \frac{\mathrm{N^i_{gen\&rec}}}{\mathrm{N^i_{gen}}}$$

A value of one corresponds to no migration. For each variable and decay channel, the binning is optimized independently to limit the migration effects to a reasonable level (s and p $> 0.4-0.5$). For the kinematic quantities of the final state leptons, migration effects are typically less pronounced (s and p > 0.9).
The differential cross sections are derived for every bin i from the number of measured events in data ($\mathrm{N^i_{data}}$) by subtracting the background ($\mathrm{N^i_{BG}}$) and dividing by the binwidth (Δ^i), the integrated luminosity (L), the efficiency (ϵ^i), and the in situ measured inclusive cross section in the same phase space σ_PS:

$$\frac{1}{\sigma}\frac{d\sigma^i}{dX} = \frac{1}{\sigma_\mathrm{PS}}\frac{\mathrm{N^i_{data}} - \mathrm{N^i_{BG}}}{\Delta^\mathrm{i} \quad \epsilon^\mathrm{i}\,\mathrm{L}}$$

The measurement is performed in the visible parton level phase space. Therefore, the efficiency, which includes migration corrections, is defined concerning events that contain leptons and quarks within the same kinematic region (p_T and $|\eta|$) as used for the event selection. This

avoids extrapolation into an unmeasured region of the phase space.
Dividing by the inclusive cross section normalizes the differential distributions to unit area and helps to reduce systematic uncertainties.

5. Systematic Uncertainties

Systematic uncertainties are evaluated separately for each source and measured distribution. Even with the 1.1 fb^{-1} subset of data recorded in 2011, the measurement is already limited by systematic effects. Due to the normalization concerning the inclusive cross section in the same phase space, all bin-by-bin correlated uncertainties cancel (at least partly). The remaining uncertainties are therefore shape uncertainties. The main model uncertainties originate from the matching threshold between matrix element calculation and parton showering and the renormalization and factorization scale Q^2 of the process. The main sources of experimental uncertainties are the jet energy scale and the b-tag efficiency.

6. Results

The $t\bar{t}$ production cross section is measured differentially as a function of the final state leptons (p_T, η), the top quarks (p_T, rapidity y) and the $t\bar{t}$ system (p_T, y, invariant mass) in all final states involving muons and electrons. For the dileptonic final states, the production cross section is additionally measured as a function of the lepton pair p_T and invariant mass. The measured points from data are compared to different SM simulations (MadGraph+Pythia, Powheg+Pythia, MC@NLO+Herwig) [8, 9, 10, 11]. Horizontal bin center corrections [12] (based on the simulation of MadGraph) have been applied to the data points in order to compare them to the continuous predictions from simulation. A selection of results can be found in Figure 1. In general, the different channels agree well with each other and the studied SM predictions. However, the transverse momentum of the top quarks (Figure 1 e, f) tend to be lower in data than predicted by the simulations. For the measurements as a function of the dilepton system (Figure 1 c, d), data prefers simulations including spin correlation (MC@NLO and Powheg) over the one without (MadGraph). In particular, this is interesting as the invariant mass of two leptons is an important variable for searches.

7. Summary

This measurement is the most detailed differential cross section measurement at the LHC with $\sqrt{s}$=7 TeV so far, involving five different final states and nine measured quantities. Within the precision of this measurement, no significant deviation from the SM is observed. However, for the transverse momentum of the top quarks, data prefers a softer spectrum than predicted by the simulations (involving QCD predictions up to NLO).

References

[1] CMS Collaboration 2008 *JINST* **0803** S08004
[2] CMS Collaboration 2012 *CMS Physics Analysis Summary* **CMS-PAS-TOP-11-013**
[3] CMS Collaboration 2009 *CMS Physics Analysis Summary* **PFT-09-001**
[4] Cacciari M, Salam G P and Soyez G 2008 *JHEP* **04** 063
[5] CMS Collaboration 2011 *CMS Physics Analysis Summary* **BTV-11-001**
[6] d'Hondt J *et al.* 2006 Fitting of Event Topologies with External Kinematic Constraints in CMS CMS Note 2006-023
[7] Abbott B *et al.* (D0) 1998 *Phys. Rev. Lett.* **80** 2063
[8] Sjöstrand T, Mrenna S and Skands P 2006 *JHEP* **05** 026
[9] Maltoni F and Stelzer T 2003 *JHEP* **02** 027
[10] Frixione S, Nason P and Oleari C 2007 *JHEP* **11** 070
[11] Frixione S and Webber B R 2002 *JHEP* **06** 29
[12] Lafferty G D and Wyatt T R 1995 *Nucl. Instrum. Meth.* **A355** 541

Journal of Physics: Conference Series **452** (2013) 012027 doi:10.1088/1742-6596/452/1/012027

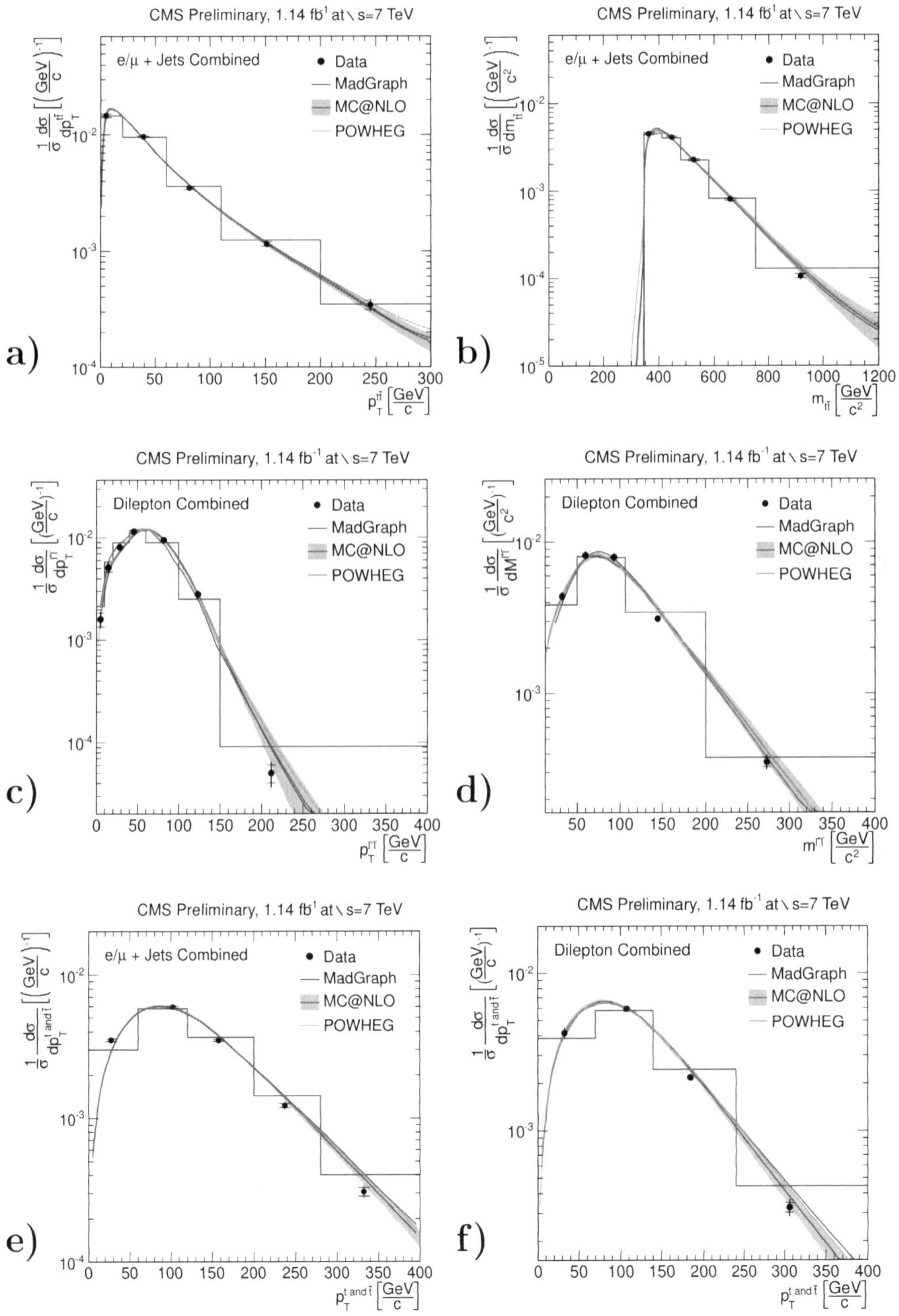

Figure 1. Normalized differential cross sections as a function of the transverse momentum (a) and invariant mass (b) of the $t\bar{t}$ system, the transverse momentum (c) and the invariant mass (d) of the lepton pair, and the top quark transverse momentum (e, f). Horizontal bin center corrections (using the MADGRAPH simulation) have been applied to the data points.

Journal of Physics: Conference Series **452** (2013) 012028 doi:10.1088/1742-6596/452/1/012028

Top quark pair production cross-section results at LHC in final states with tau or no leptons

Gia Khoriauli
On behalf of the ATLAS and CMS Collaborations
Physik. Inst., Nußallee 12, 53115 Bonn, Germany
E-mail: khoriauli@physik.uni-bonn.de

Abstract. ATLAS and CMS measured the top quark pair production inclusive cross section using proton-proton collision data collected at 7 TeV c.m. energy at the LHC during the year 2011 in several channels. This article presents the measurements of the cross section using selections of top quark pair production events with one hadronically decaying τ lepton or no leptons at all.

1. Introduction

Top quark pairs, $t\bar{t}$, are produced in proton-proton collisions via the strong interaction of colliding partons. At 7 TeV c.m. energy about 80% of $t\bar{t}$ is produced by gluon-gluon interactions. The remaining 20% is produced in quark and anti-quark annihilation processes. The theoretical cross section of the inclusive $t\bar{t}$ production within the Standard Model (SM) is calculated at the approximate NNLO level using the HATHOR tool [1], and is $\sigma^{\text{theory}} = 167^{+17}_{-18}$ pb at $\sqrt{s} = 7$ TeV and $m_{\text{top}} = 172.5$ GeV.

A top quark decays almost always into a W boson and b quark. Therefore, $t\bar{t}$ events are distinguished by the decay products of the two W bosons. W bosons can decay into a quark pair or a lepton-neutrino pair. The ATLAS [2] and CMS [3] experiments measured the $t\bar{t}$ production cross section in different final states. These final states are affected by different background processes, which makes it necessary to have unique analysis techniques developed for each particular measurement. Precise measurements of the cross section in all final states helps to test the SM against some new physics phenomena, which could influence on the measurements.

Details of cross-section measurements of both ATLAS and CMS in three final states are reviewed in this document: a) τ_{had}+lepton: one W boson decays into a τ lepton and a neutrino and the τ lepton decays hadronically. The second W boson decays leptonically including the decay into τ and neutrino, when the former decays leptonically; b) τ_{had}+jets: same as in the previous case but the second W boson decays hadronically; c) fully hadronic: both W bosons decay hadronically.

2. Data 2011 and Monte-Carlo (MC) simulation

Both ATLAS and CMS detectors recorded a large fraction of the total integrated luminosity delivered by the LHC machine during the data taking year 2011. The ATLAS and CMS studies, which are presented in this document used up to 4.7 fb^{-1} and 3.9 fb^{-1} of data, respectively.

The ATLAS signal sample was generated with MC@NLO [4] and showered and hadronised with HERWIG [5]. The single top quark background sample was also produced with these generators. In contrast, CMS used MADGRAPH [6] together with PYTHIA [7] to generate the signal as well as W/Z+jets. The single top quark background was generated with POWHEG [8] together with PYTHIA. PYTHIA was used by CMS also to generate di-bosons and QCD multi-jet samples. The ATLAS studies used di-boson samples generated with HERWIG while the QCD multi-jet and W/Z+jets background was generated with ALPGEN [9] and HERWIG. Both experiments used TAUOLA [10] to simulate τ decays. All top quark samples were generated assuming $m_{\text{top}} = 172.5$ GeV.

3. Three cross-section measurements

3.1. τ_{had}+lepton final state

The total branching ratio of all hadronic decays of τ is about 65%. The branching ratio of the τ_{had}+lepton final state is about 3.7% of the inclusive $t\bar{t}$ production. The background is any process with a τ_{had}+lepton or a lepton+jets final state, where one jet is wrongly identified as τ_{had}. The main background processes are $t\bar{t}$ with different W boson decay modes and W/Z+jets.

CMS measured the cross section separately in $t\bar{t} \to \tau_{\text{had}} + e + X$ and $t\bar{t} \to \tau_{\text{had}} + \mu + X$ final states using data samples of 2.0 fb^{-1} and 2.2 fb^{-1} respectively [11]. Signal candidate events were selected using the combined lepton (electron or muon) + dijets + $H_{\text{T}}^{\text{miss}}$ trigger, where $H_{\text{T}}^{\text{miss}}$ is the absolute value of the vectorial sum of all jets momenta. The further (offline) selection of events required an isolated electron (muon) with a transverse momentum, $p_{\text{T}} > 35\,(30)$ GeV, and an absolute pseudo-rapidity, $|\eta| < 2.5\,(2.1)$. A large missing transverse energy, $E_{\text{T}}^{\text{miss}} > 45\,(40)$ GeV, and at least two jets with $p_{\text{T}} > 35\,(30)$ GeV and $|\eta| < 2.4$ were required. At least one of the jets was required to be b-tagged. One isolated τ_{had} candidate jet was required with $p_{\text{T}} > 20$ GeV, $|\eta| < 2.4$ and a charge sign opposite to the selected electron (muon). The τ_{had} algorithm used neutral and charged hadrons in a jet as inputs in order to discriminate τ_{had} from hadronic jets. The background was estimated from MC simulation except for the misidentified τ_{had}+lepton background, which was not well modeled in simulation due to the complexity of a precise description of numerous detector effects. A data-driven method was used for the estimation of the misidentified τ_{had}+lepton background. The method assigns a variable probability to every jet to be misidentified. This probability map, which depends on p_{T}, η and a width in $\eta \times \phi$ of jets, where ϕ is the azimuthal angle around the beam axis, was obtained from a data sideband region enriched with QCD multi-jet events. Since one knows the total rate of the processes (e.g., $W(\to l\nu)$+jets), which contribute into the misidentified background one estimates this contribution by weighting jets. The cross section was measured using the "cut and count" method: the expected number of background events was subtracted from the number of selected events in data and the result was normalized to the inverse of the product of the signal acceptance (obtained from MC simulation) and the integrated luminosity of data. Systematic uncertainties on the expected number of background events, the signal acceptance and the integrated luminosity of data were considered. The results are $\sigma_{\text{t}\bar{\text{t}}}(t\bar{t} \to \tau_{\text{had}} + e + X) = 136 \pm 23(\text{stat.}) \pm 23(\text{syst.}) \pm 3(\text{lumi.})$pb and $\sigma_{\text{t}\bar{\text{t}}}(t\bar{t} \to \tau_{\text{had}} + \mu + X) = 147 \pm 18(\text{stat.}) \pm 22(\text{syst.}) \pm 3(\text{lumi.})$pb [11]. These results were combined using the Best Linear Unbiased Estimator (BLUE) method [12]. The result of the combination is $\sigma_{\text{t}\bar{\text{t}}} = 143 \pm 14(\text{stat.}) \pm 22(\text{syst.}) \pm 3(\text{lumi.})$pb [11].

The ATLAS measurement used 2.05 fb^{-1} of data [13]. Single lepton triggers were used for the event selection. The offline selection required a single isolated electron (muon) with $E_{\text{T}} > 25$ GeV ($p_{\text{T}} > 20$ GeV) and $|\eta| < 2.47\,(2.5)$ excluding the calorimeter gap regions $1.37 < |\eta| < 1.52$. Presence of a τ_{had} candidate with $20 < E_{\text{T}} < 100$ GeV and $|\eta| < 2.3$ and not overlapping with the selected lepton, $\Delta R(\text{lepton}, \tau_{\text{had}}) > 0.4$, was required. In addition, $E_{\text{T}}^{\text{miss}} > 30$ GeV and $E_{\text{T}}^{\text{miss}} + \sum |p_{\text{T}}| > 200$ GeV, where the sum is over the lepton and hadronic jets, requirements were applied. Presence of at least two hadronic jets with $p_T > 25$ GeV and $|\eta| < 2.5$ and not overlapping with the τ_{had} candidate, $\Delta R(\text{jet}, \tau_{had}) > 0.4$, was required. Selected events were split in eight categories depending on the decay type of the τ_{had} candidate: with one charged track ("1 prong") and with three charged tracks ("3 prong"); the sign of the isolated lepton and τ_{had} electric charges: opposite-signed (OS) and same-signed (SS); the number of b-jets: ≥ 1 and 0 b-jets. A track-based algorithm was used to identify "1 prong" and "3 prong" τ_{had} candidates. A Boosted Decision Tree discriminator, BDT_e was used to reject the background from misidentified electrons. Another BDT_j discriminator was used for the further selection of events. The input templates for BDT_j were obtained by subtracting the corresponding distributions, OS-SS, that canceled out the symmetric background in these two selections. Data-driven templates of the fake τ_{had} background due to jets were obtained in the 0 b-jet events and corrected for the kinematics of ≥ 1 b-jets events using weights calculated from MC simulation. Figure 1 shows the BDT_j output for the "1 prong" τ_{had}+lepton events with ≥ 1 b-jets after the template fit. Using the template fit of the BDT_j

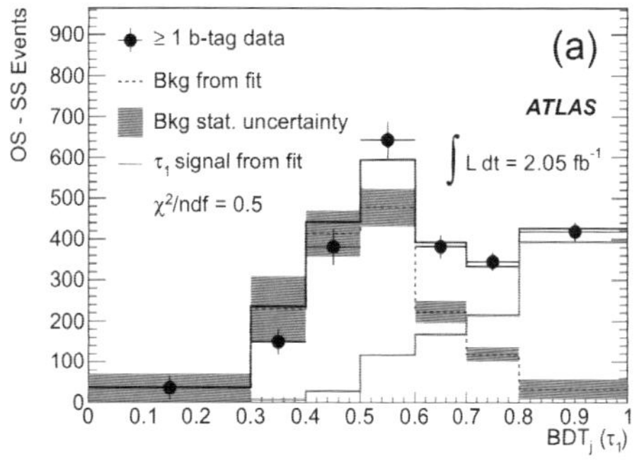

Figure 1. BDT_j (OS-SS) templates for τ_{had}+lepton events with the "1 prong" τ_{had} and ≥ 1 b-jets after the fit to data. The fitted contributions are shown as the light/red (signal), dashed/blue (background) and dark/black (total) histograms. Shaded/blue bands are the statistical uncertainty of the background template [13].

output for the "1 prong" and "3 prong" τ_{had}+lepton events with ≥ 1 b-jets the cross section was measured to be $\sigma_{t\bar{t}}(t\bar{t} \to \tau_{had} + e + X) = 186 \pm 15(\text{stat.}) \pm 20(\text{syst.}) \pm 7(\text{lumi.})\text{pb}$ and $\sigma_{t\bar{t}}(t\bar{t} \to \tau_{had} + \mu + X) = 187 \pm 18(\text{stat.}) \pm 20(\text{syst.}) \pm 7(\text{lumi.})\text{pb}$ [13]. These results were combined using the BLUE method, $\sigma_{t\bar{t}} = 186 \pm 13(stat.) \pm 20(syst.) \pm 7(\text{lumi.})\text{pb}$ [13].

3.2. τ_{had}+jets final state

The branching ratio of the τ_{had}+jets final state is about 9.8%. QCD multi-jet events are one of the main backgrounds due to fake τ_{had} leptons. The combinatorial background, where in a signal event the τ_{had} candidate is misidentified as a hadronic jet, is important, too.

The ATLAS measurement used 1.67 fb^{-1} of data [14]. The event selection was done with a multi-jet trigger. The offline selection required the presence of at least five jets with $p_T > 20$ GeV, $|\eta| < 2.5$ and at least two of them tagged as b-jets. One of the untagged jets had to be identified as τ_{had}. A veto on isolated leptons was applied. To reduce the QCD multi-jet background large E_T^{miss} significance, $E_T^{miss}/(0.5 \times \sqrt{\text{GeV}} \times \sqrt{\sum E_T}) > 8$, was required, where $\sum E_T$ is the sum of transverse energies of all objects contributing to the calculation of E_T^{miss}. The τ_{had} candidate jet was selected by excluding other jets using the kinematics and b-tagging information. A template fit method was used to measure the signal cross section, exploring the number of tracks, n_{track}, of the τ_{had} candidate jet. MC simulation predicted a template for misidentified electrons as τ_{had} candidates similar to the signal template. Therefore, the two were joined in a single template. A multi-jet template was derived using a sideband region while a template for the combinatorial background was obtained from MC simulation. In the template fit the number of events of the combinatorial background was constrained to the number of the signal events since they both originate from $t\bar{t}$ production. Figure 2 shows the fit result. Contamination of the fitted signal+electron template by the single top quark and

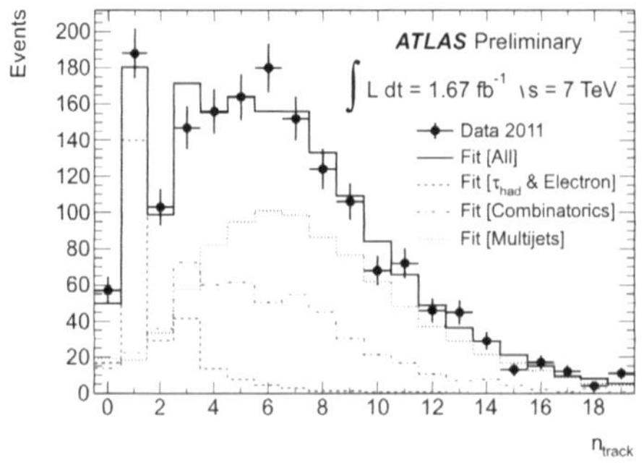

Figure 2. The $n_{\rm track}$ distribution for $\tau_{\rm had}$ candidates after all selection cuts. The solid histogram is the result of the fit. The red, blue and magenta dashed curves show the fitted contributions from the $\tau_{\rm had}$+electron "signal", the multi-jet and combinatorial backgrounds respectively [14].

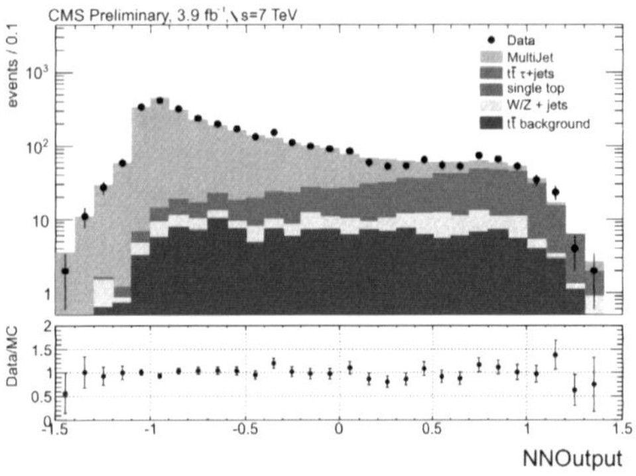

Figure 3. The distribution of the Neural-Network output after the template fit of the signal (red) and background to data. The multi-jet template (cyan) is derived from data. The combinatorial $t\bar{t}$ (blue), W/Z+jets (yellow) and single top quark (pink) templates are obtained from MC simulation [15].

W+jets background was subtracted using the predictions from MC simulation. The known (MC simulation) fraction of $\tau_{\rm had}$ events, $N_{\tau_{\rm had}}/(N_{\tau_{\rm had}} + N_{\rm electron}) = 0.77 \pm 0.03\,({\rm stat.}) \pm 0.03\,({\rm syst.})$, was used to estimate the number of $\tau_{\rm had}$ events in the fitted signal+electron template. The result is $\sigma_{t\bar{t}} = 200 \pm 19({\rm stat.}) \pm 43({\rm syst.} + {\rm lumi.})$pb [14].

The CMS measurement used 3.9 fb^{-1} of data [15]. A multi-jet trigger was used for the event selection. A jet identified as a $\tau_{\rm had}$ candidate with $p_{\rm T} > 45$ GeV and $|\eta| < 2.3$ was required to match to one of the trigger jets. The event selection required the presence of at least three jets with $p_{\rm T} > 45$ GeV and $|\eta| < 2.4$ among which exactly one jet is tagged as b-jet. Additional jets were counted if $p_{\rm T} > 20$ GeV. A veto on isolated leptons was also applied. Finally, it was required that $E_{\rm T}^{\rm miss} > 20$ GeV. A Neural-Network discriminator output was built using the MultiLayerPerceptron (MLP) algorithm. The input distributions were derived from MC simulation except for the multi-jet background. For the latter, the same signal selection of events was used but with a 0 b-jet requirement. Correction factors as derived from MC simulation were applied to the distributions in order to take the difference between the kinematics of the 0 and 1 b-jet events into account. A result of the template fit of the MLP output is shown in Figure 3. The final event selection was done by imposing a cut, NNOutput > 0.5, on these distributions after the fit. The measured cross section is $\sigma_{t\bar{t}} = 156 \pm 12({\rm stat.}) \pm 33({\rm syst.}) \pm 3({\rm lumi.})$pb [15].

3.3. Fully hadronic final state

The fully hadronic final state has the largest branching ratio, 46%. A cross-section measurement suffers from the overwhelming QCD multi-jet background. The ATLAS and CMS measurements were based on kinematic fit methods described below.

The CMS measurement used 1.09 fb^{-1} of data [16]. The event selection was done using a multi-jet trigger. The offline selection of events required at least six jets with $|\eta| < 2.4$ among which four leading jets had to fulfill a $p_{\rm T} > 60$ GeV requirement. The fifth and sixth jets were required to have $p_{\rm T} > 50$ GeV and $p_{\rm T} > 40$ GeV, respectively. Additional jets with $p_{\rm T} > 30$ GeV were also counted. At least two jets were required to be b-tagged. The jets were used to reconstruct the $t\bar{t}$ system. All possible permutations of the jets were used to reconstruct two W bosons and two top quarks but only the b-tagged jets were used as candidates for the true b quark jets. For each permutation a least-square fit was done. Gaussian resolutions as derived from MC simulation were used for the energies of the jets. The true masses of the W bosons were constrained to the known value, $m_W = 80.4$ GeV, while both top quark true masses were the free parameters of the fit, but constrained to be equal to each other. The permutation leading to the smallest χ^2 value was taken and if its probability was greater than 0.01, the event was selected. The fitted mass in the selected events was used to create templates for a likelihood fit with the signal fraction, $f_{\rm sig}$, as a free parameter. The QCD multi-jet template was derived from a sideband selection of events, which was similar to the signal selection but requiring no b-jets. A two dimensional map ($\eta \times p_{\rm T}$) of weights, aimed at taking into account the difference in kinematics of light jets in events with 0 and ≥ 2 b-jets, was derived using the whole data sample (after trigger selection). Since this sample is dominated by events with only light jets, the impact of the true b-jets on the weights is negligible. The likelihood fit result is shown in Figure 4. The fitted value of the signal fraction was converted to a cross section, $\sigma_{\rm t\bar{t}} = 136 \pm 20({\rm stat.}) \pm 40({\rm syst.}) \pm 8({\rm lumi.})$pb [16].

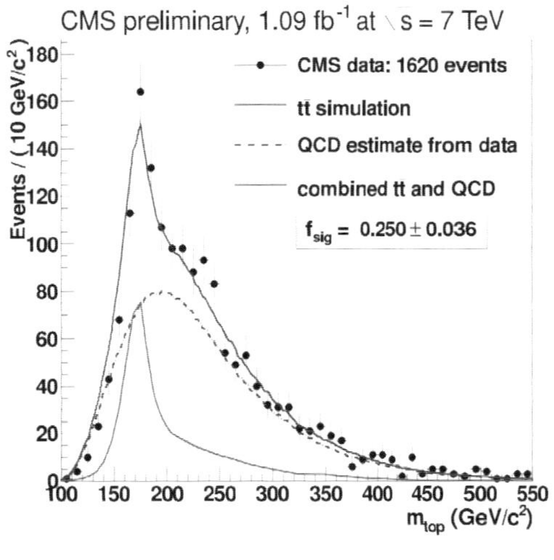

Figure 4. The result of the likelihood fit of the $t\bar{t}$ signal (solid red curve) and QCD multi-jet background (dashed blue curve) templates of the top quark mass (from the kinematic fit) to data. The uncertainty stated on the signal fraction, $f_{\rm sig}$, is only statistical [16].

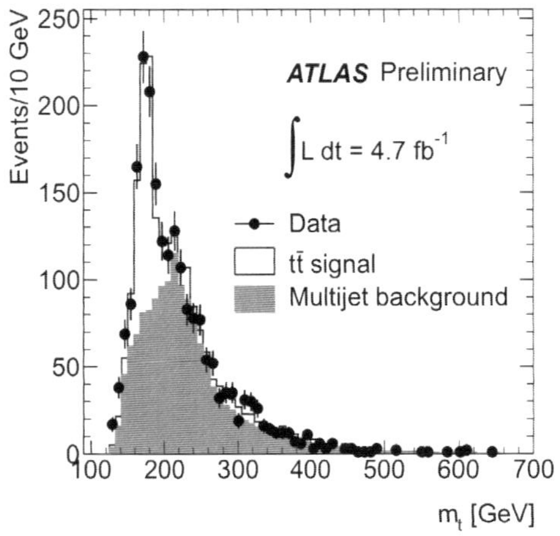

Figure 5. The result of the unbinned likelihood fit of the signal (black histogram) and background (green filled histogram) templates of the top quark mass, m_t, to data. m_t is calculated in the kinematic fit (see text) [17].

The ATLAS measurement used 4.7 fb^{-1} of data [17]. Events were selected with a multi-jet trigger. The offline selection required at least six jets with $|\eta| < 2.5$ and among them at least five jets with $p_{\rm T} > 55$ GeV and a sixth jet with $p_{\rm T} > 30$ GeV. At least two out of the leading five jets were required to be b-tagged. The jets were required to not overlap, $\Delta R({\rm jet}_i, {\rm jet}_k) > 0.6$, where i and k are the jet indices but $i \neq k$. A more stringent cut was applied to the two leading b-jets, $\Delta R({\rm bj}_1, {\rm bj}_2) > 1.2$. It was required that the $E_{\rm T}^{\rm miss}$ significance be $S_{\rm T} = E_{\rm T}^{\rm miss}/(0.5\sqrt{\rm GeV} \times \sqrt{H_{\rm T}}) < 6$, where $H_{\rm T}$ is the scalar sum of $p_{\rm T}$ of all jets. A veto on isolated leptons was applied. All possible permutations of at least six and up to ten jets were used to construct a likelihood function for every selected event. The free parameters of the likelihood function were the top quark true mass and the six quark energies. Breit-Wigner constraints were introduced in the function for the masses of the two reconstructed W bosons with the fixed true mass and width. Breit-Wigner constraints were also used for the reconstructed top quark masses. Both true top quark masses (free parameters) were required to be equal to each other. The true width of top quarks was constrained in the fit to vary according to expectation. The likelihood function included a product of six transfer functions of quark energies to energies of jets as obtained from MC simulation. Also, the likelihood function had factors, which accounted for the b-tagging efficiency and the mistagging rate of jets depending on a particular permutation of these jets. These additional factors helped to give higher probability to those permutations, where b-tagged jets were used as true b-jets. An event was selected if the permutation with a maximum likelihood value, $L^{\max}$, fulfilled the requirement, $L^{\max}/\sum L > 0.8$, where the sum is over all permutations, and the fitted mass, m_t, was greater than 125 GeV. Using the best permutation of jets and fitted values of free parameters, a χ^2 value was calculated for the reconstructed $t\bar{t}$ system in a given event. The event was finally selected if $\chi^2 < 30$. The distributions of the fitted top quark mass in the selected events were used in a template fit in order to measure the signal fraction in data. The fit result is shown in Figure 5. The QCD multi-jet template was derived using the same signal selection but requiring no b-tagged jet. Differences in kinematics of jets in the events with 0 and ≥ 2 b-tagged jets were studied in a MC simulation sample of QCD multi-jet production and taken into account as a systematic uncertainty. The measured signal fraction was converted into a measurement of the cross section, $\sigma_{t\bar{t}} = 168 \pm 12(\text{stat.})^{+60}_{-57}(\text{syst.}) \pm 7(\text{lumi.})\text{pb}$ [17].

4. Summary

The $t\bar{t}$ production cross section was measured by the ATLAS and CMS experiments in three different final states independently: τ_{had}+lepton, τ_{had}+jets and fully hadronic. The obtained results are consistent with the Standard Model prediction for the $t\bar{t}$ production cross section.

Acknowledgments

I gratefully acknowledge the support of the Deutsche Forschungsgemeinschaft (DFG) through the Emmy-Noether grant CR-312/1-2.

References

[1] Aliev M et al 2011 *Comput. Phys. Commun.* **182** 1034
[2] ATLAS Collaboration 2008 *JINST* **3** S08003
[3] CMS Collaboration 2008 *JINST* **3** S08004
[4] Frixione S and Webber B R 2002 *JHEP* **06** 029; Frixione S et al 2003 *JHEP* **08** 007; Frixione S et al 2006 *JHEP* **03** 092
[5] Corcella G et al 2001 *JHEP* **01** 010; Corcella G et al *Preprint* hep-ph/0210213
[6] Alwall J et al 2007 *JHEP* **09** 028
[7] Sjöstrand T et al 2006 *JHEP* **05** 026
[8] Frixione S et al 2007 *JHEP* **11** 070
[9] Mangano M L et al 2003 *JHEP* **07** 001
[10] Davidson N et al *Preprint* hep-ph/1002.0543
[11] CMS Collaboration 2012 *Phys. Rev. D* **85** 112007
[12] Lyons L et al 1988 *NIM* A **270** 110
[13] ATLAS Collaboration 2012 *Phys. Lett.* B **717** 89-108
[14] ATLAS Collaboration ATLAS-CONF-2012-032 `http://cdsweb.cern.ch/record/1432198`
[15] CMS Collaboration CMS-PAS-TOP-11-004 `https://cdsweb.cern.ch/record/1446652`
[16] CMS Collaboration CMS-PAS-TOP-11-007 `http://cdsweb.cern.ch/record/1371755`
[17] ATLAS Collaboration ATLAS-CONF-2012-031 `http://cdsweb.cern.ch/record/1432196`

5th International Workshop on Top Quark Physics (TOP2012) IOP Publishing
Journal of Physics: Conference Series **452** (2013) 012029 doi:10.1088/1742-6596/452/1/012029

Measurement of the $\mathrm{t\bar{t}}$ production cross section in the dilepton channel in proton-proton collisions at $\sqrt{s} = 8$ TeV with the CMS experiment

Jan Kieseler for the CMS Collaboration

Deutsches Elektronen-Synchrotron (DESY), Notkestraße 85, 22607 Hamburg, Germany

E-mail: jan.kieseler@desy.de

Abstract. The cross section for top quark pair production is measured in proton-proton collisions at $\sqrt{s} = 8$ TeV in a data sample corresponding to 2.4 fb^{-1} of integrated luminosity collected by the CMS experiment in 2012. The measurement is performed with events with two leptons (electrons or muons), as well as with identified b-quark jets in the final state. The combined measured cross section is $\sigma_{t\bar{t}} = 227 \pm 3$ (stat.) $\pm$ 11 (syst.) $\pm$ 10 (lumi.) pb, in agreement with theoretical predictions.

1. Introduction

A precise measurement of the $\mathrm{t\bar{t}}$ production cross section in proton-proton (pp) collisions at the Large Hadron Collider (LHC) offers a benchmark for the production mechanism of other processes also dominated by gluon-gluon fusion. Moreover, it can be used to validate calculations in quantum chromodynamics or to provide constraints on parton distribution functions. An accurate understanding of the $\mathrm{t\bar{t}}$ production and decay is also crucial for searches for new physics processes that show a similar event topology.

Here, the measurement [1] of the $\mathrm{t\bar{t}}$ production cross section in pp collisions at $\sqrt{s} = 8$ TeV with the CMS experiment [2] is described. It is performed in the dileptonic decay mode using data corresponding to an integrated luminosity of 2.4 fb^{-1} delivered by the LHC in 2012 and compared to standard model predictions at approximate next-to-next-to-leading order (NNLO).

2. Event Selection and Simulation

The event selection is based on the decay topology of the $\mathrm{t\bar{t}}$ pair in the dilepton channel, where both W-bosons from the decay $t\bar{t} \to W^+W^-b\bar{b}$ subsequently decay to a lepton and a neutrino. Therefore, dilepton triggers are used. These are based on the presence of two high transverse momentum (p_T) leptons, in particular electrons or muons.

Proton-proton collision events are selected with at least two oppositely charged isolated leptons and two jets of which one is identified as a b-jet. The leptons are required to have $p_T > 20$ GeV and a pseudorapidity $|\eta| < 2.5$ (e) or $|\eta| < 2.4$ (μ). Jets are selected with $p_T > 30$ GeV and $|\eta| < 2.5$. Jets originating from the b-quarks are identified using combined secondary vertex and track-based lifetime information. Events with heavy flavor resonance decays with a dilepton mass $M_{ll} < 20$ GeV are removed in all channels. Contributions from Z production in the e^+e^- and $\mu^+\mu^-$ decay modes are further reduced by requiring M_{ll} to be outside a Z mass window of 91 ± 15 GeV and $\not{E}_T > 40$ GeV. Figure 1 shows the multiplicity of b-tagged jets after applying the aforementioned selection.

The simulation of signal and background events is performed using Monte Carlo (MC) event generators. The signal events are modeled by MADGRAPH. The background contributions from Drell-Yan (DY) production, W+jets events and WW processes are generated with MADGRAPH, WZ and ZZ contributions with PYTHIA, and the single top-quark background with POWHEG. In all cases, the subsequent hadronization and parton showering are simulated with PYTHIA.

Among all of the simulated background contributions, only those from single-top-quark production and diboson (VV, where V=W or Z) are used to estimate the absolute number of background events. The remaining background events are estimated using data-based methods.

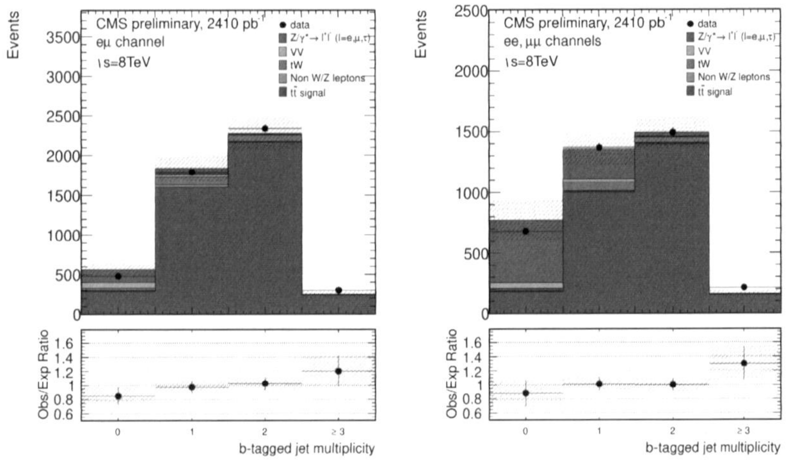

Figure 1. Multiplicity of b-tagged jets after the full event selection except the b-tag requirement for the $e^{\pm}\mu^{\mp}$ (left) and the summed e^+e^- and $\mu^+\mu^-$ channels (right). The corresponding data-to-simulation ratios are also shown.

3. Background Determination

DY processes are the dominant background in the e^+e^- and $\mu^+\mu^-$ final states. Their contribution is estimated using the Z invariant mass interval as a control region to rescale the DY contamination in the signal region. The contribution of non-DY background in the control region is subtracted using the $e^{\pm}\mu^{\mp}$ channel taking into account the corresponding lepton efficiencies. The DY background in the $e^{\pm}\mu^{\mp}$ final state is mainly dominated by Z processes with subsequent τ decays. Here, the DY contribution is estimated by fitting the dilepton invariant mass distribution in data taking into account the remaining backgrounds.

The contamination from events with non-W/Z leptons is extrapolated from an estimate at 7 TeV with a data-based method by comparing the yields of opposite-sign (N_{OS}) and same-sign (N_{SS}) dilepton events. The ratio $R_{OS/SS}$ of these yields is assumed to be independent of the center-of-mass energy. Thus, the number of background events with non-W/Z leptons can be estimated by counting the number of same-sign dilepton events at 8 TeV as

$$N_{OS}(8\text{ TeV}) = N_{SS}(8\text{ TeV}) \cdot R_{OS/SS}(7\text{ TeV}).$$

4. Cross Section Determination

The $t\bar{t}$ production cross section $\sigma_{t\bar{t}}$ is determined by a cut-and-count approach in each dilepton channel separately. The combination of all three channels, using the BLUE method, results in a cross section of $\sigma_{t\bar{t}} = 227 \pm 3$ (stat.) $\pm$ 11 (syst.) $\pm$ 10 (lumi) pb. The dominant systematic uncertainties contributing to the combined measurement come from the jet energy scale (2.5%), the trigger and lepton efficiencies (1.8%), the jet energy resolution (1.7%), the branching ratios (1.7%), and the determination of the non-W/Z background (1.4%).

The measured $t\bar{t}$ cross section is combined with the corresponding measurement [3] in the lepton plus jets channel (l+jets) and compared to different theory predictions, see Figure 2.

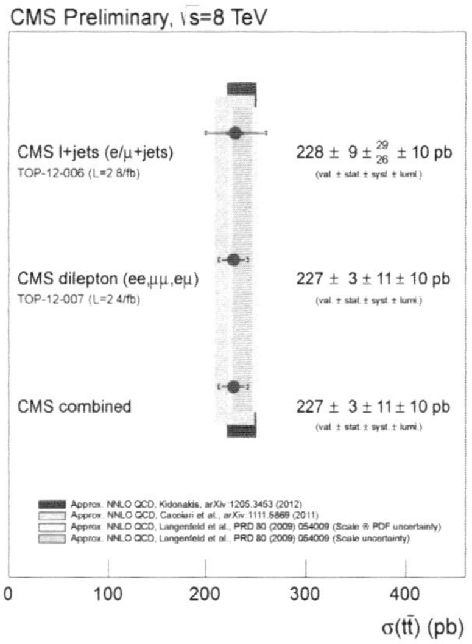

Figure 2. The $t\bar{t}$ cross section measurements using different decay modes, l+ jets and dilepton, and their combination. The vertical band shows different theory predictions at approximate NNLO.

5. Summary

CMS has performed a measurement of the $t\bar{t}$ cross section in the dilepton final states using event candidates containing two opposite sign leptons, two jets, and requiring the presence of missing transverse energy. To further disentangle signal and background, b-tagging information is used as an additional selection requirement on at least one of the jets. The DY contribution is estimated from data as well as the non-W/Z contribution. For a top-quark mass of 172.5 GeV, the $t\bar{t}$ cross section is measured to be $\sigma_{t\bar{t}} = 227 \pm 3$ (stat.)$\pm$11 (syst.)$\pm$10 (lumi) pb, in agreement with standard model predictions.

References

[1] The CMS Collaboration, *Measurement of the* $t\bar{t}$ *production cross section in the dilepton channel in pp collisions at* $\sqrt{s} = 8$ *TeV*, PAS-TOP-12-007 (2012)
[2] The CMS Collaboration, *The CMS experiment at the CERN LHC*, JINST 03 (2008) S08004
[3] The CMS Collaboration, *Top pair cross section in e/mu+jets at 8 TeV*, PAS-TOP-12-006 (2012)

Measurement of the top quark pair production cross section with ATLAS in pp collisions at $\sqrt{s} = 7\,\text{TeV}$ in the single-lepton channel using semileptonic b decays

Morris JD

Queen Mary University of London, Mile End Road, London, E1 4NS, UK

E-mail: j.d.morris@qmul.ac.uk

Abstract. The cross section for top quark pair production in pp collisions at $\sqrt{s} = 7\,\text{TeV}$ is measured using data recorded with the ATLAS detector at the Large Hadron Collider. Events are selected requiring a single high transverse momentum isolated electron or muon, missing transverse momentum and jets. Semileptonic b decays are identified by a lower momentum muon close to a jet, leading to substantially different sources of systematic uncertainty compared with other measurements. With a data sample corresponding to 4.66 fb^{-1} of integrated luminosity, the top quark pair production cross section is measured to be

$$\sigma_{t\bar{t}} = 165 \pm 2(\text{stat.}) \pm 17(\text{syst.}) \pm 3(\text{lumi.})\ \text{pb}$$

in agreement with theoretical predictions based on perturbative QCD and with other measurements which used different techniques or decay channels.

1. Introduction

The experimental determination of the $t\bar{t}$ cross section is an important measurement and a test of the Standard Model(SM). Full details of this analysis are to be found in [1]. The predicted SM $t\bar{t}$ cross section for pp collisions at a centre-of-mass energy of $\sqrt{s} = 7\,\text{TeV}$, for a top quark mass of 172.5 GeV is $\sigma_{t\bar{t}} = 167^{+17}_{-18}$ pb, calculated at approximate next-to-next-to-leading order (NNLO) in QCD with Hathor 1.2 [2] using the MSTW2008 NNLO parton density function (PDF) sets [3, 4] and cross checked with the next-to-leading order (NLO) and next-to-next-to-leading log (NNLL) calculation of Cacciari et al. [5] as implemented in Top++ 1.0 [6].

2. Data and simulated samples

This measurement uses data recorded by the ATLAS detector [7] between March and November 2011 at the LHC [8] corresponding to an integrated luminosity of 4.66±0.08 fb^{-1} [9]. The $t\bar{t}$ signal Monte Carlo (MC) has been simulated using the MC@NLO v4.01 [10, 11] generator interfaced to Herwig [12] for parton showering and hadronisation and Jimmy [13] for the underlying event simulation. The PDF set is CT10 [14] and the top quark mass is set to 172.5 GeV. The W/Z+jets MC samples were generated using Alpgen [15] interfaced to Herwig+Jimmy. The single top quark MC samples used MC@NLO interfaced to Herwig+Jimmy for the s-channel and Wt-channel, while the t-channel was generated using the AcerMC [16] generator interfaced to Pythia [17]. The diboson ($WW/WZ/ZZ$) MC samples were generated using Herwig.

3. Soft Muon Tagging (SMT) algorithm

The SMT algorithm used in this analysis relies on the presence of a reconstructed muon within a jet. The total $b \to \mu X$ branching ratio, including cascade $b \to c \to \mu X$ decays, is about 20%. The tagger uses the quality of the match between the inner detector and muon spectrometer tracks of the muon (χ^2_{match}) as a discriminating variable to separate heavy flavour (b, c) jets from light flavoured (u, d, s or gluon) ones.

Muons considered by the tagger are required to pass a series of quality cuts on the number of hits present in the various tracking subdetectors [19]. Muons are required to be with $\Delta R(\mu, \mathrm{jet}) < 0.5$[1] with $|d_0| < 3$ mm, $|z_0 \sin(\theta)| < 3$ mm, a transverse momentum $p_{\mathrm{T}} > 4\,\mathrm{GeV}$ and pseudorapidity $|\eta| < 2.5$. Muons are required to have a $\chi^2_{\mathrm{match}} < 3.2$ which has been optimized to provide a b-jet identification efficiency of 10% whilst providing a light jet rejection factor of about 200 per jet. Correction factors for simulation have been calculated as a function of p_{T} and η and are between 0.97 and 1 within an uncertainty $\leq$1%. The mistag rate is found to be 1.4$\pm$0.20 (stat. $\oplus$ syst.).

4. Event selection

The analysis requires collision data selected by an inclusive single electron or muon trigger with offline-reconstructed candidates satisfying $p_{\mathrm{T}} \geq 25\,\mathrm{GeV}$ for electrons and $p_{\mathrm{T}} \geq 20\,\mathrm{GeV}$ for muons. Electrons are required to have $|\eta| < 2.47$ and not be in the transition region between the barrel and the endcap calorimeters ($1.37 \leq |\eta| \leq 1.52$). Muon candidates are obtained by combining track segments from the layers of the muon chambers with tracks found in the inner detector within $|\eta| < 2.5$.

Leptons from W decays are isolated and thus, to reduce backgrounds, both calorimeter and track isolation are required. For the electron, the cut value on the calorimeter and track variables depend on the electron p_{T} and η such that the isolation cut efficiency for each variable is maintained at 90%. For the muon, the calorimeter isolation is required to be $< 4\,\mathrm{GeV}$ while the track isolation is required to be $< 2.5\,\mathrm{GeV}$. Leptons are also required to be separated further than an angular distance of $\Delta R = 0.4$ from any jet. Muons tagged by the SMT algorithm are not considered. Events are required to contain only one selected electron or muon.

Jets are reconstructed with the anti-k_{t} algorithm [20, 21] with a distance parameter $R = 0.4$, calibrated at the electromagnetic (EM) scale [22]. Jets are required to have a $p_{\mathrm{T}} \geq 25\,\mathrm{GeV}$, $|\eta| < 2.5$. Jets that are within $\Delta R < 0.2$ from a selected electron are rejected.

The e+jets channel requires that the missing transverse energy $E_{\mathrm{T}}^{\mathrm{miss}} \geq 30\,\mathrm{GeV}$ and that the transverse W mass, denoted $m_{\mathrm{T}}(W) \geq 30\,\mathrm{GeV}$, while the μ+jets channel requires that $E_{\mathrm{T}}^{\mathrm{miss}} \geq 20\,\mathrm{GeV}$ and a triangular cut is performed by requesting that $E_{\mathrm{T}}^{\mathrm{miss}} + m_{\mathrm{T}}(W) \geq 60\,\mathrm{GeV}$. In the analysis, events passing all selections described above with at least 3 jets are defined as "pretag" events. Pretag events with at least one jet identified as coming from a semileptonic b-quark decay by the SMT algorithm are referred to as "tagged". In the μ+jets channel, events for which the invariant mass of the W-decay muon(μ_W) and the $b-$tagged SMT muon(μ_{SMT}) is in the ranges $8\,\mathrm{GeV} \leq m_{\mu\mu} \leq 11\,\mathrm{GeV}$ and $80\,\mathrm{GeV} \leq m_{\mu\mu} \leq 100\,\mathrm{GeV}$ are vetoed to remove contributions from Υ and Z dimuon decays. Additionally, to ensure that the tagged muon is not the same as the W-decay muon, a cut of $\Delta R(\mu_W, \mu_{\mathrm{SMT}}) > 0.01$ is applied.

The signal selection efficiencies, found by applying the full selection described above on the simulated $t\bar{t}$ signal sample, are 1.42% in the e+jets channel and 2.15% in the μ+jets channel.

5. Data-driven background estimation

The W+jets background is the dominant background since these events already contain a real lepton and significant $E_{\mathrm{T}}^{\mathrm{miss}}$. The W charge asymmetry method is used to provide an overall

[1] $\Delta R(i,j) = \sqrt{\Delta\eta(i,j)^2 + \Delta\phi(i,j)^2}$, where $\Delta\eta(i,j) = \eta_i - \eta_j$ and $\Delta\phi(i,j) = \phi_i - \phi_j$

normalisation. This relies on the observation that the ratio of positively charged W bosons to negatively charged W bosons suffers from relatively small theoretical uncertainty [3, 23]. The final contributions are detailed in Table 1.

Multijet background events enter into the final event selection. Events in the e+jets analysis come mainly from photon conversions and from jets with high EM fractions. Events in the μ+jets analysis come from the semileptonic decay of heavy quarks, the decay in flight of pions and kaons and from "punch-through" hadrons that are not fully absorbed within the hadronic calorimeter. Multijet estimates are obtained by considering control regions close to the signal region [24] and obtaining an SMT tagging rate. The final contributions are detailed in Table 1.

6. Systematic uncertainties and Results

For each uncertainty listed in Table 2, the combined $t\bar{t}$ cross section is recalculated by summing the e+jets and μ+jets event yields, providing an uncertainty on the combined $t\bar{t}$ cross section. This procedure takes into account any correlation that may be present across the e+jets and μ+jets analyses. The individual systematic uncertainties on the $t\bar{t}$ cross section are summed in quadrature to provide an overall systematic uncertainty. The dominant systematic uncertainties come from the uncertainty on the multijet and W+jets backgrounds and from the jet energy scale.

The number of observed and expected jets is shown in Figure 1. The cross section is calculated using:

$$\sigma_{t\bar{t}} = \frac{N_{\text{data}} - N_{\text{bkg}}}{\int L\,dt \cdot \epsilon \cdot BR(\text{noFullHad})}$$

where N_{data} and N_{bkg} are the data and background yields respectively, ϵ is the signal efficiency for single-lepton and dilepton channels and $BR(\text{noFullHad}) = 0.543$ is the single-lepton branching ratio [18]. The combined $t\bar{t}$ cross section is calculated by summing the e+jets and μ+jets event yields and is measured to be:

$$\sigma_{t\bar{t}} = 165 \pm 2(\text{stat.}) \pm 17(\text{syst.}) \pm 3(\text{lumi.})\ \text{pb}$$

Figure 1. Number of observed and expected jets in the e+jets(left) and μ+jets(right) analysis [1].

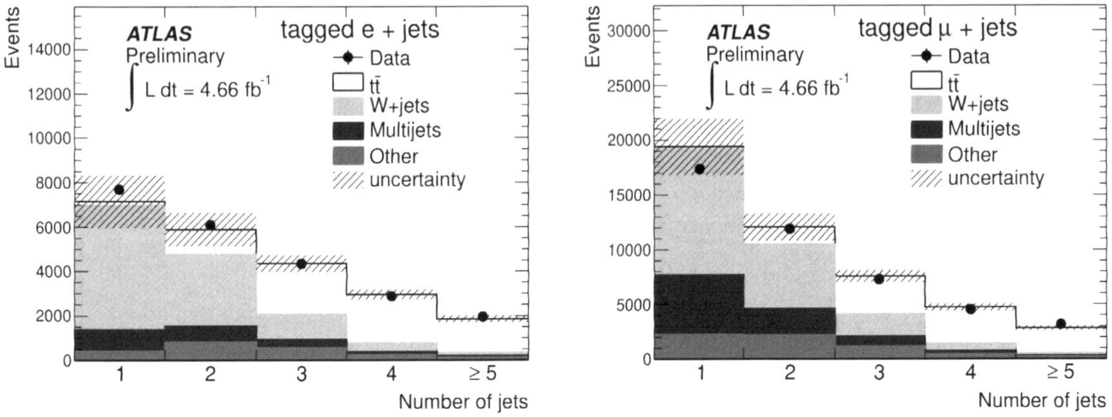

Table 1. Observed and estimated event yields in the pretag and tagged samples. Backgrounds are evaluated with Data Driven (DD) techniques or with MC simulation. Uncertainties are quoted as the quadratic sum of statistical and systematic contributions [1].

Sample	Pretag	Tagged
Data (4.66 fb^{-1})	351742	24105
$t\bar{t}$ MC	84000±2060	15060±590
W+jets DD	176500±10750	4520±600
Multijet DD	43200±9700	1930±470
Z+jets MC	21400^{+3470}_{-2130}	1050^{+145}_{-105}
Single Top MC	11430±680	1630±100
DiBoson MC	3220^{+410}_{-350}	100±15
$t\bar{t}$ MC + Backgrounds	340000± 15000	24300±980
Measured $t\bar{t}$		14900±800
Selection Efficiency (%)		3.57 ± 0.03

Table 2. Summary of individual relative systematic uncertainty contributions to the cross section measurement, in percent [1].

Source	Uncertainty [%]
Statistical Uncertainty	±1.0
Object selection	
Lepton energy resolution	+0.2 /-0.1
Lepton reco, identification, trigger	+1.7 /-1.8
Jet energy scale	+3.5 /-3.8
Jet energy resolution	±0.2
Jet reconstruction efficiency	±0.06
Jet vertex fraction	+1.2 /-1.4
$E_{\mathrm{T}}^{\mathrm{miss}}$ uncertainty	±0.07
SMT muon reco, identification	±1.3
SMT muon χ^2_{match}efficiency	±0.6
Background estimates	
Multijet normalisation	± 4.4
W+jet normalisation	± 5.5
Other bkg normalisation	± 0.1
Other bkg systematics	+2.2 /-1.8
Signal simulation	
$b \to \mu X$ Branching ratio	+2.9 /-3.1
Initial/Final state raditaion	± 1.5
PDF	±3.1
NLO generator	± 3.2
Parton shower	± 2.2
Total systematics	±10.5
Integrated luminosity	± 1.8

References

[1] ATLAS Collaboration 2012, ATLAS-CONF-2012-131, http://cds.cern.ch/record/1478370
[2] Aliev M et al. 2011 *Comput. Phys. Commun.* **182** 1034-46
[3] Martin A D et al. 2009 *Eur. Phys.* J **C** **63** 189-285
[4] Martin A D et al. 2009 *Eur. Phys.* J **C** **64** 653-680
[5] Cacciari M el al. (*Preprint hep-ph/1111.5869*)
[6] Czakon M and Mitov, A (*Preprint hep-ph/1112.5675*)
[7] ATLAS Collaboration 2008, *JINST* 3 S08003
[8] Evans L and Bryant P (editors) 2008, *JINST* 3 S08001
[9] ATLAS Collaboration 2012, ATLAS-CONF-2012-080, http://cds.cern.ch/record/1460392
[10] Frixione S and Webber B R 2002, *JHEP* **06** 029
[11] Frixione S, Nason P and Webber B R 2003, *JHEP* **08** 007
[12] Corcella G el al. 2001, *JHEP* **01** 010
[13] Butterworth J M, Forshaw J F and Seymour M H 1996, *Z. Phys.* **C** **72** 637-646
[14] Lai H L et al. 2010, *Phys. Rev.* D **82**
[15] Mangano M L el al. 2003, *JHEP* **07** 0013
[16] Kersevan B P and Richter-Was E (*Preprint hep-ph/0405247*)
[17] Sjöstrand T, Mrenna S and Skands P 2006, *JHEP* **05** 26
[18] Beringer J el al. 2012, *Phys. Rev.* D **86** 010001
[19] ATLAS Collaboration 2010, *JHEP* **1012** 060
[20] Cacciari M and Salam G P 2006 *Phys. Lett.* B **641** 57-61
[21] Cacciari M, Salam G P and Soyez, G 2008 *JHEP* **04** 063
[22] ATLAS Collaboration (*Preprint hep-ex/1112.6426v1*)
[23] Kom S H and Stirling J, 2010 *Eur. Phys.* J **C** **69** 67
[24] ATLAS Collaboration 2011, *Eur. Phys.* J **C** **71** 1577

Journal of Physics: Conference Series **452** (2013) 012031 doi:10.1088/1742-6596/452/1/012031

Top pair production cross-section results at Tevatron

Gianluca Petrillo, for the CDF and DØ collaborations
Fermi National Accelerator Laboratory, MS 352, 60510 BATAVIA, IL, U.S.A.
E-mail: petrillo@fnal.gov

Abstract. We present the latest measurements of the cross section of the production of the top quark in $t\bar{t}$ pairs from $p\bar{p}$ collisions at $\sqrt{s} = 1.96$ TeV.
Measurements are performed using up to 9.1 fb^{-1} of data collected by the CDF and DØ experiments at the Tevatron.
The first combination of measurements from the two experiments is presented. From six different measurements, the cross section is measured to be $\sigma\left(p\bar{p} \to t\bar{t}, \sqrt{s} = 1.96\,\text{TeV}\right) = 7.65\ \pm 0.42\,\text{pb}$.

1. Introduction

According to the Standard Model, the top quark can be produced in association with a b quark via electroweak interaction, or in a $t\bar{t}$ pair via strong interaction. The rate of the two production mechanisms are predicted to be of the order of 3 and 7 pb, respectively, at $\sqrt{s} = 1.96$ TeV.
In a recent achievement, the computation of the production cross section from strong interaction has been completed at Next-to-Next-to-Leading Order (NNLO), achieving an uncertainty of about 3.5% (evaluated as $7.24^{+0.15}_{-0.24}\,\textit{(scale)}^{+0.18}_{-0.12}\,\textit{(PDF)}$ pb using Top++[1] with $\sqrt{s} = 1.96$ TeV, MSTW2008nlo68cl PDF set and $m_t = 172.5\ \text{GeV}/c^2$).
The experimental precision required for the falsification of this prediction is a challenge for the experiments at Tevatron, that can be won only by the combination of their results.

The most recent measurements of $t\bar{t}$ cross section ($\sigma_{t\bar{t}}$) from the CDF collaboration are described, together with the first combination of $\sigma_{t\bar{t}}$ measurements from the CDF and DØ experiments. For a detailed description of the detectors, see [2] and [3].

2. Statistical methods for cross section measurements

The simplest way to measure the $t\bar{t}$ production cross section employs the observation of a number of events (N_{obs}) and the estimation of the expected background events (N_{bkgr}) and of the fraction of $t\bar{t}$ events reconstructed by the detector and selected by the analysis (ε). The cross section is then $\sigma = \frac{N_{\text{obs}} - N_{\text{bkgr}}}{\mathcal{BR} \cdot \varepsilon \cdot \mathcal{L}\,dt}$ (1), where $\mathcal{L}\,dt$ is the integrated luminosity corresponding to the analysed sample and $\mathcal{BR}$ is the branching fraction of the final state.
If the prediction of the background is dependent on σ, the cross section can be extracted from the maximization of a likelihood $L(\sigma) = \mathcal{P}\left(N_{\text{obs}}, N_{\text{pred}}\right)$ (2), where $\mathcal{P}$ typically describes the Poisson distribution. Methods based on the likelihood also provide an estimation of the statistical uncertainty.
This method can be extended by splitting the sample in independent subsamples and building a joint likelihood as the product $L(\sigma) = \prod_k \mathcal{P}\left(N^{(k)}_{\text{obs}}, N^{(k)}_{\text{pred}}\right)$ (3) of one likelihood from each

subsample, or *bin* ("binned likelihood").

3. New cross section measurements

The most precise measurements at $\sqrt{s} = 1.96\,\text{TeV}$ to date are based on $\approx 5\,\text{fb}^{-1}$, on events with one reconstructed light lepton ("ℓ", either electron or muon), and they achieve singularly the precision of about 7 to 9%.
The measurements presented here focus on different final states. While their branching fraction is too small for them to compete with the former in precision, they provide independent information on the decay of the top quark.

3.1. Measurement from τ+jets *final state* ($\tau \to$ hadrons)

The measurement of $t\bar{t}$ cross section in final states including τ leptons is particularly interesting in that it could reveal the presence of intermediate particles, such as a charged Higgs boson, whose coupling to fermions is proportional to their mass. The final state is characterized by four jets, two of which from b quarks, a neutrino and a τ lepton.

This measurement [4] is performed on 2.2fb^{-1} of CDF data. The section of the events includes at least four jets, at least one of which is identified as coming from a b quark ("b-tagging"), a minimum imbalance of transverse momentum in the event and a τ lepton. The latter is identified as a narrow hadronic jet associated to either one or three tracks, representing the lepton decay into one or three charged mesons. The tracks are required to be isolated: no additional track and less than 10% of the total jet energy must fall in a cone of radius 30° around the jet axis.

The main source of background is from events with many reconstructed jets ("multijet"), where one of them can mimic the signature of a τ lepton. The second largest source is from events with a W boson and jets ("W+jets"), where W decays into a real τ lepton. To describe the multijet background, events are selected out of the data sample. These events are required to fulfill all the criteria of the analysis, except that the jet candidate as τ must be *non*-isolated. A discriminant based on a Neural Network is built to separate the signal from the multijet background.

The shape of the distribution of the discriminant is used to quantify the contribution of the two dominant backgrounds (Fig. 1). The size of W+jets contamination is determined without the requirement of b-tagging. The contributions from all the processes are fixed to the predicted values, except for the one from the multijet and the W+jets processes. Their size is estimated by the maximization of a Poisson likelihood on the binned discrimnant distribution. The value obtained for the W+jets fraction is used, together with the knowledge of b-tagging efficiency, for the estimation of the W+jets background in the data sample used for the measurement. The estimation of the multijet background does not come from this maximization. Instead, a similar procedure is performed on the samples *with* the b-tagging requirement, after fixing all the contributions from the processes other than multijet to their expectation (including the one from $t\bar{t}$ signal, assumed $\sigma_{t\bar{t}} = 7.4\,\text{pb}$). By this second maximization, the *fraction* of data coming from multijet processes is obtained.

The expected number of $t\bar{t}$ events and of multijet events can be computed for each given $\sigma_{t\bar{t}}$, using the multijet fraction from the previous step. The expectation of the contributions from all the other processes does not depend on $\sigma_{t\bar{t}}$. The cross section is extracted from a signal region defined by the requirement of the discriminant value being above 0.85. This region encompasses 41 candidate events, with an expectation, for $\sigma_{t\bar{t}} = 7.4\,\text{pb}$, of 20.4 background events.

By maximizing a Poisson likelihood Eq. 2, the measured value of cross section is $\sigma_{t\bar{t}} = 8.8 \pm 3.3\,(stat) \pm 2.2\,(syst{+}lumi)\,\text{pb}$, compatible with the theoretical prediction. The dominant systematic uncertainty is from the modelling of the multijet background.

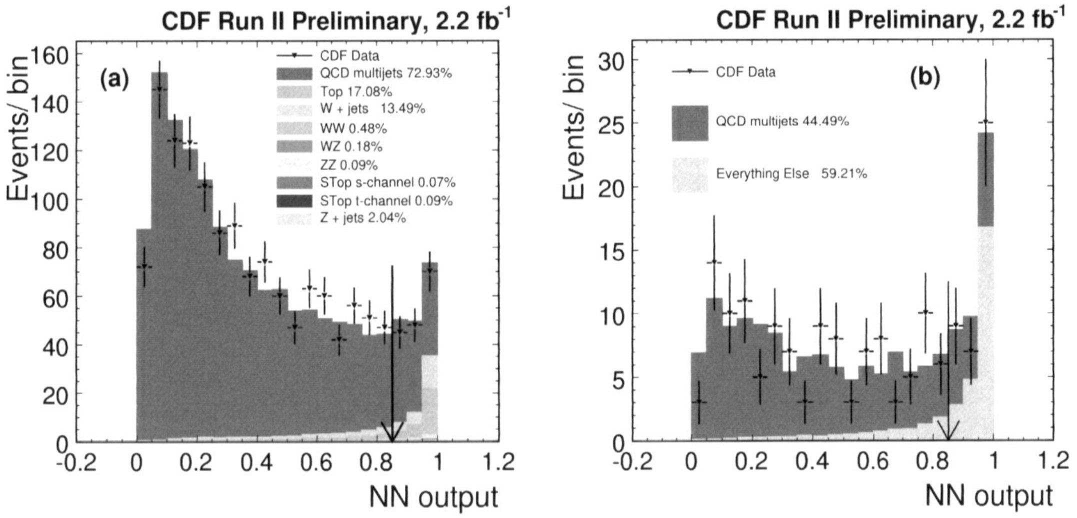

Figure 1. Distribution of the discriminant of $t\bar{t}$ signal vs. multijet background for τ+jets events, *(a)* before and *(b)* after the b-tagging requirement. The arrows mark the lower bound (0.85) of the signal region for the final measurement.

3.2. Measurement from $\tau + \ell$+jets final state ($\tau \rightarrow$ hadrons)

CDF has performed a cross section measurement on events with signature including one light lepton (electron or muon) and one τ lepton decaying hadronically [5]. The light lepton can either come from the prompt W decay or from a cascade decay $W \rightarrow \tau\nu_\tau$, $\tau \rightarrow \ell\nu_\ell\nu_\tau$. This final state can give the same insight on additional charged-Higgs-like t decay channels as the one described in Sect. 3.1. The signature is cleaner, because of the presence of the light lepton and large missing energy, but it is suppressed by the lower branching fraction of the additional $W \rightarrow \ell\nu_\ell$ decay. This is the first analysis presented here which uses the complete CDF RunII dataset (9.0 fb^{-1}).

There are two main sources of background for this signature. The first is from Z+jets events, when the Z boson decays into two τ leptons, one decaying into hadrons, the other into leptons. The second background comes from W+jets events, where the light lepton comes from the $W \rightarrow \ell\nu_\ell$ decay and one of the additional jets is misidentified as a τ lepton decaying into hadrons.

After the first kinematic selection, requiring a light lepton and a narrow jet (the τ candidate) with opposite charge, two or more additional jets, and moderate transverse momentum imbalance (10 GeV/c), the signal-over-background ratio (S:B) is estimated to be 2:9. The Z+jets events can be effectively rejected by requiring one jet to be identified as coming from a b quark. After this requirement, S:B becomes 2:1.

To further increase the signal purity, a log-likelihood-ratio discriminant (LLR) is designed to resolve the signal from the W+jets background. The quantities employed in the LLR include the transverse momentum imbalance ($\not{\vec{p}}_T$), the transverse mass of the system composed by the light lepton and the $\not{\vec{p}}_T$, and the transverse energy (E_T) of the jet with the third largest E_T.

The selection of events with $LLR \geq 0$ yields to a final sample of 36 candidate events, with an expected S:B ratio of 5:1 and an expected background of 6.9 events. The cross section is measured to be 8.2 ± 2.3 *(stat)* $^{+1.2}_{-1.1}$ *(syst)* ± 0.5 *(lumi)* pb, also compatible with the theoretical prediction. This measurement is affected by large statistical uncertainty. Nevertheless, the

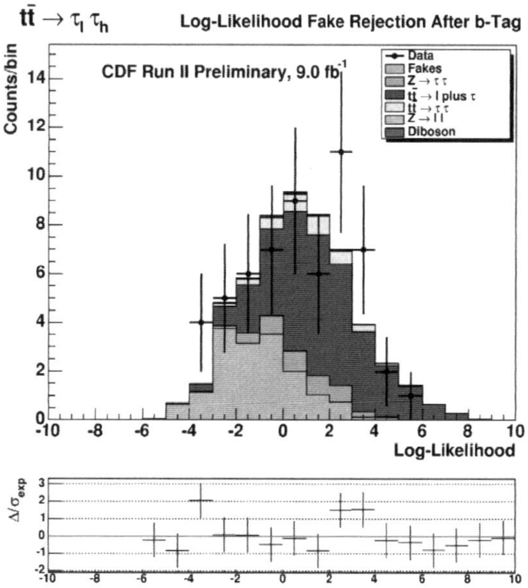

Figure 2. Distribution of the discriminant of $t\bar{t}$ signal vs. W+jets background for $\tau + \ell$+jets events.

aggressive pursue of a pure sample allows the resolution of prompt $t\bar{t} \to b\bar{b}\,\tau\nu_\tau\,\ell\nu_\ell$ and cascade $t\bar{t} \to b\bar{b}\,\tau\nu_\tau\,\tau\nu_\tau \to b\bar{b}\,\tau\nu_\tau\,\ell\nu_\ell\nu_\ell\nu_\tau$ decays, which was described by A. Jung in this same conference.

3.3. Measurement from 2ℓ+jets final state

The last single measurement by CDF presented here [6] is also performed on the complete RunII dataset. The final state is defined by the presence of two light leptons and jets. It is characterized by large transverse momentum imbalance, two jets from b quarks and light leptons of opposite charge.

This final state is fairly pure due to the presence of the two light leptons. The most relevant background sources are the production of Z+jets, and W+jets events with one real lepton and one jet misidentified as the other lepton.

The measurement is performed twice, using two different sets of selection criteria. The first one includes the requirement for large $\not{p}_T$, very large H_T (sum of the transverse energy of all the reconstructed objects, $H_T \geq 200$ GeV/c) and at least two jets. In addition, other cuts are aimed to reduce specific backgrounds (e.g. the exclusion of events with the invariant mass of the leptons in the [76; 106] GeV/c^2 range). This selection yields to 625 candidate events for 9.1 fb^{-1} of CDF data, achieving a signal-over-background ratio of 2:1. The measured cross section is $\sigma_{t\bar{t}} = 7.66\ \pm 0.46\,(stat)\ \pm 0.66\,(syst)\ \pm 0.47\,(lumi)$ pb. The main systematic uncertainty comes from the modelling of the background processes.

The second selection adds the requirement for at least one of the jets to be identified as coming from a b quark. This last requirement roughly halves the signal, but annihilates the background and achieves S:B of 14:1. The selection includes 254 events out of 8.8 pb^{-1} of CDF data (the smaller size of the dataset is due to the stricter data quality requirements needed by the b-jet identification algorithm). This yields to a measured cross section of $\sigma_{t\bar{t}} = 7.47\ \pm 0.50\,(stat)\ \pm 0.53\,(syst)\ \pm 0.46\,(lumi)$ pb. The background modelling is not a factor anymore in a measurement on a sample with less than 10% of background contamination. Unfortunately the additional uncertainty from the b-tagging undermines this gain. Nevertheless this measurement turns to be more precise.

4. Combination of measurements

The combination of different results can sensibly increase the precision of a measurement. In the case of two different experiments like CDF and DØ, not only the doubled size of the sample reduces the statistical uncertainty, but also many of the systematic uncertainties are largely independent due to the different detectors and algorithms. The first combination of measurements of top quark properties was for its mass (the latest of which can be found in [7]). The experience and techniques of that combination have been employed now, for the first time, to the measurements of the production rate. A *Best Linear Unbiased Estimator* is used to quantify the cross section.

A detailed knowledge of the uncertainties is required. The uncertainties from each measurement have been split in 10 categories in order for each of them to be defined consistently between the analyses. These uncertainties are considered either completely correlated within each experiment (as for the ones relates to detector modelling) or across the two experiments (as for the integrated luminosity), or completely independent (as for the statistical uncertainty). The uncertainties are assumed to follow a normal distribution. All the measurements from CDF and all the ones from DØ are combined into a CDF and a DØ combination result on the first step, then the two experiment-wide combinations are combined again into the Tevatron combination.

The choice of the measurements has been performed with the goal to achieve the largest independence. There are four measurements contributed by CDF: one from 8.8 fb^{-1} of di-lepton events (see Sect. 3.3); one from 4.6 fb^{-1} of events with one electron or muon, which is a relative measurement of $\sigma(t\bar{t})/\sigma(Z)$; another from the same dataset (4.3 fb^{-1}), with the additional identification of jets from b quarks; and one from 2.9 fb^{-1} of all-jet events. There are two measurements contributed by DØ: one based on 5.3 fb^{-1} of events with one electron or muon, and one based on 5.4 fb^{-1} of events with two leptons chosen between electrons and muons.

The combined result yields to a cross section at $\sqrt{s} = 1.96$ TeV

$$\sigma\,(p\bar{p} \to t\bar{t}) = 7.65\ \pm 0.20\,(stat)\ \pm 0.29\,(syst)\ \pm 0.22\,(lumi)\,\text{pb} = 7.65\ \pm 0.42\,\text{pb}$$

assuming the mass of the top quark to be 172.5 GeV/c^2. This corresponds to a precision of 5.5%, to be compared with the 3.5% of the current state-of-the-art prediction from theory [1].

5. Conclusion

We described here the first two measurements of the production cross section of $t\bar{t}$ using the complete dataset available to CDF. Other analyses will follow in order to cover all the main observable signatures of a $t\bar{t}$ event.

While it is important to have all these separate measurements, the main goal is to achieve the best possible precision in the characterization of the $p\bar{p} \to t\bar{t}$ process at $\sqrt{s} = 1.96$ TeV. The challenge of the precision can be won only by the combination of heterogeneous measurements relying on different datasets, experimental apparatuses and analysis techniques. The combination presented here is even more important as preparation to such final $\sqrt{s} =$ 1.96 TeV combination, and as an example for similar efforts that could be undertaken in the future by the collaborations working on LHC as the preferred path to the reduction of systematic uncertainties.

References

[1] For an updated version see: Czakon M and Mitov A NNLO corrections to top pair production at hadron colliders: the quark-gluon reaction *Preprint* hep-ph/1210.6832

[2] Abulencia A *et al.* (CDF Collaboration) 2007 Measurements of inclusive W and Z cross sections in $p\bar{p}$ collisions at $\sqrt{s} = 1.96$ TeV *J. Phys. G: Nucl. Part. Phys.* **34**, 245

[3] Abazov V M *et al.* 2006 The upgraded DØ detector *Nucl. Instr. Meth.* A **565**, 463

[4] Aaltonen T *et al.* (CDF Collaboration) 2012 Measurements of the Top-quark Mass and the $t\bar{t}$ Cross Section in the Hadronic τ + Jets Decay Channel at $\sqrt{s} = 1.96$ TeV *Phys. Rev. Lett.* **109**, 192001
[5] The CDF Collaboration 2012 Measurement of the top pair cross section in dileptonic channels with a hadronic tau and branching ratio $t \to \tau\nu b$, with $9.0\,\mathrm{fb}^{-1}$ *CDF note 10915*
[6] The CDF Collaboration 2012 Measurement of Top Dilepton Cross Section with CDF Full data the DIL selection *CDF note 10878*
[7] The CDF and DØ Collaborations 2012 Combination of the top-quark mass measurements from the Tevatron collider *Phys. Rev.* D **86**, 092003
[8] The CDF and DØ Collaborations, 2012 Combination of the $t\bar{t}$ production cross section measurements from the Tevatron Collider *DØ note 6363, CDF note 10926*

Measurement of Missing Transverse Energy in Top Pair Events

Philip Symonds
Brunel University
Kingston Ln, Uxbridge, Middlesex, UB8 3PH, United Kingdom
On behalf of the CMS Collaboration

Abstract. A differential cross section measurement of top quark pair production with missing transverse energy is presented using 5.1 fb^{-1} of data collected using the CMS detector at the LHC at a centre of mass energy of 7 TeV. The analysis selects events with a single isolated high energy electron or muon, which is assumed to come from one of the W bosons produced in the decays of the top quark and antiquark. The differential cross section is measured in bins of missing transverse energy. The results are consistent with the predictions of simulation.

1. Introduction

During 2011 the LHC produced almost 800000 top quark pairs and is therefore often referred to as a top quark factory. This allows a detailed investigation of the missing transverse energy (E_T^{miss}) in top quark pair events. This can potentially be used not only to verify the current models of top quark production, but also to measure rare Standard Model (SM) processes with additional E_T^{miss}. Top pair production is a major background for new physics searches involving missing transverse energy signatures, so it is important that the distribution of E_T^{miss} in top events is well understood. In addition there are many new physics scenarios that produce additional E_T^{miss} in association with a $t\bar{t}$ pair.

Top quarks decay with close to 100% probability into a W boson and a bottom quark. In $t\bar{t}$ pair production the semileptonic channel involves one of the W bosons decaying into a lepton along with its associated neutrino while the other W boson decays hadronically.

New results from the CMS experiment at the LHC are presented, based on data collected at a centre-of-mass energy of 7 TeV during 2011, measuring the distribution of E_T^{miss} in the production of top quark pairs and comparing with the predictions from event generator codes. The CMS detector is described in detail elsewhere [1].

2. Analysis Strategy

The analysis is designed to select semileptonic top pair decays, those in which one of the W bosons has decayed leptonically and the other has decayed into a quark pair. We include decays to electrons or muons, plus an accompanying neutrino. Additional leptons are vetoed to avoid contamination from dilepton processes. W boson decays to τ leptons are not explicitly searched for in the analysis although the selected samples include events where the τ decays to muons and electrons. The final state includes the charged lepton and neutrino, and four jets of hadrons from the decays of the top quarks. Two of these come directly from the decays of top to bW, and contain b-flavoured hadrons; the other two jets come from the decay of a W. The requirement of two b tagged jets enables the number of background events selected to be minimised. A sizeable fraction of events in simulation contain extra jets due to hard QCD radiation.

In semileptonic top pair decays an imbalance is expected in the measured momentum, in the plane perpendicular to the beam direction, due to the presence of a neutrino in the final state. Subsamples are divided according to the measured missing transverse momentum, and study the distribution of the polar angle of leptons to the beam axis in each subsample. As a consequence of the large mass of the top quark, the leptons in top pair events are produced predominantly at large angles to the beam axis, whereas the main background processes give rise to leptons produced at smaller angles. A template fit of the lepton pseudorapidity is used to estimate the composition of each subsample, and hence the number of top pair events in the range of missing transverse energy concerned.

The analysis is initially carried out separately for event samples where the charged lepton is a muon or an electron. The final result is based on a combination of the two channels, and is compared against predictions from a range of Monte Carlo event generator codes.

3. Simulation

For the simulation of the $t\bar{t}$ signal sample the MadGraph event generator [2] is used, which implements the relevant matrix elements up to three additional partons. The generated events are subsequently processed with PYTHIA [3] for parton showering and hadronisation. The PYTHIA parton shower is matched to the jets from the hard QCD matrix element via the MLM prescription [4] with a threshold of 40 GeV.

All of the samples are generated at the centre of mass energy of 7 TeV, using the CTEQ6L parton distribution functions [5]. PYTHIA is used for the radiation and fragmentation using the prescriptions in [6] and the underlying event tuning described in reference [7].

For comparison with the measured lepton $|\eta|$ and $E_{\mathrm{T}}^{\mathrm{miss}}$ distributions (Figure 1), the events in the simulated samples are normalised to the integrated luminosity of the data sample, 5.05 fb^{-1}. For the SM top-quark pair production cross section, the next-to-next-to-leading (NNLO) order value [8] of 163 pb is used. For the vector boson production, the inclusive NNLO calculation from [9] is used.

The Monte Carlo samples used are reweighted to match the pileup conditions of the data. A further reweighting takes account of known small differences between the performance of the b-tagging algorithm on data and Monte Carlo.

4. Cross Section Measurement

4.1. Fit Procedure

The observed events are divided into five $E_{\mathrm{T}}^{\mathrm{miss}}$ bins. In each bin, a template fit of the observed distribution of lepton rapidity $|\eta_\ell|$ is used to subtract the main backgrounds and estimate the number of $t\bar{t}$ events.

Three separate $|\eta_\ell|$ distributions are produced for each $E_{\mathrm{T}}^{\mathrm{miss}}$ bin as input to the template fit. The first template is the $|\eta_\ell|$ shape for top-like events, including $t\bar{t}$ and single top. This is taken from simulation, where the relevant event samples all give similar shapes. The $|\eta_\ell|$ distributions for the W+jets and Z+jets samples are summed together to give the second template. The distribution for QCD multijet events is estimated from data, looking at control samples where one of the lepton selection cuts is inverted.

A binned maximum likelihood fit is then used to estimate the contribution of each template to the overall $|\eta_\ell|$ distribution seen in data. The fit yields an estimate of the number of top-like events observed in the $E_{\mathrm{T}}^{\mathrm{miss}}$ bin. By subtracting the Monte Carlo estimate for the contribution from single-top production, the number of $t\bar{t}$ events is obtained.

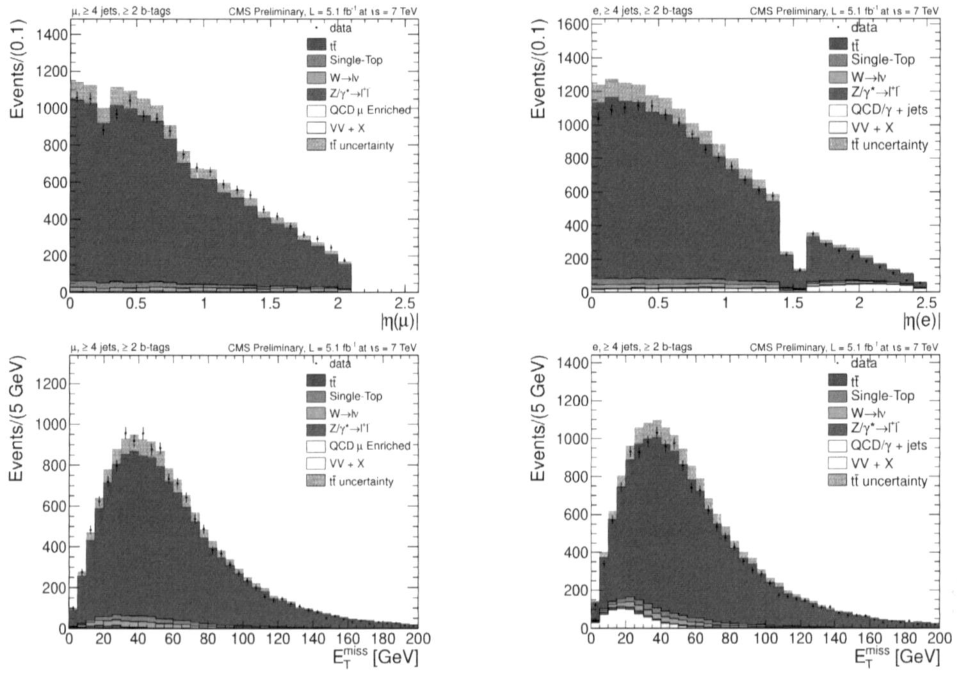

Figure 1. Kinematic variable distributions seen in the event samples after all cuts. Points with error bars are the 2011 data, stacked histograms the Monte Carlo expectations, the grey error bars the uncertainty on the measured $t\bar{t}$ cross section. Left plots: muon+jets; right plots: electron+jets. Top: lepton $|\eta_\ell|$; Bottom: $E_{\mathrm{T}}^{\mathrm{miss}}$.

4.2. Cross Section Corrections

The differential cross section for semileptonic $t\bar{t}$ production is measured as a function of $E_{\mathrm{T}}^{\mathrm{miss}}$, correcting for the effect of $E_{\mathrm{T}}^{\mathrm{miss}}$ bin migration due to finite detector resolution; the experimental acceptance and selection efficiency; and the contribution of other $t\bar{t}$ decay channels. The required correction factors are derived from simulation.

5. Systematic Errors

The uncertainties are estimated by varying the input quantities according to their theoretical and/or experimental uncertainties and observing the change in the final result. The propagated uncertainties are set to be symmetric, conservatively taking the larger of the upward and downward variation. Correlations between input variables are taken into account.

The dominant sources of systematic error are the uncertainty in the measurement of $E_{\mathrm{T}}^{\mathrm{miss}}$, and the theoretical uncertainty in the W-plus-jets background sample. Other non-negligible sources of uncertainty are the energy and momentum scales for electrons, muons and jets. The error in the QCD estimation is relatively large, but this is mitigated by the small QCD contamination of the sample after requiring two b-tagged jets. Uncertainties that affect the overall rate (e.g. the luminosity error) contribute a negligible amount as they cancel in the normalisation.

6. Results

The procedure described in section 4 has been carried out for both muon and electron samples, to give measurements of $\frac{d\sigma_{t\bar{t}}}{dE_{\mathrm{T}}^{\mathrm{miss}}}$ in five bins of $E_{\mathrm{T}}^{\mathrm{miss}}$. The total $\sigma_{t\bar{t}}$ measured with the same analysis is used to arrive at a normalised differential cross section for each channel.

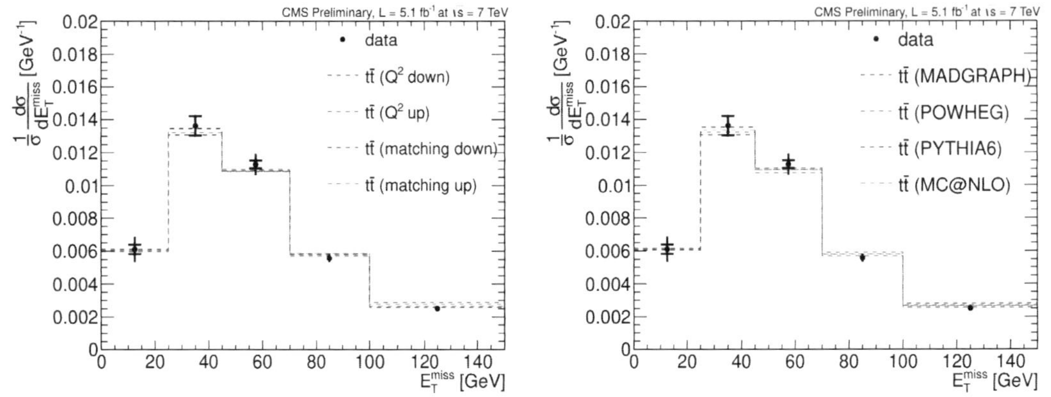

Figure 2. Comparison of normalised differential cross section result to matching threshold and Q^2 scale up and down samples (left) and different generators (right) for the combination of both channels. The inner error bars indicate the fit uncertainty and the outer error bars the c ombined fit and systematic uncertainty.

The two channels have also been combined by summing the two cross section measurements in each $E_{\mathrm{T}}^{\mathrm{miss}}$ bin, after all corrections. The summed total cross section is then used to normalise the results. A comparison to other generators and different theory inputs is shown in figure 2.

7. Summary

A measurement of the differential cross section of top quark pair production with respect to $E_{\mathrm{T}}^{\mathrm{miss}}$ has been performed in pp collisions at a 7 TeV centre of mass energy, in 5.1 fb^{-1} of data collected by the CMS experiment at the LHC in 2011. The analysis selects events with a single isolated high energy electron or muon, which is assumed to come from one of the W bosons produced in the decays of the top quark and antiquark. The differential cross section is measured in five bins of missing transverse energy. The results are consistent with the predictions of Monte Carlo generators.

References

[1] CMS Collaboration 2008 *JINST* **0803** S08004

[2] Maltoni F and Stelzer T 2011 *JHEP* **1106** 128 (*Preprint* tt̄ 1106.0522)

[3] Sjöstrand T, Mrenna S and Skands P 2006 *JHEP* **05** 026 (*Preprint* tt̄ hep-ph/0603175)

[4] Hoche S *et al.* 2006 *hep-ph/0602031*

[5] Pumplin J *et al.* 2002 *JHEP* **07** 012 (*Preprint* tt̄ hep-ph/0201195)

[6] Bartalini P, Chierici R and de Roeck A 2005 Guidelines for the estimation of theoretical uncertainties at the lhc Tech. Rep. CMS-NOTE-2005-013 CERN Geneva

[7] Acosta D, Ambroglini F, Bartalini P, de Roeck A, Fano L, Field R D and Kotov K 2006 The underlying event at the lhc Tech. Rep. CMS-NOTE-2006-067 CERN Geneva

[8] Kidonakis N 2010 *Phys. Rev. D* **82**(11) 114030 URL tt̄ http://link.aps.org/doi/10.1103/PhysRevD.82.114030

[9] Melnikov K and Petriello F 2006 *Phys. Rev.* **D74** 114017 (*Preprint* tt̄ hep-ph/0609070)

Journal of Physics: Conference Series **452** (2013) 012033 doi:10.1088/1742-6596/452/1/012033

Differential top pair cross section and top anti-top plus jets Physics [1]

Malgorzata Worek

Theoretische Physik, Fachbereich C, Bergische Universität Wuppertal, D-42097 Wuppertal, Germany

Institut für Theoretische Teilchenphysik und Kosmologie, RWTH Aachen University, D-52056 Aachen, Germany

E-mail: worek@physik.rwth-aachen.de

Abstract. A brief summary of the current status of the next-to-leading order QCD corrections to top quark pair production and the associated production of $t\bar{t}$ with jet(s) in different configurations, *i.e.* with one jet, two jets and another $t\bar{t}$ pair, is presented.

1. Introduction

By the end of 2012 a successful run of the Large Hadron Collider (LHC) with proton-proton collisions at 8 TeV delivered 23.3 fb^{-1} of collected data. This large available statistics has opened a window on entirely new measurements of more complex final states. Reducing theoretical uncertainties for correct interpretation of these data is among the highest priorities of the theoretical high energy community. A theoretical accuracy at least at next-to-leading order (NLO) level is desirable and for most analyses even demanded.

The major purpose of the LHC is a deeper understanding of the interactions among fundamental constituents of matter. To this end, the top quark plays a special role. At the LHC it is copiously produced via strong interactions in top anti-top pairs, so its production cross section, decays and properties are studied with high precision. Apart from being studied as a signal process, the top quark constitutes the main background in analyzes of the recently discovered Standard Model (SM) like Higgs boson. Distinguished by its large mass, the top quark is also potentially sensitive to physics beyond the SM (BSM), which concerns among others searches for non-SM Higgs bosons and supersymmetric particles. In order to understand and control top quark background processes to the SM and the BSM physics, precise predictions for these reactions at the differential level are indispensable.

In this contribution, a brief report on the recent activities in the calculations of the NLO QCD corrections to the differential top pair cross section and the associated production of $t\bar{t}$ with jet(s) in different configurations, *i.e.* with one jet, two jets and a $t\bar{t}$ pair, is given.

2. Top Anti-Top

The NLO QCD corrections to the production of the top anti-top pairs followed by top decays in the so called narrow-width approximation (NWA) have already been available for some

[1] Preprint number: WUB/13-02

time. In this approach the full amplitude is factorized into production of the unstable particles and their subsequent decays neglecting non-resonant and non-factorizable contributions to the amplitude. The NWA significantly simplifies calculations of higher order corrections on the one hand at the same time preserving gauge invariance. First calculations of the NLO QCD corrections to the differential cross section of top quark pair production in hadronic collisions taking only double resonance contributions into account and top and anti-top spin degrees of freedom in the production and decay have been presented in [1,2]. Mixed QCD and electroweak NLO corrections have been also calculated see e.g. [3] and references therein. In addition, several charge asymmetries have been computed for the $t\bar{t}$ production at the LHC and at the Tevatron [4]. Recently independent implementations of the NLO QCD corrections to top quark production and decay at fully differential level with all spin correlations appeared [5,6]. Moreover, NLO QCD corrections matched to parton shower for heavy flavor hadroproduction are obtainable via publicly available numerical programs like e.g. MC@NLO [7] and POWHEG [8], however these programs do not fully include spin correlations through NLO QCD.

If resonant top production dominates, the NWA approach is very well motivated. Nevertheless the accuracy of the NWA needs to be tested, which requires a full NLO QCD calculation of off-shell effects, i.e. double-, single- and non-resonant top quark contributions of the order $\mathcal{O}(\alpha_s^3\alpha^4)$ need to be taken into account. This has been done by two independent groups that have calculated NLO QCD corrections to the production of top anti-top pairs including interferences, off-shell effects, non-resonant contributions and spin correlations [9–11]. This required the introduction of a complex-mass scheme for unstable top quarks. Moreover, the intermediate W bosons were treated off-shell. These NLO calculations provide the most complete description of top anti-top pair production with leptonic decays of both $W^\pm$ gauge bosons. For the inclusive cut selection finite-width effects on $\sigma_{t\bar{t}}$ have been found to be around 1% both at the Tevatron and the LHC. In addition, the asymmetries for the top quark and the charged lepton in top anti-top production at the Tevatron and the LHC have been studied [10,11].

3. Top Anti-Top Plus Jet

For the $p\bar{p}$ and pp collisions at the TeVatron and the LHC a substantial number of events in the inclusive top anti-top sample is accompanied by an additional jet. Since top quarks are produced with large energies and at high p_T, the probability for a top quark to radiate gluons considerably increases. Depending on the p_T of the additional jet the fraction of events with an additional jet can easily be of the order of 30% or more [12]. Sample values for two different $p_T(j)$ cuts are given in Table 1, both for the TeVatron and the LHC, where results for $\sigma_{t\bar{t}j}^{\rm NLO}$ have been taken from [12], while $\sigma_{t\bar{t}}^{\rm NLO}$ have been calculated with the help of TOP++ [13]. The

Table 1. *Fraction of events in the inclusive top anti-top sample that is accompanied by an additional jet depending on the p_T of the additional jet at the TeVaron and the LHC.*

TEVATRON $\sqrt{s} = 1.96$ TeV	$\sigma_{t\bar{t}j}^{\rm NLO}/\sigma_{t\bar{t}}^{\rm NLO}$	LHC $\sqrt{s} = 14$ TeV	$\sigma_{t\bar{t}j}^{\rm NLO}/\sigma_{t\bar{t}}^{\rm NLO}$
$p_T \geq 20$ GeV	30%	$p_T \geq 50$ GeV	47%
$p_T \geq 40$ GeV	11%	$p_T \geq 100$ GeV	22%

production of a top anti-top pair together with an additional jet is therefore crucial for a more precise understanding of the topology of top anti-top events. The $t\bar{t}j$ production is, however, also a dominant background to various new physics searches. Among others, the analysis of the

SM Higgs boson in the vector boson fusion process is a prominent example. In this case, precise theoretical predictions are indispensable.

NLO QCD corrections to the on-shell production of $t\bar{t}j$ have been first calculated in [14] and subsequently confirmed in [16,32]. Furthermore in [16] top quark decays at leading order (LO) in the NWA were included. Quite recently NLO QCD corrections to the production and top quark decays in the NWA together with jet radiation in top quark decays have been presented in [17]. Moreover, spin correlations were preserved throughout the entire decay chain. This state-of-the-art calculation of the NLO QCD corrections to the $t\bar{t}j$ process has shown that NLO QCD corrections and jet radiation in decays can lead to significant changes in shapes of distributions. Therefore, they need to be included for a fully consistent description of top anti-top plus jet production.

First results for the top anti-top plus jet process at NLO combined with parton shower via POWHEG method [18, 19] are also available [20, 21]. However, in both cases only NLO QCD corrections to the on-shell production are calculated and LO decays in NWA are included through the parton shower programs. Consequently full spin correlations at NLO are omitted.

4. Top Anti-Top Plus Two Jets

Even thought a fraction of events in the inclusive top anti-top sample that is accompanied by two additional jets is only at the few percent level [22], see Table 2, $t\bar{t}$ plus two jets is an important background for the SM Higgs boson studies at the LHC. Two noticeable examples include:

(i) $H \to WW^* \to \ell\ell'\nu\bar{\nu}$ where the Higgs boson is produced via weak boson fusion [23, 24],

(ii) $H \to b\bar{b}$ where the Higgs boson is produced via associated production with a $t\bar{t}$ pair [25,26].

In both cases the invariant mass of the Higgs decay products cannot be directly reconstructed. Either because of the two missing neutrinos in the decay of the two W gauge bosons or because the $b\bar{b}$ pair can be chosen incorrectly within the complex $W^+W^-b\bar{b}b\bar{b}$ final state. In the latter case also the b-tagging efficiency plays a crucial role since two b-jets can arise from mistagged light jets. Consequently, a very precise knowledge of QCD backgrounds, i.e. $t\bar{t}jj$ as well as $t\bar{t}b\bar{b}$ is essential. Lately however, a new strategy for the $Ht\bar{t}$ channel, based on a boosted Higgs boson and a boosted top quark, has been explored that should help to reduce complicated QCD backgrounds and resolve a multi b-jet combinatorial problem [27]. Fortunately, the NLO

Table 2. *Fraction of events in the inclusive top anti-top sample that is accompanied by two additional jets depending on the p_T of the additional jets at the TeVaron and the LHC.*

TEVATRON $\sqrt{s} = 1.96$ TeV	$\sigma^{\rm NLO}_{t\bar{t}jj}/\sigma^{\rm NLO}_{t\bar{t}}$	LHC $\sqrt{s} = 7$ TeV	$\sigma^{\rm NLO}_{t\bar{t}jj}/\sigma^{\rm NLO}_{t\bar{t}}$
$p_T \geq 20$ GeV	4%	$p_T \geq 50$ GeV	6%
$p_T \geq 40$ GeV	1%	$p_T \geq 100$ GeV	1%

QCD corrections have already been calculated for both $t\bar{t}b\bar{b}$ [28–31] and $t\bar{t}jj$ [22,32] background processes at the TeVatron and the LHC. In the former case two independent calculations exist and the per-mille level agreement between them have been obtained. Nevertheless, due to complexity of the NLO calculations for the $2 \to 4$ processes only corrections to the on-shell top production have been evaluated in all cases. Furthermore, they have not yet been matched to the parton shower programs.

Figure 1. *Differential cross section distributions as a function of the invariant mass of the $t\bar{t}t\bar{t}$ system (left panel) and the $t\bar{t}$ pair (right panel) for $pp \to t\bar{t}t\bar{t} + X$ production at the LHC with $\sqrt{s} = 14$ TeV. The dash-dotted (blue) curve corresponds to the LO, whereas the solid (red) one to the NLO result. The scale choice is $\mu_F = \mu_R = H_T/4$. The uncertainty bands depict scale variation. The lower panels display the differential $\mathcal{K}$ factor.*

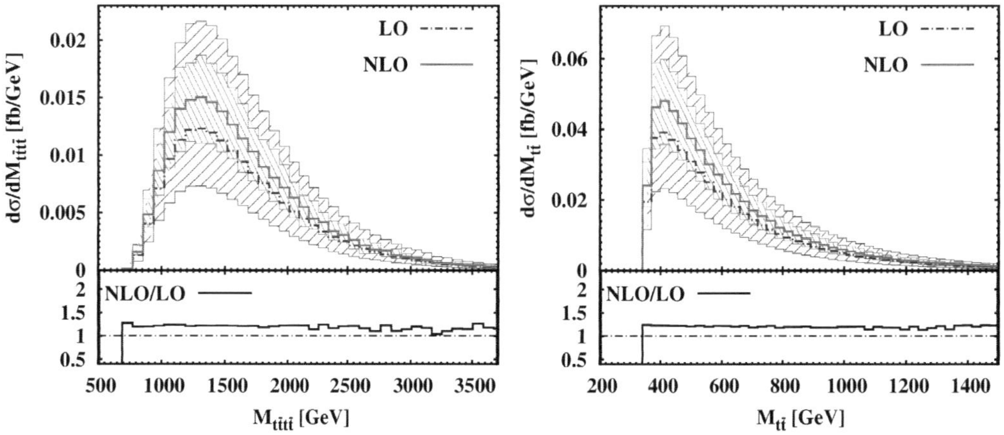

5. Top Anti-Top Plus Top Anti-Top

At the LHC the energy is sufficient to produce a four top final state at a sensible rate, see Table 3, where results obtained with the HELAC-DIPOLES Monte Carlo program [33], for four different center of mass energies are given. The four top final state is an interesting channel to probe

Table 3. *Inclusive leading order cross section in fb for the $pp \to t\bar{t}t\bar{t}$ production at the LHC with $\sqrt{s} = 7$ TeV, $\sqrt{s} = 8$ TeV, $\sqrt{s} = 13$ TeV and $\sqrt{s} = 14$ TeV. Results for the MSTW2008 LO PDF set are presented with the scale choice $\mu = 2m_t$. For the top quark mass the $m_t = 173.2$ GeV value is used. Also given is a number of expected events assuming an integrated luminosity of 5.6^{-1} fb, 23.3^{-1} fb, 100^{-1} fb and 300^{-1} fb respectively.*

LHC	$\sigma^{\rm LO}_{t\bar{t}t\bar{t}}$ [fb]	NUMBER OF EVENTS
$\sqrt{s} = 7$ TeV	0.624(1)	3
$\sqrt{s} = 8$ TeV	1.173(3)	27
$\sqrt{s} = 13$ TeV	9.08(3)	908
$\sqrt{s} = 14$ TeV	12.07(4)	3621

several realizations of BSM Physics. The most noticeable models being [34]

(i) Higgs and top compositeness,

(ii) new resonances from the Randall-Sudrum warped extra dimensions,

(iii) effective four-top quark interactions.

In addition, $t\bar{t}t\bar{t}$ is a major background for many processes arising from supersymmetric extensions of the SM, among others, the production of a heavy Higgs boson or long cascade decays of colored new particles like squarks or gluinos, see e.g. [34] and references therein. Therefore, a precise theoretical description of the four-top production rate in the SM might help to constrain new physics scenarios.

The NLO QCD corrections to the four top quark final state have been recently computed for on-shell tops [35]. Despite its relatively small cross section at NLO of the order of

$$\sigma^{NLO}_{t\bar{t}t\bar{t}}(\mathrm{LHC}_{14\mathrm{TeV}}, m_t = 173.2 \text{ GeV}, \mathrm{MSTW2008NLO}) = 17 \pm 4 \text{ [scales]} \pm 1 \text{ [PDF] fb} \quad (1)$$

a theoretical control over $pp \to t\bar{t}t\bar{t}$ is relevant if we take into account that typical predictions of various new physics scenarios are set in the range of 1-100 fb for m_{new}= 1-3 TeV [34], where m_{new} is a mass of the new heavy particle or in general the energy scale that is associated with new physics.

In addition, a judicious choice of a dynamical scale has been presented, that allowed to obtain nearly constant $\mathcal{K}$-factors in most differential cross sections. Two examples, the invariant mass of the $t\bar{t}t\bar{t}$ system and the averaged invariant mass of the $t\bar{t}$ pair are shown in Figure 1.

6. Summary

Driven by the LHC needs, a tremendous development in the NLO QCD calculations for top quark physics has recently been achieved. Currently, $2 \to 4$ processes are scrutinized at NLO. In many cases dynamical scales that depend on the event structure have been applied that allow for a better understanding of the high p_T tails of distributions. For the top anti-top pair production with leptonic decays the situation is the most satisfying, since complete off-shell and finite width effects of top quark have been calculated at NLO. Studies to match these calculations to the parton shower programs via POWHEG method are currently ongoing [36]. For the semi-leptonic decay channel and more complex final states progress is still needed, an ultimate goal being a description of a fully realistic final state such as $W^+W^-b\bar{b} + X$ with $X = j, jj, H, Z, W^\pm, \gamma$ matched to parton shower with higher than LL accuracy.

Meanwhile a huge progress in attempts to calculate the $t\bar{t}$ cross section beyond NLO has been attained. An outstanding example being a first genuine calculation of next-to-next-to-leading order QCD corrections to top anti-top pair [37–39].

Acknowledgments

We would like to thank the organizers of the 5th International Workshop on Top Quark Physics, TOP2012, for a kind invitation and a very pleasant atmosphere during the conference.

References

[1] W. Bernreuther, A. Brandenburg, Z. G. Si and P. Uwer, Phys. Rev. Lett. **87** (2001) 242002.
[2] W. Bernreuther, A. Brandenburg, Z. G. Si and P. Uwer, Nucl. Phys. B **690** (2004) 81.
[3] W. Bernreuther and Z. G. Si, Nucl. Phys. B **837** (2010) 90.
[4] W. Bernreuther and Z. G. Si, Phys. Rev. D **86** (2012) 034026.
[5] K. Melnikov and M. Schulze, JHEP **0908** (2009) 049.
[6] J. M. Campbell and R. K. Ellis, arXiv:1204.1513 [hep-ph].
[7] S. Frixione, P. Nason and B. R. Webber, JHEP **0308** (2003) 007.
[8] S. Frixione, P. Nason and G. Ridolfi, JHEP **0709** (2007) 126.
[9] A. Denner, S. Dittmaier, S. Kallweit and S. Pozzorini, Phys. Rev. Lett. **106** (2011) 052001.
[10] G. Bevilacqua, M. Czakon, A. van Hameren, C. G. Papadopoulos and M. Worek, JHEP **1102** (2011) 083.
[11] A. Denner, S. Dittmaier, S. Kallweit and S. Pozzorini, JHEP **1210** (2012) 110.
[12] S. Dittmaier, P. Uwer and S. Weinzierl, Eur. Phys. J. C **59** (2009) 625.
[13] M. Czakon and A. Mitov, arXiv:1112.5675 [hep-ph].
[14] S. Dittmaier, P. Uwer and S. Weinzierl, Phys. Rev. Lett. **98** (2007) 262002.
[15] G. Bevilacqua, M. Czakon, C. G. Papadopoulos and M. Worek, Phys. Rev. Lett. **104** (2010) 162002.
[16] K. Melnikov and M. Schulze, Nucl. Phys. B **840** (2010) 129.
[17] K. Melnikov, A. Scharf and M. Schulze, Phys. Rev. D **85** (2012) 054002.

[18] P. Nason, JHEP **0411** (2004) 040.
[19] S. Alioli, P. Nason, C. Oleari and E. Re, JHEP **1006** (2010) 043.
[20] A. Kardos, C. Papadopoulos and Z. Trocsanyi, Phys. Lett. B **705** (2011) 76.
[21] S. Alioli, S. -O. Moch and P. Uwer, JHEP **1201** (2012) 137.
[22] G. Bevilacqua, M. Czakon, C. G. Papadopoulos and M. Worek, Phys. Rev. D **84** (2011) 114017.
[23] N. Kauer, T. Plehn, D. L. Rainwater and D. Zeppenfeld, Phys. Lett. B **503** (2001) 113.
[24] S. Asai *et al.*, Eur. Phys. J. C **32S2** (2004) 19.
[25] G. L. Bayatian *et al.* [CMS Collaboration], J. Phys. G **34** (2007) 995.
[26] G. Aad *et al.* [ATLAS Collaboration], arXiv:0901.0512 [hep-ex].
[27] T. Plehn, G. P. Salam and M. Spannowsky, Phys. Rev. Lett. **104** (2010) 111801.
[28] A. Bredenstein, A. Denner, S. Dittmaier and S. Pozzorini, Phys. Rev. Lett. **103** (2009) 012002.
[29] G. Bevilacqua, M. Czakon, C. G. Papadopoulos, R. Pittau and M. Worek, JHEP **0909** (2009) 109.
[30] A. Bredenstein, A. Denner, S. Dittmaier and S. Pozzorini, JHEP **1003** (2010) 021.
[31] M. Worek, JHEP **1202** (2012) 043.
[32] G. Bevilacqua, M. Czakon, C. G. Papadopoulos and M. Worek, Phys. Rev. Lett. **104** (2010) 162002.
[33] M. Czakon, C. G. Papadopoulos and M. Worek, JHEP **0908** (2009) 085.
[34] G. Brooijmans *et al.* [New Physics Working Group Collaboration], arXiv:1005.1229 [hep-ph].
[35] G. Bevilacqua and M. Worek, JHEP **1207** (2012) 111.
[36] M. V. Garzelli, A. Kardos, C. G. Papadopoulos and Z. Trocsanyi, in preparation.
[37] P. Baernreuther, M. Czakon and A. Mitov, Phys. Rev. Lett. **109** (2012) 132001.
[38] M. Czakon and A. Mitov, JHEP **1212** (2012) 054.
[39] M. Czakon and A. Mitov, JHEP **1301** (2013) 080.

Journal of Physics: Conference Series **452** (2013) 012034 doi:10.1088/1742-6596/452/1/012034

Boosted top quark techniques and searches for $t\bar{t}$ resonances at the LHC

S Fleischmann
On behalf of the ATLAS and CMS Collaborations
Bergische Universität Wuppertal, Wuppertal, Germany
E-mail: Sebastian.Fleischmann@cern.ch

Abstract. In many models of beyond the Standard Model physics (BSM) the top quark plays a special role. The ATLAS and CMS experiments at the Large Hadron Collider (LHC) have performed dedicated searches for resonances decaying into top quark pairs, which may occur in BSM models. Especially at high $t\bar{t}$ masses the top quarks can have strongly boosted decay products. In recent years many new techniques have been developed to reconstruct and identify such decays. Both ATLAS and CMS use jet substructure to identify hadronic W boson and top quark decays. Some of those techniques are summarised in this article and the recent results of searches for $t\bar{t}$ resonances are presented. In the lepton+jets channel a leptophobic top-colour Z' with a small width is excluded at the 95% CL in the range 0.5 TeV $< m_{Z'} <$ 1.7 TeV as a benchmark scenario. Kaluza-Klein gluons and Z' bosons with larger widths are excluded in similar mass ranges.

1. Introduction

In 2011 the ATLAS [1] and CMS [2] experiments both recorded about 5fb^{-1} of pp collisions at a centre-of-mass energy of $\sqrt{s} = 7$ TeV. This data set allowed one to start to explore the multi-TeV regime. The top quark plays a special role in many models of physics beyond the Standard Model (BSM), therefore top quark pairs with high invariant mass can be sensitive probes of new physics. Both experiments performed searches for deviations from the Standard Model in the $t\bar{t}$ mass spectrum. The results are usually interpreted in terms of benchmark scenarios for $t\bar{t}$ resonances which differ in width and spin.

Experimentally, the width of the resonance compared to the experimental resolution of the invariant $t\bar{t}$ mass has a big impact on the obtained sensitivity of the searches. The spin of the resonance on the other hand plays only a minor role experimentally. Two benchmark scenarios are used by the LHC experiments. A leptophobic topcolour Z' boson [3] is used as an example for a resonance with width below or of the order of the experimental $m_{t\bar{t}}$ resolution of about 10%. Kaluza-Klein excitations of gluons are predicted by models with warped extra dimensions and used as a benchmark model with Γ/m larger than the experimental resolution [4].

One can estimate the angular separation in η-ϕ space of decay products from heavy particle (2-body) decays by $\Delta R \equiv \sqrt{(\Delta\eta)^2 + (\Delta\phi)^2} \approx 2\frac{M}{p_{\mathrm{T}}}$, where M is the mass of the decaying particle and p_{T} its transverse momentum. For example, the W and the b-quark from the top decay ($t \to bW$) are usually separated by $\Delta R(W, b) < 1.0$ for $p_{\mathrm{T}}^t > 350$ GeV. Correspondingly, the hadronic decays of W bosons ($t \to bW \to bqq$) may not be fully resolved anymore by usual jets with a radius parameter of $R = 0.4$ or $R = 0.5$, if the transverse momentum p_{T}^W of the

W boson is larger than about 200 GeV [5]. As a consequence, the top decay products start to merge and parton-level objects cannot be identified with individual jets.

One can use a different approach in the case of overlapping decay products from decays of heavy particles. Instead of using several narrow ($R = 0.4$ or $R = 0.5$) jets and constructing the hadronic decay of the top quark from them, one can use a single large radius jet (e.g. $R = 1.0$ or $R = 1.5$) and investigate its substructure. Several different methods have been proposed recently [6, 7] and section 2 will summarise a few of them. In section 3 the searches for $t\bar{t}$ resonances in different channels using 7 TeV data are presented. Finally, section 4 summarises the results.

2. Jet substructure techniques

The usage of jets with large radii (e.g. $R = 1.0$ or $R = 1.5$), so-called *fat jets*, instead of many fixed-size, narrow jets has the advantage that the fat jet can be decomposed into subjets of varying sizes and more information about the substructure of the fat jet can be taken into account. However, fat jets suffer more than narrow jets from contamination by the underlying event, initial/final state radiation and pile-up. These additional energy deposits do not only introduce larger systematic uncertainties in the jet energy and jet mass scales, but also dilute the substructure from the decays of heavy particles. Therefore, *grooming* techniques have been developed, which allow one to remove contributions from soft radiation. The most commonly used grooming techniques are *filtering* [8], *pruning* [9] and *trimming* [10]. The CMS collaboration applies jet pruning in their searches for $t\bar{t}$ resonances [11, 12]. The ATLAS collaboration has performed extensive comparisons of filtering, pruning and trimming including validations of these methods with data at $\sqrt{s} = 7$ TeV [5]. The impact of pile-up on groomed jets was evaluated in data and good performance of grooming was observed [13].

Jet trimming takes the constituents of a fat jet and applies a $k_\perp$ jet algorithm with smaller radius R_{sub} to obtain sub-jets of the original jet. All constituents belonging to sub-jets that carry only a transverse momentum fraction $p_{\text{T}}^{\text{subjet } i}/p_{\text{T}}^{\text{fat jet}}$ below a given cut value are discarded in the trimmed jet. Trimming significantly improves the mass resolution for heavy particles and pushes the mass for jets from light quarks and gluons lower, therefore giving a better separation power between jets stemming from heavy particles and multi-jet background [5].

Similarly, pruning tries to discard soft particles and wide-angle radiation. In contrast to trimming, jet constituents are selected at each recombination step of a $k_\perp$ or Cambridge/Aachen (C/A) jet algorithm. When merging two constituents or proto-jets into a new proto-jet one requires the softer component to contribute a significant fraction of the transverse momenta to the merged proto-jet or to be close-by in ΔR, otherwise the softer component is discarded.

In order to separate jets from heavy particle decays from the overwhelming background of jets induced by light quarks and gluons one takes variables into account which are sensitive to the substructure of the jets. The invariant mass of a jet is a very obvious variable. Another widely used class of variables are the last splitting scales of a $k_\perp$ jet algorithm. One defines the splitting scale $\sqrt{d_{12}} = \min(p_{\text{T}1}, p_{\text{T}2})\Delta R_{12}$, where $p_{\text{T}1}$ and $p_{\text{T}2}$ are the transverse momenta of the last two proto-jets in the recombination process of the $k_\perp$ algorithm and ΔR_{12} their distance in η-ϕ space. One expects $\sqrt{d_{12}} \approx m_t/2$ for a jet containing all products of a hadronic top decay $t \to bW \to bqq$.

Top tagging algorithms aim for the reconstruction and identification of hadronic top decays. The *HEPTopTagger* [14] is applied by the ATLAS collaboration and applies mass-drop filtering to remove contamination. It tries to reconstruct "tree level" decay products ($t \to bqq$) and uses mass relations between those to identify top decays. For this purpose the fat jet is decomposed with a mass drop requirement, i.e. the jet clustering is run backward and at each step one requires the two subjets to have significantly lower mass than their combination. The jet is decomposed until all subjets have a mass less than 30 GeV. From those subjets three subjets are selected and their constituents investigated further. In the next step only the selected constituents are clustered using the C/A algorithm with radius parameter $R = \min(0.3, \min_{ij} \Delta R(j_i, j_j)/2)$. The five hardest subjets of this step are processed further and exactly three subjets are built from their constituents, which are then interpreted as the tree-level objects.

The CMS top tagger [11] is based on the *Johns Hopkins tagger* [15] and decomposes the initial fat jet in two steps, where also grooming techniques are applied. This follows the decay chain $t \to bW, W \to qq$. Afterwards top candidates are tagged using variables like the jet mass, the number of subjets found and the minimum pairwise mass of the three hardest subjets. CMS also uses a special W boson tagger which applies jet pruning and requires a mass drop of the last two subjets to account for the components of the heavy particle decays.

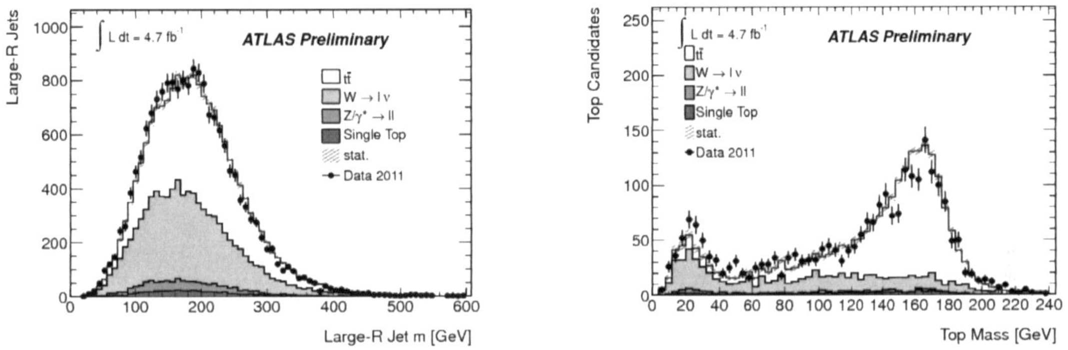

(a) $R = 1.5$ C/A jets before HEPTopTagger [5] (b) after HEPTopTagger [5]

Figure 1: Mass distribution of top candidates with a basic lepton+jets selection.
(a) Cambridge/Aachen jets ($R = 1.5$), (b) reconstructed top mass from the HEPTopTagger.

Figure 1 shows the mass distributions of hadronic top candidates for the $t\bar{t}$ lepton+jets channel. A broad peak is obtained for the mass of large radius ($R = 1.5$) C/A jets when no grooming is applied (figure 1a). In this selection without b-tagging the fraction of $t\bar{t}$ events is only about 50%. The dominant background consists of W+jets events, where the hadronic top candidate stems from additional light quarks or gluons. After applying the HEPTopTagger the background from W+jets events is already strongly reduced and the mass peak of the top quark is much more narrow (figure 1b). Background events can be further suppressed by applying a mass window cut, giving a fraction of $t\bar{t}$ events of about 86% within 140 GeV $< m_t <$ 200 GeV.

3. Searches for $t\bar{t}$ resonances

The searches for $t\bar{t}$ resonances are performed in all possible $t\bar{t}$ decay channels. The all-jets channel, where both top quarks are considered to decay hadronically, mostly aims at high di-top masses $m_{t\bar{t}}$. Only for significantly boosted top decays this final state can be distinguished from the overwhelming multijet background at the LHC. The di-leptonic channel is considered for low $m_{t\bar{t}}$. No analyses specifically designed for the high $m_{t\bar{t}}$ regime have been performed so far by the ATLAS and CMS collaborations in this channel. The highest sensitivity for $t\bar{t}$ resonances is reached in the lepton+jets channel, where a single electron or muon and additional jets are used as final state. Both ATLAS and CMS use distinct event selections optimised for the low and the high mass regime in this channel.

Standard Model $t\bar{t}$ production is obviously an irreducible background to all searches for $t\bar{t}$ resonances and the signal only occurs as deviations in the invariant mass spectrum. W boson production with additional jets is the dominant non-$t\bar{t}$ background in the lepton+jets channel. The normalisation of the Monte Carlo prediction for W+jets events is usually estimated directly from data, e.g. using the charge asymmetry. Z/γ bosons yield an important background in the di-leptonic channel and are usually also estimated by data driven methods in this channel. Diboson production (WW, WZ) is only relevant in the di-leptonic channel. Especially in the all-jets channel multijet events are an important background which needs to be estimated with data.

In the searches all important systematic uncertainties which affect the shape of the $m_{t\bar{t}}$ spectrum or give yield variations are taken into account. Prominent examples are the jet energy scale and resolution. Furthermore, reconstruction and identification efficiencies play an important role. Especially b-tagging uncertainties can be very large for high-p_{T} jets.

3.1. All-jets channel

While the ATLAS collaboration uses rather large C/A jets with radius $R = 1.5$ in combination with the HEPTopTagger [16], the CMS analysis [12] takes C/A jets with radius $R = 0.8$, but distinguishes two different types of top decays: Type-1 decays are top decays where all products of the hadronic top decay are included in a jet tagged by the CMS top tagger, while type-2 decays include only the hadronic W decay in the fat jet while the additional jet from the b-quark is separated from it. The analysis considers events with two type-1 top tagged jets and events with one type-1 and one type-2 decay, where one top tag as well as a W-tag and an additional jet is required.

Figure 2a shows the $t\bar{t}$ invariant mass spectrum obtained by the ATLAS collaboration in the all-jets channel. Both ATLAS and CMS do not observe any significant excess in the $m_{t\bar{t}}$ distributions and set limits for the benchmark scenarios (table 1).

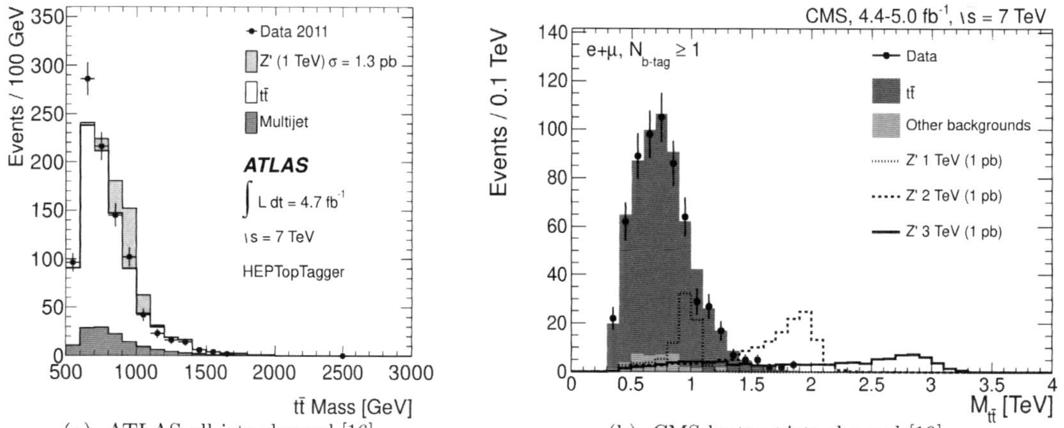

(a) ATLAS all-jets channel [16] (b) CMS lepton+jets channel [19]

Figure 2: $t\bar{t}$ invariant mass spectra at $\sqrt{s} = 7$ TeV (about 5fb^{-1}). (a) ATLAS all-jets channel with HEPTopTagger. (b) CMS lepton+jets channel using a boosted selection with b-tag.

3.2. Di-leptonic channel

The di-leptonic channel has only low non-$t\bar{t}$ background, but suffers from the two neutrinos in the final state, which complicate the reconstruction of the invariant di-top mass. In the ATLAS analysis [17] $H_{\mathrm{T}} + E_{\mathrm{T}}^{\mathrm{miss}}$ with $H_{\mathrm{T}} = \sum p_{\mathrm{T}}^{\ell} + \sum p_{\mathrm{T}}^{j}$ ($\ell = e^{\pm}, \mu^{\pm}$) is used as a proxy for $m_{t\bar{t}}$ and the search for deviations from the Standard Model prediction is done in this variable. In simulation one finds a decent correlation between this variable and the simulated true $t\bar{t}$ mass. The corresponding CMS analysis [18] simply estimates $m(2j2\ell2\nu)$ by setting the z-component of the neutrinos to zero. In addition to a bump hunting in the $m(2j2\ell2\nu)$ spectrum a Bayesian neural network is used for a multivariate event selection to obtain a better discrimination power between Standard Model $t\bar{t}$ production and the signal. Kinematic variables like the momenta of the leptons and jets as well as various angles between them and the missing transverse momentum $E_{\mathrm{T}}^{\mathrm{miss}}$ are used.

3.3. Lepton+jets channel

In the lepton+jets channel one assumes a final state with only one neutrino which allows one to estimate the longitudinal component p_z^{ν} of the neutrino momentum by applying a W boson mass constraint on the lepton+$E_{\mathrm{T}}^{\mathrm{miss}}$ system by different methods. In the event selection and the reconstruction of the $t\bar{t}$ system both ATLAS and CMS use specialised analyses for the "resolved" case, where the hadronically decayed top is reconstructed as three (or two) narrow jets and the "boosted" case, where the decay products merge. This affects not only the hadronic top decay, but also the leptonic decay, such that the isolation criteria of leptons need to be adapted.

The CMS analysis [19] uses narrow ($R = 0.5$) jets. In the resolved case leptons with standard isolation criteria are used. The boosted selection replaces isolation by a cut on $p_{\mathrm{T}}^{\mathrm{rel}}(\ell, j)$, defined as the lepton momentum perpendicular to the closest jet axis. This allows for a certain overlap of leptons and jets, while keeping the multijet background from fake leptons low. In both cases a χ^2 method including kinematic constraints on the leptonic and hadronic top decay is applied to find the most likely combination of jets for the calculation of the $t\bar{t}$ mass. The $m_{t\bar{t}}$ spectrum for the boosted selection with b-tagging is shown in figure 2b. In the calculation of exclusion limits the resolved analysis is used at low $m_{t\bar{t}}$ and the boosted selection at high $m_{t\bar{t}}$ (figure 3a).

In the ATLAS search [20] the default lepton isolation was replaced by a definition with a shrinking cone size of the isolation region. Tracks in a cone with size $\Delta R(\ell, \mathrm{track})$ proportional to $1/p_{\mathrm{T}}^{\ell}$ are used in the isolation criteria. This gives a strong improvement in the selection efficiency for leptons from boosted top quarks compared to isolation criteria with fixed cone sizes. In the resolved selection a single lepton trigger is used and a χ^2 method is applied on narrow jets to find the best jet combination for the $m_{t\bar{t}}$ calculation. The boosted selection uses a dedicated fat jet trigger on jets with $R = 1.0$. For high $t\bar{t}$ masses this trigger has a much better efficiency of nearly 100% than the single lepton triggers. Fat jets with a high p_{T} threshold and cuts on their mass and $k_{\perp}$-splitting scale $\sqrt{d_{12}}$ represent the hadronic top decay products in the reconstruction of the $t\bar{t}$ system. However, both selection strategies are not used exclusively in the limit calculation at a certain $t\bar{t}$ mass, but events from both are taken into account, where the boosted reconstruction is used, if a single event is selected by the boosted and the resolved strategy (figure 3b).

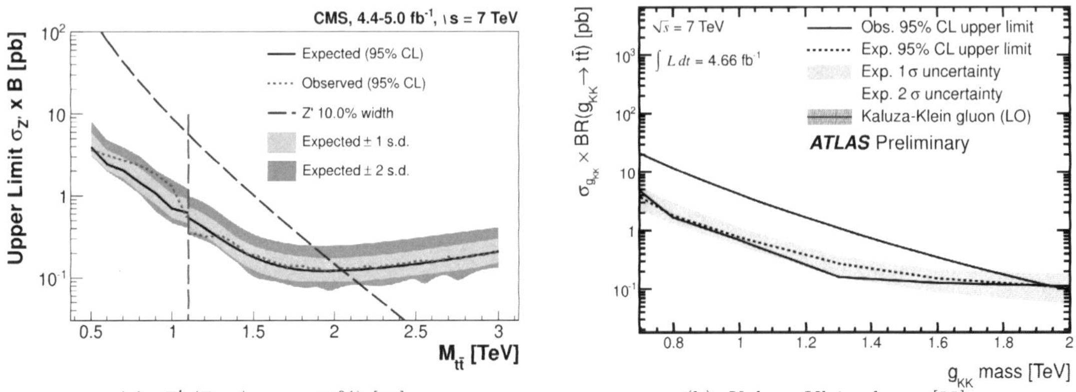

(a) Z' ($\Gamma_{Z'}/m_{Z'} = 10\%$) [19]   (b) Kaluza-Klein gluons [20]

Figure 3: 95% CL expected and observed upper limits on the cross section times the $t\bar{t}$ branching fraction (a) for a medium width Z' ($\Gamma_{Z'}/m_{Z'} = 10\%$) and (b) for Kaluza-Klein gluons ($\Gamma_{g_{\mathrm{KK}}}/m_{g_{\mathrm{KK}}} = 15.3\%$) obtained in the lepton+jets channel.

4. Results and Summary

Many new techniques to reconstruct and identify boosted top decays were developed in recent years. Both ATLAS and CMS could show their feasibility at the LHC and improve their searches for $t\bar{t}$ resonances especially at high $m_{t\bar{t}}$ by their application. Boosted techniques are expected to gain even more importance with higher centre-of-mass energies and more integrated luminosity as higher mass scales get accessible. Grooming techniques can help to reduce the pile-up dependence and effects from the underlying event on jets with large radii.

In none of the channels any significant excess in the $t\bar{t}$ mass spectra was observed. Therefore, exclusion limits for certain benchmark scenarios were set. The expected and observed mass limits for the three different benchmark models considered by the ATLAS and CMS collaborations are summarised in table 1. The lepton+jets channel has the highest sensitivity and a leptophobic top-colour Z' with a small width of $\Gamma_{Z'}/m_{Z'} = 1.2\%$ is excluded at the 95% CL in the range 0.5 TeV $< m_{Z'} <$ 1.7 TeV by the ATLAS experiment. Z' bosons with larger widths (figure 3a) and Kaluza-Klein gluons (figure 3b) are excluded in similar mass ranges.

Table 1: Summary of the observed and expected exclusion ranges at 95% CL of the resonance mass in TeV for three different benchmark models.

		Z' ($\Gamma_{Z'}/m_{Z'} = 1.2\%$)		Z' ($\Gamma_{Z'}/m_{Z'} = 10\%$)		g_{KK} ($\Gamma_g/m_g = 15.3\%$)	
		obs.	exp.	obs.	exp.	obs.	exp.
ATLAS ℓ+jets	[20]	0.5—1.7	0.5—1.6			0.7—1.9	0.7—1.9
CMS ℓ+jets	[19]	0.5—1.5	0.5—1.5	0.5—2.0	0.5—2.0	1.0—1.8	1.0—1.8
ATLAS all-jets	[16]	0.7—1.0 1.3	0.7—1.2			0.7—1.5	0.7—1.5
CMS all-jets	[12]	1.0 1.3—1.5		1.0—2.0	1.0—2.0	1.0 1.4—1.5	1.0—1.5
ATLAS $\ell\ell$	[17]					0.5—1.1	0.5—1.1
CMS $\ell\ell$	[18]	0.5—1.1	0.5—1.1				

References

[1] ATLAS Collaboration 2008 *JINST* **3** S08003
[2] CMS Collaboration 2008 *JINST* **3** S08004
[3] Harris R M, Hill C T and Parke S J 1999 *Preprint* hep-ph/9911288
[4] Agashe K, Belyaev A, Krupovnickas T, Perez G and Virzi J 2008 *Phys. Rev.* D **77** 015003
[5] ATLAS Collaboration 2012 ATLAS-CONF-2012-065, `https://cds.cern.ch/record/1459530`
[6] Abdesselam A et al 2011 *Eur. Phys. J.* C **71** 1661
[7] Altheimer A et al 2012 *J. Phys.* G **39** 063001
[8] Butterworth J M, Davison A R, Rubin M and Salam G P 2008 *Phys. Rev. Lett.* **100** 242001
[9] Ellis S D, Vermilion C K and Walsh J R 2009 *Phys. Rev.* D **80** 051501
[10] Krohn D, Thaler J and Wang L T 2010 *JHEP* **1002** 084
[11] CMS Collaboration 2011 CMS-PAS-JME-10-013, `https://cds.cern.ch/record/1333700`
[12] CMS Collaboration 2012 *JHEP* **1209** 029
[13] ATLAS Collaboration 2012 ATLAS-CONF-2012-066, `http://cds.cern.ch/record/1459531`
[14] Plehn T, Spannowsky M, Takeuchi M and Zerwas D 2010 *JHEP* **1010** 078
[15] Kaplan D E, Rehermann K, Schwartz M D and Tweedie B 2008 *Phys. Rev. Lett.* **101** 142001
[16] ATLAS Collaboration 2012 *Preprint* 1211.2202, accepted by *JHEP*
[17] ATLAS Collaboration 2012 *Eur. Phys. J.* C **72** 2083
[18] CMS Collaboration 2012 CMS-PAS-TOP-11-010, `http://cds.cern.ch/record/1438739`
[19] CMS Collaboration 2012 *JHEP* **1212** 015
[20] ATLAS Collaboration 2012 ATLAS-CONF-2012-136, `http://cds.cern.ch/record/1478974`

5th International Workshop on Top Quark Physics (TOP2012) IOP Publishing
Journal of Physics: Conference Series **452** (2013) 012035 doi:10.1088/1742-6596/452/1/012035

Measurements of top-quark properties at the Tevatron

Andreas Werner Jung
Fermilab, MS 357, P.O. Box 500, Batavia, IL, 60510, USA
FERMILAB-CONF-13-044-PPD

E-mail: ajung@fnal.gov

Abstract. Recent measurements of top-quark properties at the Tevatron are presented. CDF uses data corresponding up to 9.0 fb^{-1} to measure the ratio R of the branching fractions $\mathcal{B}(t \rightarrow Wb)/\mathcal{B}(t \rightarrow Wq)$, the branching fraction for top-quarks decaying into τ leptons and the cross section for the production of an additional γ in $t\bar{t}$ production. The results from all these measurements agree well with their respective Standard Model expectation. DØ uses 5.3 fb^{-1} of data to measure the $t\bar{t}$ cross section as a function of the time. A time dependency would imply Lorentz invariance violation as implemented by the Standard Model extension. No time dependency is observed and DØ sets first limits in the top-quark sector for Lorentz invariance violation. DØ also determines indirectly the top quark width using the results of earlier measurements at DØ. The measured top quark width is in agreement with the SM expectation and does not show any hints for new physics contributions.

1. Introduction

The top quark is the heaviest known elementary particle and was discovered at the Tevatron $p\bar{p}$ collider in 1995 by the CDF and DØ collaboration [1, 2] with a mass around 173 GeV. The production is dominated by the $q\bar{q}$ annihilation process with 85% as opposed to gluon-gluon fusion which contributes only 15%. The top quark has a very short lifetime, which prevents any hadronization process of the top quark. Instead bare quark properties can be observed by measuring top quark properties.
The measurements presented here are performed using either the dilepton ($\ell\ell$) final state or the lepton+jets (ℓ+jets) final state. Within the ℓ+jets final state one of the W bosons (stemming from the decay of the *top* quarks) decays leptonically, the other W boson decays hadronically. For the dilepton final state both W bosons decay leptonically. The branching fraction for top quarks decaying into Wb is almost 100%. Jets originating from a beauty quark (b-jets) are identified by means of a neural network (NN) built by the combination of variables describing the properties of secondary vertices and of tracks with large impact parameters relative to the primary vertex.

2. Measurement of the ratio R of branching fractions (CDF)

CDF uses all available data corresponding to 8.7 fb^{-1} in the ℓ+jets decay channel to measure the ratio R of the branching fractions $\mathcal{B}(t \rightarrow Wb)/\mathcal{B}(t \rightarrow Wq)$ [3]. The data is selected by requiring a lepton, missing transverse energy $\not{E}_T$, at least three jets with exactly one or two b-jets. Figure

1(a) shows the number of data events with their respective number of background events for different jet and b-jet bins. As an example three different templates with different values of R are shown: $R = 0.1$ (green dashed histogram), $R = 0.5$ (blue dashed histogram) and $R = 1.0$ (red dashed histogram). A simultaneous log-likelihood fit to $R = \mathcal{B}(t \to Wb)/\mathcal{B}(t \to Wq)$ and

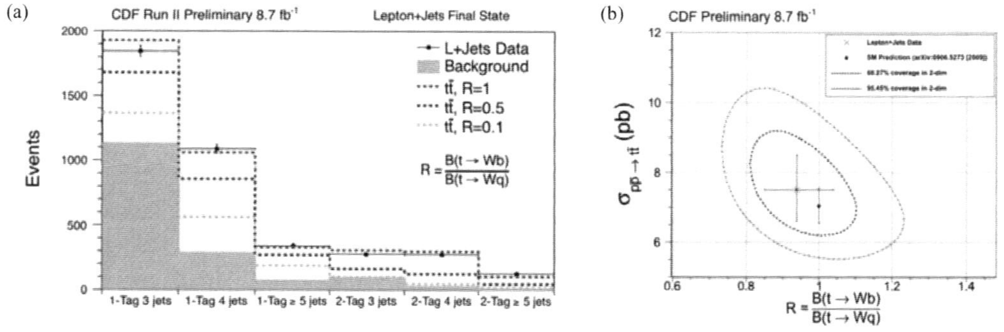

Figure 1. (a) shows the number of data events with their respective number of background events for different jet and b-jet bins compared to three different R templates: $R = 0.1$ (green dashed histogram), $R = 0.5$ (blue dashed histogram) and $R = 1.0$ (red dashed histogram). (b) shows the result of a simultaneous log-likelihood fit to the $t\bar{t}$ cross section and the ratio R with the 2 and 3 standard deviation (s.d.) contours.

to the $t\bar{t}$ cross section $\sigma_{p\bar{p}\to t\bar{t}}$ is applied. The resulting log-likelihood is shown in Figure 1b). An inclusive $t\bar{t}$ cross section of $\sigma_{p\bar{p}\to t\bar{t}} = 7.5 \pm 1.0$ (stat. + syst.) is measured in good agreement with earlier measurements at the Tevatron. The ratio R of the branching fractions is measured to $R = 0.94 \pm 0.09$ (stat. + syst.). The presented measurement of R agrees with the SM expected value and with measurements from DØ and CMS: an earlier measurement of R by DØ results in $R = 0.90 \pm 0.04$(stat. + syst.) [4] and the currently most precise R measurement from CMS presented at this conference yields $R = 0.98 \pm 0.04$ [5]. Assuming 3 generations and unitarity of the CKM matrix CDF also measured V_{tb} to 0.97 ± 0.05(stat. + syst.), which is in good agreement with the SM expectation.

3. Measurement of the branching fraction to τ leptons (CDF)

The measurement of the branching fraction for top-quarks decaying into τ leptons $\mathcal{B}(t \to Wb \to \tau\nu_\tau b)$ is measured by CDF using 9.0 fb^{-1} of data in the dilepton decay channel [6]. It relies on the precise measurement of the τ cross section in $t\bar{t}$ production, which selects an e or μ originating from the W decay or the leptonic τ decay and thus allows for one hadronic τ decay (more details can be found in [7]). In addition to the usual kinematic cuts an requirements on identified b-jets events are selected by imposing a certain log-likelihood cut enhancing τ fraction of events. The likelihood combines the following variables: $\not{E}_T$, transverse mass of lepton plus $\not{E}_T$ and the transverse energy of the third highest E_T jet. Figure 2(a) shows the log-likelihood distribution before applying any cut. Imposing a minimum log-likelihood value of 0 yields the final event selection with 36 observed events to be compared with the SM expectation of $33.6 \pm 0.7^{+3.7}_{-3.8}$ events. Figure 2(b) shows the final event selection after applying the log-likelihood cut of 0; data are nicely described by the signal and background contributions. A branching fraction from decays to single and di-τ events of $\mathcal{B}(t \to Wb \to \tau\nu_\tau b) = 0.120 \pm 0.030(\text{stat.})^{+0.022}_{-0.019}(\text{syst.}) \pm 0.007(\text{lumi})$ is measured. A further tightened likelihood cut using additional information from the transverse energy of the lepton $E_T(\ell)$ and $\Delta\phi(\not{E}_T, \ell)$ allows for a measurement of the branching fraction

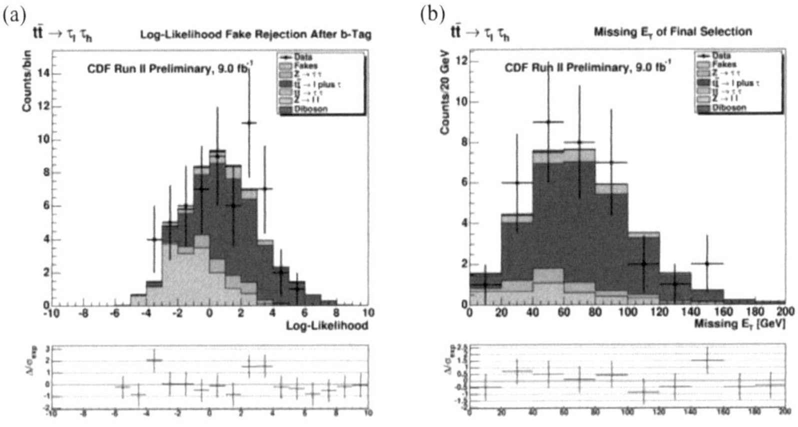

Figure 2. (a) shows the log-likelihood distribution before applying any cut, whereas (b) shows the final event selection after applying the log-likelihood cut of 0.

using only di-τ events: $0.098 \pm 0.022(\text{stat.}) \pm 0.014(\text{syst.})$. The results from these measurements agree well with the Standard Model expectation of $\mathcal{B}(t \to Wb \to \ell\nu_\ell b) = 0.108 \pm 0.009$ (averaged over e, μ and τ decay modes.

4. Measurement of the $t\bar{t} + \gamma$ production cross section (CDF)

In addition to determining decay branching fractions CDF also measures the cross section for the production of an additional particle in $t\bar{t}$ production: $p\bar{p} \to t\bar{t} + \gamma$ [8]. The measurement is based on the ℓ+jets decay channel and thus events are selected by requiring a lepton, $\not\!\!E_T$, at least three jets with one identified b-jet and a photon. The photon candidate is required to have $E_T^\gamma > 10$ GeV and no track with $p_T > 1$ GeV and at most one track with $p_T < 1$ GeV, pointing at the calorimeter cluster; and minimal leakage into the hadronic calorimeter. Figure 3 (left) shows that pre-selected data events are well described by all the various background contributions. A tight χ^2 cut on the photon identification suppresses largely the background

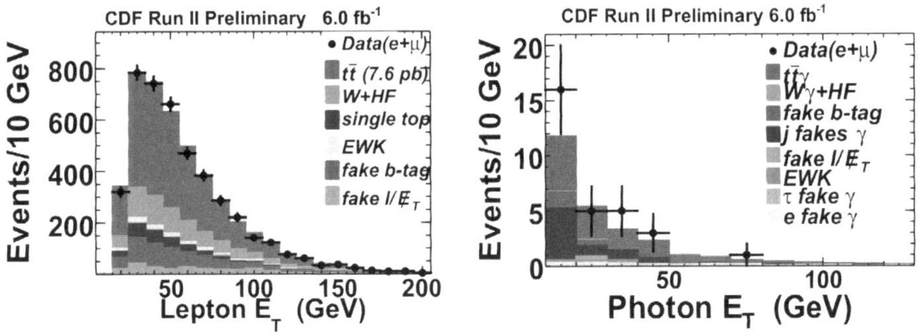

Figure 3. (left) shows that pre-selected data events compared to the various background contributions, whereas (right) shows the $t\bar{t}$ $+\gamma$ contribution after requiring a tight photon identification criteria.

and the signal contribution (red) is clearly visible in Figure 3 (right). The most dominant remaining background contribution originates from miss-identification of jets as photons. In order to get a handle on this contribution the isolation cuts are not applied and $Z \to ee$ events are used to extrapolate the isolation shape to the used cuts at higher isolation values. Taking into account this fake photon contribution, all the other backgrounds and the signal contribution the expected amount of events is 26.9 ± 3.4 (e and μ channel combined). The probability of the background alone to mimic the observed signal of 30 events is: 3 standard deviations. The cross section for $p\bar{p} \to t\bar{t} + \gamma$ is measured to $0.18 \pm 0.07(\text{stat.}) \pm 0.04(\text{syst.}) \pm 0.01(\text{lumi})$, which is a factor of 40 lower than the inclusive $t\bar{t}$ cross section! The cross section and its ratio to the $t\bar{t}$ cross section of $0.024 \pm 0.009(\text{stat.}) \pm 0.001(\text{syst.})$ are in good agreement to the SM.

5. Lorentz Invariance Violation (DØ)

DØ searches for a time dependent $t\bar{t}$ production cross section employing 5.3 fb^{-1} of data [9]. For the analysis $t\bar{t}$ events in the ℓ+jets final state are selected with a lepton (e/μ), at least four jets, exactly one jet identified as a b-jet and $\not{E}_T$. In addition the analysis relies on the timestamp of the data at production time. The Standard Model Extension (SME) [10] is an effective field theory and implements terms that violate Lorentz and CPT invariance. The modified SME matrix element adds Lorentz invariance violating terms for the production and decay of $t\bar{t}$ events to the Standard Model terms. The SME predicts a cross section dependency on siderial time as the orientation of the detector changes with the rotation of the earth relative to the fixed stars. The luminosity-corrected relative $t\bar{t}$ event rate (R) is expected to be flat within the Standard Model, i.e. no time dependency of the $t\bar{t}$ production cross section. Figure 4 shows this ratio as a function of the siderial phase, i.e. 1 corresponds to one siderial day. There is no indication of a

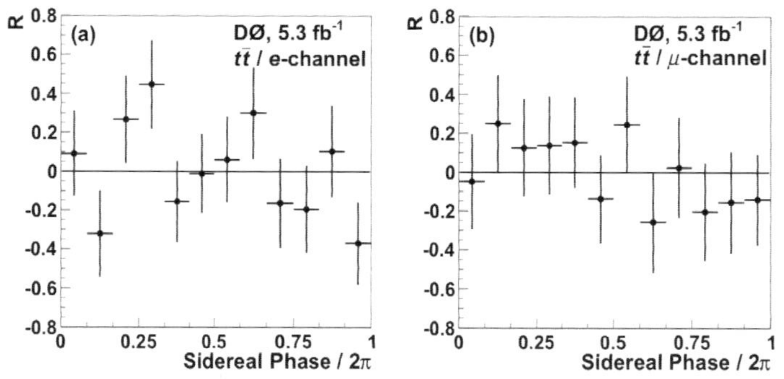

Figure 4. (a) shows the background and luminosity corrected ratio R as a function of the siderial phase (one siderial day) for events containing electrons, whereas (b) shows the same ratio R for the muon case.

time dependent $t\bar{t}$ production cross section. Instead this measurement sets the first constraints on Lorentz invariance violation in the top quark sector. As the top quark decays before it can hadronize the constraints are also the first ones for a bare quark.

6. Measurement of the top-quark width (DØ)

An indirect determination of the top quark width at DØ [11] is presented using the results of earlier measurements at DØ. The total width Γ_t is determined from the ratio of the partial decay with $\Gamma(t \to Wb)/\mathcal{B}(t \to Wb)$ under the assumption of the same tWb couplings in production and decay. The partial decay width is derived from the single top t-channel cross section measurement of $2.90 \pm 0.59(\text{stat.} + \text{syst.})$ [12] corrected by the ratio of the SM expected values for partial decay width and single top t-channel cross section. To obtain the total decay width Γ_t the derived partial decay width is corrected by the branching fraction $\mathcal{B}(t \to Wb) = 0.90 \pm 0.04$

mentioned also earlier in this text [4]. With 5.3 fb^{-1} of data the currently most precise indirect

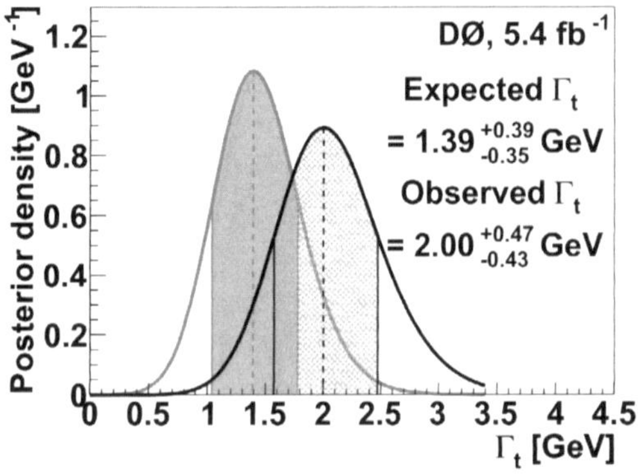

Figure 5. Probability density for the expected and observed total width Γ_t. The shaded areas represent one standard deviation around the most probable value.

determination of the top quark width results $\Gamma_t = 2.00^{+0.47}_{-0.43}$ GeV, which corresponds to a lifetime of $\tau_t = (3.29^{+0.90}_{-0.63}) \times 10^{25}$ s. Figure 5 shows the probability density for the expected and measured total width Γ_t. As the partial decay width is determined using SM predicted values any deviation from the SM expected value of the total decay width would indicate contributions due to new physics. The result does not indicate any hints for new physics contributions.

7. Conclusion

Various recent measurements of top quark properties at the Tevatron are discussed. The presented results are in good agreement with the Standard Model expectations and do not show any hints for new physics. More details and results are given at the DØ and CDF webpage [13]. CDF and DØ continue to provide unique results in the top sector and more top quark measurements using the full data sample are expected to come out soon.

Acknowledgments

The author thanks the organizers of the TOP 2012 workshop for the invitation and for the hospitality of the conference venue.

References

[1] Abe, F. *et al.* (CDF), *Phys. Rev. Lett.* **74**, 2626 (1995) [arXiv:hep-ex/9503002].
[2] Abachi, S. *et al.* (DØ), *Phys. Rev. Lett.* **74**, 2632 (1995) [arXiv:hep-ex/9503003].
[3] Aaltonen, T. (CDF), *Measurement of the ratio* $R = \mathcal{B}(t \to Wb)/\mathcal{B}(t \to Wq)$ *with 8.7 fb*$^{-1}$ *of CDF data*, Conf. Note 10887, 2012.
[4] Abazov, V.M. *et al.* (DØ), *Phys. Rev. Lett.* **107**, 121802 (2011) [arXiv:hep-ex/1106.5436].
[5] CMS, *First measurement of* $\mathcal{B}(t \to Wb)/\mathcal{B}(t \to Wq)$ *in the dilepton channel in pp collisions at* $\sqrt{s} = 7$ *TeV*, CMS-PAS-TOP-11-029, 2012.
[6] Aaltonen, T. (CDF), *Measurement of the top pair cross section in dileptonic channels with a hadronic tau and* $\tau \to b\tau\nu$ *branching ratio, with 9.0 fb*$^{-1}$, Conf. Note 10915, 2012.
[7] Petrillo, G., Talk at this conference *Top pair production cross-section results at Tevatron*, 2012.
[8] Aaltonen, T. (CDF), *Phys. Rev.* **D 84** 031104 (2011) [arXiv:hep-ex/1106.3970].
[9] Abazov, V.M. *et al.* (DØ), *Phys. Rev. Lett.* **108**, 261603 (2012) [arXiv:1203.6103].
[10] Colladay, D. and Kostelecky, V.A. *Phys. Rev.* **D 58**, 116002 (1998); Kostelecky, V.A., *Phys. Rev.* **D 69**, 105009 (2004).
[11] Abazov, V.M. *et al.* (DØ), *Phys. Rev.* **D 85**, 091104 (2012) [arXiv:1201.4156].
[12] Abazov, V.M. *et al.* (DØ), *Phys. Rev.* **D 84**, 112001 (2011) [arXiv:1108.3091].
[13] CDF: http://www-cdf.fnal.gov/physics/new/top/top.html;
DØ: http://www-d0.fnal.gov/Run2Physics/top/top_public_web_pages/top_public.html

Journal of Physics: Conference Series **452** (2013) 012036 doi:10.1088/1742-6596/452/1/012036

Top Quark Asymmetry Results at the LHC

Tae Jeong Kim for the ATLAS and CMS collaborations
Korea University, Seoul, Korea
E-mail: Tae.Jeong.Kim@cern.ch

Abstract. In 2011, an integrated luminosity of more than 5 fb^{-1} at 7 TeV has been delivered by the Large Hadron Collider (LHC). The measurements of the top-antitop quark charge asymmetry in top quark pair production have been performed at the ATLAS and CMS experiments. The measured charge asymmetry values from both experiments are within the uncertainties in agreement with the SM theory predictions. An overview of the latest results of these measurements by the time of TOP2012 conference will be presented.

1. Introduction

Top-antitop quark pair production is symmetric under charge conjugation at leading order in the standard model (SM). However, the top-antitop charge asymmetry can appear in $t\bar{t}$ pair production through quark and anti-quark annihilation at next-to-leading order (NLO) in perturbative QCD through the interference between Initial State Radiation (ISR) and Final State Radiation (FSR), which can contribute to negative asymmetry, and the interference between the Born and box diagrams that enhances the asymmetry [1]. Tevatron has observed significant deviations in forward-backward measurements at CDF and D0 [2]–[3]. In particular, CDF has observed 3.4σ deviation with respect to SM above 450 GeV for $t\bar{t}$-based forward-backward asymmetry (A_{FB}) and D0 for A_{FB} using final state leptons from top decays. These latest results have had large attention and triggered many theoretical explanations. For example, the deviation could be explained by possible exchange of new particles in t-channel and the charge asymmetry variable is sensitive to this additional production mode.

Unlike Tevatron, it is hard to measure at the LHC due to the symmetric initial state of pp collisions since the rapidity distributions of top and anti-top quarks are symmetrically distributed around zero. However, it is feasible when we consider the fact that the top quark (valence quark) width in rapidity distribution is broader than anti-top quark (see quark) width. At ATLAS [4] and CMS [5], the analyses were previously performed with around 1 fb^{-1} showing no deviations [6]–[7].

In 2011, an integrated luminosity of more than 5 fb^{-1} at 7 TeV has been delivered by the LHC. The measurements of the top-antitop quark asymmetry in top-antitop quark pair production have been performed with this data at the ATLAS and CMS experiments. In addition to $t\bar{t}$-based top-antitop quark asymmetry measurement, since the charge asymmetry from $t\bar{t}$ production can be transmitted to leptons from top decays in the dilepton decay mode, the purely lepton-based asymmetry measurement was performed in ATLAS, taking advantage of the high resolution for leptons. The lepton-based asymmetry measurement is also sensitive to top polarization. The variables sensitive to the $t\bar{t}$ and lepton-based asymmetries used at the LHC are following:

$$A_C^{t\bar{t}} = \frac{N(\Delta|y| > 0) - N(\Delta|y| < 0)}{N(\Delta|y| > 0) + N(\Delta|y| < 0)}, \quad A_C^{ll} = \frac{N(\Delta|\eta| > 0) - N(\Delta|\eta| < 0)}{N(\Delta|\eta| > 0) + N(\Delta|\eta| < 0)} \tag{1}$$

where $\Delta|y| \equiv |y_t| - |y_{\bar{t}}|$ ($\Delta|\eta| \equiv |\eta_{l^+}| - |\eta_{l^-}|$) represents the difference of absolute values of top and anti-top rapidities (positively and negatively charged lepton pseudorapidities) and N is the number of events with $\Delta|y|$ ($\Delta|\eta|$) being positive or negative. The rapidity y is defined as $y = \frac{1}{2}\ln\frac{E+P_Z}{E-P_Z}$, where E is the particle energy and P_Z its momentum component along the beam direction while the pseudorapidities η is defined as $\eta = -\ln[\tan(\theta/2)]$, where the polar angle θ is measured with respect to the beamline.

In order to confirm the Tevatron results, it is also important to measure the asymmetry as a function of variables that can enhance the charge asymmetry in certain regions. Therefore, the differential top-antitop quark asymmetry measurements are performed in CMS. In this proceedings, an overview of the latest results of these measurements by the time of TOP 2012 conference will be presented.

2. Charge asymmetry measurement at ATLAS

At ATLAS, with around 5 fb^{-1} data, the inclusive top charge asymmetry using the sensitive variable y_t and the lepton-based charge asymmetry using the sensitive variable η_l, are measured in the dilepton decay mode [8].

The event selection is following. Muons (Electrons) are required to have p_T larger than 20 (25) GeV and $|\eta|$ smaller than 2.5 (2.5). The events which have the dilepton invariant mass from opposite-sign isolated leptons within 10 GeV respect to m_Z are rejected and the missing transverse momentum E_T^{miss} should be larger than 60 GeV in the ee and $\mu\mu$ decay modes. The scalar sum of energy of two leptons and jet should be larger than 130 GeV in the $e\mu$ decay mode. The events also should have at least 2 jets with $p_T > 25$ GeV and $|\eta| < 2.5$. The contribution from fake leptons such as semi-leptonic b decays is estimated using the so called Matrix method [8]. Other backgrounds from single top production, $Z \rightarrow ll$+jets and diboson processes are evaluated using Monte Carlo simulated samples. Top quark four-momentum is reconstructed using kinematic reconstruction method constraining top quark mass and W boson mass together with missing transverse energy and transverse momentum of $t\bar{t}$ system, which yields the true combination matches in 47% of the cases where the solution is found.

In order to compare with theory predictions or with other experiments, the results at reconstruction level need to be corrected for bin migration and efficiency. These corrections are estimated using $t\bar{t}$ samples simulated with MC@NLO. The $t\bar{t}$ samples are reweighted to vary the size of the asymmetry at the generated level. Dependency of the reconstructed A_C on the generated A_C is described by a straight line. The background-subtracted data raw A_C is corrected using both the slope and the offset of the fitted straight line. The slope of the correction for the lepton-based asymmetry is very close to unity while around 0.5 for the $t\bar{t}$-based asymmetry since top quark kinematic quantities are more affected by the detector effect.

After applying unfolding procedure, the combined value for the lepton charge asymmetry is $A_C^{ll} = 0.023 \pm 0.012$ (stat.) $\pm$ 0.008 (syst.) and for the top quark charge asymmetry, $A_C^{t\bar{t}} = 0.057 \pm 0.024$ (stat.) $\pm$ 0.015 (syst.). Figures 1–2 show the measured $\Delta|\eta|$ distributions for lepton pairs and top quark pairs, respectively after the final selection for the ee, $e\mu$ and $\mu\mu$ decay modes. The $t\bar{t}$-based asymmetry in this dilepton analysis is further combined with the previous lepton+jets analysis [6]. The combined value for the top quark charge asymmetry is $A_C^{t\bar{t}} = 0.029 \pm 0.018$ (stat.) $\pm$ 0.014 (syst.). These results are compatible with the SM predictions (from the MC@NLO Monte Carlo generator) of $A_C^{ll} = 0.004 \pm 0.001$ and $A_C^{ll} = 0.006 \pm 0.002$, respectively for the lepton-based and the $t\bar{t}$-based asymmetries.

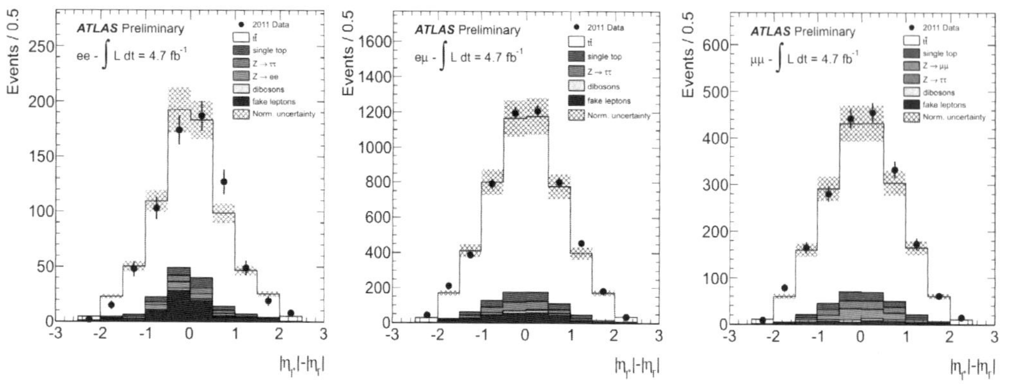

Figure 1. The distributions of $\Delta|\eta|$ for leptons in the ee, $e\mu$ and $\mu\mu$ decay modes at ATLAS [8].

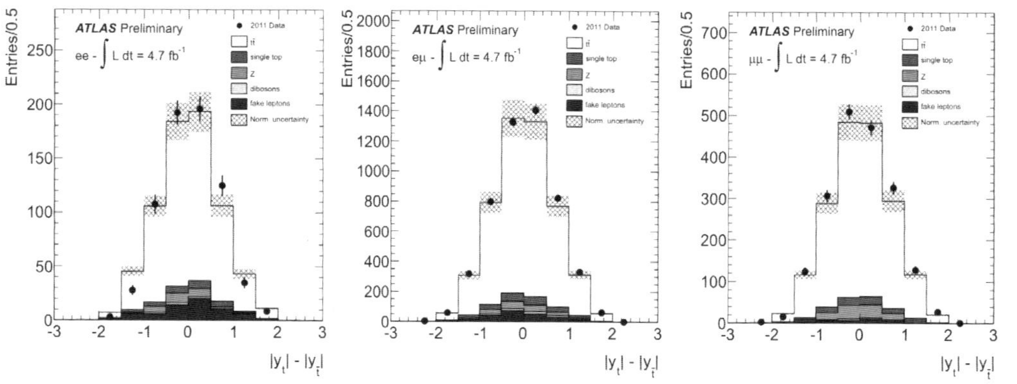

Figure 2. The distributions of $\Delta|y|$ for top quarks in the ee, $e\mu$ and $\mu\mu$ decay modes at ATLAS [8]

3. Charge asymmetry measurement at CMS

At CMS, the previous analysis [7] in the lepton+jets decay mode is updated with an integrated luminosity of 5.0 fb^{-1} [9]. Event selection follows the lepton+jets typical analysis requiring 4 jets and one b-tagging with one exclusive isolated muon (electron) with p_T larger than 20 (30) GeV and $|\eta|$ smaller than 2.1 (2.5). The contribution from multi-jet process is obtained from data events with loosened lepton isolation requirement. The four-momenta of top and anti-top quark are reconstructed by reconstructing the missing z component of neutrino four-vector considering the W boson to be on shell and then associating the measured jets in $t\bar{t}$ decays with a likelihood ratio function of masses of the reconstructed top quarks and the hadronically decay W boson and using b-tagging information. The sensitive variables related to the top quark are reconstructed with correct sign in 72% of all events.

In the SM, it is possible to increase or decrease the asymmetry by changing the kinematic boundary conditions. Therefore, it is crucial to measure the asymmetry differentially in kinematic variables that are sensitive to the asymmetry. Three different variables are used in CMS, namely the rapidity, the transverse momentum and the invariant mass of $t\bar{t}$ system, the vectorially summed four-vector of top quark and anti-quark in the laboratory frame. The reasons why these three variables are chosen are the following. For $y_{t\bar{t}}$, the charge asymmetry occurs only in the $q\bar{q}$ initial state and $t\bar{t}$ production through $q\bar{q}$ annihilation is more prominent

Table 1. The measured inclusive asymmetry at the different stages of the analysis and the corresponding theory prediction from the SM.

Uncorrected	0.003 ± 0.004 (stat.)
BG-subtracted	0.002 ± 0.005 (stat.) $\pm$ 0.003 (syst.)
Final corrected	0.004 ± 0.010 (stat.) $\pm$ 0.011 (syst.)
Theory prediction (SM)	0.0115 ± 0.0006

at larger values of rapidity $y_{t\bar{t}}$. For $p_T^{t\bar{t}}$, the presence of ISR or FSR implies on average higher $p_T^{t\bar{t}}$ so the negative contribution from the ISR or FSR interference is enhanced with large values of $p_T^{t\bar{t}}$. For $\mathrm{m}_{t\bar{t}}$, if heavy particles with different couplings to top quarks and anti-quarks exist, one can expect an effect on the charge asymmetry with rising invariant mass of the $t\bar{t}$ system.

The results at reconstruction level are corrected for bin migration and efficiency. For bin migration correction, the regularized unfolding method is applied. For the response matrix, POWHEG which preserves the asymmetry in NLO is used. For the differential measurement, in order to take into account the bin migration not only between the sensitive variable $\Delta|y|$ but also between differential sensitive variables, 2D regularized unfolding method is applied. The bin width is determined in a way that the statistical uncertainty in each bin becomes approximately the same. The 2D response matrix and efficiency for the measurement differential in $\mathrm{m}_{t\bar{t}}$ are shown in Fig. 3.

Table 1 shows the values of the measured inclusive asymmetry at the different stages of the analysis. Figure 4 shows the distributions of the inclusive and three differential measurements comparing with predictions from the SM calculations and also from an effective field theory [10]. The results from the inclusive measurement and differential cross section measurement show a good agreement with the SM predictions.

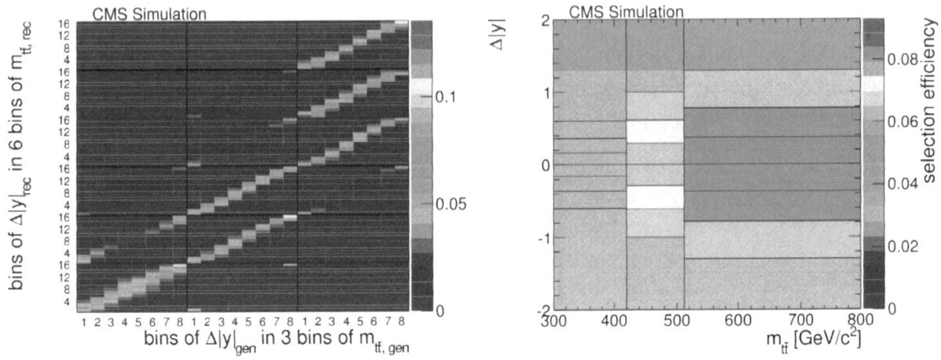

Figure 3. The 2D response matrix (left) and efficiency (right) for the measurement differential in $\mathrm{m}_{t\bar{t}}$ variable at CMS [9].

4. Conclusions

The top-antitop quark charge asymmetry performed by ATLAS and CMS with 2011 data have been presented. The lepton-based A_C in the dilepton decay mode as well as differential asymmetry in the lepton+jet decay mode have been also measured. All measured charge asymmetry values are within the uncertainties in agreement with the SM theory predictions.

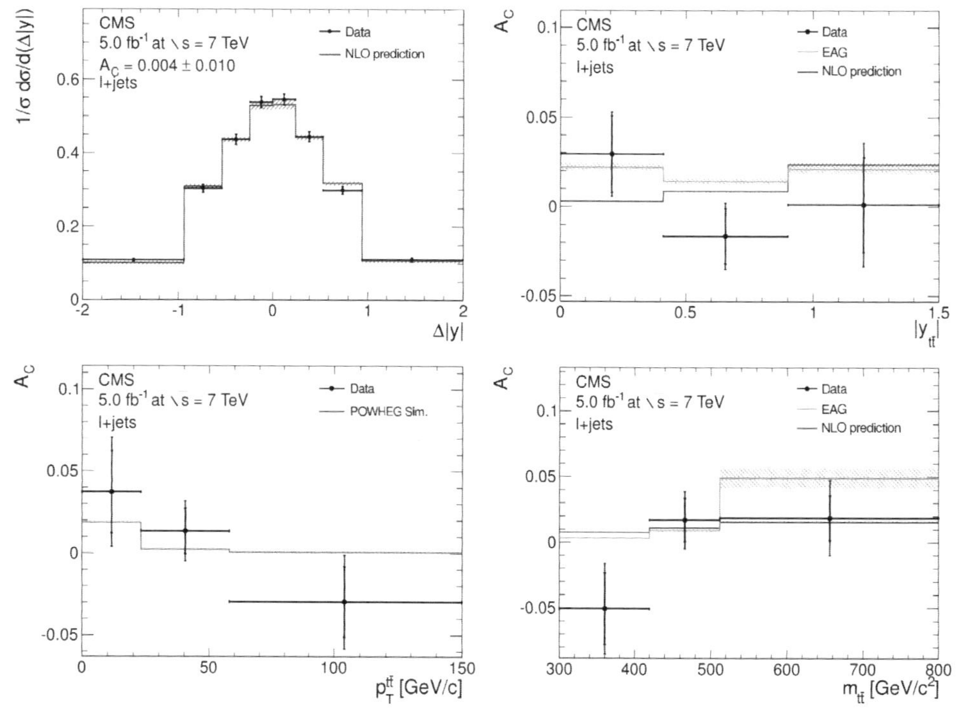

Figure 4. Unfolded inclusive $\Delta|y|$ spectrum (upper-left) and corrected asymmetry as a function of $|y_{t\bar{t}}|$ (upper right), $p_T^{t\bar{t}}$ (lower left) and $m_{t\bar{t}}$ (lower right). The NLO SM calculation and the prediction featuring an effective axial-vector coupling of the gluon (EAG) [10] are also shown [9].

The LHC results do not confirm the Tevatron observation of charge asymmetry deviation: more statistics is needed to validate or exclude the deviation.

References

[1] J. H. Kühn and G. Rodrigo, "Charge Asymmetry in Hadroproduction of Heavy Quarks", Phys. Rev. Lett., **81** (1998) 49.
[2] CDF Collaboration, "Evidence for a mass dependent forward-backward asymmetry in top quark pair production", Phys. Rev. D **83** (2011) 112003.
[3] D0 Collaboration, "Forward-backward asymmetry in top quark-antiquark production", Phys. Rev. D **84** (2011) 112005.
[4] ATLAS Collaboration, "The ATLAS experiment at the CERN LHC", JINST 3:S08003,2008.
[5] CMS Collaboration, "The CMS experiment at the CERN LHC", JINST 3:S08004,2008.
[6] ATLAS Collaboration, "Measurement of the charge asymmetry in top quark pair production in pp collisions at $\sqrt{s}$ = 7 TeV using ATLAS detector", Eur.Phys.J **C72** (2012) 2039.
[7] CMS Collaboration, "Measurement of the charge asymmetry in top-quark pair production in proton-proton collisions at $\sqrt{s}$ = 7 TeV", Phys. Lett. B **709** (2012) 28.
[8] ATLAS Collaboration, "Measurement of the charge asymmetry in dileptonic decay of top quark pairs in pp collisions at $\sqrt{s}$ = 7 TeV using ATLAS detector", ATLAS-CONF-2012-057, https://cds.cern.ch/record/1453785.
[9] CMS Collaboration, "Inclusive and differential measurements of the $t\bar{t}$ charge asymmetry in proton-proton collisions at 7 TeV", Phys. Lett. B **717** (2012) 129.
[10] E. Gabrielli, M. Raidal, and A. Racioppi, "Implications of the effective axial-vector coupling of the gluon on top-quark charge asymmetry at the LHC", Phys. Rev. D **85** (2012) 074021.

5th International Workshop on Top Quark Physics (TOP2012) IOP Publishing
Journal of Physics: Conference Series **452** (2013) 012037 doi:10.1088/1742-6596/452/1/012037

Top couplings and top partners

J A Aguilar Saavedra and M Pérez-Victoria

CAFPE and Departamento de Física Teórica y del Cosmos
Universidad de Granada, E-18071, Spain

E-mail: jaas@ugr.es,mpv@ugr.es

Abstract. We review the model-independent description of the couplings of the top quark to the Higgs and gauge bosons in theories beyond the Standard Model. Then we examine these couplings in the case of arbitrary heavy vector-like quarks mixing with the third family. We also discuss the couplings of these top partners, and comment on implications for LHC searches.

1. Introduction

The experiments at Tevatron and the LHC have allowed to determine with remarkable precission many properties of a resonance with mass around 173 GeV and width about 2 GeV, discovered at Tevatron in 1995. The properties measured so far are consistent with the ones of the particle called top quark: a spin-1/2 fermion with electric charge $Q_t = 2/3$ that is a colour triplet and whose left-handed component is the weak isospin partner of the left-handed b quark. A particle with these quantum numbers is actually required for anomaly cancellation, and furthermore the measured mass is compatible with electroweak precision tests. Therefore, it is safe to assume that the top-like resonance observed in hadron colliders is actually the top quark, as defined two sentences above. This by no means implies that the couplings of this particle are equal to the ones in the Standard Model.[1] Indeed, the simple form of the SM top couplings follows not only from the SM gauge invariance and top quantum numbers, but also from renormalizability and the complete SM field content. In this talk we will examine top couplings in the absence of these last two hypotheses. Especifically, we first consider corrections to the top couplings from non-renormalizable operators constructed with the SM fields, and then study a renormalizable extension of the SM with heavy vector-like quarks. We will keep the requirement of invariance under the (spontaneously broken) SM gauge group, since the consequences of breaking this symmetry are rather disastrous for our understanding of nature. To keep the discussion as simple as possible, we will restrict ourselves to new physics that couples only to the third family in the interaction basis, and assume that electroweak breaking is induced by a light Higgs doublet, just as in the SM. Moreover, we focus on the trilinear top interactions.

2. General top couplings

Let us study the trilinear couplings of the top quark in the presence of arbitrary heavy new physics at a scale Λ, which we assume to be of the decoupling kind. The effective Lagrangian

[1] Still, we call it "top quark". Analogous experimental, theoretical and linguistic considerations may be applied to the recently discovered Higgs-like resonance, which will be called "Higgs boson" hereafter.

describing processes at energies smaller than the scale Λ can be expanded in a power series

$$\mathcal{L}^{\text{eff}} = \mathcal{L}^{(4)} + \frac{1}{\Lambda}\mathcal{L}^{(5)} + \frac{1}{\Lambda^2}\mathcal{L}^{(6)} + \ldots, \tag{1}$$

where each $\mathcal{L}^{(n)}$ is a finite sum of local operators $\mathcal{O}_i^{(n)}$ of scaling dimension n, made of SM fields. These operators are required to be invariant under the full $SU(3)_C \times SU(2)_L \times U(1)_Y$ gauge group. Note that if we imposed only color and electromagnetic symmetries, we would introduce unnecessary parameters and miss some model-independent relations between the couplings. The lowest-order term, $\mathcal{L}^{(4)}$, is just the SM Lagrangian. The first corrections to top couplings arise from $\mathcal{L}^{(6)}$ and change substantially the SM pattern: flavour-changing neutral currents (FCNC) and right-handed charged currents are allowed at the tree-level, the left-handed CKM matrix needs not be unitary and extra CP phases can appear [1]. With our assumption of third-family-philic new physics, the mixing with the light families of quarks is suppressed by small CKM entries and can be neglected. $\mathcal{L}^{(6)}$ also contains four-fermion operators, which give important contributions to top pair production—see [2, 3, 4, 5] and the talk by J. Kamenik in this conference—but are not studied here. A convenient minimal set of operators contributing to trilinear interactions has been given in Refs. [6, 7]. Using these operators, we can write the most general effective interactions of the top quark with the gauge bosons and the Higgs, to order Λ^{-2}, in the following fashion:

$$\mathcal{L}_{gtt} = -g_s\bar{t}\frac{\lambda^a}{2}\gamma^\mu t G^a_\mu - g_s\bar{t}\lambda^a\frac{i\sigma^{\mu\nu}q_\nu}{m_t}\left(d_V^g + i d_A^g\gamma_5\right)tG^a_\mu. \tag{2}$$

$$\mathcal{L}_{\gamma tt} = -eQ_t\bar{t}\gamma^\mu t A_\mu - e\bar{t}\frac{i\sigma^{\mu\nu}q_\nu}{m_t}\left(d_V^{t\gamma} + i d_A^{t\gamma}\gamma_5\right)tA_\mu. \tag{3}$$

$$\mathcal{L}_{Wtb} = -\frac{g}{\sqrt{2}}\bar{b}\gamma^\mu\left(V_L P_L + V_R P_R\right)tW^-_\mu - \frac{g}{\sqrt{2}}\bar{b}\frac{i\sigma^{\mu\nu}q_\nu}{M_W}\left(g_L P_L + g_R P_R\right)tW^-_\mu + \text{h.c.}. \tag{4}$$

$$\mathcal{L}_{Ztt} = -\frac{g}{2c_W}\bar{t}\gamma^\mu\left(X^t_L P_L + X^t_R P_R - 2s_W^2 Q_t\right)tZ_\mu - \frac{g}{2c_W}\bar{t}\frac{i\sigma^{\mu\nu}q_\nu}{M_Z}\left(d_V^{tZ} + i d_A^{tZ}\gamma_5\right)tZ_\mu. \tag{5}$$

$$\mathcal{L}_{Htt} = -\frac{1}{\sqrt{2}}\bar{t}\left(Y_V + iY_A\gamma_5\right)tH. \tag{6}$$

$$\mathcal{L}_{\gamma bb} = eQ_b\bar{b}\gamma^\mu bA_\mu + e\bar{b}\frac{i\sigma^{\mu\nu}q_\nu}{m_b}\left(d_V^{b\gamma} + i d_A^{b\gamma}\gamma_5\right)bA_\mu. \tag{7}$$

$$\mathcal{L}_{Zbb} = \frac{g}{2c_W}\bar{b}\gamma^\mu\left(X^b_L P_L + X^b_R P_R + 2s_W^2 Q_b\right)bZ_\mu + \frac{g}{2c_W}\bar{b}\frac{i\sigma^{\mu\nu}q_\nu}{M_Z}\left(d_V^{bZ} + i d_A^{bZ}\gamma_5\right)bZ_\mu. \tag{8}$$

We emphasize that this parametrization is completely general and valid not only on-shell, but also when the particles are off-shell or inside loops.[2] We have also written the trilinear interactions of the b quark with the Z boson and the photon because, as we will see shortly, they are connected to the top interactions. At order Λ^0, the effective couplings above are as in the SM: $g_{L,R}$, V_R, $X_R^{t,b}$, $d_{V,A}^{g,t\gamma,tZ,b\gamma,bZ}$ and Y_A vanish, while $X_L^t = X_L^b = 1$, $V_L = V_{tb}^{\text{SM}} \simeq 1$ and $Y_V = \sqrt{2}m_t/v$, with $v \simeq 246\,\text{GeV}$ the Higgs vaccuum expectation value. At order Λ^{-2}, all the effective couplings, which we represent generically by κ_j, receive corrections of the form

$$\delta\kappa_j = \kappa_j - \kappa_j^{\text{SM}} = A_j\frac{v^2}{\Lambda^2}, \tag{9}$$

where A_j are dimensionless linear combinations of the operator coefficients. These corrections are called anomalous couplings. The gauge invariance of the operators implies the following

[2] The effect in the amplitudes of other possible trilinear terms can be reproduced by the ones we have written plus four-fermion interactions, see [6] for more details.

relations:

$$\delta X_L^t + \delta X_L^b = 2\delta V_L, \tag{10}$$

$$c_W d^{tZ} + e s_W \frac{v}{m_t} d^{t\gamma} = g_R, \tag{11}$$

$$c_W d^{bZ} + e s_W \frac{v}{m_b} d^{b\gamma} = g_L^*, \tag{12}$$

where we have defined $d^{t,b\,Z,\gamma} = d_V^{t,b\,Z,\gamma} + i d_A^{t,b\,Z,\gamma}$.

Let us make two simple observations with important consequences in the analysis of top data. First, even though the anomalous coupling δV_L, which accounts for the loss of unitarity of a CKM matrix beyond the SM, can be taken to be real [7], its sign is not determined *a priori*. Actually, we give below examples of SM extensions giving rise to either sign. Therefore, V_L can be smaller than, equal to or larger than 1. Requiring $V_L \leq 1$ (or $V_L \geq 1$) is a prior without any model-independent theoretical justification.

The second observation comes from relation Eq. (10). Electroweak precision data (R_b at the Z pole, in particular) put a tight limit on the anomalous coupling δX_L^b. Using this piece of information, we can approximate $\delta X_L^b \simeq 0$ [3] and conclude that

$$\delta X_L^t \simeq 2\delta V_L. \tag{13}$$

This means that, to a good approximation, the zero-momentum couplings of the top quark to the W and Z bosons depend on just one real parameter. In this way, the limit on V_L from single top production can be translated into a limit on the coupling of the left-handed top to the Z boson. Using the results in Ref. [8], we find $X_L^t \in [0.7, 1.8]$.

To finish this section, we discuss the relevance of the different anomalous couplings and their expected size. Because the $\delta\kappa_j$ are of order Λ^{-2}, their contribution to observables will be $O(\Lambda^{-2})$ when the amplitudes these insertions interfere with the SM ones, and $O(\Lambda^{-4})$ otherwise. For this reason, some authors drop non-interfering anomalous couplings from their analyses. However, it should be noted that these couplings contribute to observables that vanish to $O(\Lambda^{-2})$. Hence, they give the leading contribution to some genuinely new effects with small backgrounds. This, in many cases, compensates for the extra suppression. One example is given by the top helicity fraction F_+, which is very small in the SM. The leading corrections to F_+ arise from the anomalous couplings V_R and g_L, and are $O(\Lambda^{-4})$ (plus $O(m_b^2\Lambda^{-2})$) [9]. Dimension-8 operators only contribute to such observables at order Λ^{-8} and can be safely neglected. An additional reason for keeping all the anomalous couplings is that some extensions of the SM only produce the non-interfering type.

On the other hand, it is easy to show that none of the vertices above with a $\sigma^{\mu\nu}$ can be generated by any new physics at the classical level. Therefore, the corresponding couplings d are expected to have a $1/(16\pi^2)$ suppression relative to the other anomalous couplings, as long as i) they arise from new physics at the same scale and ii) the couplings of the new particles to the SM ones are perturbative. These conditions are met, for instance, in the SM extensions considered in the next section. But once again, to keep full generality in a model-independent analysis all the anomalous couplings above should be included.

3. Top partners

The anomalous top couplings in the previous section can be generated at the classical level in theories that contain extra bosons mixing with the SM Higgs or gauge bosons and/or extra quarks that mix with the top or bottom quark. We study here extensions with extra quarks that

[3] This condition occurs naturally in some models, see the examples below.

couple to the third family. New chiral quarks, such as the ones in a 4th generation of fermions, are all but excluded by the new Higgs data [10, 11] (besides having severe difficulties with electroweak precision observables and perturbativity). Therefore, we concentrate on the case of extra vector-like quarks, whose left-handed and right-handed components have the same gauge quantum numbers. These heavy particles appear at the TeV scale in several well motivated extensions of the SM, such as extra dimensions, composite Higgs and little Higgs theories. In order to modify the top couplings at the tree level, the new quarks must mix with the top or bottom quarks. This is only possible for the following multiplets [12]:

$$\begin{aligned} &\mathbf{1}_{\frac{2}{3}} = T, \qquad \mathbf{1}_{-\frac{1}{3}} = B, \qquad \mathbf{2}_{\frac{1}{6}} = \begin{pmatrix} T \\ B \end{pmatrix}, \qquad \mathbf{2}_{\frac{7}{6}} = \begin{pmatrix} X \\ T \end{pmatrix}, \\ &\mathbf{2}_{-\frac{5}{6}} = \begin{pmatrix} B \\ Y \end{pmatrix}, \quad \mathbf{3}_{\frac{2}{3}} = \begin{pmatrix} X \\ T \\ B \end{pmatrix}, \quad \mathbf{3}_{-\frac{1}{3}} = \begin{pmatrix} T \\ B \\ Y \end{pmatrix}. \end{aligned} \tag{14}$$

The electric charge is 2/3, -1/3, 5/3 and -4/3 for the components denoted by T, B, X and Y, respectively. These vector-like fermions can have a gauge invariant Dirac mass M, and decouple in the limit $M \to \infty$. They can couple to the SM quarks via Yukawa interactions of the form

$$\begin{aligned} &-\lambda_Q \bar{Q}_R \phi q_L + \text{h.c.}, \quad \text{if } Q \text{ is a singlet or triplet}, \\ &-\lambda_Q \bar{Q}_L \phi q_R + \text{h.c.}, \quad \text{if } Q \text{ is a doublet}. \end{aligned} \tag{15}$$

$Q_{L,R}$ are the two chiral components of the vector-like multiplet Q, $q_{L,R}$ denote, respectively, the SM (third-family) left-handed quark doublet and right-handed quark singlets, ϕ is the Higgs doublet, and the components in the three field multiplets must be combined in such a way that the term is an $SU(2)_L$ singlet. When hypercharge conservation is also imposed, only one of these Yukawa terms is allowed for each vector-like multiplet (with a coupling λ that can be taken real), except for the doublet $\mathbf{2}_{\frac{1}{6}}$, which can couple to both t_R and b_R (with two real λ's and a relative phase). In extensions with more than one type of multiplet, it is also possible to write Yukawa terms coupling heavy doublets with heavy singlets or triplets.

Table 1. Leading corrections to the couplings induced by each vector-like quark multiplet. Up, down and up-down arrows indicate positive, negative or indefinite corrections, respectively, whereas "—" means no corrections.

	T	B	$\begin{pmatrix} T \\ B \end{pmatrix}$	$\begin{pmatrix} X \\ T \end{pmatrix}$	$\begin{pmatrix} B \\ Y \end{pmatrix}$	$\begin{pmatrix} X \\ T \\ B \end{pmatrix}$	$\begin{pmatrix} T \\ B \\ Y \end{pmatrix}$
V_L	↓	↓	—	—	—	↑	↑
V_R	—	—	↕	—	—	—	—
X_L^t	↓	—	—	—	—	↓	↑
X_R^t	—	—	↑	↑	—	—	—
X_L^b	—	↓	—	—	—	↑	↓
X_R^b	—	—	↑	—	↑	—	—
Y_V^t	↓	—	↓	↓	—	↓	↓

Upon electroweak breaking, the interactions in Eq. Eq. (15) give rise to mass terms that mix the heavy and SM quarks. When this matrix is diagonalized, the trilinear couplings of the t and b quarks are modified, giving rise to some of the anomalous couplings in Eqs. (2–8). Equivalently, these corrections can be obtained integrating out the new fields to find $\mathcal{L}^{\text{eff}}$, with scale $\Lambda = M$. In table 1 we show which anomalous couplings are generated by each individual vector-like multiplet at the tree level, and indicate the sign of the corrections. Note that the singlets decrease the value of V_L while the triplets increase it. We also see that three of the multiplets correct some top couplings but do not modify X_L^b. This is enforced, without any fine tuning, by gauge invariance. The anomalous couplings Y_A and d appear at the loop level and have extra $1/(16\pi^2)$ suppressions. There are also exact relations between the different anomalous couplings, which are given in Ref. [12].

The very same parameters that modify the top couplings enter in the couplings of the heavy quarks to the gauge and Higgs bosons. This allows to connect the searches of new heavy quarks to the precision study of top couplings. On the other hand, the couplings of each extra quark to the W, Z and Higgs are also related. These relations imply that their decay branching ratios into different final states depend only on the type of multiplet they belong to, except for the case of the doublet $\mathbf{2}_{1/6}$, in which the branching ratios depend on one continuous parameter [13]. This is illustrated in Fig. 1, where we represent the possible branching ratios of heavy T and B

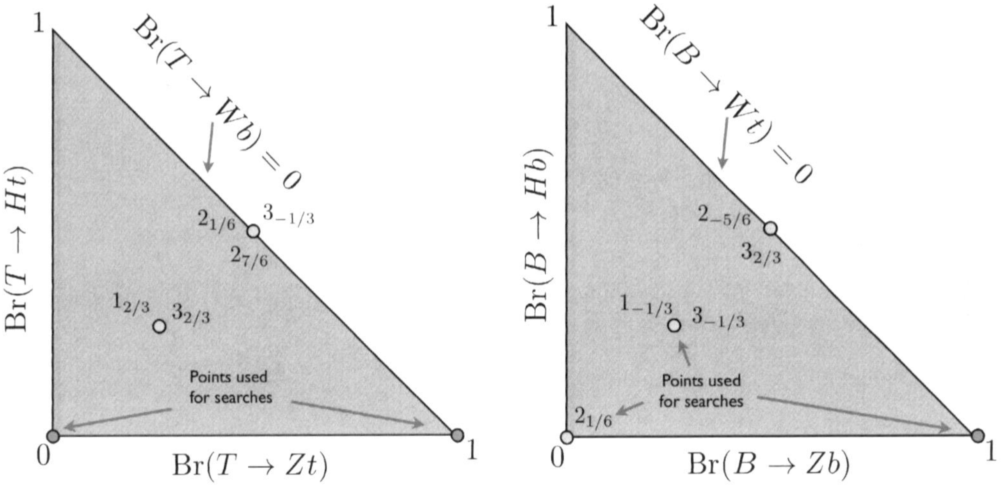

Figure 1. Branching ratios of $T \to Ht$ and $T \to Zt$ (left) and $B \to Ht$ and $B \to Zt$ (right). The green points indicate the branching ratios in the limit of large M for each multiplet, whereas the arrows point to the branching ratios usually assumed in experimental searches.

quarks decaying into final states with Z and Higgs bosons. The branching ratio into W bosons is determined from these by the requirement that $\text{Br}(T \to Zt) + \text{Br}(T \to Ht) + \text{Br}(T \to Wb) = 1$, and analogously for B. The green points indicate the branching ratios in the limit of large M for each multiplet. For $\mathbf{2}_{1/6}$, we have assumed that the mixing with the top is much larger than the mixing with the b quark (in general this multiplet would be represented by a line). The arrows point to the branching ratios usually assumed in experimental searches. We see that in many cases these assumptions do not correspond to the allowed branching ratios, so the results of these searches need to be reinterpreted. [4]

[4] The Atlas collaboration has recently interpreted their searches as constraints on the allowed points of these triangles [14].

Finally, let us comment on the contribution of these top partners to Higgs production by gluon fussion and Higgs decay into $\gamma\gamma$. The SM predictions are corrected in two ways: by loops of the heavy quarks and by modifications of the top (and b quark) couplings. It is clear that both types of corrections vanish in the limit $M \to \infty$, since the mass of these vector-like fermions does not arise from a Yukawa coupling to the Higgs doublet. Furthermore, it turns out that the corrections from T loops and t anomalous couplings cancel out against each other to a good approximation, for all types of multiplets, as long as there are no Yukawa couplings between the heavy quarks. On the other hand, the loops with B and b quarks are suppressed by m_b. To obtain larger corrections, it is necessary to include more than one type of multiplet, as in the models in Ref. [15]. Note also that in the effective Lagrangian formalism the effect of the heavy quark loops is reproduced by additional local operators that contribute to these amplitudes.

Acknowledgments
We thank the organizers of Top 2012 for a very pleasant and interesting conference. This work has been supported by the MICINN project FPA2010-17915.

References
[1] del Aguila F, Pérez-Victoria M and Santiago J 2000 *Phys. Lett.* B **492** 98
[2] Degrande C, Gerard J M, Grojean C, Maltoni F and Servant G 2011 *JHEP* **1103** 125
[3] Delaunay C, Gedalia O, Hochberg Y, Perez G and Soreq Y 2011 *JHEP* **1108** 031
[4] Aguilar-Saavedra J A and Pérez-Victoria M 2011 *JHEP* **1105** 034
[5] Aguilar-Saavedra J A and Pérez-Victoria M 2011 *Phys. Lett.* B **701** 93
[6] Aguilar-Saavedra J A 2009 *Nucl. Phys.* B **812** 181
[7] Aguilar-Saavedra J A 2009 *Nucl. Phys.* B **821** 215
[8] Aad G *et al.* [ATLAS Collaboration] 2012 *Phys. Lett.* B **717** 330
[9] Aguilar-Saavedra J A and Bernabeu J 2010 *Nucl. Phys.* B **840** 349
[10] Djouadi A and Lenz A 2012 *Phys. Lett.* B **715** 310
[11] Eberhardt O, Herbert G, Lacker H, Lenz A, Menzel A, Nierste U and Wiebusch M 2012 *Phys. Rev. Lett.* **109** 241802
[12] del Aguila F, Pérez-Victoria M and Santiago J 2000 *JHEP* **0009** 011
[13] Aguilar-Saavedra J A 2009 *JHEP* **0911** 030
[14] Aad G *et al.* [ATLAS Collaboration] 2013 *Phys. Lett.* B **718** 1284
[15] Bonne N and Moreau G 2012 *Phys. Lett.* B **717** 409

5th International Workshop on Top Quark Physics (TOP2012) IOP Publishing
Journal of Physics: Conference Series **452** (2013) 012038 doi:10.1088/1742-6596/452/1/012038

SM Top properties at LHC: top charge, Branching Ratio and top pair production associated with gauge bosons

Mara Senghi Soares, on behalf of the CMS and ATLAS Collaborations

Centro de Investigaciones Energéticas, Medioambientales y Tecnológicas (CIEMAT), Madrid, Spain

E-mail: mara@cern.ch

Abstract. In this contribution, we review some of the properties of the top quark in LHC measurements by the CMS and ATLAS Collaborations using pp collisions at 7 TeV. Firstly, measurements of two intrinsic properties of the top quark are revisited: its charge and branching ratio. Then, the first measurements of associated production of top pairs with electroweak gauge bosons W, Z and photons are presented. Associated production cross-sections provide information on the structure of the electroweak couplings to top quarks. All measurements are consistent with the expectations from the standard model top properties.

1. Introduction

With the LHC data from pp collisions at 7 TeV collected in 2011 using ATLAS and CMS detectors a new level of understanding on the properties of the top quark was achieved. Some of the experimental results from the Tevatron collider could be verified at much better precision, and new measurements which were not accessible at the Tevatron became possible with these data.

We review a few results from top measurements by the CMS and ATLAS Collaborations. A detailed description of the ATLAS and CMS detectors can be found elsewhere [1, 2]. The measurements of special interest are those which can potentially indicate the existence of physics beyond the standard model (SM).

One of the intrinsic properties of the quark top, its electric charge, was investigated under the possibility of an exotic scenario in disagreement with the SM.

Properties of the top decay are also an excellent testing ground to the SM. In particular, deviations from the SM prediction of top quarks decaying almost exclusively in the mode $t \rightarrow Wb$ can be indicative of a fourth quark generation. We present measurements by the CMS Collaboration on the decay rates of top quarks to b and light quarks. Experimental results concerning other properties of the top quark decay are covered in other contributions to this conference.

The production of $t\bar{t}$ associated to gauge bosons are of special interest in the LHC physics program since their cross-sections are too low to be measured with high precision at the Tevatron. Properties of $t\bar{t}$ couplings to bosons can be inferred from the cross-sections and angular distributions observed in such processes. Measurements of $t\bar{t}\gamma$ from the ATLAS and $t\bar{t}V$ (where V=W, Z) from the CMS Collaborations are presented. The search for $t\bar{t}$ associated to the Higgs boson was presented in another session in this conference, and is not reported here.

2. Top electric charge

The top quark decays mainly to a W^+ boson and a b-quark. As the electroweak isospin partner of the b-quark, its expected electric charge in the SM is +2e/3. In an exotic scenario where this hypothesis does not hold, the top charge could assume the value of -4e/3, decaying instead to a W^- boson and a b-quark.

The measurements of the top electric charge [3, 4] at the LHC were performed using semileptonic $t\bar{t}$ decays. If one of the W bosons decays leptonically (i.e. to a neutrino and an electron or muon), its charge can be directly measured as the charge of the electron or muon. The challenging aspects of the measurement are, then, to perform the correct association of the b-quark and the W to its parent top, and determine the electric charge of the other top decay product, the b-quark.

The identification of jets initiated by b-jets (b-tagging) is based on the properties of the B hadrons. Due to the long B lifetime, a b-jet is produced typically with large impact parameter as compared to a light quark- or gluon-initiated jet. For this reason, information related to the impact parameter or the tracks from secondary vertices inside the jet can be combined into variables with discriminating power to separate b- from light quarks- or gluon- initiated jets. B-hadrons also decay into leptons, thus some discriminators rely on the detection of non-isolated low-p_T ("soft") muons, and use parameters such as the soft muon p_T relative to the jet axis.

Two techniques were used to determine the charge of the b-quark. The first relies on the charge of the soft muon coming from the B-hadron decay. In direct B-hadron decays, the charge of the soft muons is the same as the b-quark. However, in the cases of flavour oscillations and B-hadron cascade decays, such as b→D+X→μ + ν+X, the charges of the muon and b-quark are not necessarily the same, degrading the measurement. In the second method, a "jet charge" $Q_{b\ jet}$ is assigned according to the momentum-weigthed sum of the charges (q_i) of all jet tracks i, $Q_{b\ jet} = \frac{\sum_i q_i |\vec{j}\cdot\vec{p}_i|^\kappa}{\sum_i |\vec{j}\cdot\vec{p}_i|^\kappa}$, where $\vec{j}$ is the jet axis direction and the parameter κ (=0.5 and =0.7 for the analysis by ATLAS and CMS, respectively) is optimized for the best separation between b and $\bar{b}$ jets.

The charge of the b-jet estimated with one of the described methods is combined with that of the lepton from the W decay to determine the charge of the top quark, and compared to the expectations from the SM and the exotic scenario. Both experiments exclude an exotic scenario at 5 σ, confirming at the LHC the top charge constraints found previously at the Tevatron [5].

3. Branching Ratio

CMS has measured [6] the ratio of the Branching Ratios (BR) R=B(t→Wb)/B(t→Wq) where B(t→Wb) is the BR of the top decaying to b's and B(t→Wq) the BR of the top decaying to light quarks. A high-purity $t\bar{t}$ sample in the full-leptonic decay mode, corresponding to 2.2 fb^{-1}, was used. As in the case of the electric charge measurement, the measurement requires the correct identification of b and light quarks, and correct association to its parent top.

The ratio was measured using a model based on 3 parameters: the b-tagging efficiency, the efficiency for the b-tagging criteria to accept light jets (mistags), and the fractions of events with two, one or no top quarks reconstructed and selected.

Events were none or only one top is correctly reconstructed correspond to background events or to $t\bar{t}$ events where at least one of the jets is missed and the top is reconstructed with a wrong choice of jet. Single-top events correctly reconstructed also contribute to the "one top" category. The number of top quarks with misassigned jets was estimated in data using the distribution of the invariant mass of the lepton-jet pairs. Distributions of wrongly reconstructed jets were built rotating the components of the b-jet momentum, or using in the reconstruction a b-jet of another event. For top quarks correctly reconstructed, the distribution is expected to fall-off around $\sqrt{m_t^2 - m_W^2} = 156$ GeV/c^2; therefore, the wrongly reconstructed distribution can be normalized on the tails of the data distribution. This technique was applied in data and cross-checked in simulation, as shown in Fig. 1 (a). The two remaining parameters in the model, the efficiencies for b-tagging and mistags, were measured using a multi-jets (QCD) sample as described elsewhere [7].

The model for the observation of different b-tag multiplicities as a function of R and the multiplicities observed in the data, compared to the simulation, are shown in Fig. 1 (b) and (c), respectively. The measurement yields R=0.98±0.04, in agreement with the SM. If the requirement R≤1 is imposed, the limit R>0.85 is obtained at 95% C.L.

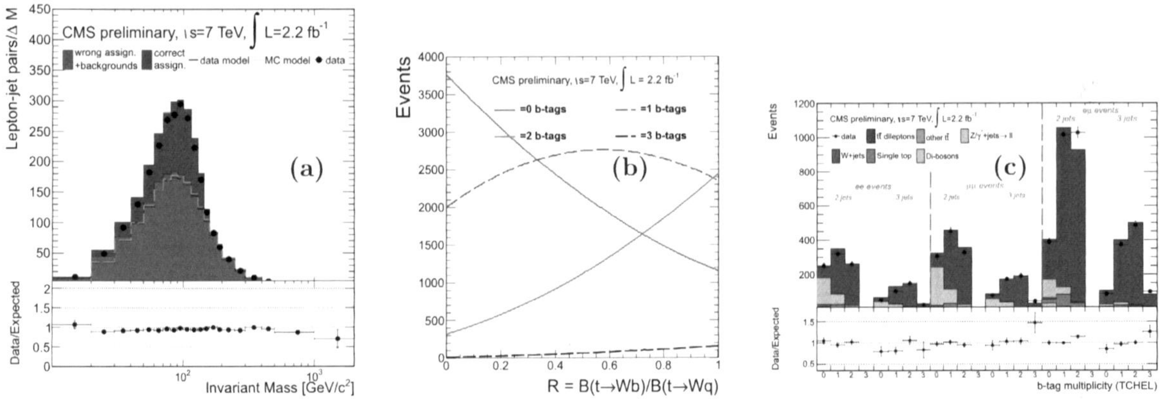

Figure 1. (a) Invariant mass of the lepton-jet pairs in the inclusive dilepton channel: number of correct jet assignments (simulation), model for wrongly assigned jets and the sum (data and simulation). (b) Model for R as a function of the b-jet multiplicities. (c) Observed b-tag multiplicity in data and simulation.

4. Associated top pair production

ATLAS Collaboration has measured [8] the cross-section times branching ratio of the inclusive $t\bar{t}+\gamma$ process using 1.04 fb^{-1}. The sample is selected to contain semi-leptonic $t\bar{t}$ decays to both electrons and muons. Additionally, a photon with $p_T > 15$ GeV and selected according to a set of shower shape criteria was required. Photons irradiated both at top production and decay were considered as part of the signal. The signal to background separation is performed by fitting the data using templates for prompt photons and "fake photons", which are hadrons in $t\bar{t}$ events passing the full γ selection criteria. Promptly produced photons are mostly isolated and have similar shower shapes than electrons. Hadrons faking photons, on the other hand, are non-isolated. Signal templates are built from the data using the isolation variable as measured in Z→ee events. The templates for "fake photons" are obtained on a sample selected using jet triggers. To avoid the presence of real photons, the events entering the "fake" template are required to fail one of the shower shape criteria applied for the signal. The results of the template fit is shown in Fig. 2. Although the fit was performed simultaneously to the electron and muon channels, results are shown separately for each channel.

The measured cross-section times BR for the $t\bar{t}+\gamma$ process, for photons with transverse momenta $p_T > 8$ GeV is 2.0±0.5(stat.)±0.7(syst)±0.08(lumi) pb. The SM predicts 2.1±0.4

pb for the same phase space and BR. Since the electromagnetic coupling of a fermion to photon depends on the fermion electric charge, this measurement can be interpreted as an alternative estimator of the top charge.

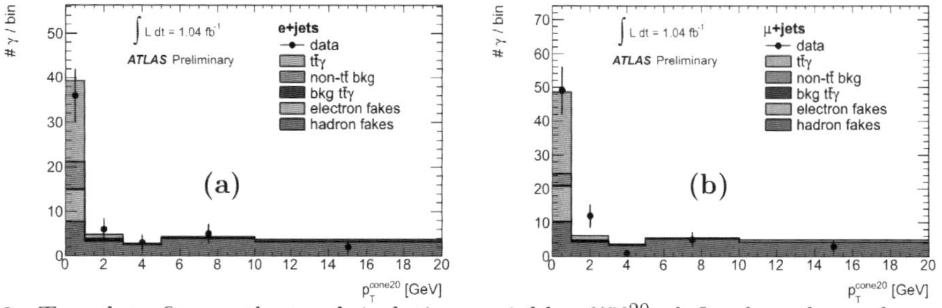

Figure 2. Template fits on the track isolation variable p_T^{cone20}, defined as the scalar sum of all tracks in a cone with $\Delta R < 0.20$ around the photon candidate. Results for (a) electrons and (b) muons.

In CMS [9], the associate production $t\bar{t}$+V, with V=W or Z was measured using two independent final states: firstly, trilepton final states with 4 jets, at least 2 of them b-tagged, and 3 leptons, at least 2 of them of the same flavor and opposite charge were considered. These requirements aimed to select events from the process pp$\rightarrow t\bar{t}$Z$\rightarrow$(t$\rightarrow$bjj)(t$\rightarrow$b$l\nu$)(Z$\rightarrow ll$), where l is an electron or muon. Secondly, a sample containing at least 3 jets, one of which b-tagged, and two leptons of opposite charge was selected. With these criteria, events from both pp$\rightarrow t\bar{t}$Z$\rightarrow$(t$\rightarrow$bjj)(t$\rightarrow$b$l\nu$)(Z$\rightarrow ll$) and pp$\rightarrow t\bar{t}$W$\rightarrow$(t$\rightarrow$bjj)(t$\rightarrow$b$l\nu$)(W$\rightarrow l\nu$) processes are selected.

Both analyses are based on 5 fb^{-1} of data. In order to guarantee statistical independence of the two analyses, events fulfilling the trilepton selection are vetoed in the dilepton analysis.

Multilepton final states are rare in the SM, so a substancial fraction of the background in both analyses are originated from misreconstruction effects, such as "fake" leptons and misidentified b-jets. The estimation of such background sources was perfomed on control samples of data with looser selection criteria and extrapolated to the fiducial region of each analysis. The number of observed events compared to background predictions in each channel is shown in Fig. 3(a) and (b) for the trilepton and dilepton selections, respectively.

The $t\bar{t}$+Z cross section measured using the dilepton selection is $0.30^{+0.14}_{-0.11}(\mathrm{stat})^{+0.04}_{-0.02}(\mathrm{syst})$ pb, with a significance of 3.66 standard deviations and compatible with the NLO prediction of 0.1387 pb. In the trilepton channel, the measured $t\bar{t}$+V cross section (where V=Z or W) was $0.45^{+0.17}_{-0.15}(\mathrm{stat})^{+0.06}_{-0.05}(\mathrm{syst})$ pb, with a significance of 2.99 standard deviations.

The $t\bar{t}$+Z cross section from the dilepton selection was extrapolated to an inclusive $t\bar{t}$+V cross section, and combined to the cross-section measured using the trilepton selection, yielding a $t\bar{t}$+V cross section of $0.51^{+0.15}_{-0.13}(\mathrm{stat})^{+0.05}_{-0.04}(\mathrm{syst})$ pb. The result is slightly above but still compatible with the NLO expectation of 0.308 pb. The results on $t\bar{t}$+V from the individual and combined measurements is summarized in Fig. 3(c).

5. Conclusions

Measurements of the top properties using LHC pp data taken at 7 TeV in 2011 were presented.

Top electric charge and flavour content of top decays were probed and found in agreement with the expectations from the SM. The low number of associated $t\bar{t}+\gamma$ and $t\bar{t}$+Z events in 2011 data does not yet allow careful experimental study of the structure of the electroweak couplings to top quarks. In spite of the large uncertainties, the measurements performed by CMS and

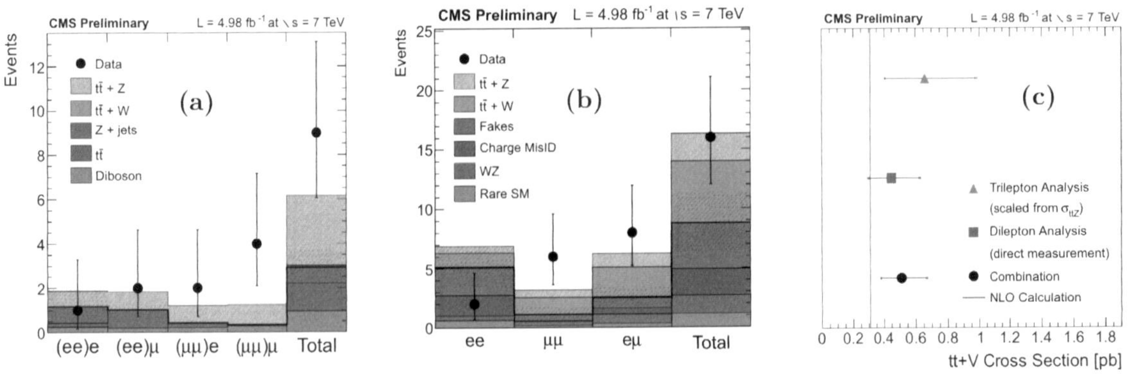

Figure 3. Number of events observed and predicted using (a) the trilepton and (b) the dilepton selection criteria. (c) Summary of the $t\bar{t}$+V cross section measurements, compared to the NLO prediction.

ATLAS are still of interest in top physics, since they pave the way for more precise measurements in the near future. Moreover, these processes, as well as $t\bar{t}$+W, are important backgrounds on searches for new physics, as for instance SUSY searches, and to the measurement of associated $t\bar{t}$+Higgs production.

Prospects for the top physics program in the near future are promising. CMS and ATLAS have arrived at the end of 2012 with reconstructed data corresponding to more than 20 fb^{-1} of integrated luminosity per experiment, taken at a higher collision energy than in 2011, $\sqrt{s} = 8$ TeV. Even more data was collected and stored (via Data Parking at CMS and the Delayed Data Stream at ATLAS), that should be available to be analyzed in 2013. This unprecedent amount of data from high-energy pp collisions will bring important improvements to the precision of the top quark properties measurements. Direct improvements will be obviously seen in the studies involving rare processes (statistically limited in 2011) such as the associated productions of $t\bar{t}$+(γ,W,Z and Higgs). In all top measurements, the reduction on some of the systematic uncertainties, such as background estimation using data, is also expected.

References

[1] ATLAS Collaboration, "The ATLAS experiment at the CERN Large Hadron Collider", J. Instrum. 3, S08003, 2008.

[2] CMS Collaboration, "The CMS experiment at the CERN LHC", J. Instrum. 3, S08004, 2008.

[3] CMS Collaboration, "Constraints on the top-quark charge from top-pair events", CMS PAS TOP-11-031, 2011.

[4] ATLAS Collaboration, "Measurement of the top quark charge in pp collisions at $\sqrt{s}$=7 TeV in the ATLAS experiment", ATLAS-CONF-2011-141.

[5] CDF Collaboration, "Exclusion of an exotic top quark with -4/3 electric charge using soft lepton tagging", Phys. Rev. Lett. 105, 101801, 2010; D0 Collaboration, "Experimental discrimination between charge 2e/3 top quark and charge 4e/3 exotic quark production scenarios", Phys. Rev. Lett. 98, 041801, 2007.

[6] CMS Collaboration, "First measurement of B(t→WB)/B(t→Wq) in the dilepton channel in pp collisions at $\sqrt{s}$=7 TeV", CMS PAS TOP-11-029, 2011.

[7] CMS Collaboration, "b-Jet identification in the CMS experiment", CMS PAS BTV-11-004, 2011.

[8] ATLAS Collaboration, "Measurements of the inclusive $t\bar{t}\gamma$ cross section at $\sqrt{s}$=7 TeV with the ATLAS detector", ATLAS-CONF-2011-153, 2011.

[9] CMS Collaboration, "First measurement of vector-boson production associated with top-antitop pairs at 7 TeV", CMS PAS TOP-12-014, 2012.

Journal of Physics: Conference Series **452** (2013) 012039 doi:10.1088/1742-6596/452/1/012039

Signal modelling systematics at ATLAS

Ivor Fleck on behalf of the ATLAS Collaboration

Universität Siegen, Naturwissenschaftliche-Technische Fakultät, 57068 Siegen, Germany

E-mail: ivor.fleck@uni-siegen.de

Abstract. For the analysis of various properties of the top quark, parameters like selection efficiency are taken from simulated event samples. These samples need several input parameters, like the top quark mass, the strength of initial and final state radiation or parton distribution functions. The presented results will depend on the choice of these input parameters. Therefore it is mandatory to determine a range of parameters which is consistent with current measurements and is able to determine the systematic uncertainty on the measurement result due to the input of certain parameters in the signal generation. In these proceedings the procedures used by the ATLAS experiment at the LHC are given.

1. Introduction

At the ATLAS experiment [1] at the LHC several modelling parameters were studied and their influence on the results of various measurements for top quarks were evaluated. The signal modelling uncertainties that are covered in this article are the following:

- Monte Carlo generator
- parton shower model
- parton distribution function
- initial and final state radiation
- top quark mass
- colour reconnection

2. Event generators

In ATLAS the default generator for events with top quarks is MC@NLO 4.0x[1] [2] with HERWIG 6.520 [3] for the parton shower and JIMMY 4.31 [4] for the underlying event. As the name suggests this generator simulates events at next-to-leading order in perturbation theory.

Two more generators, POWHEG-hvq-patch4(BOX 1.0.x) [5] and ALPGEN 2.1x [6] with the same settings, are also used to generate events within the ATLAS event generation environment.

For any analysis the largest discrepancy observed for these three generators is taken as a systematic uncertainty. Typical discrepancies are 5 % [7] for MC@NLO compared to POWHEG and up to 10 % for MC@NLO compared to ALPGEN. Even though ALPGEN is a leading order generator, in some parts of the phase space the data are better described by ALPGEN than by the other generators which are next-to-leading order generators.

[1] Version numbers greater or equal to 4.01 are currently used while older analyses were using MC@NLO 3.41.

In general one cannot say that one generator describes the data better than any of the other generators as for each generator there is some phase space where it agrees better with the data than the other generators. The largest discrepancies are observed in the acceptance. For the future it is therefore foreseen to reduce this acceptance dependency by applying looser cuts or to use variables that depend less on the boost of objects.

Data from the jet veto [8] analysis is compared to the prediction from the Monte Carlo generators mentioned above and is shown in Fig. 1. In this analysis events are vetoed if they contain in addition to the jets from the top quark pair production decay an additional jet with transverse momentum above a threshold Q_0 in a central rapidity interval. The fraction of events surviving the jet veto is presented as a function of Q_0. For the central rapidity region $|y| < 0.8$ the result is shown in Fig. 1. Discrepancies of several percent are observed between the different generators but all are in agreement with the data within the measurement uncertainties.

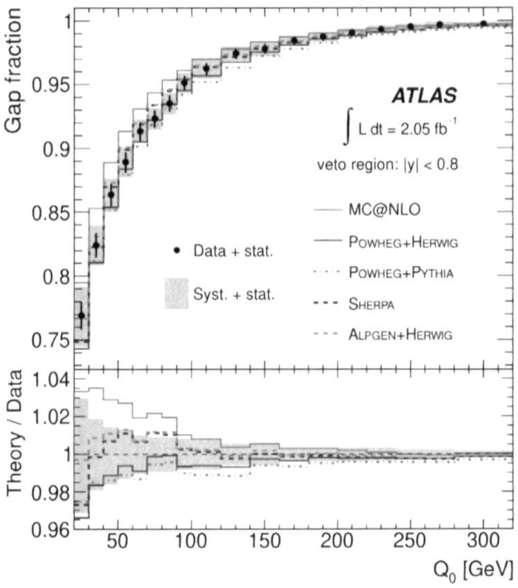

Figure 1. Gap fraction, i.e. the fraction of the number of top quark pair production events with no jet above the threshold Q_0 and the total number of top quark pair production events as a function of Q_0 for ATLAS data and several generators [8]. This plots shows the gap fraction for the central detector region with rapidity $|y| < 0.8$. It can be seen that the data and the Monte Carlo predictions agree within the uncertainty of the data.

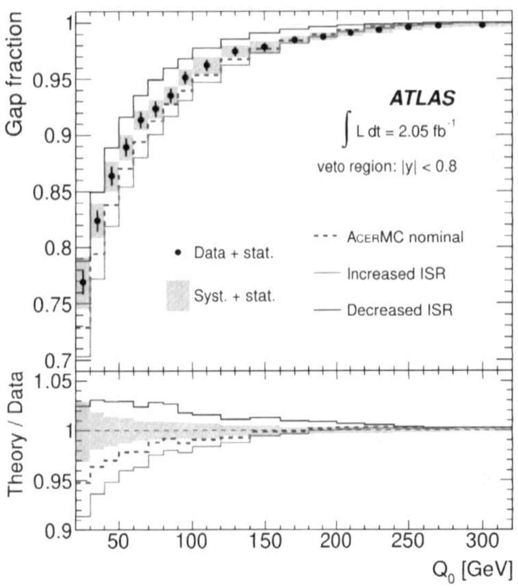

Figure 2. Gap fraction, i.e. the fraction of the number of top quark pair production events with no jet above the threshold Q_0 and the total number of top quark pair production events as a function of Q_0 for ATLAS data and several generators [8]. This plots shows the gap fraction for the central detector region with rapidity $|y| < 0.8$. It can be seen that the Monte Carlo predictions for different values of ISR parameters vary stronger than the uncertainty from this measurement.

3. Parton shower

Parameters in PYTHIA [9] for parton shower (PS) and underlying event (UE) are tuned to match data from the LHC [10]. There exist separate tunes for minimum bias (MB) and underlying event parameters. The tuning is done in several steps, beginning with flavour parameters, then using final state radiation (FSR) and hadronisation, followed by initial state radiation (ISR) and multiple-parton interactions (MPI). For HERWIG and JIMMY the UE parameters are also tuned. Tuning is done using Rivet [11] and the professor tool [12].

Tunes for PYTHIA6 were done using ATLAS data taken at a centre-of-mass energy of $\sqrt{s} = 7$ TeV using several sets of PDFs. Theses tunes are called AUET2B (ATLAS underlying event tune) and AMBT2B (ATLAS minimum bias tune). Tunes for PYTHIA8 are already available but are not yet extensively used for top quark production process simulation.

For HERWIG and JIMMY a tune is done only for MPI parameters of JIMMY, as HERWIG does not have MPI model. The minimum bias data cannot be used is this case and consequently only a UE event tune using several PDFs exists.

For the determination of the systematic uncertainties due to the modelling of the parton shower the following two models are using in combination with the POWHEG event generator:

- HERWIG with JIMMY and AUET2B
- PYTHIA with AUET2B

In Fig. 3 ATLAS data is compared to the event tune AUET2B for PYTHIA, using three different parton distribution functions. Within the uncertainty of the data there is good agreement between data and the simulated events.

The comparison between the different tunes results in an uncertainty of 2% in the measurement of the top quark pair-production cross section [7] or an uncertainty of 0.15 GeV in the measurement of the top quark mass [13].

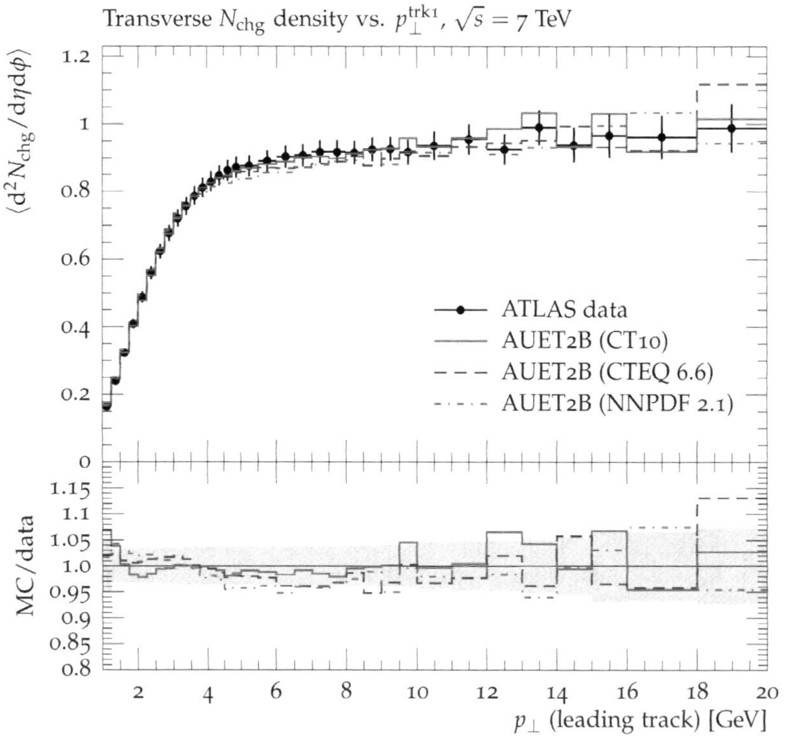

Figure 3. Mean number of selected tracks (N_{chg}) per unit of rapidity (η) and polar angle (ϕ) as a function of the transverse momentum ($p_\perp$)of the track with the largest transverse momentum [10]. Shown are data and simulated events with PYTHIA and the ATLAS tune AUET2B for three different parton distribution functions.

4. Initial and final state radiation

At hadron colliders ISR and FSR of gluons is the most common process. This results in additional jets in the event except those from of the hard scattering. As this process is so common it can be rather well measured in data. ATLAS has performed a study of such events. When using the gap fraction analysis [8] the data can be compared with different settings for initial state radiation. This is shown in Fig. 2 for the central detector region with a rapidity $|y| < 0.8$. As can be seen the parameter variation used gives larger variations than is compatible with the measured data and its uncertainties. Consequently the ISR variations in future analyses will be smaller than the ones used so far and will reduce the dependency of results on the parameters for ISR.

Taking into account the uncertainties on the ISR and FSR parameters in the simulation of events, results in uncertainties for the top quark pair production cross section in the range of $2 - 5$ % and in an uncertainty of $0.5 - 1.1$ GeV for the top quark mass [13].

5. Parton distribution functions

For the evaluation of uncertainties due to the use of a certain parton distribution function (PDF) a recommendation of the PDF4LHC group [14] exists. In ATLAS these recommendations are used comparing results for three different PDFs, i.e. MRSTW08 [15], CTEQ6.6 (or CT10) [16] and NNPDF2.2 [17]. All available sets for each PDF and the strong coupling constant α_s uncertainty are used. For all results using the different inputs an envelope is calculated and taken as the systematic uncertainty. This corresponds to an uncertainty on the top quark pair production cross section of less than 3% [7] and of 0.1 GeV for the top quark mass [13].

6. Top quark mass

The top quark mass is known with a precision of 0.9 GeV from the Tevatron experiments [18]. Consequently this uncertainty is used in the variation of the top quark mass in the analyses. Event samples are produced with a value for the top quark mass of 172.5 GeV. Additional samples with ± 2.5 GeV and ± 5 GeV are also produced. From these samples results for a variation of the top quark mass by the world average uncertainty value of 0.9 GeV are derived. For the cross section measurement for top quark pair production this results in an uncertainty of less than 1 % [19].

7. Colour reconnection

In recent analyses, especially in the top quark mass measurements, colour reconnection effects are also taken into account. In ATLAS the MC generator AcerMC [20] with Pythia and the Perugia2011 [21] tune with and without colour reconnection with the new PS/MI model in Pythia6.4 are used. In addition also the Tevatron tune A-Pro [22] and A-Pro with colour reconnection (ACR-Pro) with the old PS/MI model in previous versions of Pythia are used.

Depending on the analysis, uncertainties due to colour reconnection of up to 1.2 GeV [23] on the top quark mass are calculated.

8. Conclusions

Signal modelling is still one of the major systematic uncertainties for many analyses related to top quarks. The main contributions are the Monte Carlo generator and parton shower model uncertainties. By using the large amount of data that the LHC has delivered and the ATLAS experiment has recorded these uncertainties will be reduced in the near future.

In the case of initial and final state radiation the measurements have already constrained the uncertainties stronger than the uncertainties assumed so far. This will lead to strongly reduced uncertainties for future analyses.

References

[1] ATLAS Collaboration, JINST 3 S08003, 2008.
[2] S. Frixione and B.R. Webber, JHEP 0206:029, 2003.
[3] G. Corcella et al., JHEP 0101:010, 2001.
[4] J. M. Butterworth, J. R. Forshaw, and M. H. Seymour, Z. Phys. C72 637646, 1996.
[5] S. Frixione, P. Nason and C. Oleari, JHEP 0711:070, 2007.
[6] M.L. Mangano et al., JHEP 0307:001, 2003.
[7] ATLAS Collaboration, ATL-CONF-2012-024, 2012, http://cds.cern.ch/record/1400677.
[8] ATLAS Collaboration, Eur.Phys.J. C72 2043, 2012.
[9] T. Sjöstrand, S. Mrenna, P. Skands, JHEP 05:026, 2006.
[10] ATLAS Collaboration, ATL-PHYS-PUB-2011-014, 2011, http://cds.cern.ch/record/1430733.
[11] A. Buckley et al., arXiv:1003.0694 [hep-ph], 2010.
[12] DELPHI Collaboration, Z. Phys. C 73 11-60, 1996.
[13] ATLAS Collaboration, Eur.Phys.J. C72 2046, 2012.
[14] M. Botje et al., arXiv:1101.0538v1 [hep-ph], 2011.
[15] A.D. Martin, W.J. Stirling, R.S. Thorne and W. Watt, Eur.Phys.J. C63:189-285, 2009.
[16] Hung-Liang Lai et al., Phys. Rev. D82:074024, 2010.
[17] R.D. Ball et al., Nucl.Phys. B838:136-206, 2010.
[18] The CDF and D0 Collaborations, arXiv:1207.1069 [hep-ex], 2012.
[19] ATLAS Collaboration, ATL-CONF-2011-121, 2011, http://cds.cern.ch/record/1376413.
[20] P.B. Kersevan, E. Richter-Was, arXiv:0405.247 [hep-ph], 2004.
[21] P.Z. Skands, arXiv:1005.3457 [hep-ph], 2010.
[22] R. Field and R. C. Group, arXiv:0510.198 [hep-ph], 2005.
[23] ATLAS Collaboration, ATL-CONF-2012-082, 2012, http://cds.cern.ch/record/1460394.

Journal of Physics: Conference Series **452** (2013) 012040 doi:10.1088/1742-6596/452/1/012040

Signal modeling systematic uncertainties at Tevatron

Viatcheslav Shary
for the CDF and D0 collaborations
CEA, IRFU, SPP
Centre de Saclay,
91191 Gif-sur-Yvette, France

Abstract. An overview of the signal modeling systematic uncertainties in the top quark measurements at the Tevatron is given. We discuss approaches developed by the D0 and CDF collaborations for the estimation of these systematic uncertainties. Uncertainties related to the initial and final state radiation, higher order corrections, choice of the hadronization model, color reconnection and choice of the parton distribution functions set are considered in this contribution.

1. Introduction

Since the top quark discovery at the Tevatron in 1995, the top quark measurements reach an extremely high precision. The most important example is a top quark mass measurement with a precision of 0.56 % [1]. The improvement in the uncertainty is explained by the large statistics accumulated both by D0 [2] and CDF [3] experiments and by the advance in the analysis techniques. Limitation in the precision comes mainly from the systematic uncertainties and the largest component is often the uncertainty of the top quark production simulation or simply *signal modeling uncertainty*. The following effects are considered when this uncertainty is estimated at the Tevatron.

- Uncertainty due to the initial and final-state radiation simulation.
- Uncertainty related to the higher-order QCD corrections.
- Uncertainty due to the color reconnection effects.
- Uncertainty due to the simulation of jet hadronization and underlying events.
- Uncertainty due to the choice of the parton distribution functions.
- Uncertainty due to the multiple interactions model.
- Uncertainty due to the b-quark jet hadronization simulation.

In the following sections we discuss in details these sources of uncertainties as well as approaches developed by D0 and CDF experiments to evaluate the size of uncertainties.

2. Simulation of the $t\bar{t}$ Production and Uncertainties Estimation

The $t\bar{t}$ production is simulated differently at D0 and CDF experiments. For many analyses CDF use PYTHIA [4] generator both for the matrix element simulation and simulation of the showering and hadronization effects together with the CTEQ5L parton density function (PDF) set [5]. D0 in most of the cases use ALPGEN [6] for the ME simulation and PYTHIA for the showering and hadronization. CTEQ6L1 PDF set [7] is used by D0. CDF set the factorization and renormalization scale for events generator to $Q^2 = m_t^2 + p_\perp^2 + (p_1^2 + p_2^2)$, where $p_\perp$ is the transverse momentum characterizing the scattering process, p_1^2, p_2^2 are the virtualities of the incoming partons. D0 set both scales to $Q^2 = m_t^2 + \langle p_\perp^2 \rangle$, where $\langle p_\perp^2 \rangle$ is the average of the square of transverse momentum of all light partons produced in association with the $t\bar{t}$ pair. Some analysis use different generators. For example, for the recent single top analysis CDF uses the next-to-leading order (NLO) POWHEG [8] ME generator and PYTHIA for showering. The single top analysis at D0 uses the COMPHEP-based effective NLO generator SINGLETOP [9]. Simulated "underlying events" are added to each hard collision. For underlying event generation CDF uses the PYTHIA tune A [10], while D0 uses a modified version of the same tune.

2.1. Uncertainties due to Initial and Final State Radiation

Additional radiation in the initial state (ISR) or final state (FSR) affects the simulation of the number of jets as well as a the $t\bar{t}$ pair transverse momentum (p_T) simulation. Estimation of this uncertainty is based on the study made by CDF using Drell-Yan dilepton events [11]. These events produced by the quark-antiquark annihilation, as the most of $t\bar{t}$ events, but have no FSR radiation. The distribution of the mean p_T of the dilepton pairs in data shows the logarithmic rise with the squared invariant mass and compatible with the variation of Λ_{QCD} between 73 and 292 MeV and variation of the Q^2 scale between 0.5 and 2, after extrapolation to the $t\bar{t}$ scale. In PYTHIA the ISR and FSR radiation is controlled by the same equation, so variation of Λ_{QCD} and Q^2 parameters modifies both ISR and FSR radiation in $t\bar{t}$ events simulation. In top quark mass measurement this uncertainty is found to be small 0.09% for D0 and 0.15% for CDF and overlap with other uncertainties (out-of-cone jet energy correction or higher order QCD correction) is not excluded.

Both experiments compare the reconstructed p_T distribution of the $t\bar{t}$ pairs (extremely sensitive to ISR radiation) between data and MC in the context of the $t\bar{t}$ forward-backward asymmetry measurement, where it has a large effect on the measured asymmetry values. For example, D0 founds that this distribution is different between default MC@NLO simulation and data [12], while CDF comparison between POWHEG simulation and data shows that they are in a good agreement [13]. Unfortunately in both cases the reconstructed $t\bar{t}$ p_T distributions are not sensitive to the parton level p_T, because reconstructed effect are too large and distribution is "smeared" and couldn't be use to constrain the ISR uncertainty.

2.2. Higher Order Uncertainties

For most of the CDF and D0 analyses use tree level leading order MC generators, so it is necessary to estimate uncertainties related to the higher order QCD corrections. CDF estimates this uncertainties by reweighting the gluon fusion fraction in PYTHIA from the value 5% default for the $t\bar{t}$ production to the NLO estimation of the gluon fraction 20% [14]. For the top quark mass measurement this uncertainties found to be 0.02%. D0 uses different approach. It compares the ALPGEN+HERWIG [15] simulation of $t\bar{t}$ pairs with the simulation using MC@NLO+HERWIG. In top quark mass measurement this estimation leads to the uncertainty of 0.14% for D0. CDF also measure the difference in the top quark mass using the MC@NLO+HERWIGsimulation, but concludes that this uncertainties overlap with the uncertainties in ISR and FSR.

2.3. Color Reconnection Uncertainties

"Color reconnection" refers to the process of interaction between final state partons of *ttbar* hard scattering event and color remnants of proton and antiproton. This effect has poor constrains from the experimental data, that is why both collaborations use two different versions of PYTHIA generator to estimate this effect. First is the version with angular (Q^2) ordering for jet showers with the A-pro underlying-event model (modified version of the tune A). Second is the ACR-pro version. This version is identical to A-pro, but it includes the color reconnection effects in the model [16]. For the top quark mass measurement the estimated uncertainty is 0.32% for CDF and 0.16% for D0.

2.4. Hadronization and Underlying Events Uncertainties

In order to estimate hadronization and underlying events effect D0 and CDF compare two generators with different hadronization models, namely PYTHIA which uses the string "Lund model" [4] and HERWIG with hadronization based on the "cluster model" [15]. CDF compares PYTHIA with tune A for the underlying event model with HERWIG with tuned implementation of the underlying event generator JIMMY. D0 compare the set of ALPGEN events hadronized with PYTHIA to the same set of ALPGEN events, but hadronized with HERWIG. Hadronization and underlying event model uncertainty is one of the main source of the systematic uncertainty for the top quark mass measurement. CDF estimation gives 0.40% while D0 measure 0.33%. An attempt has been made to choose "the best" generator or tune using analysis of the jet shapes in QCD events [17], but it was not possible to have and additional constrains from this study to the hadronization uncertainty. In the same time this approach is revealed to be extremely useful to detect the "pathological" behaviour of the new tunes as it happened for the difference in the reconstructed top quark mass observed between Q^2-ordered and p_T-ordered models of parton showers in [16]. It has been shown in [18] that these new p_T-ordered models don't reproduce shape of the light-quarks and gluon jets. So additional efforts were made in tunning p_T-ordered models. Further studies of jet shape using the full Tevatron statistics may be useful to constrain better hadronization systematic uncertainty directly from data. The new p_T-ordered tunes (labeled Perugia-2011) [19] need to be compare with "default" tunes used by D0 and CDF.

2.5. Uncertainties Related to Parton Distribution Functions

CDF estimates an uncertainty due to the choice of the parton distribution functions by variation of 20 eigenvectors in CTEQ6M set. Differences due to these variations are summed up in quadrature. Additionally a difference between two different set of PDFs, MRST98L [20] and the default one CTEQ5L is estimated. The largest uncertainty between these two estimations is added in quadrature to the variation due to the different Λ_{QCD} value in MRST PDF set. D0 estimates this uncertainty by reweighting PYTHIA simulation using the quadratic sum of 20 eigenvectors variations in CTEQ6M. In the top quark mass measurement this uncertainty is estimated to be 0.08% for CDF and 0.14% for D0.

2.6. Multiple Interaction Model Uncertainties

At the Tevatron D0 and CDF experiments have a significant probability that additional $p\bar{p}$ interactions happen simultaneously with the hard scattered one and a typical average value of number of primary vertices is around 2. To simulate this effect CDF overlays the Poisson-distributed events simulated with PYTHIA. D0 overlays the real data events registered with "zero-bias" trigger. The mean value of number of multiple interactions increases with the instantaneous luminosity increase. This mean value chosen at the moment of MC events generation may be different (usually smaller) than the mean value of the analysis data set. Reweighting of the simulation distribution of number of primary vertices to the distribution observed in data, allows to estimate the uncertainty due to the imperfect simulation of this effect. Such approach is adopted by the CDF. D0 uses the similar approach, but instead of number of primary vertices distribution, D0 reweights the the instantaneous luminosity spectrum. Usually, uncertainty due to the multiple interaction model is small. For example, in the top quark mass measurement this uncertainty estimated to be 0.06% for CDF and 0.03% for D0.

2.7. b-quark Jet Hadronization Uncertainty
CDF estimates the uncertainty due to the b-quark jets simulation by varying the parameters of the default PYTHIA b-quark jets fragmentation based on the Bowler model [21], where the Bowler fragmentation-function parameter $r_q = 1.0$ and Lund fragmentation function parameters $a = 0.3$, $b = 0.58$. The SLD parametrization ($r_q = 0.980 \pm 0.010$, $a = 1.30 \pm 0.09$, $b = 1.58 \pm 0.09$) and LEP parametrization ($r_q = 0.897 \pm 0.013$, $a = 1.03 \pm 0.08$, $b = 1.31 \pm 0.08$) [22] is used to estimated the uncertainty. D0 uses the default parametrization from LEP and compares to the simulation with the SLD parametrization. The resulting uncertainties are usually small. For example, for the top quark mass measurements they found to be 0.09% for CDF and 0.03% for D0.

3. Conclusion
In conclusion, we want to remark that signal modeling uncertainties play an important role in the precise top quark measurement. The estimation of these uncertainties often based on the comparison between different generators or generator tunes. The validity of such comparison need to be verified with data when possible. This is especially valid for the case of the new Perugia 2011 p_T-ordered PYTHIA tunes, where the bias in the reconstructed top quark mass need to be compared with the bias observed in tunes traditionally used by the D0 and CDF. Further improvement of the signal modeling uncertainties estimation could be achieved by using the new tunes for the better estimation of the modeling effects (for example, new PYTHIA tunes for the color reconnection effects). Systematic uncertainties should be constrained with Tevatron data, for example, ISR/FSR estimation could be improved by using the full statistics of Drel-Yan events, the jet hadronization uncertainties could be constrained by the jet shape studies. Additionally, the overlap between different systematic uncertainties need to be studied and reduced when possible. As an example, the higher order QCD uncertainties estimation may overlap with ISR/FSR, ISR/FSR uncertainty estimation may overlap with out-of-cone jet energy scale systematic uncertainty.

Those improvements are still possible and need to be implemented in the final set of the Tevatron measurements, in particular in the top quark mass and cross section measurements, in studies of the top quark properties: forward-backward charge asymmetry, spin correlation, etc.

References
[1] Aaltonen T *et al.* (CDF and D0 Collaborations) 2012 *Phys. Rev. D* **86** 092003
[2] Abazov V *et al.* (D0 Collaboration) 2006 *Nucl.Instrum.Meth.* **A565** 463–537 (*Preprint* `physics/0507191`)
[3] Acosta D *et al.* (CDF Collaboration) 2005 *Phys. Rev. D* **71** 052003 (*Preprint* `hep-ex/0410041`)
[4] Sjostrand T, Mrenna S and Skands P Z 2006 *JHEP* **0605** 026 (*Preprint* `hep-ph/0603175`)
[5] Lai H *et al.* (CTEQ Collaboration) 2000 *Eur.Phys.J.* **C12** 375–392 (*Preprint* `hep-ph/9903282`)
[6] Mangano M L *et al.* 2003 *JHEP* **0307** 001 (*Preprint* `hep-ph/0206293`)
[7] Pumplin J, Stump D, Huston J, Lai H, Nadolsky P M *et al.* 2002 *JHEP* **0207** 012 (*Preprint* `hep-ph/0201195`)
[8] Alioli S, Nason P, Oleari C and Re E 2009 *JHEP* **0909** 111 (*Preprint* `0907.4076`)
[9] Boos E, Bunichev V, Dudko L, Savrin V and Sherstnev A 2006 *Phys.Atom.Nucl.* **69** 1317–1329
[10] Affolder T *et al.* (CDF Collaboration) 2002 *Phys.Rev.* **D65** 092002
[11] Abulencia A *et al.* (CDF Collaboration) 2006 *Phys.Rev.* **D73** 032003 (*Preprint* `hep-ex/0510048`)
[12] Abazov V M *et al.* (D0 Collaboration) 2011 *Phys.Rev.* **D84** 112005 (*Preprint* `1107.4995`)
[13] The CDF Collaboration 2012 (*Preprint* `CDF note 10807`)
[14] Aaltonen T *et al.* (CDF Collaboration) 2009 *Phys.Rev.* **D79** 031101 (*Preprint* `0807.4262`)
[15] Corcella G *et al.* 2001 *JHEP* **01** 010 (*Preprint* `hep-ph/0011363`)
[16] Skands P Z and Wicke D 2007 *Eur.Phys.J.* **C52** 133–140 (*Preprint* `hep-ph/0703081`)
[17] Acosta D *et al.* (CDF Collaboration) 2005 *Phys.Rev.* **D71** 112002 (*Preprint* `hep-ex/0505013`)
[18] Galtier A B March 2010 *MC4LHC workshop*, URL `http://indico.cern.ch/conferenceDisplay.py?confId=74601`
[19] Skands P Z 2010 *Phys.Rev.* **D82** 074018 for the description of Perugia 2011 tunes, see arxive version. (*Preprint* `1005.3457`)

[20] Martin A D, Roberts R, Stirling W J and Thorne R 1998 *Eur.Phys.J.* **C4** 463–496 mRST98 (*Preprint* hep-ph/9803445)
[21] Bowler M 1981 *Z.Phys.* **C11** 169
[22] Peters Y, Hamacher K and Wicke D (D0 Collaboration) 2006

Journal of Physics: Conference Series **452** (2013) 012041 doi:10.1088/1742-6596/452/1/012041

Status of the ATLAS Experiment

Martin Aleksa on behalf of the ATLAS Collaboration
CERN, Geneva 23, CH-1211 Geneva, Switzerland
E-mail: martin.aleksa@cern.ch

Abstract.
This document presents an overview of the status of the ATLAS experiment at the CERN Large Hadron Collider at the time of the 5th International Workshop on Top Quark Physics in Winchester (September 2012). The status of the detector and data taking are presented together with some selected physics results and plans for the first long shutdown.

1. The ATLAS Detector

The ATLAS detector [1] is a multipurpose particle physics apparatus with forward-backward symmetric cylindrical geometry at the CERN Large Hadron Collider (LHC). The entire detector (Fig. 1) weighs 7000 tons. It is 44 m long and 25 m in diameter. It is located in an underground cavern at a depth of 100 m, where it surrounds one of the collision points around the 27-km-long LHC ring. The first proton-proton (*pp*) collisions at the LHC were in 2009, and since then the collider has operated at several different center-of-mass energies.

ATLAS is composed of several distinct subdetectors in order to identify and measure the energy and momentum of a variety of particles and so reconstruct the dynamics of the collision. The inner tracking detector (ID) consists of a silicon pixel detector, a silicon microstrip detector, and a straw-tube transition radiation tracker. The ID is surrounded by a thin superconducting solenoid which provides a 2 T magnetic field, and by high granularity liquid-argon (LAr) sampling electromagnetic calorimetry. The electromagnetic calorimeter is divided into a central barrel (pseudorapidity[1] $|\eta| < 1.475$) and end-cap regions on either end of the detector ($1.375 < |\eta| < 2.5$ for the outer wheel and $2.5 < |\eta| < 3.2$ for the inner wheel). In the region matched to the ID ($|\eta| < 2.5$), it is radially segmented into three layers. The first layer has a fine segmentation in η to facilitate e/γ separation from π^0 and to improve the resolution of the shower position and direction measurements. In the region $|\eta| < 1.8$, the electromagnetic calorimeter is preceded by a presampler detector to correct for upstream energy losses. An iron-scintillator/tile calorimeter gives hadronic coverage in the central rapidity range ($|\eta| < 1.7$), while a LAr hadronic end-cap calorimeter provides coverage over $1.5 < |\eta| < 3.2$. The forward regions ($3.2 < |\eta| < 4.9$) are instrumented with LAr calorimeters for both electromagnetic and hadronic measurements. The muon spectrometer surrounds the calorimeters and consists of three large air-core superconducting magnets providing a toroidal field, each with eight coils, a system of precision tracking chambers, and fast detectors for triggering. The combination of all these systems provides charged particle measurements together with efficient and precise lepton and photon measurements in the pseudorapidity range $|\eta| < 2.5$. Jets and missing transverse

[1] The pseudorapidity η is defined in terms of the polar angle θ measured from the beam line as $\eta = -\ln\tan(\theta/2)$.

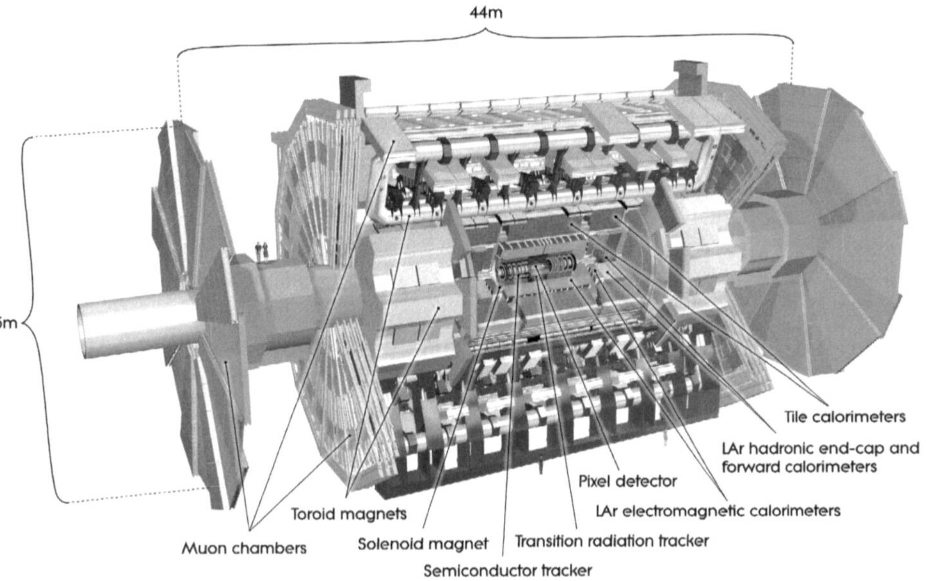

Figure 1. Cutaway drawing of the ATLAS detector showing its main components.

energy $E_{\mathrm{T}}^{\mathrm{miss}}$ are reconstructed using energy deposits over the full coverage of the calorimeters, $|\eta| < 4.9$.

2. Data taking and trigger

Physics data-taking with pp collisions at $\sqrt{s} = 7\,\mathrm{TeV}$ started in March 2011 and ended in October 2011. Between April and December 2012 pp collisions at $\sqrt{s} = 8\,\mathrm{TeV}$ have been recorded. During these periods, LHC delivered to ATLAS an integrated luminosity of $5.6\,\mathrm{fb}^{-1}$ at $\sqrt{s} = 7\,\mathrm{TeV}$ and $23.3\,\mathrm{fb}^{-1}$ at $\sqrt{s} = 8\,\mathrm{TeV}$. At the time of the conference in September 2012, only a fraction of the 2012 data-set ($14.6\,\mathrm{fb}^{-1}$) was delivered. The exact amount of data used is indicated for each analysis presented here. During both years the data-taking efficiency, signifying the ratio of the recorded and the delivered luminosity, was very high above 93.5 %. After all data quality cuts approximately 90 % of the delivered LHC luminosity was used for physics analysis. A very high fraction of $> 96\,\%$ of all ATLAS read-out channels was operational during both data-taking periods.

The LHC delivered physics fills with peak instantaneous luminosity of up to $7.7 \times 10^{33}\,\mathrm{cm}^{-2}\mathrm{s}^{-1}$, corresponding to an average of 37 superimposed minimum bias pp collisions per bunch crossing at the start of fills. The ATLAS trigger and offline reconstruction algorithms were extensively optimised during the 2011-12 winter shutdown to make them robust against such high levels of pile-up, beyond the design specifications of the detector. Special care was taken to develop robust algorithms showing flat performance versus pile-up and which minimize CPU usage. The trigger menu was optimised to give a 65 kHz level-1 trigger rate at the start of fill, and a fill-averaged event filter output rate of 400 Hz. Typical trigger thresholds at the event filter are 24-25 GeV transverse momentum for single isolated muons and electrons, 8-18 GeV for di-leptons, 20-35 GeV for diphotons, 20-29 GeV for di-taus and 80 GeV for missing transverse momentum $E_{\mathrm{T}}^{\mathrm{miss}}$.

For example, the main triggers for top physics in the di-lepton and single-lepton (plus jets) channels were single-lepton triggers with slightly increased thresholds and isolation requirements

Journal of Physics: Conference Series **452** (2013) 012041 doi:10.1088/1742-6596/452/1/012041

in 2012 compared to 2011. For the all hadronic channel, 5 jet triggers and 1 b-jet + 3 jet triggers were used.

The offline event reconstruction at Tier-0 takes from 15-40 s/event, and was run on up to 7500 CPU cores in parallel to keep up with the data-taking rate. The Worldwide LHC Computing Grid supporting ATLAS continues to be vital, with 150000 simultaneous Monte Carlo simulation and user analysis jobs being run routinely.

3. Some selected physics results

3.1. Standard model cross-sections

A summary of electroweak physics in ATLAS is shown in Fig. 2 where the measured cross-sections of the main electroweak processes are presented together with theoretical predictions [2]. The W and Z vector-boson inclusive cross-sections were measured with $35\,\text{pb}^{-1}$ of integrated luminosity from the 2010 dataset. All other measurements were performed using the 2011 or the 2012 datasets. The top quark pair production cross-section is based on a statistical combination of measurements in the single-lepton, di-lepton and all-hadronic channels. The dark error bar represents the statistical uncertainly. The red error bar represents the full uncertainty, including systematics and luminosity uncertainties. All theoretical expectations were calculated at NLO or higher. The remarkable success of NNLO pQCD calculations together with our present knowledge of the proton parton structure is demonstrated.

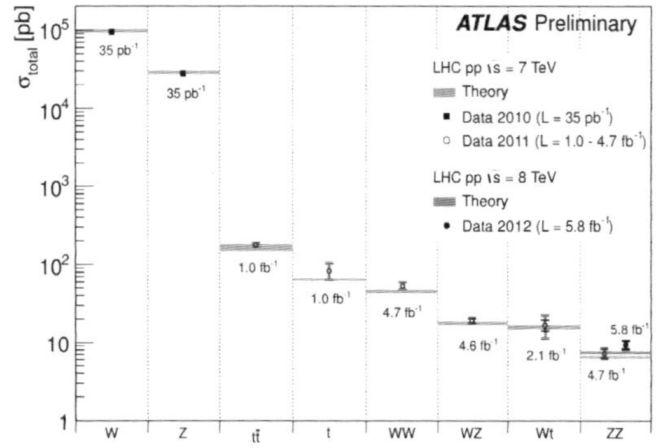

Figure 2. Summary of several standard model total production cross-section measurements compared to the corresponding theoretical expectations [2].

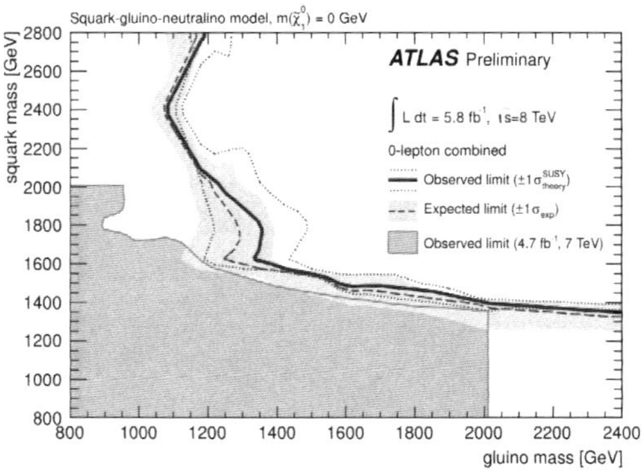

Figure 3. Exclusion limits for the squark and gluino mass in a simplified MSSM scenario with only strong production of gluinos and first- and second-generation squarks, with direct decays to jets and neutralinos [3].

3.2. SUSY searches with 2012 data

Weak scale supersymmetry (SUSY) is an extension of the standard model (SM) that provides a solution to the hierarchy problem by introducing SUSY partners of all SM particles. In the framework of a generic R-parity conserving minimal supersymmetric extension of the SM, SUSY particles are produced in pairs, and the lightest supersymmetric particle is stable and can be a dark matter candidate.

The 2012 dataset has been used to further extend the region of supersymmetric parameter space excluded by previous measurements with the ATLAS detector. First results using the 2012 dataset were presented at the conference. In Fig. 3 a search for squarks and gluinos in

Higgs boson decay	Subsequent decay	Sub-channels	m_H range [GeV]	$\int Ldt$ [fb^{-1}]
2011 $\sqrt{s}=7$ TeV				
$H \to ZZ^{(*)}$	4ℓ	$\{4e, 2e2\mu, 2\mu 2e, 4\mu\}$	110–600	4.8
	$\ell\ell\nu\bar{\nu}$	$\{ee, \mu\mu\} \otimes$ {low, high pile-up}	200–280–600	4.7
	$\ell\ell q\bar{q}$	{b-tagged, untagged}	200–300–600	4.7
$H \to \gamma\gamma$	–	10 categories $\{p_{\mathrm{Tt}} \otimes \eta_\gamma \otimes$ conversion$\} \oplus$ {2-jet}	110–150	4.8
$H \to WW^{(*)}$	$\ell\nu\ell\nu$	$\{ee, e\mu/\mu e, \mu\mu\} \otimes$ {0-jet, 1-jet, 2-jet} $\otimes$ {low, high pile-up}	110–200–300–600	4.7
	$\ell\nu qq'$	$\{e, \mu\} \otimes$ {0-jet, 1-jet, 2-jet}	300–600	4.7
$H \to \tau\tau$	$\tau_{\mathrm{lep}}\tau_{\mathrm{lep}}$	$\{e\mu\} \otimes$ {0-jet} $\oplus \{\ell\ell\} \otimes$ {1-jet, 2-jet, VH}	110–150	4.7
	$\tau_{\mathrm{lep}}\tau_{\mathrm{had}}$	$\{e, \mu\} \otimes$ {0-jet} $\otimes \{E_{\mathrm{T}}^{\mathrm{miss}} < 20$ GeV, $E_{\mathrm{T}}^{\mathrm{miss}} \geqslant 20$ GeV$\}$ $\oplus\{e, \mu\} \otimes$ {1-jet} $\oplus \{\ell\} \otimes$ {2-jet}	110–150	4.7
	$\tau_{\mathrm{had}}\tau_{\mathrm{had}}$	{1-jet}	110–150	4.7
$VH \to Vbb$	$Z \to \nu\nu$	$E_{\mathrm{T}}^{\mathrm{miss}} \in \{120\text{–}160, 160\text{–}200, \geqslant 200$ GeV$\}$	110–130	4.6
	$W \to \ell\nu$	$p_{\mathrm{T}}^{W} \in \{< 50, 50\text{–}100, 100\text{–}200, \geqslant 200$ GeV$\}$	110–130	4.7
	$Z \to \ell\ell$	$p_{\mathrm{T}}^{Z} \in \{< 50, 50\text{–}100, 100\text{–}200, \geqslant 200$ GeV$\}$	110–130	4.7
2012 $\sqrt{s}=8$ TeV				
$H \to ZZ^{(*)}$	4ℓ	$\{4e, 2e2\mu, 2\mu 2e, 4\mu\}$	110–600	5.8
$H \to \gamma\gamma$	–	10 categories $\{p_{\mathrm{Tt}} \otimes \eta_\gamma \otimes$ conversion$\} \oplus$ {2-jet}	110–150	5.9
$H \to WW^{(*)}$	$e\nu\mu\nu$	$\{e\mu, \mu e\} \otimes$ {0-jet, 1-jet, 2-jet}	110–200	5.8

Table 1. Summary of the individual Higgs decay channels entering the combination [5].

final states containing jets, missing transverse momentum and no high-p_{T} electrons or muons is presented [3]. No excess above the SM expectation is observed. When the neutralino is massless, gluino masses below 1100 GeV are excluded at the 95 % confidence level in a simplified model with only gluinos and the lightest neutralino.

3.3. Standard model Higgs searches

The standard model of particle physics has been tested by many experiments over the last four decades and has been shown to successfully describe high energy particle interactions. However, the mechanism that breaks electroweak symmetry in the SM has not been verified experimentally. This mechanism [4], which gives mass to massive elementary particles, implies the existence of a scalar particle, the SM Higgs boson. The search for the Higgs boson is one of the highlights of the LHC physics programme. The different production and decay modes give rise to a rich experimental phenomenology, with many different analysis channels being used [5] to cover the full mass range of interest, as shown in Tab. 1.

An excess of events is observed near a Higgs mass $m_H = 126$ GeV in the $H \to ZZ^{(*)} \to 4\ell$ and $H \to \gamma\gamma$ channels, both of which provide fully reconstructed candidates with high resolution in invariant mass. These excesses are confirmed by the highly sensitive but low-resolution $H \to WW^{(*)} \to \ell\nu\ell\nu$ channel. The observed local p_0 values from the combination of channels, using the asymptotic approximation, are shown as a function of m_H in Fig. 4 for the low mass range. The significance of the observed excess of events at $m_H = 126$ GeV corresponds to $5.9\,\sigma$. The observed 95 % confidence level exclusion regions are 111 – 122 GeV and 131 – 559 GeV.

The observed excess corresponds to a best-fit signal strength $\mu = \sigma/\sigma_{\mathrm{SM}} = 1.4 \pm 0.3$ for $m_H = 126$ GeV, which is consistent with the SM Higgs boson hypothesis (σ is the observed cross-section and σ_{SM} signifies the expected SM Higgs boson cross-section) . A summary of the individual and combined best-fit values of the strength parameter for a SM Higgs boson mass hypothesis of 126 GeV is shown in Fig. 5. The mass of the observed new particle is estimated using a profile likelihood ratio for the two channels with highest mass resolution. The resulting mass estimate of the observed particle is $126.0 \pm 0.4(\mathrm{stat}) \pm 0.4(\mathrm{sys})$ GeV.

4. Plans for the long shutdown 2013 and 2014

The main goal of the long shutdown starting mid February 2013 is to consolidate the LHC magnet inter-connects to be able to operate the LHC at a center-of-mass energy of $\sqrt{s} = 13$ TeV

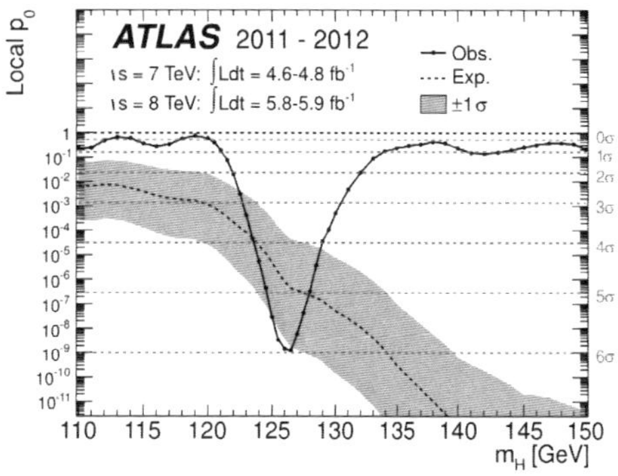

Figure 4. The observed (solid) local p_0 as a function of m_H in the low mass range. The dashed curve shows the expected local p_0 under the hypothesis of a SM Higgs boson signal at that mass with its $\pm 1\sigma$ band [5].

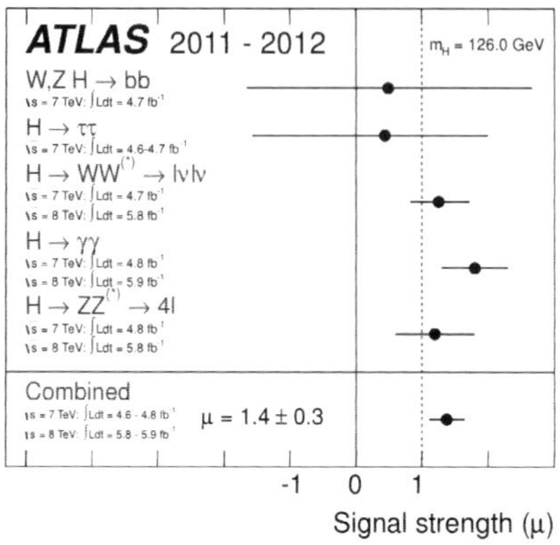

Figure 5. Measurements of the signal strength parameter μ for $m_H = 126\,\text{GeV}$ for the individual channels and their combination [5].

in the years 2015 – 2017. The ATLAS plans during this first long shutdown are the following:

- Adding a new insertable pixel b-layer (IBL) to preserve current physics performance at very high pile-up. Together with the installation of the IBL, a new beam pipe and new pixel services will be installed.
- New evaporative cooling plant for the present inner detector silicon.
- Consolidation of detector elements (e.g. calorimeter power supplies).
- Addition of specific neutron shielding.
- Completion of the installation of endcap extra muon chambers staged in 2003.
- Upgrade of magnet cryogenics.

5. Summary and conclusion

The ATLAS experiment has been performing very well during the LHC pp runs in 2011 and 2012. The standard model Higgs boson is excluded at 95 % Confidence Level in the mass range 111 – 559 GeV, except for the narrow region 122 – 131 GeV. In this region, an excess of events with a significance of $5.9\,\sigma$ is observed. Many interesting physics analyses have been reported and are presented at the ATLAS Collaboration public page [2].

6. Acknowledgements

We thank CERN for the very successful operation of the LHC, as well as the support staff from our institutions without whom ATLAS could not be operated efficiently.

References

[1] ATLAS Collaboration, J. Instrum., 3 S08003 (2008)
[2] ATLAS Collaboration, https://twiki.cern.ch/twiki/pub/AtlasPublic/
[3] ATLAS Collaboration, ATLAS-CONF-2012-109 (2012)
[4] P.W. Higgs, Phys. Lett. 12 (1964) 132; F. Englert, R. Brout, Phys. Rev. Lett. 13 (1964) 321; G.S. Guralnik, C.R. Hagen, T.W.B. Kibble, Phys. Rev. Lett. 13 (1964) 585.
[5] ATLAS Collaboration, Phys. Lett. B716 (2012)

Journal of Physics: Conference Series **452** (2013) 012042 doi:10.1088/1742-6596/452/1/012042

Vector boson plus multijet production

IPPP/13/07, DCPT/13/14, LPN13-019

Marek Schönherr

Institute for Particle Physics Phenomenology, Durham University, Durham DH1 3LE, UK

E-mail: marek.schoenherr@durham.ac.uk

Abstract. In this contribution the developments in the description of vector boson plus jets signatures at hadron colliders in recent years are summarised. Particular focus is put on its relevance as background to top physics.

1. Introduction

Within the first years of running of the LHC at centre-of-mass energies of 7 and 8 TeV top physics has been at the core of its physics programme. It serves both as a signal to be measured as precisely as possible and as a background to new physics, Higgs and Standard Model processes with many jets.

When considering measuring top quark production the production of W and Z bosons in association with many jets present the major backgrounds. In particular, the process $pp \to W+ \geq 4$ jets constitutes an irreducible background to the semileptonic $t\bar{t}$ production channel. It therefore needs to be known with as large a precision as possible. Many advances have been made in recent years for this class of processes in their own right, reaching now a stage where they start to decrease the theoretical uncertainties for the relevant background channels below the leading order.

While higher order calculations for $V(=W, Z)$ plus multiple jets become accessible they improve the description of QCD at large scales, leading to a stabilisation of the respective cross sections at the same time. But only when they are matched to parton showers a simultaneous and observable independent, reliable description of QCD dynamics at low scales is achieved. Such a matching offers additional benefits as the connection of low scale perturbative dynamics of the parton shower to the non-perturbative dynamics of hadronisation models and their subsequent hadron decays can be used to arrive at particle level descriptions that are directly comparable to experimental data. Multijet merging techniques can then be evoked to arrive at inclusive descriptions, combining successive multiplicities of fixed-order matrix elements matched to parton showers with their respective accuracies preserved and, at the same time, resumming multiscale logarithm associated with hierarchical multijet production.

2. LO calculations and MePs merging

Multijet merging techniques at leading order accuracy are known for more than ten years [1, 2, 3, 4, 5, 6, 7, 8, 9]. Combining tree-level matrix elements of successive parton multiplicities with parton showers into an inclusive description, preserving the respective accuracies, they are by now the work horses of the LHC experiments for multijet topologies. Their ability to describe data has been tested in various analyses [10, 11, 12, 13, 14].

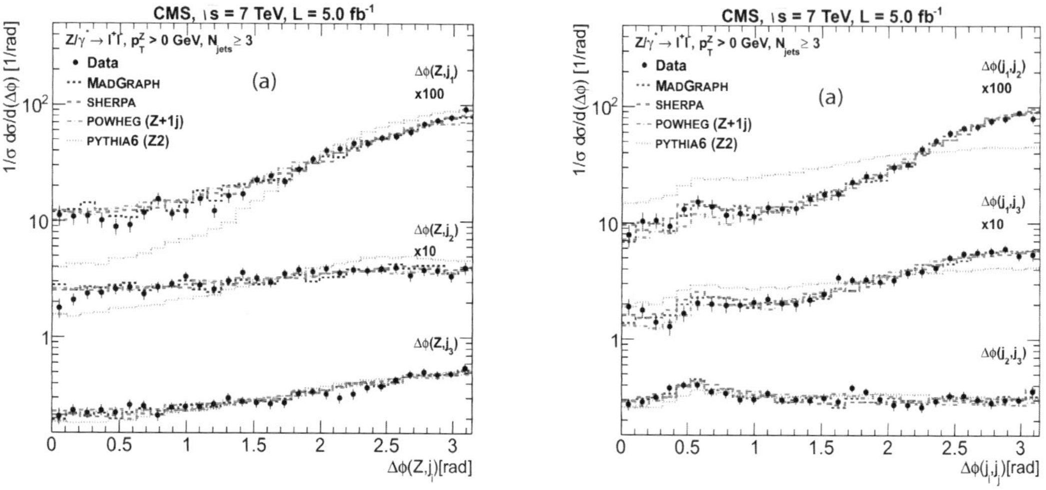

Figure 1: Azimuthal decorrellation of the Z and the 1st, 2nd and 3rd jet **(left)** and of the 1st and 2nd, 1st and 3rd, and 2nd and 3rd jet in $pp \to Z+ \geq 3$ jets at the LHC as measured by CMS. Figures are taken from [14].

Fig. 1 presents the most recent of these analyses wherein the ability of the MEPS methods to describe the radiation pattern of the jets in $pp \to Z+ \geq 3$ jets events is examined and good agreement is found. Nonetheless, the theoretical uncertainties of these methods are large owing to the leading order accuracy of the description of hard and/or wide angle parton emission, including quantum interference effects, only. By elevating their description to next-to-leading order the theoretical uncertainties on the respective observables can be reduced.

3. NLO calculations, NLOPS and MEPS@NLO

NLO

While next-to-leading order calculations for W and Z production with up to two jets are known for some time [15], such calculation for $pp \to W + 3$ jets [16, 17] and $pp \to W + 4$ jets [18], as are relevant as top backgrounds, have only become available recently. Similarly, $pp \to Z + 3$ jets [19] and $pp \to Z + 4$ jets [20] are available.

Fig. 2 shows a calculation of the transverse momenta of four leading jets in $pp \to W + 4$ jets production at the LHC using BLACKHAT+SHERPA [18, 21] and a clear reduction of the theoretical uncertainty can be seen. However, possibly large logarithms due to scale hierarchies are not taken into account by such a calculation and also small scale dynamics are absent.

NLOPS

To enhance next-to-leading order calculations with the resummation of large logarithms associated with the production of the softest jet it can be matched to a parton shower, either via the MC@NLO [22, 23, 24] or the POWHEG [25, 26, 27, 28] technique. Such a calculation has the added benefit that it can take advantage of the parton shower's infrared continuation with the non-perturbative dynamics of hadronisation models with subsequent hadron decays, and multiple parton interactions. Thus, such calculation can directly be compared to experimental data.

Fig. 3 shows the results of implementations of these methods by various groups, reaching up to multiplicities of $pp \to V + 3$ jets described at NLO. For all, good agreement with data is

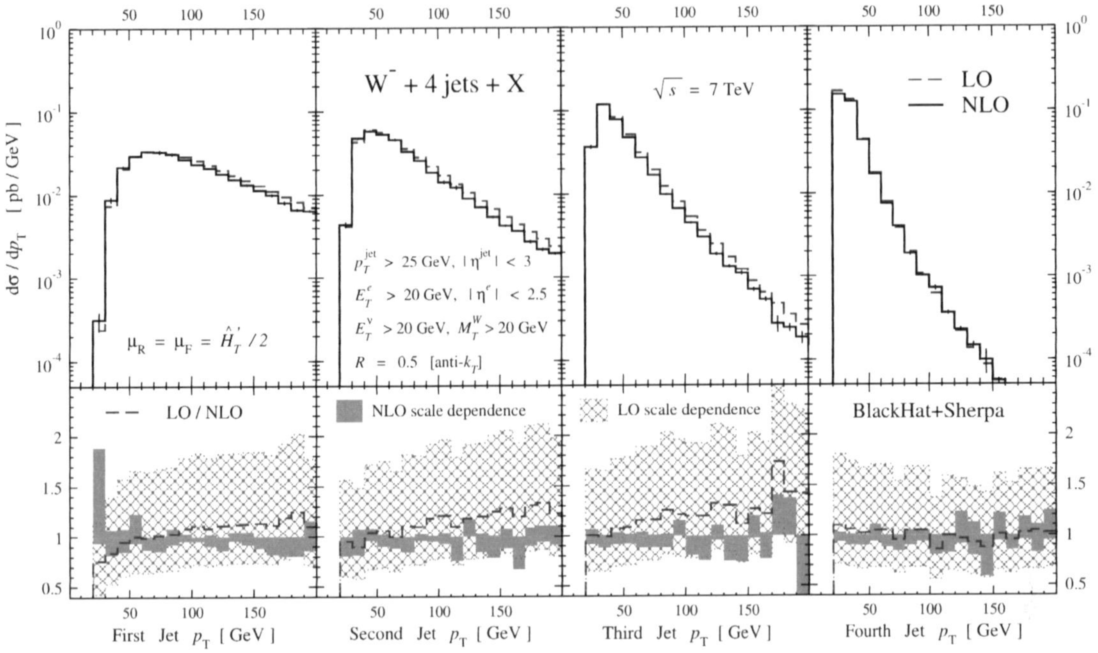

Figure 2: Transverse momenta of the four leading jets in $pp \to W^- + 4$ jets at the LHC calculated at NLO accuracy. Figures taken from [18].

found and the respective uncertainties are reduced compared to LO calculations.

MEPS@NLO

NLOPS calculations, like fixed-order NLO calculations, are lacking any resummation, of scale hierarchies of jet emissions that are described at NLO accuracy, e.g. of the emission scales of the four leading jets in $pp \to V + 4$ jets. Such a resummation of scales with respect to the inclusive sample, however, are present in MEPS methods. Therefore, as a first step, NLOPS calculations were combined with MEPS merging methods to the called MENLOPS merging method [32, 33]. Therein the most inclusive process is described by a NLOPS matched calculation while leading order matrix elements are merged on top of it.

MEPS@NLO merging [34, 35, 36] now aims at merging NLOPS matched calculations of successive jet multiplicities into such an inclusive sample. Therein, not only are the respective jet multiplicities described at next-to-leading order accuracy, but also the overall resummation of the parton shower is undisturbed. Thus, both large and small scale dynamics are accurately described.

Fig. 4 displays the inclusive n-jet cross sections and the azimuthal decorrelation of the two leading jets, probing both relative production rates and interjet dynamics. Comparing the uncertainties of the MEPS@NLO merging method (merging $pp \to W + 0, 1, 2$ jets at NLO and $pp \to W + 3, 4$ jets at LO) to those of the MENLOPS method (merging $pp \to W + 0$ jets at NLO and $pp \to W + 1, 2, 3, 4$ jets at LO) one clearly seas the added precision of including more NLO matrix elements. The predictions are compared to ATLAS data [11] and also the more accurate method is favoured. Fig. 5 presents the transverse momenta of the two leading jets in events with at least one, two or three jets. A similar reduction of the uncertainties is found.

Finally, with these methods at hand, a full assertion of all perturbative and non-perturbative uncertainties in particle level Monte-Carlo predictions following [37] can be done.

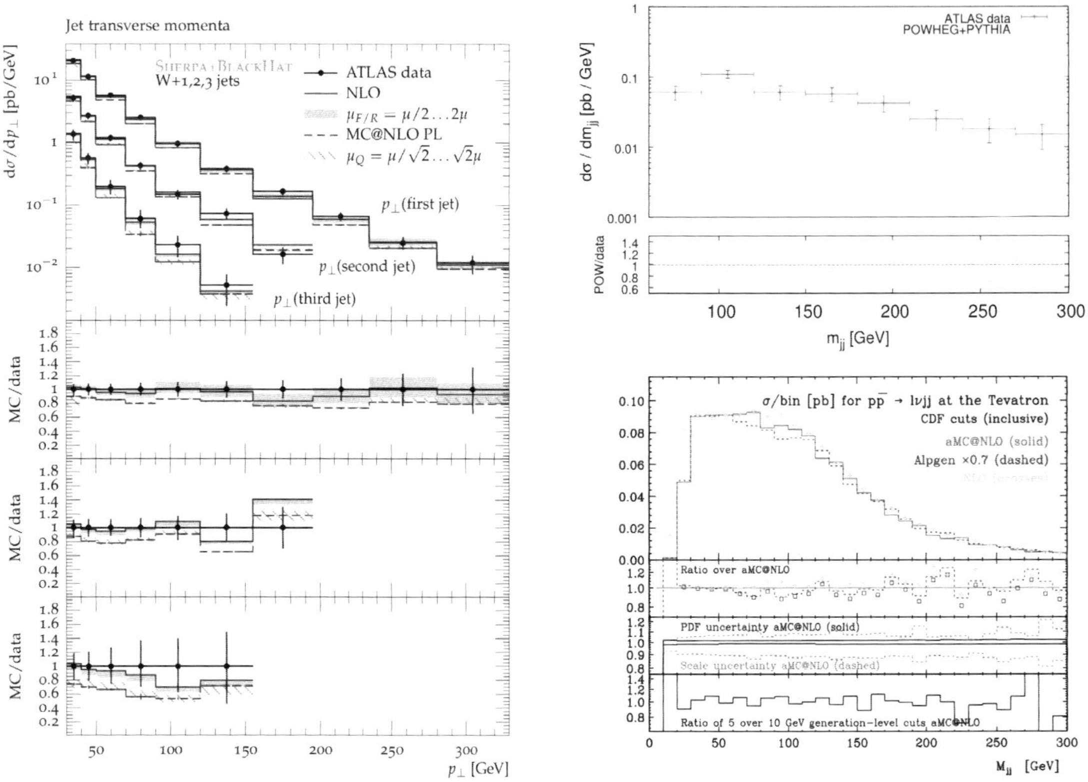

Figure 3: Left: Transverse momenta of the nth jet in MC@NLO simulations of $pp \to W + n$ jets at the LHC compared to ATLAS data [11]. **Right top:** Invariant mass of the leading jet pair in a POWHEG simulation of $pp \to Z + 2$ jets at the LHC compared to ATLAS data [12]. **Right bottom:** Invariant mass of the leading jet pair in a aMC@NLO simulation of $p\bar{p} \to W + 2$ jets at the Tevatron. Figures taken from [29, 30, 31]

References

[1] Catani S, Krauss F, Kuhn R and Webber B R 2001 *JHEP* **11** 063 (*Preprint* hep-ph/0109231) URL http://www.slac.stanford.edu/spires/find/hep/www?eprint=hep-ph/0109231

[2] Mangano M L, Moretti M and Pittau R 2002 *Nucl. Phys.* **B632** 343–362 (*Preprint* hep-ph/0108069) URL http://www.slac.stanford.edu/spires/find/hep/www?eprint=hep-ph/0108069

[3] Lönnblad L 2002 *JHEP* **05** 046 (*Preprint* hep-ph/0112284) URL http://www.slac.stanford.edu/spires/find/hep/www?eprint=hep-ph/0112284

[4] Krauss F 2002 *JHEP* **0208** 015 (*Preprint* hep-ph/0205283) URL http://www.slac.stanford.edu/spires/find/hep/www?eprint=hep-ph/0205283

[5] Mangano M L, Moretti M, Piccinini F and Treccani M 2007 *JHEP* **01** 013 (*Preprint* hep-ph/0611129) URL http://www.slac.stanford.edu/spires/find/hep/www?eprint=hep-ph/0611129

[6] Alwall J *et al.* 2008 *Eur. Phys. J.* **C53** 473–500 (*Preprint* 0706.2569) URL http://www.slac.stanford.edu/spires/find/hep/www?eprint=arXiv:0706.2569

[7] Höche S, Krauss F, Schumann S and Siegert F 2009 *JHEP* **05** 053 (*Preprint* 0903.1219) URL http://www.slac.stanford.edu/spires/find/hep/www?eprint=arXiv:0903.1219

[8] Hamilton K, Richardson P and Tully J 2009 *JHEP* **11** 038 (*Preprint* 0905.3072) URL http://www.slac.stanford.edu/spires/find/hep/www?eprint=arXiv:0905.3072

[9] Lönnblad L and Prestel S 2012 *JHEP* **03** 019 (*Preprint* 1109.4829)

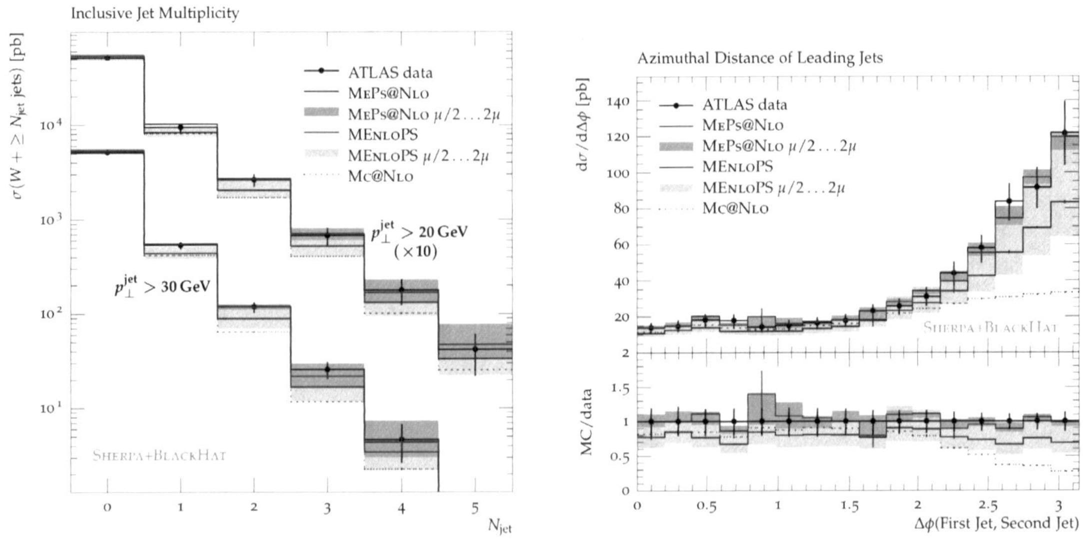

Figure 4: $W+ \geq n$-jet cross sections **(left)** and azimuthal decorrellation of the two leading jets **(right)** in $pp \to W+$ jets production at the LHC compared to ATLAS data [11]. Figures taken from [35].

[10] Aad G *et al.* (ATLAS Collaboration) 2011 *Phys.Lett.* **B698** 325–345 (*Preprint* 1012.5382)
[11] Aad G *et al.* (ATLAS Collaboration) 2012 *Phys.Rev.* **D85** 092002 (*Preprint* 1201.1276) URL http://inspirehep.net/record/1083318
[12] Aad G *et al.* (ATLAS) 2012 *Phys.Rev.* **D85** 032009 (*Preprint* 1111.2690) URL http://inspirehep.net/record/945498
[13] Chatrchyan S *et al.* (CMS) 2012 *JHEP* **1201** 010 (*Preprint* 1110.3226) URL http://inspirehep.net/record/940012
[14] Chatrchyan S *et al.* (CMS Collaboration) 2013 (*Preprint* 1301.1646)
[15] Campbell J M and Ellis R 2002 *Phys. Rev.* **D65** 113007 (*Preprint* hep-ph/0202176)
[16] Ellis R, Melnikov K and Zanderighi G 2009 *JHEP* **0904** 077 (*Preprint* 0901.4101)
[17] Berger C F, Bern Z, Dixon L J, Febres-Cordero F, Forde D, Gleisberg T, Ita H, Kosower D A and Maître D 2009 *Phys. Rev. Lett.* **102** 222001 (*Preprint* 0902.2760) URL http://www.slac.stanford.edu/spires/find/hep/www?eprint=arXiv:0902.2760
[18] Berger C F, Bern Z, Dixon L J, Febres-Cordero F, Forde D, Gleisberg T, Ita H, Kosower D A and Maître D 2011 *Phys. Rev. Lett.* **106** 092001 (*Preprint* 1009.2338) URL http://inspirebeta.net/record/867513
[19] Berger C F, Bern Z, Dixon L J, Febres-Cordero F, Forde D, Gleisberg T, Ita H, Kosower D A and Maître D 2010 *Phys. Rev.* **D82** 074002 (*Preprint* 1004.1659) URL http://www-spires.dur.ac.uk/spires/find/hep/www?eprint=arXiv:1004.1659
[20] Ita H, Bern Z, Dixon L J, Febres-Cordero F, Kosower D A and Maître D 2012 *Phys.Rev.* **D85** 031501 (*Preprint* 1108.2229) URL http://inspirehep.net/record/922868
[21] Gleisberg T, Höche S, Krauss F, Schönherr M, Schumann S, Siegert F and Winter J 2009 *JHEP* **02** 007 (*Preprint* 0811.4622) URL http://inspirebeta.net/record/803708
[22] Frixione S and Webber B R 2002 *JHEP* **06** 029 (*Preprint* hep-ph/0204244) URL http://www.slac.stanford.edu/spires/find/hep/www?eprint=hep-ph/0204244
[23] Frixione S, Nason P and Webber B R 2003 *JHEP* **08** 007 (*Preprint* hep-ph/0305252) URL http://www.slac.stanford.edu/spires/find/hep/www?eprint=hep-ph/0305252
[24] Höche S, Krauss F, Schönherr M and Siegert F 2012 *JHEP* **09** 049 (*Preprint* 1111.1220) URL http://inspirehep.net/record/944643
[25] Nason P 2004 *JHEP* **11** 040 (*Preprint* hep-ph/0409146) URL http://inspirebeta.net/record/659055
[26] Frixione S, Nason P and Oleari C 2007 *JHEP* **11** 070 (*Preprint* 0709.2092) URL http://www.slac.stanford.edu/spires/find/hep/www?eprint=arXiv:0709.2092
[27] Hamilton K, Richardson P and Tully J 2008 *JHEP* **10** 015 (*Preprint* 0806.0290) URL http://www.slac.stanford.edu/spires/find/hep/www?eprint=arXiv:0806.0290

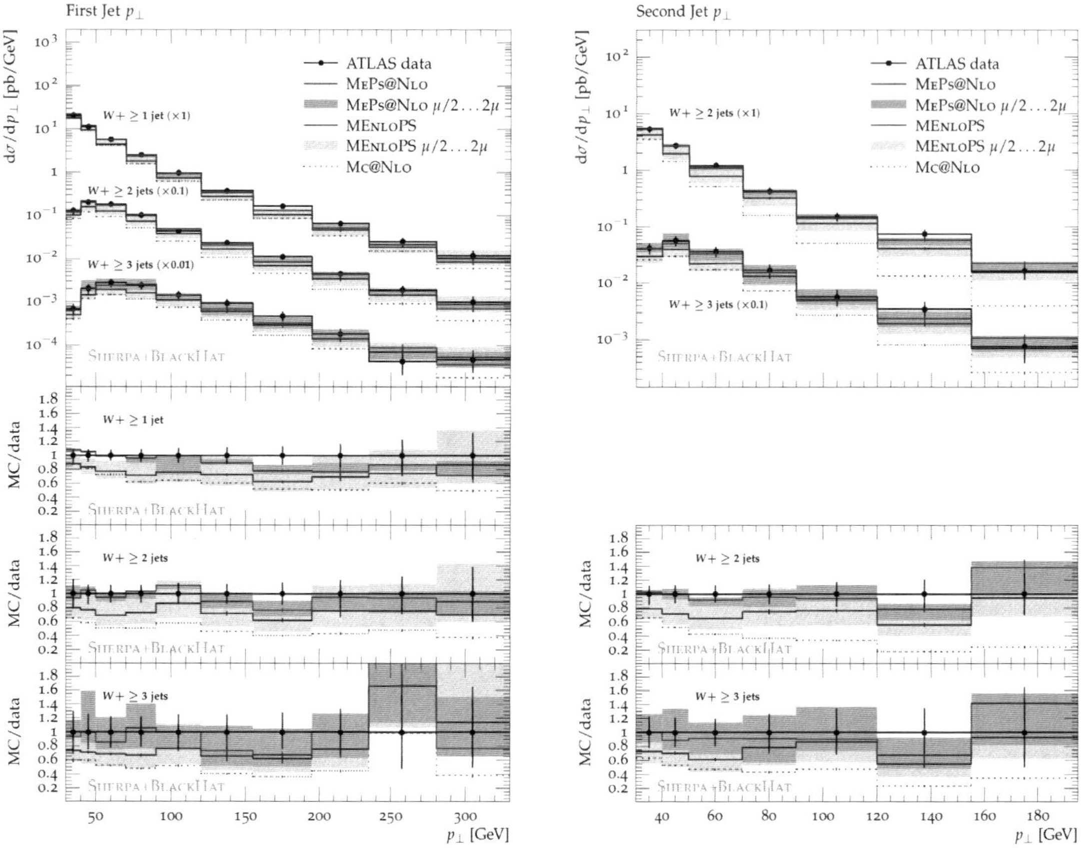

Figure 5: Transverse momentum of the leading and subleading jet in events with at least 1, 2 or 3 jets in $pp \to W+$ jets production at the LHC compare to ATLAS data [11]. Figures taken from [35].

[28] Höche S, Krauss F, Schönherr M and Siegert F 2011 *JHEP* **04** 024 (*Preprint* 1008.5399) URL http://inspirebeta.net/record/866705
[29] Höche S, Krauss F, Schönherr M and Siegert F 2013 *Phys.Rev.Lett.* **110** 052001 (*Preprint* 1201.5882) URL http://inspirehep.net/record/1086175
[30] Re E 2012 *JHEP* **1210** 031 (*Preprint* 1204.5433) URL http://inspirehep.net/record/1112186
[31] Frederix R, Frixione S, Hirschi V, Maltoni F, Pittau R *et al.* 2012 *JHEP* **1202** 048 (*Preprint* 1110.5502)
[32] Hamilton K and Nason P 2010 *JHEP* **06** 039 (*Preprint* 1004.1764) URL http://www.slac.stanford.edu/spires/find/hep/www?eprint=arXiv:1004.1764
[33] Höche S, Krauss F, Schönherr M and Siegert F 2011 *JHEP* **08** 123 (*Preprint* 1009.1127) URL http://www.slac.stanford.edu/spires/find/hep/www?eprint=arXiv:1009.1127
[34] Lavesson N and Lönnblad L 2008 *JHEP* **12** 070 (*Preprint* 0811.2912)
[35] Höche S, Krauss F, Schönherr M and Siegert F 2012 (*Preprint* 1207.5030) URL http://inspirehep.net/record/1123387
[36] Gehrmann T, Höche S, Krauss F, Schönherr M and Siegert F 2013 *JHEP* **1301** 144 (*Preprint* 1207.5031)
[37] Höche S and Schönherr M 2012 *Phys.Rev.* **D86** 094042 (*Preprint* 1208.2815) URL http://inspirehep.net/record/1127523

Physics of the Top Quark at Future Lepton Colliders

Thomas Teubner

Department of Mathematical Sciences, University of Liverpool, Liverpool L69 3BX, U.K.

E-mail: `thomas.teubner@liverpool.ac.uk`

Abstract. The physics scenario of the top quark at future lepton colliders is discussed, concentrating on the theory and simulation of $t\bar{t}$ and $t\bar{t}H$ production close to threshold. The study of these processes would allow to determine the mass and couplings of the top quark with very high precision, including the direct measurement of the top Yukawa coupling.

1. Introduction

At this conference, impressive measurements of the top quark's properties at hadron colliders have been discussed, and a lot more will come from the LHC, so why top quark physics at a lepton collider? With a Yukawa coupling y_t of order one the top plays a special role in the electroweak symmetry breaking. Its mass and couplings are key parameters for electroweak precision tests of the Standard Model (SM) and its extensions, and the accuracy of these parameters is the limiting factor in many studies [1]. Similarly, when extrapolating masses and couplings within scenarios of unified theories via renormalisation group equations ('running') over many orders of magnitude in energy, meaningful constraints require a very high accuracy of the top couplings. For such precision measurements, a linear e^+e^- (or a muon) collider would offer unique possibilities for several reasons: the centre-of mass energy is clearly defined and can be monitored, together with the sharp luminosity spectrum, with highest accuracy; the leptonic initial state has only electro-magnetic and no strong initial state radiation so no PDFs are required; the final states produced in lepton annihilation are of low multiplicity compared to hadron collisions, leading to a 'clean' environment with controllable backgrounds and measurable cross section normalisations. The top quark mass measurements at the Tevatron and the LHC have already achieved a very high relative accuracy, making this the best known quark in this respect. However, connecting a kinematical reconstruction of a coloured object to a mass parameter as used in theoretical calculations introduces systematic uncertainties of the order of the QCD scale $\Lambda_{\rm QCD}$ [2]. A threshold scan of top pair production at an e^+e^- collider offers the possibility to avoid this problem and is the only known way to determine the top quark mass with an accuracy of order 100 MeV. Similarly, studying $e^+e^- \to t\bar{t}H$ close to the kinematic threshold will, assuming that the LHC has found the Higgs at ~ 125 GeV, allow to directly measure the top Yukawa coupling. In the following these scenarios will be discussed in more detail.

2. Threshold production; theoretical description

Top quark production at threshold is strongly enhanced due to the attractive Coulomb-like gluonic interaction. The fast decays of the top and antitop prevent the formation of stable bound states, hence there will be no top quark spectroscopy. However, the remnant of the smeared-out

$1S$ peak provides a very distinctive feature in the threshold cross section $\sigma_{t\bar{t}} \equiv \sigma(e^+e^- \to t\bar{t})$, which can be related to the top quark mass without infrared ambiguities, therefore avoiding the aforementioned uncertainties of order $\Lambda_{\rm QCD}$. Note that, contrary to a kinematic reconstruction of top decay products, the threshold scan for the total cross section is basically a counting experiment. The accuracy of the extraction of the top mass and couplings will therefore rely on the precise knowledge of the cross section, i.e. of the luminosity spectrum of the e^+, e^- beams (not just the mean energy and luminosity, but see below) and, of course, on having enough statistics. In addition, the theoretical description must match the experimental accuracy.

2.1. Effective theories; observables

Due to its large mass and decay width, $t\bar{t}$ production can be described using perturbative QCD (pQCD). To achieve high accuracy of the theoretical prediction, higher order corrections for the threshold cross section are mandatory. These corrections are highly non-trivial as they (a) require the summation of certain classes of diagrams and, (b) involve several physical scales, i.e. the top mass, m_t, the top momentum, $p_t \sim m_t v$ (with v the top's non-relativistic velocity) and the 'binding' energy of the quasi-bound system, $E = \sqrt{s} - 2m_t \sim m_t v^2$. In addition the top width, Γ_t, has to be taken into account. As, close to threshold, the non-relativistic velocity v is small, these scales have a strong hierarchy, $m_t >> p_t >> E$. This makes it possible to formulate the problem in form of effective field theories. Over the last two decades several forms of such effective field theories have been developed, called Non-Relativistic QCD (NRQCD). Within this framework, $\sigma_{t\bar{t}}$ has been calculated up to next-to-next-to leading order (NNLO) in pQCD by several groups, see [3] for an early review and references therein. Close to threshold the perturbative expansion has to be organised as a series in both α_s and the small non-relativistic velocity v. Therefore the normalised cross section ratio R in the threshold region takes the form $R = \sigma_{t\bar{t}}/\sigma_{\mu^+\mu^-} \sim v \sum_n \left(\frac{\alpha_s}{v}\right)^n \left[\text{LO}\{1\}, \text{NLO}\{v, \alpha_s\}, \text{NNLO}\{v^2, \alpha_s v, \alpha_s^2\}\right]$. Already at leading order (LO) all terms of order $(\alpha_s/v)^n$ have to be summed; they correspond to ladder-type diagrams and build up the Coulombic QCD potential which is attractive for the $t\bar{t}$ colour singlet. Other potentials appear at higher orders and are taken into account using Green function techniques. The effective theory with its systematic power counting separates the dynamical from the non-dynamical degrees of freedom and allows to sum these order $(\alpha_s/v)^n$ contributions and, for certain versions of the effective field theory, also logarithmically enhanced contributions. These new theories have also improved the description of other quarkonium systems like $b\bar{b}$.

In Fig. 1 the $t\bar{t}$ production cross section is shown as a function of the centre-of-mass

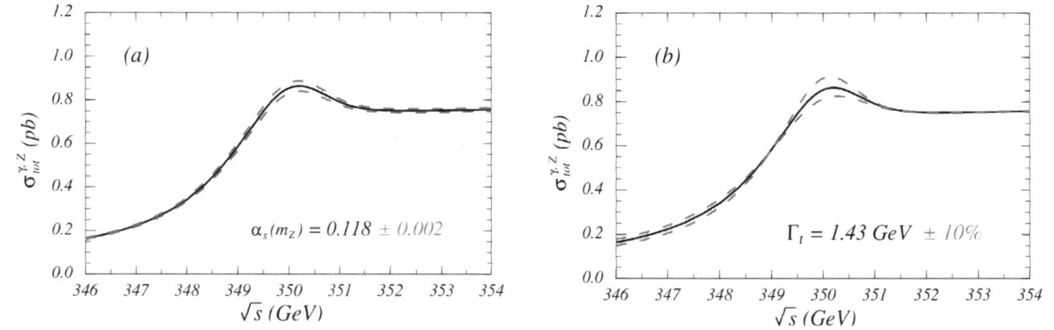

Figure 1. Dependence of the total cross section $e^+e^- \to Z^*, \gamma^* \to t\bar{t}$ as a function of $\sqrt{s}$ on (a) the strong coupling α_s and (b) the top width Γ_t. The solid (black) lines correspond to $\alpha_s = 0.118$ and $\Gamma_t = 1.43$ GeV, whereas the dashed (red) ones to variations of α_s and Γ_t by 0.002 and 10%, respectively. The predictions are from [4] and based on vNRQCD.

energy $\sqrt{s}$ in the threshold region. The results [4] are obtained within so-called velocity NRQCD (vNRQCD) and are at next-to-next-leading logarithmic order (NNLL), see also [5, 6]. They include contributions due to the summation of leading logarithms of the ratios of the different scales m_t, $p_t = m_t v$ and $E \sim m_t v^2$ through renormalisation group running [4, 5, 6]. This leads to a stabilisation of the perturbative series which then takes the form $R = \sigma_{t\bar{t}}/\sigma_{\mu^+\mu^-} \sim v \sum_{n,k} \left(\frac{\alpha_s}{v}\right)^n (\alpha_s \log v)^k \left[\mathrm{LL}\{1\}, \mathrm{NLL}\{v, \alpha_s\}, \mathrm{NNLL}\{v^2, \alpha_s v, \alpha_s^2\}\right]$. Despite the large decay width, the remnant of the $1S$ peak is clearly visible. The dashed (red) lines indicate the dependence of $\sigma_{t\bar{t}}$ on the parameters α_s (left panel) and on the width Γ_t (right panel). In addition, $\sigma_{t\bar{t}}$ also depends on the top Yukawa coupling y_t through virtual Higgs corrections and on its electroweak couplings. Although there is a correlation in the dependence on these parameters, it will be possible to measure not only m_t, but also some of the other parameters with high accuracy through a multi-parameter fit of data from a threshold scan. To help to disentangle the correlated dependencies, observables other than the total cross section can be used, such as the top momentum distribution and the forward-backward asymmetry A_{FB} which stems from the interference of the leading S wave $t\bar{t}$ production with the P wave contribution from production through a virtual Z. While the momentum distribution is mainly sensitive to m_t and less to α_s, the asymmetry A_{FB} is strongly dependent on the width Γ_t which determines the overlap of the S and P wave contributions. Other observables, like the spectra of W decay leptons, will be instrumental to measure the top polarisation and to gain sensitivity e.g. on CP violation from possible BSM contributions to the top electric dipole moment, on $V + A$ contributions or other anomalous couplings.

2.2. Status of higher order predictions

The first predictions for $\sigma_{t\bar{t}}$ were already made more than ten years ago, see e.g. [3]. They showed large corrections not only from LO to NLO, but again from NLO to NNLO. While this can be understood from the way in which certain corrections appear in the perturbative series, it means that the stability of the predictions, estimated from either the size of the corrections or from variation of the renormalisation scales, is limited. As this may threaten the accuracy of future threshold analyses, several groups have worked on further improvements. One way is the already mentioned summation of large logarithms through renormalisation group running, done within effective theories called velocity (v) or potential (p) NRQCD. This has led to a stabilisation of the corrections and hence the predictions [4, 5, 6]. Another way is to go to even higher 'fixed-order' perturbation theory, i.e. next-to-next-to-next-to leading order (NNNLO). A complete result for the cross section is not yet available at this order, but results for the $1S$ peak position and normalisation indicate an improved stability at NNNLO, see [7, 8] for short reviews and [9, 10] for recent results. As a part of this programme, the recent results for the three-loop coefficient a_3 of the QCD potential should be mentioned [11, 12]; more than 20000 Feynman diagrams contribute to this quantity.

For a reliable prediction of the top pair production cross section also electroweak corrections have to be taken into account. For the pure QCD corrections the top width Γ_t, which is due to the weak interaction, has usually been included as an imaginary part of the non-relativistic energy. However, this approach is not consistent at higher order, where, in the power counting of the effective theory, the electroweak coupling $\alpha_{\mathrm{EW}} \sim \Gamma_t/m_t \sim \alpha_s^2 \sim v^2$. In addition, electroweak interactions lead to single- and non-resonant $W^+W^-b\bar{b}$ final states which can, depending on cuts in the experimental analyses, contribute as irreducible backgrounds to the double-resonant $t\bar{t}$ signal. Recently a lot of effort has been invested in understanding these electroweak effects in the framework of effective theories which include unstable particles [13, 14, 15]. Work done so far has shown that the corrections are of the order of several percent for the total cross section and will have to be included in future complete threshold analyses.

2.3. Mass definitions

To reach an accuracy of order 100 MeV in the determination of m_t, it is crucial to go beyond the usual definition of the top mass as a pole mass. This becomes obvious when using the pole mass in the prediction of $\sigma_{t\bar{t}}$ in the threshold region: the position of the $1S$ peak, which is a physical quantity, moves significantly (by several 100 MeV) from order to order. The reason for this is understood. It is well known that, although infrared (IR) finite and gauge invariant, the pole mass itself is not an observable and suffers from ambiguities of order $\Lambda_{\rm QCD}$ which can be related to IR renormalon effects. In contrast, for the $t\bar{t}$ $1S$ peak, individual IR contributions are cancelling between the mass and potential in $E(1S) = 2m_t^{\rm pole} + V$. Hence the $1S$ peak position is IR safe and an observable from which a suitable 'short-distance mass' can be extracted. Several such mass schemes have been developed, such as the 'potential subtracted mass' [16] or the '$1S$ mass' m_t^{1S} [17] which is perturbatively defined in the $\Gamma_t \to 0$ limit, see [3] for a discussion of the application of these schemes. These masses can be translated, order-by-order in perturbation theory and *without* an IR ambiguity $\sim \Lambda_{\rm QCD}$ to other masses, like the $\overline{\rm MS}$ mass which is suitable for the high-energy regime (but not in the threshold region).

3. Simulation of the threshold scan

Several groups have studied the analysis of the top quark threshold from an experimental point of view. While backgrounds and event reconstruction should not pose a major problem for the planned machine and detector designs, the required understanding of the luminosity spectrum is challenging. The spectrum relevant for the experimentally measured threshold cross section is strongly influenced by the initial beam profile, beam-beam interactions in the interaction region and also by QED initial state radiation. It will be crucial to measure and monitor this spectrum as otherwise large effects would bias the top mass and other parameters determined via a fit of the threshold scan data. Figure 3 shows a recent simulation [18] done for CLIC and ILC

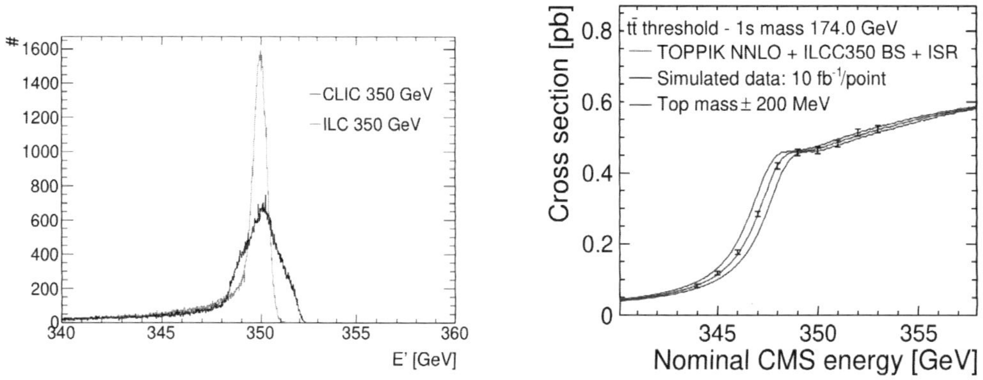

Figure 2. Luminosity spectrum and resulting $t\bar{t}$ threshold cross section as simulated by [18].

designs. The left panel shows typical beam spectra, the right one the resulting $t\bar{t}$ threshold cross section for an ILC of 350 GeV. The simulated data together with the theoretical curves used for the fit demonstrate that, while the relatively wide beam spectrum leads to a considerable smearing of the original theory curve as shown in Fig. 1, the measured threshold cross section will still be a highly sensitive probe of the top mass. This simulation of the threshold scan predicts that an experimental uncertainty of 21 MeV for m_t^{1S} will be reachable at CLIC with a one-parameter fit using a fixed input for α_s. With a two-dimensional fit $\Delta m_t^{1S} = 33$ MeV and $\Delta\alpha(M_Z^2) = 0.0009$. For an ILC the accuracy would be even better by 10 to 20%. Clearly, it will be crucial to know the beam spectrum with the best possible accuracy, see also [19] for related work on the simulation of beam effects in the $t\bar{t}$ threshold scan. The quoted fit results are in line with earlier studies by Martinez *et al* [20], who in addition predicted a relative accuracy of about 35% for the simultaneous determination of the top Yukawa coupling for a light Higgs in the mass range found at the LHC. This is possible as a relatively light Higgs leads to significant corrections at threshold from the resulting Yukawa potential.

4. Studies for $t\bar{t}$ production in the continuum

Compared to these predictions for a threshold scan, similar recent simulations for the top mass reconstruction at $\sqrt{s} = 500$ GeV at CLIC [21] or ILC indicate that the achievable statistical uncertainty, though impressive taken face value, will be about three times worse, in addition to the theoretical problem of relating such a kinematic mass to a short distance mass. However, $t\bar{t}$ production in the continuum offers unique possibilities to measure certain top couplings. One example is the Ztt coupling, which could be a sensitive probe of BSM physics but is not accessible at the LHC. Through measurements of the asymmetries $A_{\rm FB}$ and $A_{\rm LR}$ the Wtb and Ztt couplings can be determined at the 1% level of accuracy [22, 23]. When parametrising SM and possible BSM contributions in a model independent way through general dimension four and six operators, many can only be tested at lepton colliders, while for others the ILC would improve markedly on what can be done at the LHC, see e.g. [24] for recent analyses.

5. $t\bar{t}H$

To establish if the particle found at the LHC is the SM Higgs or something different one needs to measure its couplings with the best possible accuracy. At the LHC the top Yukawa coupling y_t is accessible indirectly through the cross section which is mainly induced by the ggH coupling via a top loop. Total cross section measurements at hadron colliders are plagued by various uncertainties, and the accuracy of such a determination will be about 15%. At a lepton collider, the measurement of y_t in $t\bar{t}H$ production will be possible, but also challenging due to the relatively complicated final state, low rates and high backgrounds [25]. Early studies had indicated that the prospects for such a measurement are better at higher centre-of-mass energies. However, the situation has changed with the observation that, close to the $t\bar{t}H$ threshold the final state predominantly consists of a non-relativistic $t\bar{t}$ pair, back-to-back with a Higgs boson. Therefore threshold corrections similar to the case of the $t\bar{t}$ threshold enhance the cross section and have been calculated within NRQCD to NLL order [26]. Together with possible enhancements due to e^- and possibly e^+ beam polarisation, this leads to a significantly enlarged cross section. These effects have previously been estimated to give enhancement factors of 2.4 and 2.1, respectively, and resulted in an estimated relative accuracy of 10% for the measurement of y_t [27]. Figure 5 shows the results of a recent simulation [28] of the $t\bar{t}$ mass spectrum in the

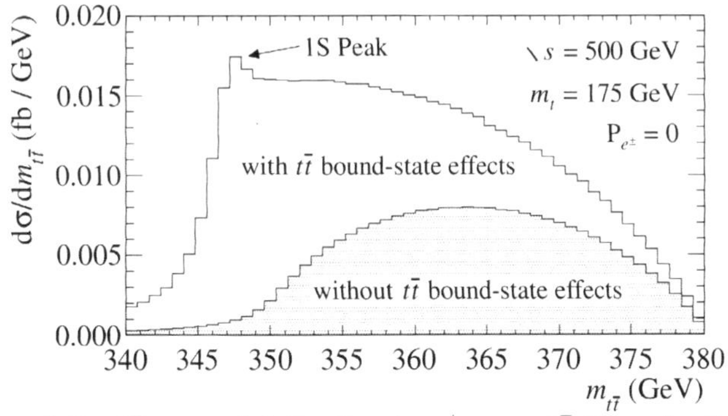

Figure 3. Simulation of the $t\bar{t}$ mass distribution in $e^+e^- \to t\bar{t}H$ production close to threshold. (From [28].)

process $e^+e^- \to t\bar{t}H$ with and without the bound-state effects due to the QCD dynamics of the $t\bar{t}$ system. Even well above the nominal threshold they lead to a significant enhancement of the cross section. This recent analysis is based on a `fastsim` simulation and has confirmed the previously estimated improvement, giving a statistical uncertainty of 10% for the measurement of y_t, if a beam polarisation of $(P_{e^+}, P_{e^-} = (+0.3, -0.8)$ is taken into account and assuming an integrated luminosity of 1000 fb^{-1} and a centre-of-mass energy $\sqrt{s} = 500$ GeV.

6. Outlook

At this conference we have seen impressive results on top quark physics from the Tevatron and the LHC, and a lot more is to come. At the same time, studies of top quark physics at a future lepton collider have moved forward tremendously compared to the first explorations of the physics scenario and have triggered important theoretical developments like effective field theories and short-distance mass definitions. The theoretical accuracy stands typically at next-to-next-to leading order or even higher, but only for inclusive quantities. To be fully prepared, more and better Monte Carlo tools will be needed, which have to include the recent theoretical developments. However, the physics case for a future linear collider running at the $t\bar{t}$ and the $t\bar{t}H$ thresholds has been clearly made: It would offer the possibility to determine the top mass and its couplings in a clean environment and with the highest accuracy, impossible to reach in hadron collisions. With the recent discovery of a Higgs-like particle at the LHC it is a good time to push for the realisation of an International Linear Collider.

References

[1] Heinemeyer S, Kraml S, Porod W and Weiglein G 2003 *JHEP* **0309** 075
[2] Moch S-O 2013 *in these proceedings*
[3] Hoang A H *et al* 2000 *Eur. Phys. J. direct* C **2** 1
[4] Hoang A H, Manohar A V, Stewart I W and Teubner T 2002 *Phys. Rev.* D **65** 014014; 2001 *Phys. Rev. Lett.* **86** 1951
[5] Stahlhofen M and Hoang A 2011 *PoS RADCOR2011* 025; 2011 *JHEP* **1106** 088; 2006 *Phys. Rev.* D **75** 054025
[6] Pineda A and Signer A 2007 *Nucl. Phys.* B **762** 67
[7] Beneke M, Kiyo Y and Schuller K 2007 *PoS RADCOR2007* 051
[8] Steinhauser M 2006 *Nucl. Phys. Proc. Suppl.* **157** 226
[9] Penin A A, Smirnov V A and Steinhauser M 2005 *Nucl. Phys.* B **716** 303
[10] Beneke M and Kiyo Y 2008 *Phys. Lett.* B **668** 143
[11] Smirnov A V, Smirnov V A and Steinhauser M 2010 *Phys. Rev. Lett.* **104** 112002; 2008 *Phys. Lett.* B **668** 293
[12] Anzai C, Kiyo Y and Sumino Y 2010 *Phys. Rev. Lett.* **104** 112003
[13] Penin A A and Piclum J H 2012 *JHEP* **1201** 034
[14] Hoang A H and Reisser C J 2005 Phys. Rev. D **71** 074022; 2006 Phys. Rev. D **74** 034002; Hoang A H, Reisser C J and Ruiz-Femenia P 2010 *Phys. Rev.* D **82** 014005
[15] Beneke M, Chapovsky A P, Signer A and Zanderighi G 2004 *Phys. Rev. Lett.* **93** 011602; 2004 *Nucl. Phys.* B **686** 205; Beneke M, Jantzen B and Ruiz-Femenia P 2010 *Nucl. Phys.* B **840** 186
[16] Beneke M 1998 *Phys. Lett.* B **434** 115
[17] Hoang A H and Teubner T 1999 *Phys. Rev.* D **60** 114027
[18] Seidel K, Simon F and Tesar M 2012 *private communication*; 2012 Prospects for the measurement of the top mass in a threshold scan at CLIC and ILC *Preprint* LCD-Note-2012-013
[19] Gournaris F and Boogert T 2008 *in '2007 International Linear Collider Workshop (LCWS07 and ILC07)'* ed Riemann S *Preprint* DESY-PROC-2008-03, http://www.slac.stanford.edu/econf/C0705302
[20] Martinez M and Miquel R 2003 *Eur. Phys. J.* C **27** 49
[21] Simon F, Seidel K and Poss S 2012 Top mass measurement at CLIC at 500 GeV *Preprint* arXiv:1202.3341
[22] Devetak E, Nomerotski A and Peskin M 2011 *Phys. Rev.* D **84** 034029
[23] Doublet P, Richard F, Poschl R, Frisson T and Rouene J 2012 Determination of top-quark asymmetries at the ILC *Preprint* arXiv:1202.6659
[24] Aguilar-Saavedra J A, Fiolhais M C N and Onofre A 2012 *JHEP* **1207** 180; Aguilar-Saavedra J A 2010 *PoS ICHEP2010* 378; 2011 *Nucl. Phys.* B **843** 638, Erratum-ibid 2011 *Nucl. Phys.* B **851** 443
[25] Gay A 2007 *Eur. Phys. J.* C **49** 489

[26] Farrell C and Hoang A H 2005 *Phys. Rev.* D **72** 014007; 2006 *Phys. Rev.* D **74** 014008
[27] Juste A 2005 Towards a precise measurement of the top quark Yukawa coupling at the ILC *Preprint* arXiv:hep-ph/0512246
[28] Yonamine R *et al* 2011 *Phys. Rev.* D **84** 014033; Yonamine R, Ikematsu K, Uozumi S and Fujii K 2010 A study of top-quark Yukawa coupling measurement in $e^+e^- \to t\bar{t}H$ at $\sqrt{s} = 500$ GeV *Preprint* arXiv:1008.1110

Measurement of the mass difference between top and antitop quarks

S. Blyweert, on behalf of the CMS Collaboration
Inter-university Institute for High Energies, Vrije Universiteit Brussel - Brussels, Belgium
E-mail: stijn.blyweert@cern.ch

Abstract. A measurement of the mass difference between the top quark and the antitop quark ($\Delta m_{\mathrm{t}} = m_{\mathrm{t}} - m_{\bar{\mathrm{t}}}$) is performed using events with a muon or an electron and at least four jets in the final state. Data collected in 2011 with the CMS detector at the LHC and corresponding to an integrated luminosity of $4.96 \pm 0.11\ \mathrm{fb}^{-1}$ are analyzed. The measured value of $\Delta m_{\mathrm{t}} = -0.44 \pm 0.46$ (stat.) $\pm\ 0.27$ (syst.) GeV is consistent with the expectation of CPT invariance, and provides a significantly improved precision relative to existing measurements.

1. Introduction

CPT symmetry is one of the fundamental symmetries in the Standard Model, which is not known to be violated. A direct measurement of a mass difference between particle and anti-particle would indicate a violation of CPT symmetry. The only quark where this measurement can be performed is the top quark, since it decays before the hadronization can take place. This mass difference between the top quark and the antitop quark has been measured before at the Tevatron and no significant deviation from zero was found [1, 2].

In this analysis, $\mathrm{t\bar{t}}$ events decaying in the e+jets and μ+jets channels are used. In these events, one of the two top quarks decays hadronically ($\mathrm{t} \rightarrow \mathrm{bW} \rightarrow \mathrm{bq\bar{q}'}$) and the other top quark decays in the lepton channel ($\mathrm{t} \rightarrow \mathrm{bW} \rightarrow \mathrm{b}\ell\nu_\ell$), where the lepton ℓ is an electron or a muon. Based on the lepton charge, the data can be split into two distinct samples consisting of hadronically decaying top or antitop quarks. For each event category, the Ideogram method [3] is used to measure the mass of the top quarks and finally both masses are subtracted from each other: $\Delta m_{\mathrm{t}} = m_{\mathrm{t}}^{hadr,\ \mathrm{Ideogram}} - m_{\bar{\mathrm{t}}}^{hadr,\ \mathrm{Ideogram}}$.

2. Event selection and event reconstruction

The measurement is performed on $4.96 \pm 0.11\ \mathrm{fb}^{-1}$ of proton-proton collisions at a center-of-mass energy of 7 TeV which were recorded in 2011 by the CMS detector [5]. These events were first selected by a trigger requiring an isolated electron or muon with $p_{\mathrm{T}} > 25$ or 17 GeV, respectively. The events are reconstructed with the particle-flow (PF) algorithm [6]. They are selected by requiring exactly one isolated electron with $p_{\mathrm{T}} > 30$ GeV and $|\eta| < 2.5$, or exactly one isolated muon with $p_{\mathrm{T}} > 20$ GeV and $|\eta| < 2.1$, and at least four jets with $p_{\mathrm{T}} > 30$ GeV and $|\eta| < 2.4$.

A kinematic fit is used to reconstruct the mass of the hadronically decaying top quark. There are 12 possible jet-quark assignments for every event when considering the four leading jets. For each of these, the kinematic fit returns the minimal χ^2 ($\chi^2_{\mathrm{min},i}$), the fitted top quark mass ($m_{\mathrm{t},i}^{Fit}$) and the uncertainty on the fitted top quark mass ($\sigma_i(m_{\mathrm{t}}^{Fit})$). Only the jet-quark assignments with $\chi^2_{\mathrm{min},i}/n_{\mathrm{dof}} < 10$ are considered for further analysis. A comparison between data and simulation of the fitted top quark mass can be found in Figure 1. A good agreement between data and simulated events is observed.

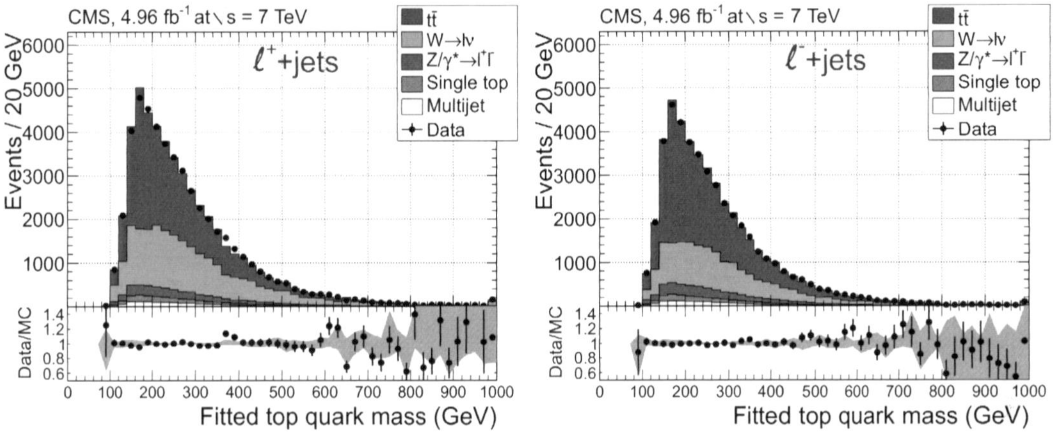

Figure 1. Fitted top quark mass for ℓ^++jets events (left) and ℓ^-+jets events (right). Only the jet combination with lowest χ^2 from the kinematic fit is shown. The simulation is normalized to the number of events observed in data. The ratio of the number of observed events in data to the number of events expected from simulation is shown at the bottom, including an error band corresponding to the systematic uncertainty.

3. The Ideogram method

The output of the kinematic fit for all the possible jet-quark assignments ($\chi^2_{\mathrm{min},i}$, $m_{\mathrm{t},i}^{Fit}$ and $\sigma_i(m_{\mathrm{t}}^{Fit})$) is used in the Ideogram method to measure the top quark mass. For every event a likelihood is calculated:

$$\mathcal{L}_{\mathrm{event}}(x; y \mid m_{\mathrm{t}}) = f_{\mathrm{t\bar{t}}}\ P_{\mathrm{t\bar{t}}}(x; y \mid m_{\mathrm{t}}) + (1 - f_{\mathrm{t\bar{t}}}) P_{\mathrm{bkg}}(x)$$

In this equation, x represents the number of b-tagged jets $n_{\mathrm{b-tag}}$, the muon charge q^{μ} and the fitted top quark masses $m_{\mathrm{t},i}^{\mathrm{Fit}}$. The quantity y is $\sigma_i(m_{\mathrm{t}}^{\mathrm{Fit}})$ and $\chi^2_{\mathrm{min,i}}$.

In the next step, $n_{\mathrm{b-tag}}$ and q^{μ} are factorized out:

$$P_{\mathrm{t\bar{t}}}(x; y \mid m_{\mathrm{t}}) = P_{\mathrm{t\bar{t}}}(n_{\mathrm{b-tag}}) \cdot P_{\mathrm{t\bar{t}}}(q^{\mu}) \cdot P_{\mathrm{t\bar{t}}}(x_{\mathrm{mass}}; y \mid m_{\mathrm{t}})$$
$$P_{\mathrm{bkg}}(x) = P_{\mathrm{bkg}}(n_{\mathrm{b-tag}}) \cdot P_{\mathrm{bkg}}(q^{\mu}) \cdot P_{\mathrm{bkg}}(x_{\mathrm{mass}})$$

Finally, the $\mathrm{t\bar{t}}$ signal probability $P_{\mathrm{t\bar{t}}}(x; y \mid m_{\mathrm{t}})$ is calculated as a weighted sum over all jet-quark assignments, containing two components: a correct jet combination probability and a wrong jet combination probability:

$$P_{\mathrm{t\bar{t}}}(x_{\mathrm{mass}}; y \mid m_{\mathrm{t}}) = \sum_{i=1}^{12} w_i \Big(f_{\mathrm{gc}} \int \mathrm{d}m' G(m_i \mid m', \sigma_i) BW(m' \mid m_{\mathrm{t}}, \Gamma_t)$$
$$+ (1 - f_{\mathrm{gc}}) WP(m_i \mid m_{\mathrm{t}}) \Big)$$

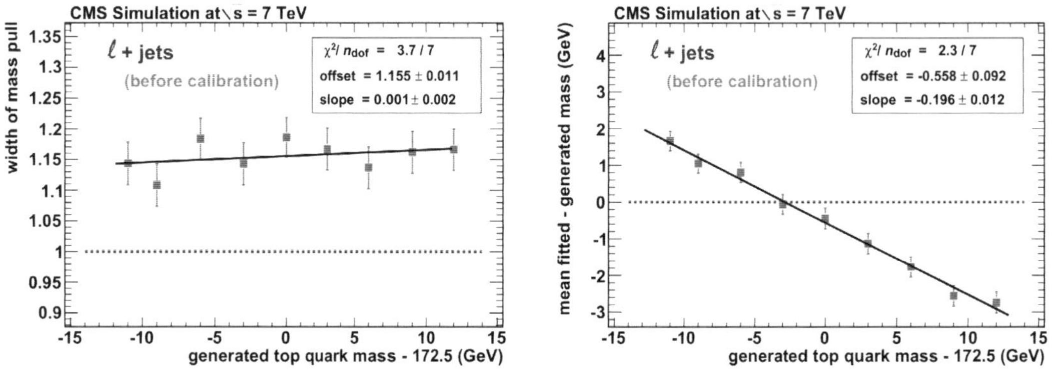

Figure 2. Width of the pull distribution (left) and bias on the estimated top quark mass (right) as a function of the generated top quark mass for ℓ+jets events.

The weights w_i of the previous equation is calculated as follows:

$$w_i = \exp\left(-\frac{1}{2}\,\chi^2_{\mathrm{min},i}\right) w_{\mathrm{b-tag}}, \text{ with } w_{\mathrm{b-tag}} = \prod_j p^j$$

Here, the probabilities p^j are equal to ε_l, $(1-\varepsilon_l)$, ε_b or $(1-\varepsilon_b)$, depending on the flavor hypothesis for each jet in the jet combination and depending on whether the jet is b-tagged or not. This weight significantly reduces the effect of wrong jet combinations and of background events.

The measured top quark mass is extracted from the combined likelihood of the full sample. This combined likelihood is calculated by combining the individual likelihoods of all the events:

$$\mathcal{L}_{\mathrm{sample}}(x; y \mid m_{\mathrm{t}}, f_{\mathrm{t\bar{t}}}) = \prod_j \mathcal{L}_{\mathrm{event,j}}(x; y \mid m_{\mathrm{t}}, f_{\mathrm{t\bar{t}}})$$

4. Calibration of the individual mass measurements

The event likelihood is a simplified representation of the underlying probability processes, which means the results need to be calibrated using pseudo-experiments with simulated events. The calibration is performed inclusively for ℓ+jets events. The bias on the estimated top quark mass as a function of the generated to quark mass is shown in Figure 2. Although the biases are withing 3 GeV, the final results need to be corrected by using the fitted linear calibration curve. The width of the pull distribution is found to be slightly larger than one, so the statistical uncertainties on the final mass measurement are scaled up by about 9%.

5. Systematic uncertainties

Many of the systematic uncertainties that are important for the measurement of m_{t} are reduced in the context of this measurement, as these systematic effects would alter the measured properties of top and antitop quarks in a similar and correlated manner. Several sources of systematic uncertainties on the modeling of the physical processes evaluated in absolute n_{t} measurements are not expected to affect the measurement of Δm_{t}. These include modeling of hadronization, underlying event, initial and final-state radiation, factorization scale used, and matching threshold between matrix elements and parton shower.

Systematic uncertainties for other effects considered are included, as listed in Table 1. For most systematic uncertainties, the statistical significance of the observed shift in Δm_{t} is small. We therefore quote the observed shift as a systematic uncertainty when it is larger than the statistical uncertainty, and otherwise we quote just the statistical uncertainty. The total systematic uncertainty is taken to be the quadratic sum of the values quoted for each source.

Table 1. Overview of systematic uncertainties on Δm_t. The total is defined by adding in quadrature the contributions from all sources, by choosing for each the larger of the estimated shift or its statistical uncertainty, as indicated by the bold script.

Source	Estimated effect (GeV)
Jet energy scale	$0.04 \pm \mathbf{0.08}$
Jet energy resolution	$0.04 \pm \mathbf{0.06}$
b vs. $\bar{\text{b}}$ jet response	$\mathbf{0.10} \pm 0.10$
Signal fraction	$\mathbf{0.02} \pm 0.01$
Difference in W^+/W^- production	$\mathbf{0.014} \pm 0.002$
Background composition	$\mathbf{0.09} \pm 0.07$
Pileup	$\mathbf{0.10} \pm 0.05$
b-tagging efficiency	$\mathbf{0.03} \pm 0.02$
b vs. $\bar{\text{b}}$ tagging efficiency	$\mathbf{0.08} \pm 0.03$
Method calibration	$0.11 \pm \mathbf{0.14}$
Parton distribution functions	**0.088**
Total	0.27

6. Final results

The final result obtained for 4.96 ± 0.11 fb^{-1} of data taken in 2011 by the CMS experiment is equal to:

$$\Delta m_t^{\text{measured}} = -0.44 \pm 0.46\ (\text{stat}) \pm 0.27\ (\text{syst})\ \text{GeV}$$

The measured value is in agreement with the expectation of the Standard Model. This is the most precise measurement to date of this quantity. More details can be found in [7].

References

[1] CDF Collaboration, "Measurement of the mass difference between t and $\bar{t}$ quarks", *Phys. Rev. Lett.* **106** (2011) 152001.

[2] D0 Collaboration, "Direct measurement of the mass difference between top and antitop quarks", *Phys. Rev. D* **84** (2011) 052005.

[3] DELPHI Collaboration, "Measurement of the Mass and Width of the W boson in e^+e^- Collisions at $\sqrt{s}$ = 161 - 209 GeV", *Eur. Phys. J. C* **55** (2008) 1-38.

[4] CMS Collaboration, "Measurement of the top quark mass in the lepton+jets channel", CMS Physics Analysis Summary **CMS-PAS-TOP-10-009** (2011).

[5] CMS Collaboration, "The CMS experiment at the CERN LHC", *JINST* **03** (2008) S08004.

[6] CMS Collaboration, "Commissioning of the Particle-Flow reconstruction in Minimum-Bias and Jet Events from pp collisions at 7 TeV", CMS Physics Analysis Summary **CMS-PAS-PFT-10-002** (2010).

[7] CMS Collaboration, "Measurement of the mass difference between top and antitop quarks", *JHEP* **1206** (2012) 109.

Journal of Physics: Conference Series **452** (2013) 012045 doi:10.1088/1742-6596/452/1/012045

Fully hadronic $t\bar{t}$ cross section with the ATLAS detector

Andrea Coccaro on behalf of the ATLAS Collaboration
Department of Physics, University of Washington, Seattle WA, United States of America
E-mail: `Andrea.Coccaro@cern.ch`

Abstract. A measurement of the $t\bar{t}$ production cross section in the all-hadronic decay mode is presented using 4.7 fb^{-1} of proton-proton collisions at a centre of mass energy of $\sqrt{s} = 7$ TeV collected by the ATLAS experiment in 2011. Events are selected using a multi-jet trigger. Kinematic and b-tagging requirements are then applied to identify $t\bar{t}$ event candidates. A kinematic fit reconstructs the event topology of the final state extracting the top-quark mass which is then used to measure the production cross section with an unbinned likelihood fit. The result is found in good agreement with the Standard Model prediction for a top-quark mass of 172.5 GeV.

1. Introduction

The measurement of the top-quark pair production cross section provides an important test of perturbative QCD calculations and top-quark pair production constitutes a major background source to many new physics scenarios predicted to be accessible at the LHC. The $t\bar{t}$ production cross section is nowadays calculated at approximate NNLO and is $\sigma_{t\bar{t}} = 167^{+17}_{-18}$ pb for a top-quark mass of 172.5 GeV at a centre-of-mass energy of $\sqrt{s} = 7$ TeV [1].

In the fully hadronic final state, the two W's decay hadronically and the experimental signature is characterized by a nominal six-jet topology with two jets stemming from the hadronization of b-quarks. The all-hadronic channel has the advantage of a large branching ratio, 46%, although it suffers from a huge background of multi-jet events.

The present result [2] is based on data recorded during the 2011 run with pp collisions at $\sqrt{s} = 7$ TeV using the ATLAS detector [3]. The total integrated luminosity is 4.7 fb^{-1} with an uncertainty of 3.9% after asking for stable beam conditions and all subsystems to be operational. This analysis is complementary to similar measurements in the single lepton and dilepton channels.

2. Trigger, event selection and simulated samples

The events considered in this analysis are triggered asking for at least five jets with $|\eta| < 3.2$ and $E_T > 30$ GeV. The calibration constants to correct for the lower hadron response of the non-compensating calorimeters were not included in the trigger system and an offline cut of $E_T > 55$ GeV is introduced at the analysis level to operate nearby the plateau of the trigger selection. The 55 GeV cut is obtained by studying the trigger dependence on the fifth leading jet and corresponds to a 90% efficiency with respect to jets reconstructed offline. The efficiency

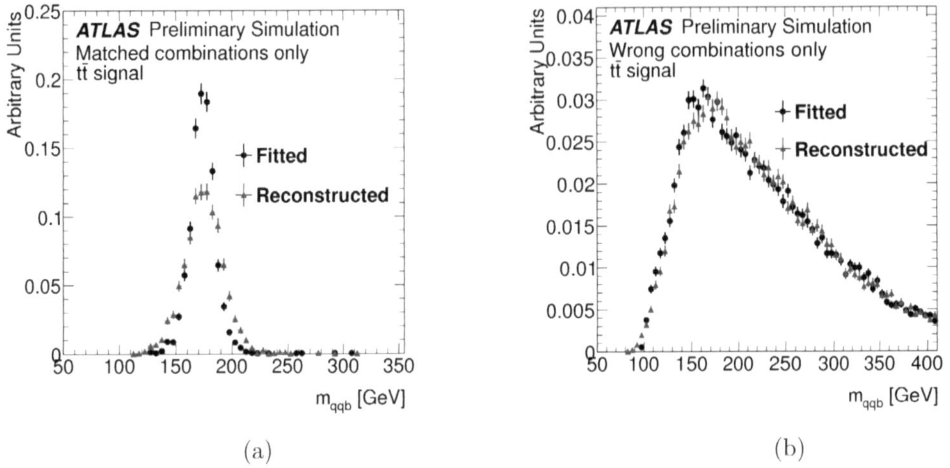

Figure 1: Top-quark mass distribution using the reconstructed or the corrected jet energies after the kinematic fit described in the text [2]. Only events where all the jets are correctly matched to quarks are shown in (a). Events where at least one jet is wrongly matched are shown in (b).

plateaus at more than 99% starting from 60 GeV and 55 GeV represents the best compromise between signal acceptance and contribution to the trigger systematic uncertainty.

To select a sample enriched in top-quark pair events, a preliminary set of cuts is applied on the triggered events: at least one reconstructed primary vertex with at least five associated tracks, no isolated leptons with $p_T > 20$ GeV, at least five jets with $E_T > 55$ GeV and $|\eta| < 2.5$ and at least one more jet with $E_T > 30$ GeV and $|\eta| < 2.5$. At least two jets, out of the five leading jets, are required to be b-tagged combining the high-performance algorithms [4] in a neural network. Jets with an unlikely association to the hard scatter in the event, using the jet vertex fraction information [5], are not considered in the analysis.

The preselection is completed requiring a minimum angular distance between the two b-tagged jets of $\Delta R > 1.2$ and a minimum angular distance between any two reconstructed jet of $\Delta R > 0.6$. The former cut reduces the background contamination from gluon splitting without compromising the signal efficiency while the latter cut is needed to avoid relying on the trigger simulation in case of close-by jets.

The modelling of the $t\bar{t}$ signal and its associated selection efficiency is derived from MC simulation. The MC@NLO generator with PDF set CT10 is used for the generation of the $t\bar{t}$ signal, assuming a top quark mass of 172.5 GeV. Due to the large uncertainty in the multi-jet cross-section prediction, a data-driven technique is employed to estimate the background, while the MC simulation is only used to estimate the uncertainty on the mass shape distribution for the background.

3. Kinematic fit

A kinematic fit is performed on the selected events to optimize the top-quark mass reconstruction. The fit is based on a likelihood approach to find the correct association of jets with the final state partons of the fully-hadronic $t\bar{t}$ decays. The likelihood function $\mathcal{L}$ includes Breit-Wigner functions used to constrain the di-jet and the tri-jet masses to the W-boson mass and the top-quark mass, respectively. The top-quark mass is treated as a free parameter but it is constrained to be identical for the top and anti-top quark candidates. The W-boson mass and width are fixed to the known values in literature. Additional terms are added taking into account the b-tagging algorithm response biasing the likelihood to select permutations in which a b-tagged jet is matched to a b-quark.

The fit is performed for all possible permutations of jets and quarks and for each permutation the function $-\ln\mathcal{L}$ is minimized with respect to the six jet energies and the reconstructed top-quark mass. The sum of all permutations is computed and the relative weight, normalized to unity, of the best computation is referred to as the *event probability*.

The method is tested on simulated events passing the selection cuts. Figure 1 (a) shows the mass distribution obtained for the events where all the jets are correctly matched to the corresponding quarks using the MC truth information. As expected, the mass resolution is improved when using the jet energies calculated during the kinematic fitting procedure compared to just using the reconstructed jet energies. The fraction of signal events fully matched by the kinematical fit is found to be 11%. For the events where at least one jet is wrongly matched there is no gain in the mass resolution by performing the kinematic fit, as shown in Figure 1 (b).

4. Cross section measurement

Few additional cuts are imposed before measuring the $t\bar{t}$ production cross section. Events are required to have $m_t > 125$ GeV and a jet multiplicity between 6 and 10. To further improve the signal to background ratio, cuts are introduced requiring the *event probability* to be greater than 0.8 and the minimal mass χ^2 to be less than 30. The χ^2 is defined as:

$$\chi^2 = \frac{(m_{j_1,j_2} - m_W)^2}{\sigma_W^2} + \frac{(m_{j_1,j_2,b_1} - m_t)^2}{\sigma_t^2} + \frac{(m_{j_3,j_4} - m_W)^2}{\sigma_W^2} + \frac{(m_{j_3,j_4,b_2} - m_t)^2}{\sigma_t^2},$$

where j_1, j_2, j_3, j_4, b_1 and b_2 are the reconstructed jets and b-jets and m_W, m_t, σ_W and σ_t are obtained by the mean and the half-width gaussian parametrization of the top-quark and W-boson mass distributions given by the all-hadronic $t\bar{t}$ MC simulation. The minimal mass χ^2 is defined as the minimum χ^2 obtained when computing its value for every permutation per event. With this final event selection, the fraction of signal events fully matched by the kinematical fit rises to 36%.

An unbinned likelihood fit is performed on the final data sample, shown in Figure 2 (a). The signal template distribution is derived from MC simulation and includes both the $t\bar{t}$ combinatorial background and the correct combinations while the background template distribution is derived from data events in the same signal region but without imposing any b-tagging requirement. The signal fraction is estimated by the fitting procedure to be $31.4 \pm 2.3\%$ and considering the inclusive $t\bar{t}$ selection efficiency, derived from MC simulation, the total cross section is measured to be $\sigma_{t\bar{t}} = 168 \pm 12(\text{stat.})$ pb.

The events passing all the cuts but dropping the b-tagging requirement are found to be dominated by multi-jet events, with a residual fraction of all-hadronic $t\bar{t}$ events estimated to be 4.6%. The top-quark mass distribution obtained with this methodology, referred as to the untagged sample, is shown in Figure 2 (b) and is used as a template for the background in the likelihood fit. Any b-tagging technique biases the jet p_T distribution and therefore most of the kinematic distributions, such as the top-quark mass spectrum, are affected by systematic uncertainties related to the b-tagging requirement. To estimate the uncertainty on the background modelling, the top-quark mass shape is studied using simulated events for both generic multi-jet and exclusive $b\bar{b}$+jets events and compared with and without the b-tagging requirement. The ratio of the top-quark mass distributions between the tagged and untagged and between the multi-jet and $b\bar{b}$+jets samples are computed and applied to the untagged data sample. The effect of these corrections on the data-driven background is shown in Figure 2 (b) and the maximum variation between the nominal and the corrected cross sections gives a systematic uncertainty on the background modelling of 4%.

The highest contribution to the total systematic uncertainty of the cross section measurement is from the jet energy scale and is estimated to be +20/-11%. Other sources of uncertainty were

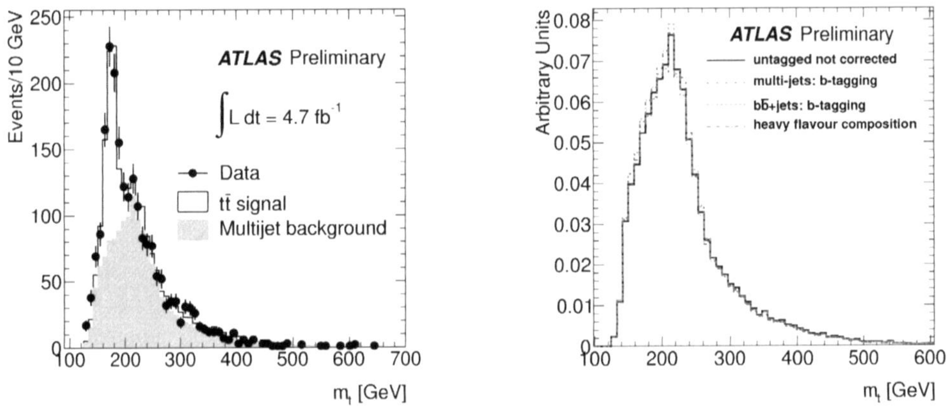

(a) (b)

Figure 2: (a) Unbinned likelihood fit of the top-quark mass distribution [2]. The signal and background templates are also displayed and the error bars associated to the data are statistical only. (b) Top-quark mass distribution after the kinematic fit using the sample selected with all the cuts described in the text but dropping the b-tagging requirement [2]. This distribution is used as a template of the multi-jet background for the likelihood fit and it is also shown in Figure 2 (a). Additional distributions obtained applying different corrections derived from MC simulations are overlaid using dashed lines.

estimated and are related to the b-tagging efficiency and mistag rate ($\pm 17\%$), the initial and final state radiation ($\pm 17\%$), the dependence on the generator and on the parton shower ($\pm 13\%$), the multi-jet trigger efficiency ($\pm 10\%$), the parton distribution function (+7/-5%), the luminosity ($\pm 3.9\%$) and the jet energy resolution ($\pm 3\%$). The total systematic uncertainty is estimated to be +36/-34%.

5. Conclusions

The production cross section of $t\bar{t}$ events in the all-hadronic decay channel was measured at the LHC using 4.7 fb^{-1} of pp collisions at the LHC with the ATLAS detector. The vastly dominant background consists of multi-jet events and its shape is modelled using a fully data-driven technique. A kinematic fit is employed to derive the top-quark mass distribution which is then used to extract the cross section measurement with an unbinned likelihood fit. The final measurement is $\sigma_{t\bar{t}} = 168 \pm (\text{stat.})^{+60}_{57}(\text{syst.}) \pm 7(\text{lum})$ pb. The measured cross section is found to be in agreement with the theoretical expectation of $\sigma_{t\bar{t}} = 167^{+17}_{-18}$ pb [1].

References

[1] M. Cacciari, M. Czakon, M. Mangano, A. Miltov, P. Nason, *Phys. Lett.* B **710** (2012)
[2] ATLAS Collaboration, ATLAS-CONF-2012-031 (2012)
[3] ATLAS Collaboration, JINST **3** S08003 (2008)
[4] ATLAS Collaboration, ATLAS-CONF-2011-102 (2012)
[5] ATLAS Collaboration, ATLAS-CONF-2012-064 (2012)

Journal of Physics: Conference Series **452** (2013) 012046 doi:10.1088/1742-6596/452/1/012046

$t\bar{t}$ pair hadroproduction in association with a heavy boson at the NLO QCD accuracy + Parton Shower

M V Garzelli[1], A Kardos[2,3], C G Papadopoulos[4,5], Z Trócsányi[2]

[1] Laboratory for Astroparticle Physics, University of Nova Gorica, SI 5000 Nova Gorica, Slovenia

[2] Institute of Physics and MTA-DE Particle Physics Research Group, University of Debrecen, H 4010 Debrecen P.O.Box 105, Hungary

[3] INFN, Sezione di Milano-Bicocca, I 20126 Milano, Italy

[4] Institute of Nuclear Physics, NCSR Δημόκριτος, GR 15310 Athens, Greece

[5] CERN, TH-Unit, CH 1211 Geneva 23, Switzerland

E-mail: `garzelli@mi.infn.it`

Abstract. The `PowHel` framework allows to make predictions of total and differential cross-sections of multiparticle hadroproduction processes at both NLO QCD accuracy and NLO QCD matched to Parton Shower, on the basis of the interface between the `POWHEG-BOX` and `HELAC-NLO` codes. It has already been applied to study several processes involving a $t\bar{t}$ pair in association with a third particle or hadronic jet. Our most recent predictions concern $t\bar{t}V$ hadroproduction (with $V = W$ or Z), at both parton and hadron level, by considering different decay channels (hadronic and leptonic) of the heavy particles. In particular, we show the results of our phenomenological analyses under the same system of cuts also recently adopted by the CMS collaboration at LHC.

1. Introduction

In recent years significant effort was devoted to increase the accuracy of theoretical predictions for p-p collision processes, in order to match the experimental capabilities of the LHC detectors. One line of development aims at providing refined automatic computations of fixed-order total and differential cross-sections with increased accuracy, leading to predictions of the final products of hard-scattering processes at the parton level. A second line of development, strongly tied to the first one, relies on the combination of these results with shower algorithms, able to track the evolution of these final-state particles to hadronization. In practice, to realize the combination between hard-scattering and parton shower, different matching schemes have been proposed, depending of the order of the perturbative expansion. At NLO accuracy two methods dominate the literature in the last ten years, the MC@NLO [1] and POWHEG [2, 3]. Here we make use of the latter, as implemented in the computer program `POWHEG-BOX` [4].

The `POWHEG-BOX` provides an almost automatic and process-independent matching framework yet it requires some process-dependent input, that has to be specified by the user, process by process. Thus, it offers a framework to implement the matching of new processes, once one is able to provide all required input. In case of processes of some complexity already at the parton level, e.g. multiparticle production, this input can be conveniently computed by external codes.

Our `PowHel` package uses the `HELAC-NLO` [5] set of codes, to provide all matrix-elements required as input by `POWHEG-BOX`, for the computation of cross-sections at the NLO + PS accuracy for $t\bar{t}$ pair hadroproduction in association with a third particle. The output of `PowHel` are Les Houches event (LHE) files [6], including events at the first radiation emission level. Additional radiation emissions can be simulated by further showering these events using Shower Monte Carlo (SMC) programs, such as `PYTHIA` [7] and `HERWIG` [8]. This can be done a-posteriori, since all essential process and event information needed at this purpose is already stored in the LHE files. Thus the full method can be exploited to produce predictions at the NLO level, at the first radiation emission level (LHE level), after parton shower (PS) and after PS + hadronization + hadron decay.

The total cross-sections as computed from LHE events are expected to have NLO accuracy, whereas the corresponding differential distributions have formally NLO accuracy up to higher order corrections, as LHE events also include the effect of Sudakov form factors embedded in the POWHEG formula. Therefore LHE predictions, as compared to NLO ones, are very useful to check the correctness of the implementation and to understand, as well, in which phase-space regions one could expect that the effects of higher-order corrections will sensibly modify the NLO distributions. On the other hand, predictions after the full SMC chain are the closest to the experimental results, allowing for an immediate comparison with the data collected in the detectors. Actually, in many experiments reconstruction procedures have been developed in order to reconstruct the intermediate unstable particles, like top quarks and heavy bosons, from which the final-state products were originated (bottom-up approach). We would like to emphasize however, that, once full high-accuracy simulations from the parton level down to the hadron level (top-down approach) are available, like in the case of `PowHel` + SMC, this reconstruction is unnecessary.

In this paper we concentrate on the case of $t\bar{t}$-pair hadroproduction in association with an heavy gauge boson, $t\bar{t}W$ and $t\bar{t}Z$, that we have recently studied extensively in Ref. [9, 10], leaving the reader interested in our predictions for other $t\bar{t}$-X associated processes, like $t\bar{t}j$, $t\bar{t}H$ and $t\bar{t}A$, to `PowHel` previous works [11, 12, 13].

2. Results

$t\bar{t}Z$ and $t\bar{t}W^{\pm}$ LO and NLO inclusive cross-sections as predicted by PowHel were already presented in Ref. [10, 12]. Here we emphasize that these predictions follow extensive checks and comparisons to the results of previous works presented by different authors in Ref. [14, 15], on the basis of different methods. Further predictions concerning several differential distributions at NLO and at NLO + PS accuracy were included, as for $t\bar{t}Z$ hadroproduction, for the first time in Ref. [16] and [10], respectively, whereas, as for $t\bar{t}W$, in Ref. [9].

Our predictions use the following parameters: m_t = 172.5 GeV, m_W= 80.385 GeV, m_Z=91.1876 GeV, $\sin^2\theta_C$=0.049284, $\mu_R = \mu_F = m_t + m_V/2$, where m_V is the mass of the vector boson. We adopted the SMC versions `PYTHIA-6.4.26` and `HERWIG-6.520`. We enforced stability of π^0 in both, whereas μ stability was enforced in `HERWIG`, to match the default in `PYTHIA`. All other hadrons and leptons were assumed to be stable or to decay with B.R. according to the default implementation of each SMC.

When making predictions at NLO + PS accuracy, we concentrated on two different channels, the trilepton and the same-sign dilepton ones, as also recently exploited by the CMS collaboration in Ref. [17]. Essentially, the trilepton channel selection aims at isolating the Z, decaying in $\ell^+\ell^-$ pairs, together with the $t\bar{t}$ pair decaying semileptonically. The selection of the dilepton channel aims at isolating the $W^{\pm}$, decaying in $\ell^{\pm}\overset{(-)}{\nu}_{\ell}$, again with the $t\bar{t}$ pair decaying semileptonically. The list of cuts we adopted for this study is extensively detailed in Ref. [9].

In the trilepton channel, requesting the invariant mass of a pair of opposite-charge isolated leptons to be close to the Z mass, the contribution of $t\bar{t}W^+$ and $t\bar{t}W^-$ is suppressed with respect to the contribution of $t\bar{t}Z$, as shown in Fig. 1.a. The invariant mass of the reconstructed Z after cuts is shown in Fig. 2.a, by cumulating the results of the three processes. Pairs of opposite charge leptons, emitted by heavy particle decays, turned out to be quite insensitive to SMC

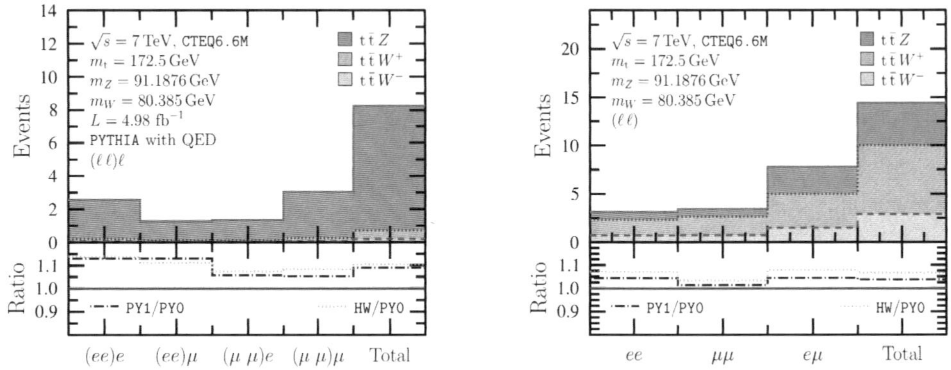

Figure 1. Number of events in a) the trilepton and b) the dilepton channels at the $\sqrt{s} = 7$ TeV LHC, as predicted by PowHel + PYTHIA, for an integrated luminosity $L = 4.98\,\text{fb}^{-1}$. The contribution in all possible channels are shown separately, as well as their sum in the last bin of each panel. The contributions due to $t\bar{t}Z$, $t\bar{t}W^+$ and $t\bar{t}W^-$ are cumulated one over the other. To be compared with the experimental data in Fig. 4 and 6 of the CMS technical report [17], respectively. In the lower inset the ratios between cumulative results using different SMC (HW/PY) and between cumulative results obtained by neglecting and including photon radiation from leptons (PY1/PY0) are also shown.

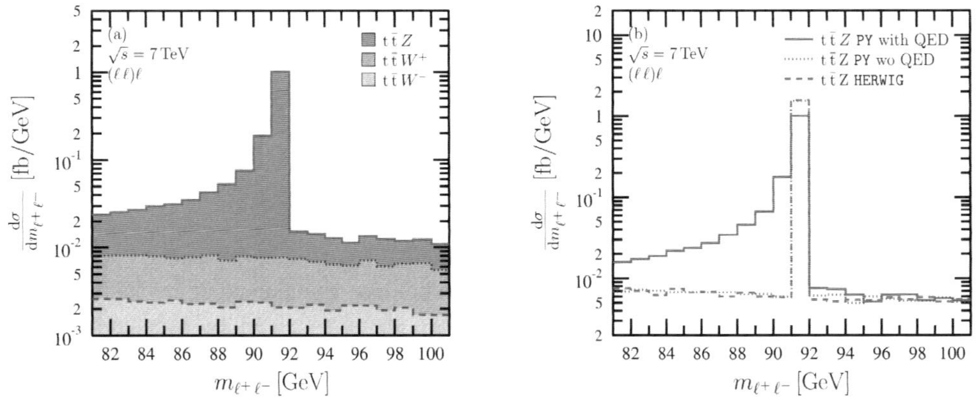

Figure 2. Invariant mass of the Z reconstructed from same-flavour (ℓ^+, ℓ^-) pairs after the trilepton analysis, as obtained by PowHel + PYTHIA at the $\sqrt{s} = 7$ TeV LHC: in panel a) the results corresponding to the different processes $t\bar{t}Z$, $t\bar{t}W^+$ and $t\bar{t}W^-$ are cumulated one over the other, whereas in panel b) distributions obtained by using different SMC (PYTHIA, HERWIG and PYTHIA without photon radiation from leptons) are also shown, limited to $t\bar{t}Z$.

effects, allowing for a good reconstruction of heavy objects. This opens good perspectives for the study, in the trilepton channel, of even other neutral boson candidates, like an exotic Z'. Even if further photon emissions from the emitted leptons smear the Z peak in the (ℓ^+, ℓ^-) invariant mass distribution, this does not disappear, as shown in Fig. 2.b. In the experimental event reconstruction electrons are often corrected for photon emission effects, but this is not always the case of muons. Thus, one could treat the two cases separately even in the shower simulation, since it is not always true that photon emission from muons leads to negligible effects.

The reconstruction of charged heavy objects like a W boson or a t-quark is complicated

channel	PowHel+SMC 7 TeV (fb)	PowHel+SMC 8 TeV (fb)	ratio
$(e,e)e$	0.516	0.782	1.515
$(e,e)\mu$	0.255	0.388	1.521
$(\mu,\mu)e$	0.273	0.420	1.538
$(\mu,\mu)\mu$	0.613	0.934	1.523
total	1.658	2.524	1.522

channel	PowHel+SMC 7 TeV (fb)	PowHel+SMC 8 TeV (fb)	ratio
(e,e)	0.631	0.907	1.437
(e,μ)	0.694	0.991	1.428
(μ,μ)	1.569	2.289	1.459
total	2.894	4.187	1.446

Table 1. Cumulative cross-sections at $\sqrt{s} = 7$ and 8 TeV, as predicted by PowHel+PYTHIA in a) all trilepton and b) all same-sign dilepton channels. The statistical error is always $< 1\%$.

because the measurement of missing energy, attributed to heavy particle decay, has large uncertainties. As a consequence, we did not introduce any explicit cut to distinguish $t\bar{t}W$ from $t\bar{t}Z$ contributions in the analysis in the same-sign dilepton channel. Thus the contribution of the $t\bar{t}Z$ process is not negligible with respect to that of the $t\bar{t}W$ ones, so all three processes contribute in a consistent way to the total amount of observed events, as shown in Fig. 1.b.

The total cumulative cross-sections after full SMC, including the contributions of $t\bar{t}Z$, $t\bar{t}W^+$ and $t\bar{t}W^-$, for all different lepton combinations in the trilepton and dilepton decay channels, at 7 and 8 TeV, are reported in Tables 1.a and 1.b. In the last column of each table, the ratio of the results at 7 and 8 TeV for each single channel is also reported showing an almost constant value. The slightly different average value of this ratio from all trilepton channels, as compared to the average from all dilepton channels, is related to the different set of cuts.

The $\sqrt{s} = 7$ and 8 TeV PowHel events used for these analyses are freely available on the web at http://www.grid.kfki.hu/twiki/bin/view/DbTheory/WebHome.

Acknowledgments

We are grateful to the TOP-2012 Conference organizers and partecipants for many interesting discussions. This work was supported by the LHCPhenoNet network PITN-GA-2010-264564, the Hungarian Scientific Research Fund grant K-101482 and the Slovenian Ministery of Work under the AD-FUTURA program.

References

[1] Frixione S and Webber B R 2002 *JHEP* **0206** 029 (*Preprint* hep-ph/0204244)
[2] Nason P 2004 *JHEP* **0411** 040 (*Preprint* hep-ph/0409146)
[3] Frixione S, Nason P and Oleari C 2007 *JHEP* **11** 070 (*Preprint* 0709.2092)
[4] Alioli S, Nason P, Oleari C and Re E 2010 *JHEP* **06** 043 (*Preprint* 1002.2581)
[5] Bevilacqua G, Czakon M, Garzelli M, van Hameren A, Kardos A *et al.* 2013 *Comput.Phys.Commun.* **184** 986–997 (*Preprint* 1110.1499)
[6] Alwall J, Ballestrero A, Bartalini P, Belov S, Boos E *et al.* 2007 *Comput.Phys.Commun.* **176** 300–304 (*Preprint* hep-ph/0609017)
[7] Sjostrand T, Mrenna S and Skands P Z 2006 *JHEP* **0605** 026 (*Preprint* hep-ph/0603175)
[8] Corcella G, Knowles I, Marchesini G, Moretti S, Odagiri K *et al.* 2002 (*Preprint* hep-ph/0210213)
[9] Garzelli M, Kardos A, Papadopoulos C G and Trócsányi Z 2012 *JHEP* **1211** 056 (*Preprint* 1208.2665)
[10] Garzelli M, Kardos A, Papadopoulos C G and Trócsányi Z 2012 *Phys.Rev.* **D85** 074022 (*Preprint* 1111.1444)
[11] Kardos A, Papadopoulos C G and Trócsányi Z 2011 *Phys.Lett.* **B705** 76–81 (*Preprint* 1101.2672)
[12] Garzelli M, Kardos A, Papadopoulos C G and Trócsányi Z 2011 *Europhys.Lett.* **96** 11001 (*Preprint* 1108.0387)
[13] Dittmaier S, Dittmaier S, Mariotti C, Passarino G, Tanaka R *et al.* 2012 (*Preprint* 1201.3084)
[14] Lazopoulos A, McElmurry T, Melnikov K and Petriello F 2008 *Phys.Lett.* **B666** 62–65 (*Preprint* 0804.2220)
[15] Campbell J M and Ellis R K 2012 *JHEP* **1207** 052 (*Preprint* 1204.5678)
[16] Kardos A, Trócsányi Z and Papadopoulos C G 2012 *Phys.Rev.* **D85** 054015 (*Preprint* 1111.0610)
[17] CMS Collaboration 2012 *CMS-PAS-TOP-12-014*.

5th International Workshop on Top Quark Physics (TOP2012) IOP Publishing
Journal of Physics: Conference Series **452** (2013) 012047 doi:10.1088/1742-6596/452/1/012047

Search for exotic same-sign dilepton signatures (b' quark, $T_{5/3}$ and four top quarks production) in 4.7 fb^{-1} of pp collisions at $\sqrt{s} = 7$ TeV with the ATLAS detector

Léa Gauthier
CEA-Saclay
On behalf of the ATLAS collaboration
E-mail: lea.gautheir@cern.ch

Abstract. We present a search for pair production of down-type heavy quarks ($b'\bar{b}'$), single and pair production of heavy top quark partners ($T_{5/3}q\bar{q}'$ and $T_{5/3}\bar{T}_{5/3}$, respectively), and production of events containing four top quarks ($tt\bar{t}\bar{t}$) in pp collisions at $\sqrt{s} = 7$ TeV recorded with the ATLAS detector at the LHC. For an integrated luminosity of 4.7 fb^{-1}, the corresponding lower limit on both the b' and $T_{5/3}$ mass is 0.67 TeV at 95% confidence level, when produced in pairs. When including the single production mechanism, limits on $T_{5/3}$ production are 0.68 TeV and 0.70 TeV for a coupling constant of the $tWT_{5/3}$ vertex equal to 1 and 3, respectively. The upper limit on the four top quarks production cross section is 61 fb at 95% confidence level.

1. Introduction

The Standard Model (SM) is extremely successful in explaining experimental measurements in particle physics, however it possesses some theoretical problems such as the Higgs boson mass fine-tuning [2] which have motivated a large number of extensions to the theory. Due to the large variety of models proposed, searches based on the signature of same-sign lepton pairs are often useful since they have a low background rate in the SM and potentially large contributions from new phenomena. In this note, three specific signal processes are considered: pair production of down-type heavy quarks, single and pair production of heavy top quark partners and production of events containing four top quarks, produced via a four-top-quark contact interaction. This note presents the first ATLAS [1] results on the experimental search on four-top-quarks production and the first search for both single and pair-production of $T_{5/3}$ [3].

2. Processes

2.1. Pair production of down-type heavy quarks (b')

A fourth generation of heavy chiral quarks is a natural extension to the SM [4]. This model is based on a new chiral generation: a $SU(2)_L$ doublet $(t', b')_L$ with the corresponding right-handed singlets t'_R, b'_R. It remains consistent with precision electroweak measurements if one assumes small mass splitting between the heavy up-type quark (t') and the heavy down-type quark (b') [4]. This model is disfavoured, however, by $m_H \sim 126$ GeV [5]. In the scenario studied in this note, the b' decays into the final states $u/c + W^-$ as well as $t + W^-$, as long as

$m_{b'} - m_t > m_W$. If one assumes $V_{u(c)b'} \ll V_{tb'}$ the dominant decay is $b' \to t + W^-$.

2.2. Single and pair production of heavy top quark partners ($T_{5/3}$)

This natural, non-supersymmetric solution to the hierarchy problem requires fermionic partners of the top quark [6]. These heavy fermions are coupled to SM quarks through a W boson preferentially to the third generation. We study here the pair and single production of the top-quark partner with electric charge $Q = 5/3$, denoted $T_{5/3}$. For single production, the cross section depends on the coupling constant λ of the $tWT_{5/3}$ vertex. In this analysis, we study the cases $\lambda = 1$, $\lambda = 3$ and $\lambda \ll 1$. In the latter case, the single production vanishes and the study corresponds to the pair production of b' quarks (same final state and acceptance).

2.3. Production of events containing four top quarks

Many new physics models couple strongly to the top quark, including a new interaction related to electroweak symmetry breaking that couples uniquely to the top quark, models where only the top quark is composite, or new coloured heavy particles that decay preferentially to top quarks. Many of these new physics models predict an enhanced rate of events containing four top quarks with respect to the SM production. In this analysis, only the contact interaction operator with right-handed top quarks is considered, as left-handed top operators are already strongly constrained by electroweak precision.

3. Background

3.1. Data-driven background

This background includes top-quark pair production, W+jets and single-top-quark production. It is divided into:

- charge mis-identification: the sign of the electric charge of the electron can be mis-reconstructed due to incorrect measurement of the sign of the track curvature or hard bremsstrahlung producing trident electrons ($e^\pm \to e^\pm\gamma^* \to e^\pm e^+ e^-$). The charge mis-identification of the muon is negligible. The rate of charge mis-identification for electrons is derived as a function of the electron pseudorapidity ($|\eta|$) from the rate of same-sign and opposite-sign electron pairs in data events from Z bosons. The dependence of this rate as a function of the transverse momentum (p_T) is much smaller than the variation with η and so is neglected in the parametrization of the rate.
- mis-reconstructed leptons: at least one of the two leptons in the selected same-sign pair is not a real isolated lepton. The leptons may come from the semi-leptonic decay of a b or a c hadron, π^0's or photons mis-reconstructed as leptons. To estimate this background, two sets of lepton selection criteria are defined, named *tight* [3] (same as standard selection) and *loose* (*tight* without isolation requirement). The probabilities r and f that a real or a fake loose lepton, respectively, passes the tight criteria are measured in dedicated control regions. The matrix method [7] is then used to estimate the number of events in the signal region with at least one fake lepton.

3.2. Monte Carlo background

Additional background estimates are derived using simulated Monte Carlo samples:

- di-boson production: WZ and ZZ take into account hard emission of up to three partons, the soft emission, showering and hadronization. Cross sections are normalised to next-to-leading order (NLO) theoretical calculations. $W\gamma$, with hard emission of up to five partons, are also considered and $W^\pm W^\pm$+2 jets is generated taken into account showering and hadronization.
- $t\bar{t}W$(+jets), $t\bar{t}Z$(+jets) and $t\bar{t}W^\pm W^\pm$ take into account showering and hadronization and

the $t\bar{t}W/Z(+\text{jet})$ are rescaled to NLO calculations.

4. Event selection

Jets are reconstructed using the anti-k_t [8] algorithm with a distance parameter of 0.4 and are required to satisfy $p_{\mathrm{T}} > 25$ GeV and $|\eta| < 2.5$. In order to avoid reconstructing electrons as jets, the latter are required to have an angular separation $\Delta\mathrm{R}(\text{jet},e)$ of at least 0.2. Jets from the decay of heavy flavour hadrons are selected by a multivariate b-tagging algorithm [9] at an operating point of 70% efficiency, which corresponds to a mis-tag rate of less than 1%. Electrons are required to have a transverse energy $E_{\mathrm{T}} > 25$ GeV and $|\eta_{\text{cluster}}| < 2.47$ excluding $1.37 < |\eta_{\text{cluster}}| < 1.52$). Electrons are also required to be well isolated from jets ($\Delta\mathrm{R}$(e, jet) > 0.4). Muons are required to have transverse momentum $p_{\mathrm{T}} > 20$ GeV and to fall within $|\eta| < 2.5$. Muons are also required to be well isolated from jets ($\Delta\mathrm{R}(\mu$, jet)> 0.4). Events that contain two muons which could originate from a cosmic muon are rejected, when the angle between them in the transverse plane is larger than 3.1 rad.

Events must contain at least two isolated leptons of the same charge with $p_{\mathrm{T}} > 25$ GeV for the leading lepton. In events with more than one same-sign pair, the pairs are sorted according to the leading lepton p_{T}, then by the sub-leading lepton p_{T}. At least one of the selected leptons must match the one that triggered the readout of the event. Events should also contain at least two jets, including at least one b-tagged jet and in the ee or $\mu\mu$ channel, the invariant mass of the two selected leptons must exceed 15 GeV and be out of the Z boson mass window, meaning $|m_{ll} - m_Z| > 10$ GeV. The transverse missing energy $E_{\mathrm{T}}^{\text{miss}}$ must be greater than 40 GeV and the total transverse energy H_{T}, defined as the scalar sum of the p_{T} of all leptons and jets, must exceed 550 GeV. This cut optimization is done with the extraction of the expected limit by taking into account the systematic uncertainties described in the following section.

5. Systematics

Several systematic uncertainties have been considered:

- Uncertainties affecting the Monte Carlo samples: uncertainties in the jet and lepton efficiency, energy or momentum calibration, and resolution lead to systematic uncertainties on the signal and background acceptance. Their effect is evaluated by varying each parameter independently within its uncertainty and recomputing the yield of each sample. The uncertainty on the luminosity has been estimated from van der Meer scans and is 3.9% [10]. Uncertainties on Monte Carlo background cross sections depend on the process.
- Uncertainties affecting the data-driven backgrounds: The uncertainty on the overall scale of the charge mis-identification background in the signal region is derived from a comparison of the mis-id rate extracted by three different methods. The uncertainties on the estimate of background coming from fake electrons was estimated by using different loose electron definitions while, for the muon case, it was estimated by changing the definition of the multi-jet enriched region.

6. Results

For each model, upper limits at 95% Confidence Level (CL) on the cross sections of the hypothetical processes are derived using the CL_s method [11]. We use single-bin counting experiments to extract the most likely signal cross section. Systematic uncertainties are included as variations in the expected signal and background yields, which are allowed to vary in the ensembles used to generate the CL_s distributions. Correlations of the systematic uncertainties across samples and channels are taken into account.

The expected and observed limits on the masses of the b' and $T_{5/3}$ and on the four top quark events production cross-section are shown in Table 1.

Table 1. Expected and observed limits on the masses of the b' and $T_{5/3}$ and on the four-top-quark events production cross-section.

Constrained parameter	Expected	Observed
b' mass or $T_{5/3}$ mass for $\lambda \ll 1$	> 0.64 TeV	> 0.67 TeV
$T_{5/3}$ mass for $\lambda = 1$	> 0.64 TeV	> 0.68 TeV
$T_{5/3}$ mass for $\lambda = 3$	> 0.66 TeV	> 0.70 TeV
Four top quark production cross-section	< 90 fb	< 61 fb

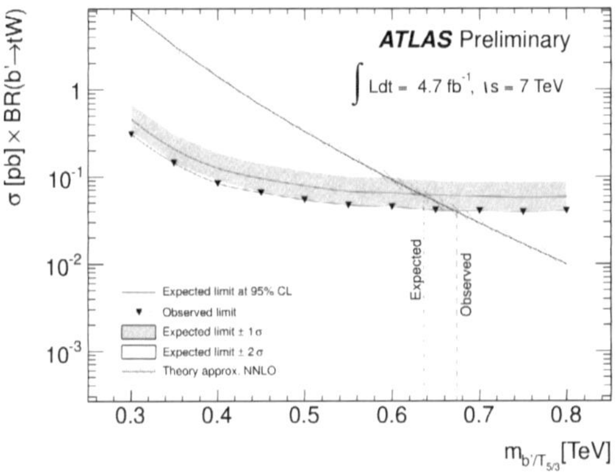

Figure 1. Expected and observed upper limits on the pair production cross section of the b' and $T_{5/3}$, as a function of their mass [3].

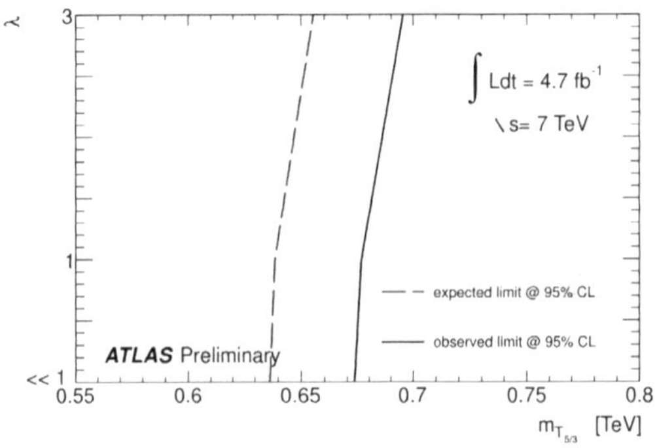

Figure 2. Expected and observed lower limits on the $T_{5/3}$ signal, as a function of the $T_{5/3}$ mass and the coupling constant λ. The shaded area is excluded at 95% CL [3].

7. Conclusion

A search for exotic processes in the same-sign dilepton plus jets signature with the ATLAS detector for an integrated luminosity of 4.7 fb^{-1} at $\sqrt{s} = 7$ TeV has been presented. The obtained limit on the b' and $T_{5/3}$ mass in the pair production is the most stringent limit to date in the same-sign dilepton channel. The limits obtained on the $T_{5/3}$ including the single production, and on four top quark event production are the first to date.

References

[1] ATLAS Collaboration, (2008) *JINST* 3 S08003
[2] E. Gildener, *Phys. Rev* D **14** (1976) 1667
[3] ATLAS Collaboration, ATLAS-CONF-2012-130 (2012)
[4] B. Holdom, *J. High Energy Phys* JHEP0608(2006)076
[5] ATLAS Collaboration, *Phys. Rev.* D **86** (2012) 074014
[6] R. Contino and G. Servant, *J. High Energy Phys* JHEP0806(2008)026
[7] ATLAS Collaboration, *Eur. Phys. J.* **C71** (2011) 1577
[8] M. Cacciari, G. P. Salam, and G. Soyez, *J. High Energy Phys* JHEP0804(2008)063
[9] ATLAS Collaboration, ATLAS-CONF-2012-043 (2012)
[10] ATLAS Collaboration, *Eur. Phys. J.* **C71** (2011) 1630
[11] T. Junk, *Nucl.Instrum.Meth.* **A434** (1999) 435443

Search for production of SUSY partners of the top quark with the CMS detector

Grégory Hammad on behalf of the CMS Collaboration
Groupe de Physique des Particules, Université de Mons, Belgium
E-mail: gregory.hammad@umons.ac.be

Abstract. The results of a search for production of supersymmetric partners of the top quark in proton-proton collisions at $\sqrt{s} = 7$ TeV at the LHC are presented. These results are based on a data sample, collected by the CMS detector in 2011, corresponding to 5 fb^{-1} of integrated luminosity. No sign of new physics is observed and exclusion limits on masses of supersymmetric particles are derived in the context of simplified models.

1. Introduction
In many supersymmetric extensions (SUSY) of the Standard Model, SUSY partners of the 3rd quark generation (stop, sbottom) are lighter than the partners of the 1st and 2nd generations, allowing production rates high enough to be observed at the LHC. Moreover, recent results on Higgs boson searches favour light stop particles [1]. As these 3rd generation SUSY partners should decay to heavy quarks (t, b), evidence for the existence of such SUSY partners may be observed by studying events with top quark pairs in the final state.

2. Experimental signature
In the context of R-parity conserving SUSY models, supersymmetric particles, which have an odd R-parity, are produced in pairs and lead to final states containing lightest SUSY particles (LSP) via cascade decays. The LSP are stable and are assumed to be weakly interacting massive particles. Experimentally, the non-detection of the LSP causes to an energy imbalance in the transverse plane with respect to the direction of the colliding particles. This imbalance, called missing transverse energy (MET), is a distinctive SUSY signature.

At the Large Hadron Collider, the CMS Collaboration [2] searched for the production of stop quark pairs either via gluino decay ($pp \to \tilde{g}\tilde{g} \to \tilde{t}_1 t \tilde{t}_1^* t$) or via direct production ($pp \to \tilde{t}_1 \tilde{t}_1^*$). As the top quark decays almost exclusively into a b quark and a W boson, the final state topology depends on the subsequent decay of the W boson ($W \to l\nu$ or $W \to q\bar{q}'$). Experimental signatures of stop production are thus characterised by:

- isolated leptons with high transverse momentum (p_T) in leptonic decay channels,
- large missing transverse energy (MET),
- jets from b-quarks.

The main algorithms used by the CMS Collaboration to identify b-jets are the simple secondary vertex tagger (SSV), the track counting tagger (TC) and the combined secondary vertex tagger (CSV) [3].

3. Search strategies

3.1. Search with same-sign dileptons (SS+b)

In the context of Supersymmetry, final states with two same-sign leptons may correspond to the production of gluino pairs decaying into $t\bar{t}\tilde{\chi}_0$ via an intermediate virtual or on-shell stop with subsequent decays of two same-sign top quarks via the leptonic channel $t \to bW \to bl\nu$. In this analysis [4], the event selection requires at least two isolated same-sign leptons with $p_T > 20$ GeV/c. In addition, at least two jets with $p_T > 40$ GeV/c are required, among which at least two are tagged as b-jets by the SSV tagger. To suppress background events from Z+jets production, events with any third isolated lepton forming an opposite-sign, same-flavour lepton pair with a mass compatible with the Z boson mass (± 15 GeV/c^2) are rejected. Remaining background events containing one or more fake leptons (i.e., mis-identified hadrons, electrons from photon conversion, etc.) are estimated from data using events with one or more leptons failing tight identification and isolation requirements but not looser ones. These events are then corrected for the probability for a fake lepton to pass the selection, obtained from events with a lepton candidate and a separated jet. The background contribution from opposite-sign dilepton events with one lepton having a mis-reconstructed charge is estimated using a control sample with opposite-sign dilepton events corrected for the charge misidentification probability derived from simulation and validated with a data sample containing lepton pairs whose invariant mass is compatible with the Z boson mass. Finally, events with genuine same-sign leptons from rare SM processes (e.g., ttW or ttZ) are estimated using simulation.

3.2. Search with single lepton final states

Two different analyses searched for stop production with a single lepton in the final state. Their search sensitivity was found to be enhanced by requiring a high b-tag jet multiplicity.

(e/μ, $\geq$ 2b, MET) analysis [5] : event selection requires one isolated lepton with $p_T > 20$ GeV/c, at least three jets with $p_T > 40$ GeV/c and MET > 100 GeV. In addition, the scalar sum of the selected jet transverse momenta, H_T, is required to be higher than 350 GeV/c. Identification of b-jets is made with the SSV algorithm. The main background sources after event selection are $t\bar{t}$ and W+jets production and are estimated separately by fitting the MET distribution with distinct exponential templates parametrized as a function of H_T. The free parameters of these templates are determined from data in a control region defined by $350 < H_T < 700$ GeV and $100 <$ MET < 400 GeV where the fraction of $t\bar{t}$ to W+jets events is derived using template fits to the observed b-tagged jet multiplicity. The fitted MET templates are then extrapolated at high value of H_T.

(e/μ, $\geq$ 3b, Y_{MET}) analysis [5] : events are selected by requiring one isolated lepton with $p_T > 20$ GeV/c, at least four jets with $p_T > 40$ GeV/c, MET > 60 GeV and $H_T > 375$ GeV/c. In addition, selected events are required to have at least one b-tagged jet, using the TC tagger. Background events, mainly $t\bar{t}$ and W+jets events, are estimated using a factorisation method in the H_T-Y_{MET} plane, with $Y_{\mathrm{MET}} = \mathrm{MET}/\sqrt{H_T}$. The signal region is defined by $H_T > 650$ GeV and $Y_{\mathrm{MET}} > 5.5\ \sqrt{\mathrm{GeV}}$. Correlation factors between H_T and Y_{MET} for the different background processes are derived from simulation. Background estimations are calculated per b-tagged jet multiplicity.

3.3. Search with fully hadronic final states

Three different analyses searched for stop production with multi-jet final states.

α_{T} analysis [6] : selected jets are combined into two pseudo-jets in such a way that their E_T difference, ΔH_T, is minimum. The α_T variable is then defined as $\alpha_T = 1/2(H_T - \Delta H_T)/(H_T^2 - \not{H}_T^2)^{1/2}$ with $\not{H}_T$ the magnitude of the vector sum of all the jet transverse momenta. For perfectly balanced dijet events with jets back-to-back in ϕ, α_T is equal to 0.5. For imbalanced events, this value becomes smaller. In principle, only dijet systems, with jets not back-to-back in ϕ, recoiling against genuine MET can have values greater than 0.5. Jets are required to have $E_T > 50$ GeV and events are selected if they have at least two jets with $E_T > 100$ GeV, the highest one in E_T being within the tracker acceptance. Events with forward jets ($|\eta| > 3$) are vetoed. The signal region is defined by $H_T > 275$ GeV and $\alpha_T > 0.55$. This region is then divided into eight variable-size bins in H_T and b-tagged jet multiplicity bins. For each bin, background event estimations are derived from control samples using transfer functions calculated with simulation.

$\mathbf{M_{T2}}$ analysis [7] : originally developed to measure the mass of the primary pair-produced sparticles, the M_{T2} variable is used in this analysis to discriminate SUSY events against SM background events. First, two *event hemispheres* are formed by merging the selected jets. Then the transverse mass m_T^i of each hemisphere i is calculated as a function of m_χ, the mass of the LSP and $p_T^{\chi(i)}$, the unknown LSP transverse momentum in hemisphere i. Finally, M_{T2} is defined as $M_{T2}(m_\chi) = \min_{p_T^{\chi(1)}+p_T^{\chi(2)}=p_T^{miss}} \left[max(m_T^{(1)}, m_T^{(2)})\right]$. A maximum difference of 70 GeV is imposed between MET and the magnitude of the vector sum of all the selected jet transverse momenta in order to reject events with large contributions from soft and/or forward jets to the momentum imbalance. Events with hard jets falling identification criteria in the central detector acceptance are also rejected. In order to reject further events with non back-to-back hemispheres, as expected in SUSY events, a minimum in azimuth ϕ between any selected jets and the MET, $\Delta\phi_{min.}(\text{jets, MET}) > 0.3$ is required. Events are selected by requiring at least four jets with $p_T > 150/100/40/40$ GeV/c, $H_T > 750$ GeV and MET > 30 GeV/c. In addition, events are required to have at last one b-tagged jet, using the SSVHE tagger. After event selection, the number of remaining multi-jet events is estimated from a side-band region defined with $\Delta\phi_{min.}(\text{jets, MET}) < 0.2$. The level of events with lost leptons are determined from single lepton data set corrected for the probability to loose a lepton, derived from simulation. Finally, the number of events with genuine MET from Z($\rightarrow \nu\nu$)+jets processes is estimated from a control sample with an isolated photon and corrected for the M_{T2}-dependent ratio of Z($\rightarrow \nu\nu$)+jets to γ+jets events obtained from simulation. The signal region is divided in several H_T and M_{T2} bins.

(MET+b) analysis [8] : events are selected by requiring at least three jets with $p_T > 50$ GeV and at least one identified b-jet using the CSV tagger. Events with at least one isolated lepton with $p_T > 10$ GeV/c are vetoed. The multi-jet background is estimated from a control sample defined by a cut on the minimum azimuthal opening angle between MET and the three jets with the highest p_T. Background from $t\bar{t}$ and W+jets processes are estimated using a MET template derived from a single lepton sample. Five signal regions are then defined based on the number of required b-tagged jets, using different minimum requirement on H_T and MET.

3.4. Inclusive search

A search for stop production was performed using the so-called *razor* variable R which is related to MET and to M_R, an event-by-event indicator of the sparticle mass [9]. It is observed that distributions of R and M_R for SM background events can be modelled by simple exponentials in the $R^2 - M_R$ plane. A dedicated control sample is then defined for each SM background process in order to get a data-driven estimation of the parameters of these exponential shapes. A complete model for the SM background processes is then built and used to fit a background dominated region in the $R^2 - M_R$ plane in order to derive the SM background yield from data. Finally, this background model is extrapolated to the signal region, at large R^2 and high M_R and the expected yields are then compared to the observed yields in different non overlapping signal regions. Events are classified according to the number of isolated lepton (0,1,2) and according to lepton flavours. Results are obtained for each category and then combined to test the compatibility of the background-only hypothesis with the data. In order to enhance the

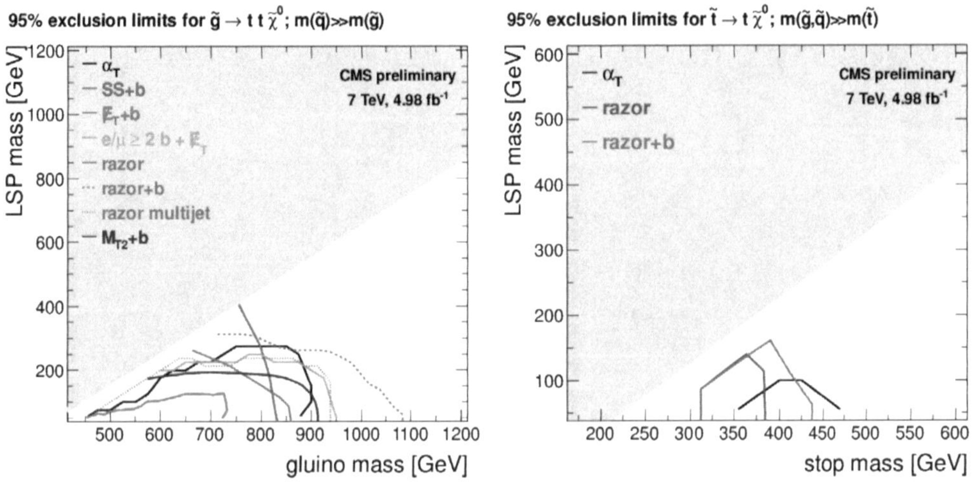

Figure 1. Exclusion limits at 95% C.L. on the mass parameters for simplified models of stop production, via gluino decay (left) and via direct production (right).

sensitivity to stop production, a second razor analysis, **razor+b** [10], used only events with at least one b-tagged jet (TCHE). Motivated by models predicting low mass stop decaying via $\tilde{t}_1 \to t + \tilde{\chi}_0$ and thus producing low to moderate MET, a third complementary analysis was developed (**razor multi-jet** [11]) for which the signal region is defined with a lower cut on R^2 compared to the other razor analyses. Only events with at least 6 jets ($p_T > 30$ GeV/c) are considered and at least one b-tagged jet (TCHE) is required for events in the signal region.

4. Results and conclusion

Using proton-proton collision data collected during 2011 at $\sqrt{s} = 7$ TeV, the CMS Collaboration searched for stop pair production in all decay channels. No significant excess of data was observed over the SM expectations. Exclusion limits on SUSY particle masses were then derived in several models, particularly on the gluino and stop masses in the context of simplified models [12]. As illustrated in Figure 1, simplified models for stop production via gluino decay are excluded at 95% C.L. for a gluino mass up to 1 TeV, assuming a LSP mass below 200 GeV. Simplified models for direct stop production are excluded at 95% C.L. for a stop mass comprised between 310 GeV and 420 GeV for a LSP mass below 100 GeV.

References

[1] Buckley M R and Hooper D 2012 *Phys.Rev.* **D86** 075008 (*Preprint* 1207.1445)
[2] Chatrchyan S *et al.* (CMS Collaboration) 2008 *JINST* **3** S08004
[3] Chatrchyan S *et al.* (CMS Collaboration) 2012 *CMS Physics Analysis Summary BTV-11-004*
[4] Chatrchyan S *et al.* (CMS Collaboration) 2012 *JHEP* **1208** 110 (*Preprint* 1205.3933)
[5] Chatrchyan S *et al.* (CMS Collaboration) 2012 *Submitted to Phys. Rev. D* (*Preprint* 1211.3143)
[6] Chatrchyan S *et al.* (CMS Collaboration) 2013 *JHEP* **1301** 077 (*Preprint* 1210.8115)
[7] Chatrchyan S *et al.* (CMS Collaboration) 2012 *JHEP* **1210** 018 (*Preprint* 1207.1798)
[8] Chatrchyan S *et al.* (CMS Collaboration) 2012 *Phys.Rev.* **D86** 072010 (*Preprint* 1208.4859)
[9] Chatrchyan S *et al.* (CMS Collaboration) 2012 *CMS Physics Analysis Summary SUS-12-005*
[10] Chatrchyan S *et al.* (CMS Collaboration) 2012 *Submitted to Phys. Rev. Lett.* (*Preprint* 1212.6961)
[11] Chatrchyan S *et al.* (CMS Collaboration) 2012 *CMS Physics Analysis Summary SUS-12-009*
[12] Alves D *et al.* (LHC New Physics Working Group) 2012 *J.Phys.* **G39** 105005 (*Preprint* 1105.2838)

Journal of Physics: Conference Series **452** (2013) 012049 doi:10.1088/1742-6596/452/1/012049

Measurement of Spin Correlations in Top Pair Events in the Dilepton Channels in pp Collisions at 7 TeV

Felix Höhle on behalf of the CMS Collaboration

III Inst. Phys. RWTH Aachen University

Abstract

We present a measurement of the spin correlation of top-antitop-pair events produced in pp collisions at $\sqrt{s} = 7\,\text{TeV}$. The data used by this analysis corresponds to an integrated luminosity of $5.0\,\text{fb}^{-1}$ collected with the CMS detector at the LHC. The spin correlation in top-antitop-pair events is extracted from a fit to the angular distribution between the two selected leptons. In the helicity basis, the correlation coefficient is found to be $\mathcal{A}_{\text{hel}} = 0.24 \pm 0.02 \pm 0.08$ and agrees with the Standard Model prediction.

1 Introduction

At the LHC, top quarks are produced abundantly, mainly in pairs by gluon fusion. The top quark mass has been measured to be $172.9 \pm 0.6 \pm 0.9\,\text{GeV}$ [1] and the width of the top quark is found to be $\Gamma_t = 2.0^{+0.7}_{-0.6}\,\text{GeV}$ [1]. This implies a lifetime much shorter than the hadronization time-scale, m_t/Λ^2_{QCD}. Therefore, the spin of the top quark at production is transferred to its decay products and can be accessed by the distribution of angular observables. This offers the opportunity to study spin correlations of the $t\bar{t}$ pairs [2]. The investigation of spin correlation in the $t\bar{t}$ system probes the bare quark at production, and by this the top pair production processes and perturbative QCD. At the LHC, in the regime of small invariant masses of the $t\bar{t}$ systems, the production is dominated by the fusion of like-helicity gluon pairs, resulting in top pairs with aligned orientations of top and anti-top spins (RR or LL), see Fig. 1. At higher invariant $t\bar{t}$ masses, the dominant production process switches to unlike-helicity gluons, producing $t\bar{t}$ pairs in anti-aligned orientations (RL or LR), which give identical configurations than the ones produced by $q\bar{q}$ annihilation at the Tevatron [3]. In the dileptonic decay ($t\bar{t} \rightarrow l^+\nu l^-\bar{\nu}b\bar{b}$), the charged leptons are correlated in azimuthal angle ($\Delta\phi_{l^+l^-}$) in the laboratory frame [3] and can be measured precisely by CMS without reconstructing the full event kinematics. CDF and D0 measured the strength of the $t\bar{t}$ spin correlation to be in agreement with the Standard Model expectation using template fit methods to angular distributions [4, 5, 6]. The ATLAS experiment reported a measurement of the spin correlation using $\Delta\phi_{l^+l^-}$ distributions and found it to be in agreement with next-to-leading order Standard Model predictions [7].

This analysis uses data corresponding to an integrated luminosity of $5.0\,\text{fb}^{-1}$ of proton-proton collisions at $\sqrt{s} = 7\,\text{TeV}$, provided by the LHC and recorded with the CMS detector [8]. The measurement of the spin correlation in $t\bar{t}$ events is compared to the Standard Model expectation [9].

2 Selection

This analysis focuses on dileptonically decaying (ee, $e\mu$, $\mu\mu$) top quark pairs. All events are triggered by double lepton triggers (e, μ) and contain two isolated high energy leptons ($p_{\text{T}} \geq 20\,\text{GeV}$). An additional Z-mass-veto of $|m_{ll} - m_{\text{Z}}| > 15\,\text{GeV}$ and a cut on $\not{E}_T \geq 40\,\text{GeV}$ is applied on same lepton flavor events to reject Drell-Yan events. All events are also required to have an invariant lepton pair mass above 20 GeV. Two jets are required with a transverse momentum above 30 GeV, and one of them has to be b-tagged. A comparison of the $\Delta\phi_{l^+l^-}$ distribution of the selected events with and without spin correlations is shown in Fig. 2.

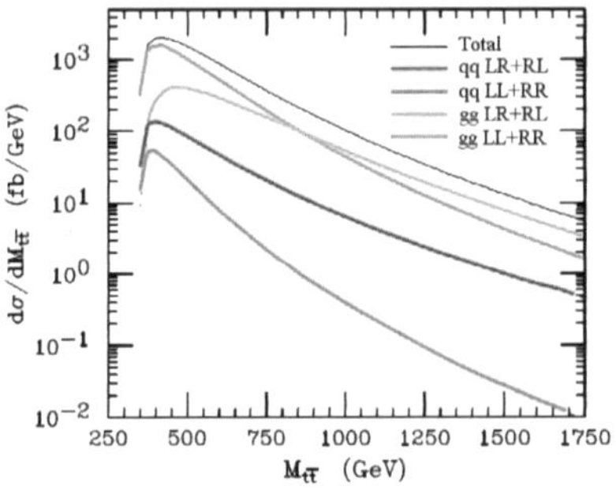

Figure 1: Differential cross sections of different helicity states and their productions mechanisms, [10].

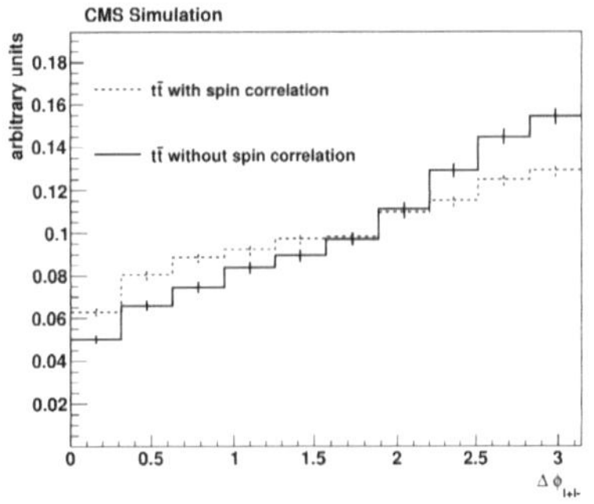

Figure 2: Simulated $\Delta\phi_{l^+l^-}$ distribution of selected events, with and without spin correlations.

3 Estimates of Background Processes

Two different kinds of background processes are estimated using data-driven methods. Firstly, the production of Drell-Yan events, which dominate the e^+e^- and $\mu^+\mu^-$ channel. It is estimated from the number of events in data inside the Z-mass-veto window, scaled by the ratio of events that fail and pass this selection. Secondly, non-prompt lepton background processes are estimated using a matrix method [11], or Barlow's event reweighting method [12]. Different dileptonic samples are constructed by loosening the lepton isolation and probabilities for prompt leptons and non-prompt leptons to be contained in these samples are introduced. The resulting system of linear equations has the number of events with two prompt-leptons ($t\bar{t}$ and Drell-Yan events) and the number of events with one or two prompt leptons as variables. Solving this system yields the expected contribution of W-like and QCD multijet events [13]. Other contributions, due to single top quark and diboson events, are estimated from simulation.

4 Systematic Uncertainties

Various systematic uncertainties are affecting the measurement. The dominant effects are listed below. Experimental systematic uncertainties include:

Lepton selection: Uncertainties on trigger and lepton selection (isolation and identification) affects the measurement and are estimated by changing the scale factors within their uncertainties.

Lepton energy scale: The uncertainty on the lepton energy scale, which affects mainly the p_{T} distribution of leptons, are estimated from comparisons between data and simulated Z events.

Jet energy scale and jet energy resolution: The uncertainty from the jet energy scale (JES) correction is estimated by variation of this scale within its uncertainties. The jet energy resolution (JER) is estimated similarly, by a variation depending on the η of the jet.

Background events: The uncertainty on the normalization of background processes is considered by varying the corresponding cross section by 100%.

Pileup (PU) modeling: The uncertainty of 8% on the proton-proton cross section at the LHC energy of $\sqrt{s} = 7\,\mathrm{TeV}$ influences the pileup modeling and is used to estimate the corresponding systematics.

FastSim vs. FullSim: The investigations of systematic uncertainties use a simplified detector simulation (FastSim). The observed discrepancies in respect to a detailed simulation (FullSim) are in the jet multiplicity distributions visible. This effect is accounted by comparing the $\Delta\phi_{l^+l^-}$ distributions between FastSim and FullSim.

The systematic uncertainties related to the modeling of $t\bar{t}$ signal events include:

Factorization and renormalization scales: The uncertainty related to factorization and renormalization scales is estimated by investigating dedicated MC@NLO predictions with in-/decreased scales.

Spin correlation and τ decay: Electrons and muons from τ decays are considered to be part of the signal, however the τ polarization is not taken into account in the simulation used in this analysis.

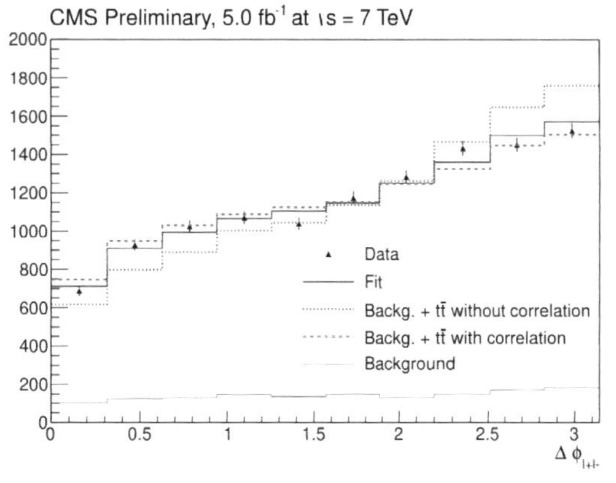

Figure 3: Simulated $\Delta\phi_{l^+l^-}$ distributions of selected events compared to data.

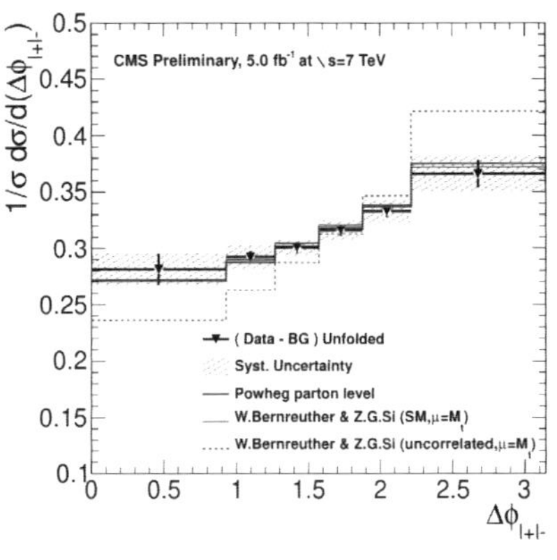

Figure 4: Unfolded $\Delta\phi_{l^+l^-}$ distributions of simulated and measured events after selection.

Therefore the effect of spin correlation is not properly propagated to its decay products. The corresponding uncertainty is estimated by comparing simulated events with and without excluding events containing at least one lepton from τ decay.

Top quark mass: The uncertainty on the top quark mass is estimated by using $t\bar{t}$ signal events with different top masses, assuming an uncertainty on the top mass of 2.5 GeV around $m_t = 172.5$ GeV.

PDF: The uncertainty related to PDF are estimated by a method described in Ref. [14].

5 Observable

The angular difference in the azimuthal plane between the two leptons coming from the W decays in $t\bar{t}$ events ($\Delta\phi_{l^+l^-} = |\phi_{l^+} - \phi_{l^-}|$) is sensitive to the spin correlation of the two top quarks. The strength of the spin correlation can be measured by the spin correlation coefficient, defined as the asymmetry between the number of $t\bar{t}$ events with spin orientation of top quarks aligned and anti-aligned, namely:

$$A = \frac{N(\mathrm{RR}) + N(\mathrm{LL}) - N(\mathrm{RL}) - N(\mathrm{LR})}{N(\mathrm{RR}) + N(\mathrm{LL}) + N(\mathrm{RL}) + N(\mathrm{LR})},$$

The spin correlation coefficient is estimated from the $\Delta\phi_{l^+l^-}$ distribution observed in data, by means of a template fit. Three different templates are considered: the simulated $\Delta\phi_{l^+l^-}$ distributions of $t\bar{t}$ events with and without Standard Model spin correlation between the top quarks, and a $\Delta\phi_{l^+l^-}$ distribution which describes the background events.

A binned likelihood fit method is used, to perform a simultaneous fit of the ee, $e\mu$ and $\mu\mu$ channels. All the events are fitted together in a single likelihood fit, but different templates are used according to the channels. The templates describing background processes are derived by simulation, and weighted according to data-driven estimates, Sec. 3, and the prediction based on simulation of single top quark and the diboson events.

Templates for $t\bar{t}$ signal events are derived from MC@NLO with and without Standard Model spin correlation. The numbers of $t\bar{t}$ signal events in each channel are parameters of the fit, as well as the fraction f of events with Standard Model spin correlation.

The fit result on data is shown in Fig. 3, combining the three channels. The number of $t\bar{t}$ signal events is measured to be 1718 ± 44 in the ee channel, 2132 ± 50 in the $\mu\mu$ channel and 6351 ± 85 in the $e\mu$ channel. The fraction of spin correlation f is measured to be 0.74 ± 0.08 (stat.).

The systematic uncertainties are estimated using pseudo-experiments based on simulated events, that are modified according to the investigated systematic effects. The average variation of the spin correlation is used to estimate the systematic uncertainty. The effects of the systematic sources on the measurement of f are presented in Tab. 1. The resulting measured value of f is 0.74 ± 0.08 (stat.) ± 0.24 (syst.).

Considering the top spin correlation predicted by MC@NLO, A^{simu}, the measured spin correlation coefficient is $f \times A^{simu}$. The spin correlation coefficient from MC@NLO $t\bar{t}$ signal events, in the helicity

Table 1: Uncertainties Δf on the fraction of events with spin correlation, as predicted by the fit.

uncertainty Δf	absolute	relative
statistical	0.08	11%
MC statistical	0.07	9%
	experimental	
lepton selection	0.01	1%
lepton energy scale	0.01	1%
JES/JER	0.02	3%
all backgrounds	0.07	9%
pileup	0.02	3%
b-tagging	0.01	1%
	$t\bar{t}$ modeling	
FastSim vs. FullSim	0.06	8%
fact. and renorm. scales	0.15	20%
τ decay	0.12	16%
top mass	0.02	3%
PDF	0.07	9%

basis, is estimated at matrix element level and found to be $A_{hel}^{simu} = 0.33$. This value of the spin correlation coefficient is close to the standard model prediction (NLO calculation) of $A_{hel}^{SM} = 0.31$ from Ref. [15], while this measurement yields a spin correlation coefficient of

$$A_{\text{hel}}^{\text{meas}} = 0.24 \pm 0.02\,(\text{stat.}) \pm 0.08\,(\text{syst.})\,.$$

References

[1] Particle Data Group Collaboration, "Review of particle physics", *J. Phys. G* **37** (2010 and 2011 partial update for the 2012 edition) 075021. doi:10.1088/0954-3899/37/7A/075021.

[2] M. Beneke et al., "Top Quark Physics", *CERN-TH/2000-100*.

[3] G. Mahlon and S. Parke, "Spin Correlation Effects in Top Quark Pair Production at the LHC", *Phys. Rev. D* **81** (2010) 074024. doi:10.1103/PhysRevD.81.074024.

[4] CDF Collaboration, "Measurement of $t\bar{t}$ Spin Correlation in $p\bar{p}$ Collisions Using the CDF II Detector at the Tevatron", *Phys. Rev. D* **83** (2011) 031104. doi:10.1103/PhysRevD.83.031104.

[5] D0 Collaboration, "Measurement of Spin Correlation in $t\bar{t}$ Production Using a Matrix Element Approach", *Phys. Rev. Lett.* **107** (2011) 032001. doi:10.1103/PhysRevLett.107.032001.

[6] D0 Collaboration, "Measurement of spin correlation in $t\bar{t}$ production using dilepton final states", *Phys.Lett.* **B702** (2011) 16–23. doi:10.1016/j.physletb.2011.05.077.

[7] ATLAS Collaboration, "Observation of Spin Correlation in $t\bar{t}$ Events from pp Collisions at $\sqrt{s} = 7$ TeV Using the ATLAS Detector", *Phys. Rev. Lett.* **108** (May, 2012) 212001. doi:10.1103/PhysRevLett.108.212001.

[8] CMS Collaboration, "The CMS experiment at the CERN LHC", *JINST* **0803** (2008) S08004. doi:10.1088/1748-0221/3/08/S08004.

[9] CMS Collaboration, "Measurement of Spin Correlations in ttbar production", *CMS Physics Analysis Summary* **TOP-12-004** (2012).

[10] G. Mahlon et al., "Angular correlations in top quark pair production and decay at hadron colliders", *Phys. Rev. D* **53** (1996) 4886–4896. doi:10.1103/PhysRevD.53.4886.

[11] D0 Collaboration, "Measurement of the $t\bar{t}$ production cross section in $p\bar{p}$ collisions at $\sqrt{s} = 1.96$ TeV using secondary vertex b tagging", *Phys.Rev. D* **74** (2006) 112004. doi:10.1103/PhysRevD.74.112004.

[12] R. J. Barlow, "Event classification using weighting methods", *J.Comput.Phys.* **72** (1987) 202–219. doi:10.1016/0021-9991(87)90078-7.

[13] CMS Collaboration, "Measurement of the $t\bar{t}$ production cross section in the dilepton channel in pp collisions at $\sqrt{s} = 7$ TeV", *CMS Physics Analysis Summary* **TOP-11-005** (2011).

[14] M. Botje et al. http://www.hep.ucl.ac.uk/pdf4lhc/.

[15] W. Bernreuther and Z.-G. Si, "Distributions and correlations for top quark pair production and decay at the Tevatron and LHC", *Nuc. Phys. B* **837** (2010) 90–121. doi:10.1016/j.nuclphysb.2010.05.001.

Journal of Physics: Conference Series **452** (2013) 012050 doi:10.1088/1742-6596/452/1/012050

Top-quark mass measurements at LHC: a new approach

S. Alioli[1], P. Fernández[2], J. Fuster[2], A. Irles[2], S. Moch[3 4], P. Uwer[5], M. Vos[2]

[1] LBNL & UC Berkeley, 1 Cyclotron Road, Berkeley, CA 94720, USA
[2] IFIC, Universitat de València and CSIC, Catedrático Jose Beltrán 2, E-46980 Paterna, Spain
[3] II. Inst. für Theoretische Physik, Universität Hamburg, Luruper Chaussee 149, D-22761 Hamburg, Germany
[4] DESY, Platanenallee 6, D-15738 Zeuthen, Germany
[5] Humboldt-Universität zu Berlin, Newtonstrasse 15, D-12489 Berlin, Germany

E-mail: adrian.irles@ific.uv.es

Abstract. We present a new method to measure the top-quark mass in high energetic hadron collisions at the LHC. We study the mass dependence of the production of top-quark pairs in association with an additional jet. The cross section of tt+1Jet production is sensitive to the top-quark mass since gluon radiation depends on the top-quark mass through threshold and cone effects. In particular we study the normalised tt +1Jet cross section differential in the invariant mass of the final state jets. We have investigated the sensitivity of the method together with the impact of various theoretical and experimental uncertainties. We find that the method has the potential to be competitive with existing methods. We emphasize that in the proposed method the mass parameter can be uniquely defined through one-loop renormalization.

1. Introduction

The top quark is the heaviest elementary particle with a mass of 173.2 ± 0.9 [1]. Due to its large mass the top quarks present the strongest coupling to the Higgs boson within the framework of the Standard Model (SM). For that, the top quark become an ideal laboratory for detailed tests of the Higgs mechanisms and of many alternative explanations of spontaneous symmetry breaking (EWSB) in Beyond Standard Models (BSM) theories. Top-quark mass comparison with the mass of the recently observed resonance [2, 3] —assuming that the resonance is the long-sought Higgs boson—can be used to test the validity of the Standard Model [4] and the structure of the electroweak vacuum [5, 6].

Quarks, in contrast to leptons, are colored objects that feel the strong interactions. As a consequence, quarks doesn't exist as free particles as they are confined into hadrons. Since quark masses are in general not observables by themselves their precise values depend on the renormalization scheme used to define them. In the case of heavy quark masses the most commonly used definitions are the pole mass m_q^{pole} and the running mass $\overline{m}_q(\mu_r)$.

The top-quark mass is presently inferred by the kinematical reconstruction of the invariant mass of its decay products (see e.g., Ref. [7]) or by its relation to NLO defined observables like the inclusive top-quark pair production cross section [8]. The extracted mass from kinematical measurements reach to precisions of ~ 1 GeV. Mass determinations using the inclusive cross section Ref. [8] presents a weak sensitivity on the top-quark mass $\frac{\Delta\sigma}{\sigma} \sim -5\frac{\Delta m_t}{m_t}$ but in this measurement the renormalization scheme is unambiguously defined in difference with the determination based on the kinematical reconstruction.

In this work we advocate a new method to measure the top-quark mass in high energetic hadron collisions at the LHC looking for a high precision measurement with a renormalization scheme

Journal of Physics: Conference Series **452** (2013) 012050 doi:10.1088/1742-6596/452/1/012050

unambiguously defined. The mass dependence of the production of top-quark pairs in association with an additional jet is exploited.

2. Top-quark pair production in association with a hard jet at NLO accuracy in QCD

The NLO QCD corrections for $t\bar{t}+1\text{-jet}+X$ have been presented in Refs. [9, 10] for Tevatron and LHC (14 TeV) conditions. The results share all the attractive features of theoretical predictions including radiative corrections at higher orders. The cross section for $t\bar{t}+1\text{-jet}+X$ is theoretically very well under control and well suited for precision measurements.

In this document we have updated the results of Refs. [9, 10] to 7 TeV proton proton collisions: see Table 1

Table 1. The $t\bar{t}+1\text{-jet}+X$ cross section using LO and NLO calculations [9, 10] for proton-proton collisions at 7 TeV and for different m_t^{pole} values. Jets are defined using the anti-kt algorithm [11] with R=0.4 as implemented in the FASTJET package [12]. The additional jet is required to have $p_T > 50$ GeV and $|\eta| < 2.5$. The uncertainty due to the limited statistics of the numerical calculation is indicated in parenthesis affecting the last digit. The scale uncertainty is also shown for some top-quark mass values ($\mu = m_t^{\text{pole}}/2 < \mu < \mu = 2m_t^{\text{pole}}$). The CTEQ6.6 [13] (CT09MC1 [14]) PDF set has been used to obtain the NLO (LO) results.

	$\sigma_{t\bar{t}+1\text{-jet}}$ [pb]	
m_t^{pole} [GeV]	LO	NLO
165	57.615(4)	52.25(9)
170	$49.910(3)^{+30}_{-17}$	$45.45(6)^{+1}_{-6}$
175	45.372(3)	39.46(6)

Finally, the uncertainties originating from the PDFs have been studied by comparing the results obtained with the PDF sets CTEQ6.6 and MSTW2008nlo90cl [15]. For $m_t^{\text{pole}} = 170$ GeV we find, for example $\sigma_{t\bar{t}+1\text{-jet}}^{\text{NLO, MSTW08}} = 49.21$ pb

Very recently predictions for $t\bar{t}+1$-jet productions matched with parton shower predictions (PS) have been provided [16, 17]. We have used the results from Refs. [18, 17] to investigate the effect of the parton shower and allowing for a more realistic study closer to what will be done in the experimental analysis. The obtained results are in good agreement with the previously mentioned.

The various studies mentioned above show, that the theoretical description of the process $t\bar{t}+1$-jet based predictions at NLO accuracy in QCD is well under control.

3. Top-quark mass measurements with $t\bar{t}+1$-jet events

The mass sensitivity of the $t\bar{t}+1$-jet cross section $\sigma_{t\bar{t}+1\text{-jet}}$ is very similar to the inclusive $t\bar{t}$ cross section as it is illustrated on Table 1 and a measurement of the top quark mass from it not lead to any significant improvement. Since inclusive cross sections are in general difficult to measure, we propose to study normalized differential distributions. Evidently distributions contain more information and may be more sensitive to the mass parameter. Furthermore, due to the normalization many experimental and theoretical uncertainties cancel between numerator and denominator. In order to enhance the mass sensitivity of $t\bar{t}+1$-jet events we need to focus on specific kinematical configurations. A natural observable to look at is the (normalized) differential distribution of the $t\bar{t}+1$-jet cross section with respect to the invariant

Journal of Physics: Conference Series **452** (2013) 012050 doi:10.1088/1742-6596/452/1/012050

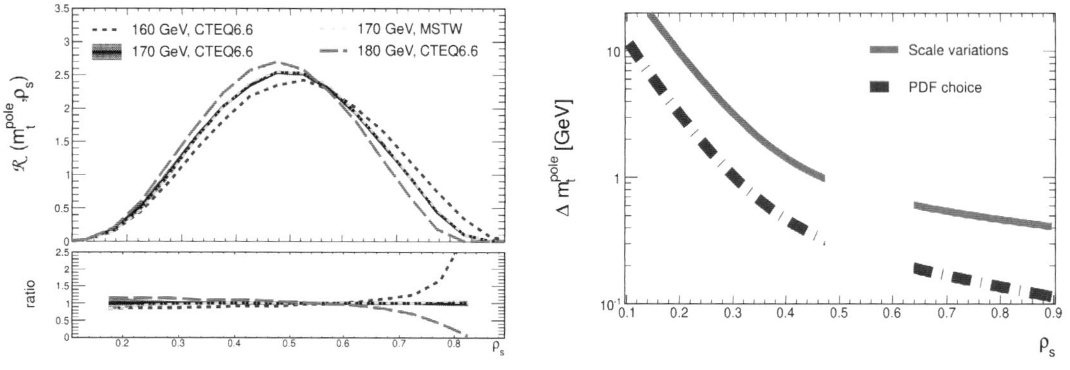

Figure 1. Left: $\mathcal{R}(m_t^{\text{pole}}, \rho_s)$ calculated at NLO accuracy for different masses $m_t^{\text{pole}} = 160$, 170 and 180 GeV. For $m_t^{\text{pole}} = 170\,\text{GeV}$ the scale and PDF uncertainties evaluated as discussed in the text are shown. The ratio with respect to the result for $m_t^{\text{pole}} = 170\,\text{GeV}$ is shown in the lower plot.
Right: Expected impact of scale (magenta line) and PDF (blue dashed line) uncertainties on the measured top-quark mass value. The region where $\mathcal{R}$ is essentially insensitive to the top-quark mass is not shown.

mass squared $s_{t\bar{t}j}$ of the final state. More precisely we study the dimensionless distribution

$$\mathcal{R}(m_t^{\text{pole}}, \rho_s) = \frac{1}{\sigma_{t\bar{t}+1\text{-jet}}} \frac{d\sigma_{t\bar{t}+1\text{-jet}}}{d\rho_s}(m_t^{\text{pole}}, \rho_s), \tag{1}$$

where ρ_s is defined as $\rho_s = \frac{2m_0}{\sqrt{s_{t\bar{t}j}}}$ with a fixed value for $m_0 = 170$ GeV.

In this study, the pole mass scheme has been chosen. The study of the convergence of the theoretical uncertainties coming from the PDFs or from the truncation of the perturbative series have been performed following the same criteria of the previous section. To investigate the sensitivity of the distribution $\mathcal{R}$ to the top-quark mass we have calculated $\mathcal{R}$ for different m_t^{pole}. The result is shown in Figure 1 (left). The three curves need to cross since the area under each curve is normalized to one. The crossing happens at $\rho_s \approx 0.55$. At this point the distribution is essentially insensitive to the top-quark mass. For ρ_s close to one we expect that the production of heavier quark masses is suppressed compared to lighter masses. For large energies we observe that the mass dependence is small as one would naively expect. From Figure 1 we conclude that a significant mass dependence can be observed for $0.3 < \rho_s < 0.5$ and $0.6 < \rho_s$.

The impact ofthe scale variations or the PDF choice on the extraction of m_t^{pole} using these distributions has been evaluated. This is summarized in Figure 1 (right). In both cases we see that $\Delta m_t^{pole} < 1$ GeV. The main source of uncertainty come from the scale variation while the impact of the PDF uncertainties is much smaller.

To investigate further the impact of higher order corrections and the effect of the parton shower we have compared the predictions for $\mathcal{R}(m_t^{\text{pole}}, \rho_s)$ using different approximations: LO, $t\bar{t}$@NLO + POWHEG [18], $t\bar{t}+1$-jet @NLO + POWHEG [17]. These comparisons show that the $\mathcal{R}$ distribution exhibit a high theoretical stability as far as different approaches give compatible results for the $\mathcal{R}$ distribution.

We are thus lead to the conclusion that from the theoretical perspective $\mathcal{R}$ provides an interesting alternative to existing methods for top-quark mass measurements.

4. Experimental viability study

We study now the stable final state particles originated from typical $t\bar{t}+1\text{-jet}+X$ events as produced in proton-proton interactions at 7 TeV center-of-mass energy. The results presented here illustrates qualitatively the viability and the potential of the method for *à la LHC* environment. In any case, a real experimental determination will necessarily need a more detailed and careful detector analysis.

This study only considers the so called semi-leptonic decay channel which assumes that one of the two W boson decays leptonically (just consider electron or muon channels) whereas the remaining W boson decays hadronically.

Here and in the Table 2 we briefly summarize effects that have been studied: the event generator and fragmentation model: POWHEG with Pythia's parton shower versus MC@NLO [19] with Herwig's parton shower [20, 21];backgrounds, mainly QCD and W+jets; the impact of a wrongly reconstructed jet energy; the unfolding procedure to correct the $\mathcal{R}$ distribution to the perturbative partonic level; different modelling of color reconnection; and the statistical error depending on the collected luminosity.

Table 2. Summary of the experimental uncertainties studied

Source of uncertainty	Result
Event Gen. & fragm. model	$\Delta m_t^{\text{pole}} \sim 0.2 \pm 0.2$ GeV $(\rho_s > 0.6)$
Backgrounds	$\sim 5\%$ of the events
Jet reconstruction	$\Delta m_t^{\text{pole}} \sim 0.8$ GeV $(\rho_s > 0.6)$ assuming a 3% for the jet energy scale uncertainty
Unfolding	$\Delta m_t^{\text{pole}} \sim 0.3$ GeV $(\rho_s > 0.6)$
Color Reconnection	$\Delta m_t^{\text{pole}} < 0.25$ GeV at $95\% CL$ $(\rho_s > 0.6)$
Statistics	$\Delta m_t^{\text{pole}} \sim 1.4$ GeV per $5\, fb^{-1}$ $(\rho_s > 0.6)$

It is important to say that a real analysis using data and detector specific tools is needed to understand the exact value of the uncertainties affecting the determination of m_t^{pole} but we can estimate that a total error around of 1 GeV is achievable.

5. Conclusions

A new method to measure the top-quark mass at hadron colliders (specifically at LHC) has been presented here. This method is based on a differential distribution of the $t\bar{t} + 1$-jet cross section with respect $\sim 1/s_{t\bar{t}j}$. In this analysis the renormalization scheme of the top-quark mass is uniquely defined. Also, this method, presents a well mass sensitivity and low uncertainties coming form uncalculated higher order corrections and from the parton distribution functions. This uncertainties are estimated to be below 1 GeV. Finally, the preliminary study of the experimental viability shows that precisions of ~ 1 GeV are achievable. This study has not been addressed to any particular experiment, so a real analysis using data and detector specific framework is needed.

6. References

[1] J. Beringer, et al., Journal of Physics G **33**, 1+ (2006)
[2] G. Aad, et al., Phys.Lett. **B716**, 1 (2012). DOI 10.1016/j.physletb.2012.08.020
[3] S. Chatrchyan, et al., Phys.Lett. **B716**, 30 (2012). DOI 10.1016/j.physletb.2012.08.021
[4] S. Heinemeyer, W. Hollik, D. Stockinger, A. Weber, G. Weiglein, JHEP **0608**, 052 (2006). DOI 10.1088/1126-6708/2006/08/052
[5] G. Degrassi, S. Di Vita, J. Elias-Miro, J.R. Espinosa, G.F. Giudice, et al., JHEP **1208**, 098 (2012). DOI 10.1007/JHEP08(2012)098
[6] S. Alekhin, A. Djouadi, S. Moch, Phys.Lett. **B716**, 214 (2012). DOI 10.1016/j.physletb.2012.08.024
[7] T. Aaltonen, et al., Phys.Rev.D (2012)
[8] U. Langenfeld, S. Moch, P. Uwer, Phys.Rev. **D80**, 054009 (2009). DOI 10.1103/PhysRevD.80.054009

[9] S. Dittmaier, P. Uwer, S. Weinzierl, Phys.Rev.Lett. **98**, 262002 (2007). DOI 10.1103/PhysRevLett.98.262002
[10] S. Dittmaier, P. Uwer, S. Weinzierl, Eur.Phys.J. **C59**, 625 (2009). DOI 10.1140/epjc/s10052-008-0816-y
[11] M. Cacciari, G.P. Salam, G. Soyez, JHEP **0804**, 063 (2008). DOI 10.1088/1126-6708/2008/04/063
[12] M. Cacciari, G.P. Salam, G. Soyez, Eur.Phys.J. **C72**, 1896 (2012). DOI 10.1140/epjc/s10052-012-1896-2
[13] P.M. Nadolsky, et al., Phys. Rev. **D78**, 013004 (2008). DOI 10.1103/PhysRevD.78.013004
[14] H.L. Lai, J. Huston, S. Mrenna, P. Nadolsky, D. Stump, et al., JHEP **1004**, 035 (2010). DOI 10.1007/JHEP04(2010)035
[15] A. Martin, W. Stirling, R. Thorne, G. Watt, Eur.Phys.J. **C63**, 189 (2009). DOI 10.1140/epjc/s10052-009-1072-5
[16] A. Kardos, C. Papadopoulos, Z. Trocsanyi, Phys.Lett. **B705**, 76 (2011). DOI 10.1016/j.physletb.2011.09.080
[17] S. Alioli, S.O. Moch, P. Uwer, JHEP **1201**, 137 (2012). DOI 10.1007/JHEP01(2012)137
[18] S. Frixione, P. Nason, G. Ridolfi, JHEP **0709**, 126 (2007). DOI 10.1088/1126-6708/2007/09/126
[19] S. Frixione, B.R. Webber, JHEP **0206**, 029 (2002)
[20] G. Corcella, I. Knowles, G. Marchesini, S. Moretti, K. Odagiri, et al., JHEP **0101**, 010 (2001)
[21] G. Corcella, I. Knowles, G. Marchesini, S. Moretti, K. Odagiri, et al., (2002)

Journal of Physics: Conference Series **452** (2013) 012051 doi:10.1088/1742-6596/452/1/012051

Mass and rapidity dependent top quark forward-backward asymmetry in the effective Lagrangian approach

Dong-Won Jung,[1] P. Ko,[2] and Jae Sik Lee[3]

[1,2] School of Physics, KIAS, Seoul 130-722, Republic of Korea

[3] Department of Physics, Chonnam National University, 300 Yongbong-dong, Buk-gu, Gwangju, 500-757, Republic of Korea

E-mail: [1]`dwjung@kias.re.kr`, [2]`pko@kias.re.kr`, and [3]`jslee@jnu.ac.kr`

Abstract. We study the invariant mass and rapidity dependent top quark forward-backward asymmetry from the effective Lagrangian viewpoint. The Wilson coefficients are constrained by the experimental observations and the concrete models that reproduce the low energy effective Lagaragians are considered. Some of them are disfavored and others relatively favored. For each cases, we estimate the appropriacy of the effective Lagrangian approach.

1. Introduction

A few years ago, the Tevatron collabroration reported the top forward-backward (FB) asymmetry ($A_{\rm FB}$) and since then it has been drawing much interest from high energy physics community. In addtion to the integrated asymmetry of $A_{\rm FB}(\rm CDF) = 0.158 \pm 0.074$ [1] [1], the mass and rapidity dependent asymmetries are also reported and the most recent results are given by the CDF collaboration [4].

In the previous publications, we performed the detailed analyses on the integrated asymmetry with the effective Lagarangian approach [5, 6] and also on the rapidity and invariant mass dependent asymmetries [7]. In this paper, we extend our analysis in Ref. [7] with updated data, also including the consideration of more general cases and their validities.

2. Effective Lagarangian

If the new physis scale is higher than the energy scale considered, we can adopt the effective Lagrangian to describe the low energy phenomena. Since no resonance has been observed at the Tevatron energy scale, we adopted the effective Lagrangian approach to draw the information on the new physics which might be responsible for top quark $A_{\rm FB}$. Most generally we can write down the dimension-6 operators for $q\bar{q} \to t\bar{t}$ [5, 6]:

$$\mathcal{L}_6 = \frac{g_s^2}{\Lambda^2} \sum_{A,B} \left[C_{8q}^{AB} (\bar{q}_A T^a \gamma_\mu q_A)(\bar{t}_B T^a \gamma^\mu t_B) \right] . \qquad (1)$$

[1] The most recent values are $A_{\rm FB}(\rm CDF) = 0.162 \pm 0.042$ [2], $A_{\rm FB}(\rm D0) = 0.196 \pm 0.065$ [3]. There are still around 2-σ deviation from the standard model calculation.

With shorthand notations of $C_1 \equiv C_{8q}^{LL} + C_{8q}^{RR}$ and $C_2 \equiv C_{8q}^{LR} + C_{8q}^{RL}$, $A_{\rm FB}$ must be expanded with $\sigma_{NLO}^{int}, \sigma_{NP}^{int}$. In that case, the FB asymmetry must be written

$$A_{FB} \simeq A_{FB}^{SM} + \Delta\sigma_{NP}^{int}/\sigma_0^{SM}. \tag{2}$$

Then the deviation from the SM are expressed as

$$\Delta\sigma_{t\bar{t}} \equiv \sigma_{t\bar{t}} - \sigma_{t\bar{t}}^{\rm SM} \propto (C_1 + C_2), \tag{3}$$
$$\Delta A_{\rm FB} \equiv A_{\rm FB} - A_{\rm FB}^{\rm SM} \propto (C_1 - C_2). \tag{4}$$

We will use $A_{\rm FB}({\rm CDF}) = 0.158 \pm 0.074$ in order to fix the effective couplings and predict the $M_{t\bar{t}}$ and Δy dependent $A_{\rm FB}$ for those C_i's within 1-σ range. We found that $C_1\,(1\,{\rm TeV}/\Lambda)^2$ and $-C_2\,(1\,{\rm TeV}/\Lambda)^2$ take values between ~ -0.5 and ~ 2.5, see Fig. 1 in Ref. [6] for updated results.

3. FB asymmetries as functions of invariant mass $M_{t\bar{t}}$ and rapidity difference Δy

It will be instructive to start this section with old wisdom of electroweak interaction. The FB asymmetry of the muon pair was measured at PETRA [8], the center of mass energy was far below the Z^0 pole mass. One can observe clear FB asymmetry by the parity violating weak interaction, which is absent in QED. More explicitly, $A_{\rm FB}$ of muon pair far below the Z^0 mass can be approximated as

$$A_{FB}(s) \simeq -\frac{3G_F}{\sqrt{2}}\frac{s}{4\pi\alpha}\,(g_L - g_R)^2 \equiv kG_F s, \tag{5}$$

which is negative definite, a generic feature of the new vector boson with universal couplings to the initial and the final fermions and antifermions. The PETRA measurement of $A_{FB}(s)$ in the region far below the Z^0 pole is that the $A_{\rm FB}(1200{\rm GeV}^2) \simeq -0.1$, which can be translated into

$$k = -7.18,$$

compared with the SM prediction: $k = -5.78$. Note that we can get the rough size of k (or $(g_L - g_R)^2/M_Z^2$) only from the interference term between the QED photon and the Z^0 boson exchanges in the limit $s \to 0$ (near threshold), if $s \ll M_Z^2$.

Similar things can be done for $q\bar{q} \to t\bar{t}$ process. In this case with the dimension-6 operator the asymmetry is

$$\widehat{A}_{\rm FB}(M_{t\bar{t}}) = \frac{\hat{\beta}_t \frac{\hat{s}}{\Lambda^2}(C_1 - C_2)}{\frac{8}{3}\left[1 + \frac{\hat{s}}{2\Lambda^2}(C_1 + C_2)\right] + \frac{16\hat{s}}{3m_t^2}\left[1 + \frac{\hat{s}}{2\Lambda^2}(C_1 + C_2)\right]} \simeq \frac{3\hat{\beta}_t \frac{\hat{s}}{\Lambda^2}(C_1 - C_2)}{8 + 16\frac{\hat{s}}{m_t^2}}. \tag{6}$$

In any case, the whole point is that the FB asymmetry near the threshold is approximately linear in $\hat{s}$ modulated by $\hat{\beta}_t = \sqrt{1 - 4m_t^2/\hat{s}}$ with a small slope parameter that could have either sign depending on $(C_1 - C_2)$, namely the underlying new physics affecting $q\bar{q} \to t\bar{t}$.

Now we are in the place to present the numerical results. In the first place, we consider the case either of C_i is zero. It is reasonable point in the parameter space since many of the models predict only one of Wilson coefficients are nonzero. See the Table(1) of Ref. [5]. The allowed ranges by the integrated FB asymmetry then become $C_1 \sim [\,0.15, 0.9\,]$ and $C_2 \sim [\,-0.67, -0.15\,]$ with C_2 or C_1 is zero for either case. The results are presented in Fig. 1. One can see that for models that produce nonzero values for either C_1 or C_2 it is impossible to fit all data altogether. They fails to generate right behaviors of FB asymmetry as functions of invariant mass and rapidity. They cannot fit the two-bin data either though the points satisfy the integrated

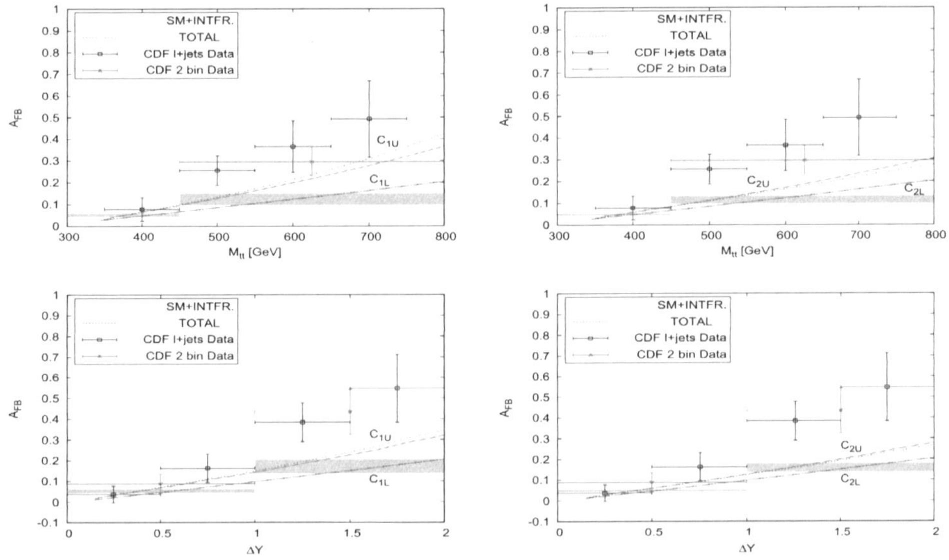

Figure 1. *Top FB asymmetry as functions of* $M_{t\bar{t}}$ *(upper) and* Δy *(low). In the left frames we are taking* C_1 *in the range between* $C_{1L} = 0.15$ *and* $C_{1U} = 0.97$ *with* $C_2 = 0$. *In the right frames, we vary* C_2 *in the range between* $C_{2L} = -0.15$ *and* $C_{2U} = -0.67$ *with* $C_1 = 0$. *We have taken* $\Lambda = 1$ *TeV in both cases. In each frame, the two bands are for* $A_{\rm FB}$ *in the lower and higher* $M_{t\bar{t}}$ *or* Δy *bins varying* C_1 *(left) and* C_2 *(right) in the ranges delimited by* $C_{1L,1U}$ *and* $C_{2L,2U}$*, respectively, and the dots for the CDF data with errors. In the solid (red) lines, we include only the SM contribution and the one from the interference between the SM and NP amplitudes while the effects of* $(NP)^2$ *term have been added in the dotted (blue) lines.*

asymmetry. So we can conclude that such models can be excluded if we trust the Tevatron observation in the sense that they fail to reconstruct the consistent prediction compared to data.

Next, let us consider the more general case of nonzero values for both C_1 and C_2. We assign the constraint that $C_1 + C_2 = 0$, which makes no contribution to the cross section. (See Eq. (3)) It is reasonable assumption since the variation of the cross section is smaller than that of $A_{\rm FB}$. In addition to that, we further assume that $C_{8q}^{RR} = C_{8q}^{LL} = \frac{1}{2}C_1$ and $C_{8q}^{RL} = C_{8q}^{LR} = -\frac{1}{2}C_1$ to minimize the the NP^2 contribution which spoil the validity of the effective Lagrangian approach.

The results are collected in Fig.(2). Unlike the previous results, now we can fit the data. Actually, this parameter set corresponds to usual axigluon models. As shown in the figure, experimental results are fit in consistent way this time. Interesting point is that this holds for rather larger values of integrated asymmetry. In other words, if we restrict the integrated asymmetry in the lower half range of 1-σ result, then the model cannot be accommodated to the observations. The parameter set which provide the consistent values for all the variables prefers the larger value of integrated asymmetry, roughly $A_{\rm FB} \sim [0.158, 0.158 + 0.074]$, to the smaller values of integrated asymmetry, namely, $A_{\rm FB} \sim [0.158 - 0.078, 0.158]$. The validity of this assertion can be tested if more experimental results are accumulated. Unfortunately it cannot be done and we only can tell about rather vague tendency instead of giving strict statements. Conservatively speaking, we can assert that the parameter set of $C_1 + C_2 = 0$ fit the data well with preference for the larger value of integrated asymmetry.

Final comments are in order. We should always be careful for treating the problem from

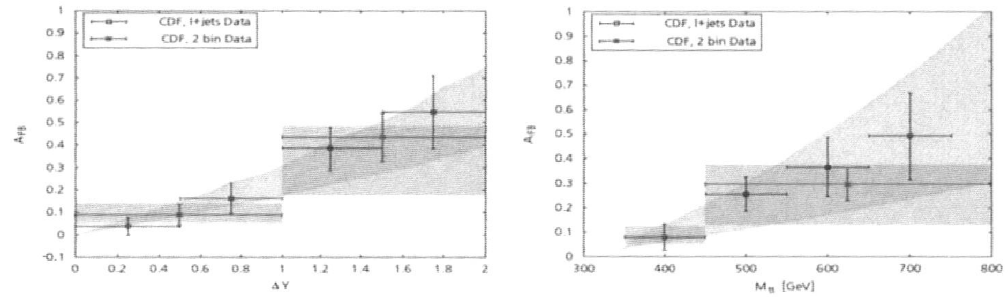

Figure 2. *Same as Fig.(1), with $C_1 + C_2 = 0$. Experimental results are added in the figures in the same ways.*

the effective Lagrangian method. In that case we should make it sure that the higher order contributions are small enough not to spoil the validity of the nonrenormalizable operators. For that purpose we estimate the relative size of the NP^2 over the SM value. The ratio goes from 0.08% up to at most 35% of the SM value in the large invariant mass region, and we can state that the analysis remains robust in most range. Similar behavior takes place in the rapidity dependent asymmetry.

4. Summary

We perform the analyses on the invariant mass and rapidity dependent top forward-backward asymmetries from the effective Lagrangian viewpoint. With the non-observation of any new resonances at the Tevatron, such method is efficient way to extract the new physics information that might be responsible for FB asymmetry. We show that generically the models that give the nonzero value for only one of C_1 and C_2 cannot reproduce the experimental observations in consistent way in the sense that they fail to fit the slope and two-bin values. If we relax the constraint and assign the axigluon-like condition as $C_1 + C_2 = 0$ then we can fit the data. Even in that case large integrated FB asymmetry are disfavored. More study including the LHC data may be necessary to clarify the experimental status of various model in relation with the information extracted by the effective Lagrangian approach at the Tevatron.

Acknowledgements

This work is supported in part by Basic Science Research Program through NRF 2011-0022996 and in part by NRF Research Grant 2012R1A2A1A01006053.

References

[1] T. Aaltonen *et al.* [CDF Collaboration], . Rev. D **83**, 112003 (2011) [arXiv:1101.0034 [hep-ex]].
[2] T. Aaltonen *et al.* [CDF Collaboration], arXiv:1211.1003 [hep-ex].
[3] V. M. Abazov *et al.* [D0 Collaboration], Phys. Rev. D **84** (2011) 112005 [arXiv:1107.4995 [hep-ex]].
[4] The CDF Collaboration, Conf. Note 10807 (2012).
[5] D. -W. Jung, P. Ko, J. S. Lee and S. -h. Nam, Phys. Lett. B **691** (2010) 238 [arXiv:0912.1105 [hep-ph]].
[6] D. -W. Jung, P. Ko and J. S. Lee, Phys. Lett. B **701** (2011) 248 [arXiv:1011.5976 [hep-ph]].
[7] D. -W. Jung, P. Ko and J. S. Lee, Phys. Lett. B **708** (2012) 157 [arXiv:1111.3180 [hep-ph]].
[8] S. L. Wu, Phys. Rept. **107** (1984) 59-324.

Measurement of the single-top-quark t-channel cross section in pp collisions at $\sqrt{s} = 7$ TeV using a multivariate analysis

Dennis Klingebiel for the CMS collaboration
III. Physikalisches Institut A, RWTH Aachen University, Germany
E-mail: klingebiel@physik.rwth-aachen.de

Abstract. Electroweak top-quark production offers a unique access to search for new physics phenomena. The measurement of the t-channel cross section allows to determine the Cabibbo-Kobayashi-Maskawa matrix element $|V_{\mathrm{tb}}|$ with high precision and without assuming its unitarity. We present the precise measurement of the t-channel single-top-quark production cross section with a multivariate analysis. Data collected with the CMS experiment in proton-proton collisions at a center-of-mass energy of 7 TeV, corresponding to 1.17 fb^{-1} for muon final states and 1.56 fb^{-1} for electron final states, are used for this measurement. The measured t-channel single-top-quark production cross section is 67.2 ± 6.1 pb, compatible with the approximate next-to-next-to-leading-order Standard Model prediction. The CKM matrix element $|V_{\mathrm{tb}}|$ is measured to be 1.020 ± 0.046 (meas.) ± 0.017 (theor.).

1. Introduction

Single top quarks can be produced via electroweak interactions. Three production modes are distinguished based on the virtuality of the exchanged W boson. In proton-proton collisions at a center-of-mass energy of 7 TeV, the dominant production mode is the t-channel with an approximate next-to-next-to-leading-order (NNLO) Standard Model (SM) cross section prediction of $64.6^{+2.1}_{-0.7}{}^{+1.5}_{-1.7}$ pb for a top-quark mass of $m_t = 172.5\,\mathrm{GeV}/c^2$ [1]. Figure 1 shows the leading order Feynman diagrams for t-channel single-top-quark production. The precise

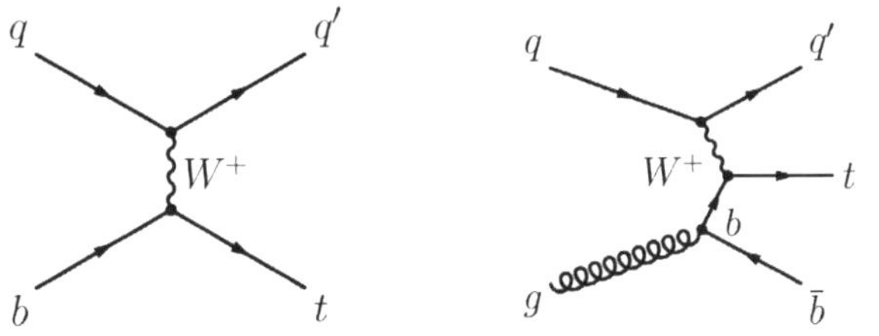

Figure 1. Feynman diagrams for t-channel single-top-quark production in leading order for the $2 \rightarrow 2$ (left) and $2 \rightarrow 3$ (right) processes.

measurement of the t-channel cross section allows a unique access to the measurement of the Cabibbo-Kobayashi-Maskawa matrix (CKM matrix) [2; 3] element $|V_{\mathrm{tb}}|$ without assuming unitarity of the CKM matrix, as the t-channel production cross section is proportional to $|V_{\mathrm{tb}}|^2$. In the CMS experiment [4], two complementary analysis approaches with three independent analyses are followed to measure the t-channel single-top-quark quark cross section in proton-proton collisions at 7 TeV [5]. One approach exploits the pseudo-rapidity distribution of the recoil jet in a signal-enriched phase space ($|\eta_{j'}|$ analysis). The other approach exploits multivariate analysis techniques and various signal- or background-enriched categories. Either Neural Networks (NN) or Boosted Decision Trees (BDT), have been used to exploit the prior knowledge of the electroweak top-quark production mechanism and to probe the SM top-quark event topology of the candidate events. Various categories are defined on the multiplicity of jets and b-tagged jets in each event are used to enhance the signal acceptance, constrain systematic uncertainties in-situ, and to gain confidence in the background modelling. This approach allows a most precise measurement of the signal cross section. All three analyses have been combined using the best linear unbiased estimator (BLUE) [6]. In this article, we summarize the analysis strategy and the results of one of the multivariate analyses, the BDT analysis, while the $|\eta_{j'}|$ analysis is described in detail elsewhere in these conference proceedings [7]. Events with leptonic decays of the top quark into a muon or electron, a W boson, and a b quark are considered ($\mathrm{t} \rightarrow \mathrm{bW} \rightarrow \mathrm{b}\ell\nu$ (ℓ = e or μ)). The data used in this analysis were collected with the CMS detector in proton-proton collisions at a center-of-mass energy of 7 TeV. The data corresponds to 1.17 fb^{-1} for the muon final states and and 1.56 fb^{-1} for the electron final states.

2. Analysis strategy

The BDT analysis was developed in a "semi-blind" way. First, the validity of the description of physics objects kinematics and properties was checked in signal-depleted control regions. Furthermore, the agreement between data and simulation in various kinematic variables, especially the input variables for the multivariate analysis and the classifier output, was confirmed and continuously checked in signal-depleted phase space. Here, both W boson plus jets and top quark dominated control regions were used. Second, the signal region was unblinded.

The BDT analysis measures the signal cross section in 12 orthogonal categories based on the flavour of the charged lepton (e, μ), the number of selected jets n-jets $= 2, 3, \geq 4$, and the number of b-tagged jets m-btags $= 1, \geq 2$. Furthermore, the 0-tag categories are used to check the modelling of input variables and multivariate discriminator output.

The t-channel signal is simulated with the POWHEG next-to-leading-order (NLO) generator [8–10] interfaced to PYTHIA 6.4.24 [11]. Data are recorded if a trigger is fired by a single isolated muon with a transverse momentum $p_{\mathrm{T}} > 17$ GeV/c or a single isolated electron with $p_{\mathrm{T}} > 27$ GeV/c, or in later runs by a single isolated electron with $p_{\mathrm{T}} > 25$ GeV/c and a b-tagged jet with $p_{\mathrm{T}} > 30$ GeV/c. The simulated events are reweighted to fit the number of additional interactions (pileup) to the conditions of collected data.

The event selection requires exactly one isolated muon with $p_{\mathrm{T}} > 20$ GeV/c and $|\eta| < 2.1$ or one isolated electron with $p_{\mathrm{T}} > 30$ GeV/c and $|\eta| < 2.5$. Charged leptons are required to be isolated to reject non-prompt electrons, i.e. to further reduce the QCD multijet background. For electrons, an additional conversion rejection is applied. Additionally, at least two jets with $p_{\mathrm{T}} > 30$ GeV/c and $|\eta| < 4.5$ are required. One striking feature of t-channel single-top-quark production is the very forward jet originating from the light quark recoiling against the top-quark. Most of the signal is expected to have two jets with one b-tagged jet, since the b-quark out of the initial gluon splitting (fig. 1 right) is collinear with the proton remnant, or has too low transverse momentum to be reconstructed. B-tagging of jets is applied using a track counting high purity algorithm at the tight working point [12]. Furthermore, a transverse W boson mass $M_T(W) > 40$ GeV/c^2 [5] is required for events with a muon in the final state and $\not{E}_{\mathrm{T}} > 35$ GeV for events with an electron in the final state.

A QCD multijet model is obtained by inverting the isolation criterion in data.

In order to discriminate between signal and background events, Boosted Decision Trees as implemented in TMVA [13] are used. A set of 11 most important variables out of the 37 variables used in Ref. [14] is chosen based on their discrimination power. For the given analysis, this set is as minimalistic as possible and provides good statistical sensitivity. The chosen variables include kinematics of reconstructed final state and composite objects, and angular correlations between them. The full list of used input variables can be found in Ref. [5]. The description of the input variables is checked in signal-depleted control regions as described in the beginning of this section. Several BDTs are constructed with the Adaptive Boosting algorithm [13], each reweighting the event weight of misclassified events by a certain factor. With the weighted majority vote of the whole forest of trees, the BDT classifier output as one powerful discriminator is calculated.

The resulting BDT discriminator output distribution for the muon-channel in the "2-jets 1-btag" signal category is shown in fig. 2 (left). For events with two jets, the main background is W boson plus jets, while the dominant background for events with three jets is top quark pair production. The resulting BDT discriminator output distribution for the sum of the "2-jets 2-btags", "3-jets 2-btags", "4-jets 1-btag", and "4-jets 2-btags" signal-depleted categories is shown in fig. 2 (right).

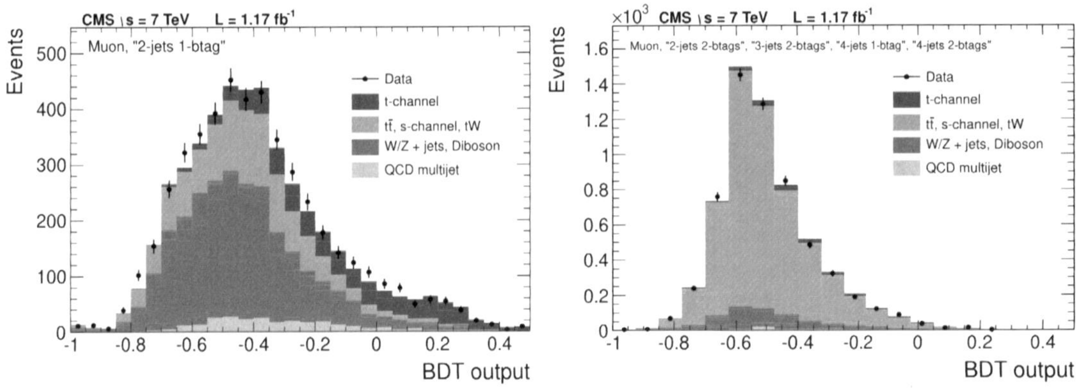

Figure 2. Distributions of the BDT discriminator output for events in a signal category "2-jets 1-btag" with a muon final state (left), and for a signal-depleted control regions, the sum of muon-final-state events in the categories "2-jets 2-btags", "3-jets 2-btags", "4-jets 1-btag", and "4-jets 2-btags" (right). Simulated signal and background contributions are scaled to the best fit results.

For the statistical evaluation, a bayesian core method is used. The implementation of the statistical model and its evaluation is done with the THETA package [15]. Experimental uncertainties and uncertainties of theoretical predictions on the background production cross sections are represented as nuisance parameters. Variations of the jet energy scale, jet energy resolution, b-tagging efficiency, muon and electron trigger and reconstruction efficiencies, pileup conditions, $\not{E}_T$ modelling and luminosity are considered as experimental uncertainties. The nuisance parameters itself are marginalized using a Markov Chain Monte Carlo technique. Correlations of nuisance parameters are taken into account. An over-constraint of the statistical model might arise from the limited knowledge of the W boson plus jets rate in the dedicated phase space of this analysis. Thus, a total of 18 nuisance parameters for the W boson plus jets processes are introduced. These depend on the flavour of the additional partons (b, c, or udsg partons) and the number of reconstructed jets in an event. Gaussian or log-normal priors are used for nuisance parameters and a flat prior for the signal strength. The median of the resulting signal strength posterior is used as the central measurement result. The [16%, 84%] quantiles are used as the uncertainty interval.

Theoretical uncertainties due to the factorization and renormalization scale, matching thresholds for samples simulated with MADGRAPH, parton distribution function, and description of the signal modelling are externalized, i.e. not modelled as additional nuisance parameters within the statistical model. Instead, their impact on the measured cross section is evaluated by generating pseudo-experiments with varied templates and re-doing the default signal extraction procedure, in order to avoid that the statistical model artificially over-constrains those systematic uncertainties. An over-constraint might arise when uncertainties cannot be fully parametrized within the statistical model, as it is often the case for theoretical uncertainties. The dominant systematic uncertainties are the signal modelling, the W boson plus jets background rate prediction, and the simulation statistics.

The CKM matrix element $|V_{\mathrm{tb}}|$ is calculated using the approach of Feldman and Cousins [16].

3. Results

The t-channel single-top-quark production cross section in the BDT analysis is measured to be

$$\sigma_{t\text{-ch.}} = 66.6^{+7.0}_{-6.6}\ (\text{stat.} + \text{syst.} + \text{lum.})^{+6.4}_{-3.5}\ (\text{theor.})\ \text{pb} \quad (\text{muons}),$$
$$\sigma_{t\text{-ch.}} = 66.4^{+8.4}_{-7.9}\ (\text{stat.} + \text{syst.} + \text{lum.})^{+5.4}_{-5.4}\ (\text{theor.})\ \text{pb} \quad (\text{electrons}),$$

and the combination of both channels is

$$\sigma_{t\text{-ch.}} = 66.6 \pm 4.0\ (\text{stat.}) \pm 3.3\ (\text{syst.})^{+3.9}_{-3.3}\ (\text{theor.}) \pm 1.5\ (\text{lum.})\ \text{pb}.$$

where "stat." refers to statistical uncertainties, "syst." to systematic uncertainties, "lum." to uncertainties of the luminosity measurement, and "theor." to externalized theoretical uncertainties as defined in the previous section.

The results of this analysis are combined with the two other analyses $|\eta_{j'}|$ and NN [5] using the BLUE method, which is applied iteratively. The obtained χ^2 corresponds to a p-value of 0.90, i.e. the individual analyses are compatible with each other. The correlations of statistical and systematic uncertainties between the three analyses are taken into account [5]. Correlations are varied, resulting in negligible changes of the combined cross section.

The combined measurement of the single-top-quark t-channel cross section is

$$\sigma_{t\text{-ch.}} = 67.2 \pm 3.7\ (\text{stat.}) \pm 3.0\ (\text{syst.}) \pm 3.5\ (\text{theor.}) \pm 1.5\ (\text{lum.})\ \text{pb}\ ,$$

which is in agreement with the approximate next-to-next-to-leading-order (NNLO) SM prediction of $\sigma^{\text{th}}_{t\text{-ch.}} = 64.6^{+2.1}_{-0.7}{}^{+1.5}_{-1.7}$ pb [1].

The CKM matrix element is measured to be $|f_{L_V} V_{\mathrm{tb}}| = \sqrt{\sigma_{t\text{-ch.}}/\sigma^{\text{th}}_{t\text{-ch.}}} = 1.020 \pm 0.046\ (\text{meas.}) \pm 0.017\ (\text{theor.})$, assuming only that the top-quark decays into a W boson and a b quark ($\mathrm{t} \to \mathrm{bW}$) with a branching ratio of 1. f_{L_V} is a general anomalous coupling form factor. Assuming SM electroweak couplings, i.e. $f_{L_V} = 1$ and $|V_{\mathrm{tb}}| \leq 1$, we measure $0.92 < |V_{\mathrm{tb}}| \leq 1$ at 95% confidence level.

4. Conclusions

Two analysis approaches have been followed to measure the t-channel single-top-quark production cross section with the CMS experiment. The results of all analyses are compatible with each other. The t-channel single-top-quark production cross section is measured in pp collisions to be 67.2 ± 6.1 pb at a center-of-mass energy of $\sqrt{s} = 7$ TeV, compatible with the approximate NNLO Standard Model prediction of $64.6^{+2.1}_{-0.7}{}^{+1.5}_{-1.7}$ pb [1]. The CKM matrix element $|V_{\mathrm{tb}}|$ is measured to be $1.020 \pm 0.046\ (\text{meas.}) \pm 0.017\ (\text{theor.})$.

Figure 3 shows the comparison of this t-channel cross section measurement [5] with dedicated measurements at the Tevatron [17; 18], ATLAS [19], and with expectations from Quantum chromodynamics. The SM expectation is computed at NLO with MCFM in the 5-flavour scheme [20] and at approximate NNLO [1].

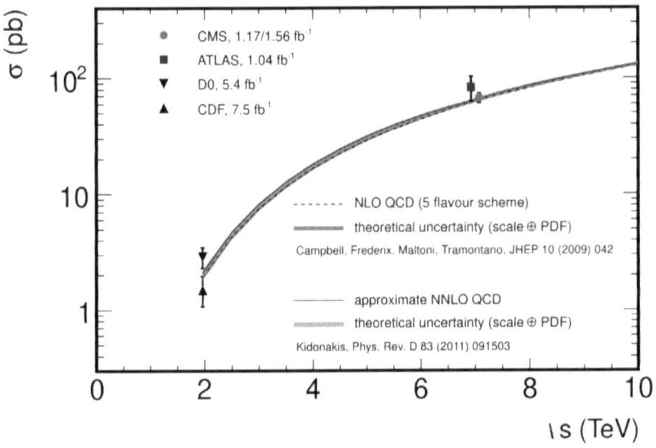

Figure 3. t-channel single-top-quark cross section vs. centre-of-mass energy. The error band (width of the curve) of the SM expectation calculation includes variations of the cross section due to the top-quark mass uncertainty [21], the parton distribution function [22], and the coherent variations of factorisation and renormalisation scales by a factor of two or 0.5.

References

[1] Kidonakis N 2011 *Phys. Rev. D* **83** 091503 (*Preprint* 1103.2792)
[2] Cabibbo N 1963 *Phys. Rev. Lett.* **10** 531
[3] Kobayashi M and Maskawa T 1973 *Prog. Theor. Phys.* **49** 652
[4] Chatrchyan S *et al.* (CMS) 2008 *JINST* **03** S08004
[5] Chatrchyan S *et al.* (CMS Collaboration) 2012 *JHEP* **1212** 035 (*Preprint* 1209.4533)
[6] Lyons L, Gibaut D and Clifford P 1988 *Nucl. Instr. and Meth. A* **270** 110
[7] Merola M 2013 *these conference proceedings*
[8] Alioli S, Nason P, Oleari C and Re E 2010 *JHEP* **06** 043 (*Preprint* 1002.2581)
[9] Alioli S, Nason P, Oleari C and Re E 2009 *JHEP* **09** 111 (*Preprint* 0907.4076)
[10] Frixione S, Nason P and Oleari C 2007 *JHEP* **11** 070 (*Preprint* 0709.2092)
[11] Sjöstrand T, Mrenna S and Skands P Z 2006 *JHEP* **05** 026 (*Preprint* hep-ph/0603175)
[12] CMS Collaboration (CMS) 2011 b-jet identification in the CMS experiment CMS Physics Analysis Summary CMS-PAS-BTV-11-004 URL http://cdsweb.cern.ch/record/1427247
[13] Hoecker A, Speckmayer P, Stelzer J, Therhaag J, von Toerne E and Voss H 2007 TMVA: Toolkit for Multivariate Data Analysis (*Preprint* physics/0703039)
[14] Chatrchyan S *et al.* (CMS) 2011 *Phys. Rev. Lett.* **107** 091802 (*Preprint* 1106.3052)
[15] Müller T, Ott J and Wagner-Kuhr J 2012 THETA – a framework for template-based statistical modeling and inference Preprint IEKP-KA-CMS/2012-1 URL http://www-ekp.physik.uni-karlsruhe.de/pub/web/cms/iekp-ka-cms2012-1.pdf
[16] Feldman G J and Cousins R D 1998 *Phys. Rev. D* **57** 3873 (*Preprint* physics/9711021)
[17] Abazov V M *et al.* (D0) 2011 *Phys. Lett. B* **705** 313 (*Preprint* 1105.2788)
[18] Aaltonen T *et al.* (CDF) 2009 *Phys. Rev. Lett.* **103** 092002 (*Preprint* 0903.0885)
[19] Aad G *et al.* (ATLAS) 2012 *Phys. Lett. B* **717** 330 (*Preprint* 1205.3130)
[20] Campbell J M, Frederix R, Maltoni F and Tramontano F 2009 *JHEP* **10** 042 (*Preprint* 0907.3933)
[21] CDF and D0 Collaborations 2009 Combination of CDF and D0 measurements of the single top production cross section (*Preprint* 0908.2171)
[22] Campbell J M, Huston J W and Stirling W J 2007 *Rept. Prog. Phys.* **70** 89 (*Preprint* hep-ph/0611148)

Journal of Physics: Conference Series **452** (2013) 012053 doi:10.1088/1742-6596/452/1/012053

Combined measurement of the $t\bar{t}$ cross section and the b-tagging efficiency using l+jet events with CMS

M. Maes on behalf of the CMS Collaboration
Vrije Universiteit Brussel, Inter-university Institute for High Energies
Brussels, Belgium

E-mail: michael.maes@cern.ch

Abstract. The Large Hadron Collider, build at CERN, is the first collider in the world able to act as a top quark factory. The large amounts of top quark events produced allow us to study the properties of this heavy quark in great detail. We propose a combined measurement of the $t\bar{t}$ cross section and the b-tagging efficiency using the lepton+jets event topology observed by the CMS experiment in the 8 TeV proton collisions of the LHC. The data collected in 2012 used for this measurement corresponds to an integrated luminosity of 2.8 fb^{-1} for muon+jets events and 2.7 fb^{-1} for electron+jets events. The obtained $t\bar{t}$ cross section is 228.4 $\pm$ 9.0 (stat.) $^{+29.0}_{-26.0}$ (syst.) $\pm$ 10.0 (lumi) pb.

1. Introduction

The measurement of the top quark pair cross section at the LHC [1] is important for multiple reasons. The measurement on its own is a test of the Standard Model as well as a benchmark for the theoretical calculations. The result is also relevant for those searches beyond the Standard Model for which the top quark pair processes are the dominant background. In the decay of top quarks bottom quarks are present as the CKM element V_{tb} is constrained to be close to unity. Therefore the top quark pair signal, two b quarks and two W bosons in the final state, can be isolated from the background in the proton collisions at the LHC by use of b-quark identification algorithms or so-called b-tagging [2]. Although the application of b-tagging will purify the selected sample, it will also introduce important systematic uncertainties due to our understanding of the performance of these b-tagging algorithms. With a combined measurement of the top quark pair cross section with the b-tagging efficiency we will be able to remove this additional uncertainty. We present the potential of performing this combined measurement in a maximally data-driven way at the CMS [3] experiment. The b-tagging efficiency will be estimated from a specific sample which is enhanced in b-quark jets. In a data-driven way the non b-quark jets will be subtracted, and the measured distribution of the b-tag variable will be obtained for true b-quark jets. The b-tagging efficiency at different working points or thresholds can be determined from this distribution. Using the same event sample, but after applying a b-tag cut in the event selection, the top quark pair cross section will be obtained from a template fit on a kinematic variable which differentiates between the top quark process and the main background.

The measurement will be applied on the semi-leptonic decaying top quark pairs, namely $t\bar{t} \rightarrow bWbW \rightarrow bqqb\ell\nu_{\ell}$. The lepton is either a muon or an electron. Both studied decay modes account for about 15% of the total branching ratio of top quark pairs.

2. Event selection and reconstruction

The results are obtained on a dataset of about $2.8 fb^{-1}$ proton-proton collisions recorded at the CMS experiment at a centre-of-mass energy of 8 TeV. A trigger is used requiring an isolated muon with a transverse momentum (p_T) exceeding 20GeV/c or an electron with a p_T exceeding 25 GeV/c. Moreover, the trigger requires the events to contain at least three jets with a p_T above 30 GeV/c.

Then the events are reconstructed using the ParticleFlow algorithm [4] and required to contain an isolated lepton. This lepton is either a muon with p_T exceeding 26GeV/c within a pseudorapidity of $|\eta| < 2.1$ or an electron with p_T exceeding 30GeV/c within a pseudorapidity of $|\eta| < 2.5$, both passing a series of tight identification criteria. Furthermore, we require at least four jets in the event with $p_T >$ 45,45,35 and 35 GeV/c respectively within the acceptance of $|\eta| < 2.5$. Finally events are rejected if they contain one or more looser leptons. No missing transverse energy ($\not{E}_T$) is used.

3. Measuring the $t\bar{t}$ cross section

The $t\bar{t}$ production cross section is measured using a template fit on a the invariant mass of the lepton and leptonic b-jet candidate ($M_{\ell b}$), shown in Figure 1. To reconstruct $M_{\ell b}$ distribution, the b quark jet in the $t \to b\mu\nu_\mu$ decay should be identified among all selected jets. A χ^2 sorting algorithm is applied using as input the top quark and W boson masses and widths from simulation to points out the b-jet candidate in each event. Since the $M_{\ell b}$ variable is sternly dependent on the $t\bar{t}$ decay kinematics it has proven to have good discriminate between $t\bar{t}$ events and the various backgrounds. The templates for $t\bar{t}$ and background distributions are obtained from simulation, after applying the event selection outlined in the previous section together with a requirement of medium b-tag on the jet associated to the leptonically decaying top quark, and are fitted on the distribution from data. A b-tag algorithm deriving a probability based on the 2D track impact parameter significance of the tracks inside the jet [2] is applied with a cut corresponding to a miss tag rate of 1%. The template fit results in a number of observed $t\bar{t}$ events. From this number, using Equation (1), the event selection efficiency and b-tagging efficiency, the cross section is obtained. The b-tagging efficiency entering the cross section calculation is measured simultaneously with the $t\bar{t}$ cross section.

$$\sigma_{t\bar{t}} = \frac{N_{t\bar{t}}}{L} = \frac{N_{t\bar{t}^{obs}}}{L} \times \frac{1}{\epsilon_{sel}\epsilon_{\chi^2_{cut}}\epsilon_{b-tag}} \tag{1}$$

The efficiencies ϵ_{sel} and $\epsilon_{\chi^2_{cut}}$ used in Equation (1) are determined on simulation and a scale factor is applied to account for differences with respect to data. The b-tagging efficiency is determined by reconstructing the b tag discriminant distribution for true b jets ($\hat{\Delta}_b$) in a fully data-driven way.

Before applying the b-tagging criteria on the b-jet associated to the leptonically decaying top quark in the events, these candidates for b-flavored jets are used to define a b candidate jet sample. This jet sample is divided into a b-flavor enriched and b-flavor depleted subsample. As shown in Figure 2 the subdivision is performed according to the $M_{\ell b}$ variable of the event, which yields the b enriched region as 50 GeV/c^2 $< M_{\ell b} <$ 140 GeV/c^2 and the b depleted region as 140 GeV/c^2 $< M_{\ell b} <$ 240 GeV/c^2. The subdivision is defined such that the $M_{\ell b}$ variable remains uncorrelated with the b-tagging discriminators in the selected enriched and depleted regions.

To obtain the true b-jet discriminator ($\hat{\Delta}_b$), the non-b jet contamination in the b flavor enriched subsample is removed by subtracting the b-tag discriminant distribution of the b-flavor depleted subsample (Δ_b^{depl}) from the b-flavor enriched subsample (Δ_b^{enr}).

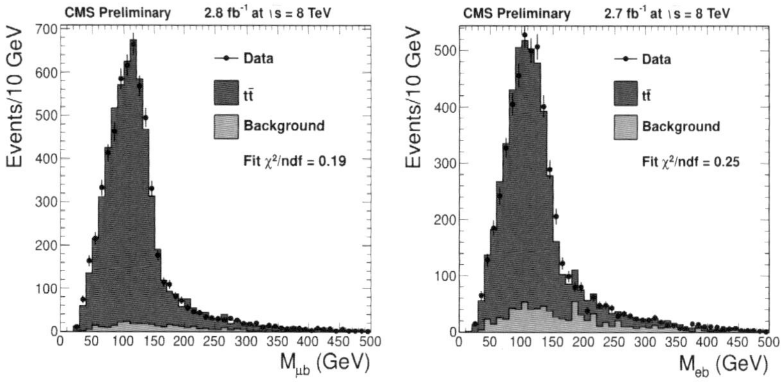

Figure 1. (left) Invariant mass distribution of the b jet candidate and the muon system and (right) Invariant mass distribution of the b jet candidate and the electron system. Signal and background distributions are normalised to the template fit results

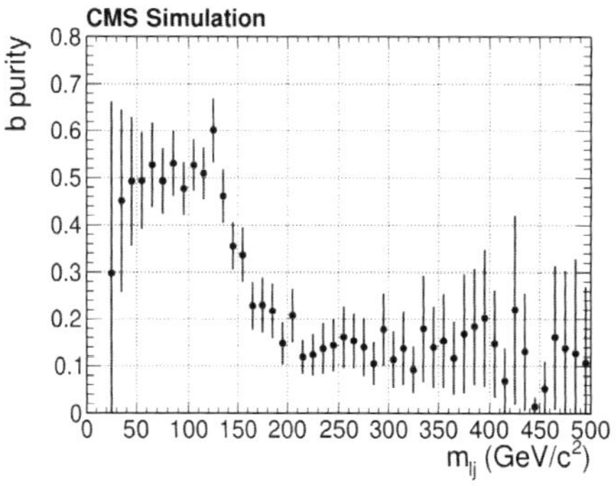

Figure 2. b-jet purity as a function of the invariant mass distribution of the b jet candidate and the muon system

$$\hat{\Delta}_b = \Delta_b^{enr} - F \times \Delta_b^{depl} \tag{2}$$

The factor F in Equation (2) represents the ratio between the number of non-b jets in the b-flavor enriched and b-flavor depleted subsamples which therefore rescales the number of non-b jets in the depleted region to the number of non-b jets in the enriched subsample. This factor F is obtained by constructing a control sample with a high purity in non-b jets. The control sample of jets is defined as those jets in the events which are associated with the hadronically decaying W boson. Additionally these jets in the control sample are requested to be anti-b-tagged to further reduce the b jet contamination. In the control sample, the same subsamples are defined as within the signal b candidate jet sample by use of the $M_{\ell b}$ variable which allows direct access to the factor F from data. The kinematics of the jets in the control sample are matched to those in the signal b candidate sample using reweighing techniques based on the observed data itself . Also in the data driven determination of F and in the application of Equation (2) the participating jets in the b-flavor depleted region are reweighed to match the kinematics of the jets in the b-flavor enriched region. From the estimated b-tag discriminator distribution, $\hat{\Delta}_b$, the b-tagging efficiency for diverse working points is estimated.

4. Results

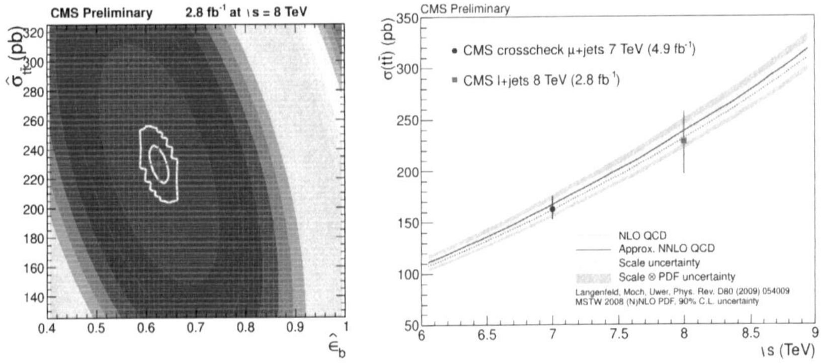

Figure 3. $\Delta\chi^2$ distribution in the $(\hat{\epsilon}_b, \hat{\sigma}_{t\bar{t}})$ plane showing the correlation between the two estimators within muon+jets events (left) and the cross section obtained with this analysis technique at 7 TeV and 8 TeV centre-of-mass energy [5, 6] (right)

On a dataset of 2.8 (2.7) fb^{-1} in the muon (electron) channel at $\sqrt{s} = 8$TeV recorded by the CMS experiment at the Large Hardron Collider during 2012, the $t\bar{t}$ cross-section was measured [5].

$$\sigma_{t\bar{t}}(\mu + jets) = 229.9 \pm 11.1(stat.) \pm^{27.6}_{29.0} (syst.) \pm 10.1(lumi.)pb$$

$$\sigma_{t\bar{t}}(e + jets) = 227.3 \pm 12.2(stat.) \pm^{35.5}_{30.0} (syst.) \pm 10.0(lumi.)pb$$

A combined fit to both channels yields the following cross section value.

$$\sigma_{t\bar{t}}(combined) = 228.4 \pm 9.0(stat.) \pm^{29.0}_{26.0} (syst.) \pm 10.0(lumi.)pb$$

Figure 3 shows the correlation between the b-tagging efficiency and the $t\bar{t}$ cross section from data which arises from the combined measurement of the two properties. The dominating systematic uncertainty in the measurements of the $t\bar{t}$ cross section arises from the uncertainty in the determination of the jet energy scale and the uncertainties on the b-tagging efficiency.

The combined fit result is also shown on the right side of Figure 3 together with the cross check result at $\sqrt{s} = 7$TeV provided by the same method in the muon+jets channel [6].

References

[1] N. Kidonakis, R. Vogt, *The theoretical top quark cross section at the Tevatron and the LHC*, **arXiv:0805.3844v1 [hep-ph]** (2008)
[2] CMS Collaboration, *Identification of b-quark jets with the CMS experiment*, **arXiv:1211.4462 [hep-ex]** (2012)
[3] CMS Collaboration **JINST, 3 S08004** (2008), doi=10.1088/1748-0221/3/08/S08004
[4] CMS Collaboration, *Commissioning of the Particle-Flow reconstruction in Minimum-Bias and Jet Events from pp collisions at $\sqrt{s} = 7$ TeV*, **CMS-PAS-PFT-10-002** (2010)
[5] CMS Collaboration, *Top pair cross section in e/mu+jets at 8 TeV*, **CMS-PAS-TOP-12-006** (2012)
[6] CMS Collaboration, *Measurement of the t t-bar production cross section in pp collisions at sqrt(s) = 7 TeV with lepton + jets final states*, **arXiv:1212.6682 [hep-ex]** (2012)

Journal of Physics: Conference Series **452** (2013) 012054 doi:10.1088/1742-6596/452/1/012054

Search for Single Production of Vector-like Quarks Coupling to Light Generations in $4.64\ fb^{-1}$ of ATLAS Data at $\sqrt{s} = 7$ TeV

Samuel Meehan on behalf of the ATLAS Collaboration

The Enrico Fermi Institute, The University of Chicago, Chicago, Illinois, USA

E-mail: samuel.meehan@cern.ch

Abstract. This analysis presents a search for singly-produced vector-like quarks (VLQs) coupling to light quark generations in events with a W or Z boson produced in association with two or more high transverse momentum jets. The vector boson produced in the VLQ decay is reconstructed in the $W \to \ell\nu$ or $Z \to \ell\ell$ modes, where ℓ is an electron or a muon. No evidence of VLQs is observed above the expected background in $4.64\ fb^{-1}$ of data taken in 2011 by the ATLAS experiment at the LHC at a centre-of-mass energy of 7 TeV. A fit to the distribution of the invariant mass of the system composed of the vector boson and the highest transverse momentum jet is used to set limits on the heavy quark production cross section times branching ratio, and mass bounds are obtained for three flavors of VLQs.

1. Introduction

Many extensions of the Standard Model (SM) predict the existence of vector-like quarks (VLQs) in Refs. [1,2], defined as spin 1/2 quarks for which both chiralities have the same transformation properties under $SU(2) \times U(1)$. Notably, VLQs are often introduced as a top partner to cancel quadratic divergence of the top loop in the Higgs propagator. Although VLQs are generally assumed to mix primarily with third-generation SM quarks in order to satisfy constraints from electroweak (EWK) precision measurements and flavor observables, mixing to first generations is not excluded. In some scenarios, including the benchmark model considered in this search from Ref. [3], corrections to the quark mixings can cancel and relax these constraints, thus producing sizeable cross sections for the production of multiple new VLQs at the LHC in which single production is found to be the most likely channel for discovery.

Searches for single production of VLQs, in these channels have previously been reported by ATLAS [4], CDF [5] and D0 [6]. This search, more fully documented in Ref. [7], is performed for three flavors of VLQs, one up-type U quark of charge $+2/3$, one down-type D quark of charge $-1/3$, and one X quark of charge $+5/3$ in which the decays are assumed to contain W and Z bosons 100% of the time. In this analysis, performed using data collected with the ATLAS detector [8] at $\sqrt{s} = 7$ TeV, these VLQs are searched for in events with a W or Z boson produced in association with at least two jets, and in which the vector boson decays leptonically with electrons or muons in the final state.

2. Selection and Optimization

The VLQs searched for in this analysis are assumed to decay to a W or Z boson, referred to as CC charged current or NC neutral current channels respectively, and a light quark. Thus, this new physics signal can be detected as an excess in the rate of production of W/Z in association with a jet. This excess will manifest itself as a resonance near the VLQ mass, and so the sensitivity of this search is increased and it is made model-independent by "bump-hunting" for W/Z+jet resonant signals. However, background processes producing a W or Z boson and a high-transverse momentum (high-p_T) jet, whether fake or irreducible, are large and so a set of selection criteria are applied to select a subset of signal-like events used for the search. This selection is optimized by using Monte Carlo simulation of the primary standard model backgrounds (W/Z+jets, $t\bar{t}$, single top, and diboson) and proceeds in two stages. First, an initial baseline preselection is made to ensure that well understood events are being used in the search. Then, a multivariate optimization procedure is implemented to exploit kinematic features of the signal to define a set of selection criteria that increase the sensitivity to the VLQ signal.

The initial baseline event selection is performed by selecting events using a set of preselection criteria that ensure events to have been recorded during data-taking periods during which all subsystems of the detector are operational. Events in both data and Monte Carlo are then required to pass a high-p_T lepton trigger fired by one of the leptons used to reconstruct the leptonic W/Z boson decay. In each event, selection criteria are applied to all the physics objects necessary to reconstruct the full topology of the production and decay of the VLQ. Electrons are reconstructed using an algorithm to select high quality, well isolated leptons with E_T >25 GeV and $|\eta| < 2.47$[1] (but vetoing electrons reconstructed in the transition region of $1.37 < |\eta| < 1.52$). Muons are reconstructed using an algorithm to veto muons coming from heavy flavor decay and cosmic rays, and are required to to be well-isolated and have $p_T > 25$ GeV and $|\eta| < 2.7$. Jet candidates are reconstructed using the anti-kt algorithm [10] with a radius parameter of 0.4, using topological calorimeter clusters as constituents. These jets are required to pass a set of selection criteria defined to suppress jets coming from pileup or problematic detector regions and are further required to have $p_T > 25$ GeV and $|\eta| < 4.5$. Lastly, the missing transverse energy ($E_{\mathrm{T}}^{\mathrm{miss}}$), used as the candidate object for the neutrino in CC channel events, is calculated as the negative vector sum of the transverse momenta of all reconstructed objects in an event within $|\eta| < 4.5$.

The four-vector of the VLQ is then reconstructed in the CC (NC) channel as the sum of the four vectors from the W (Z) boson and the highest p_T jet in the event. In the CC channel, the W boson is reconstructed in events with exactly one lepton and $E_{\mathrm{T}}^{\mathrm{miss}}$ >50 GeV and is required to have a transverse mass greater than 40 GeV. Furthermore, for full reconstruction of the VLQ in the CC channel the longitudinal momentum of the neutrino candidate is determined using a W mass constraint with the lepton and $E_{\mathrm{T}}^{\mathrm{miss}}$. In the NC channel, events with a Z boson are selected as having two oppositely charged leptons of the same flavor whose combined invariant mass is between 66 GeV and 116 GeV.

In addition to these VLQ signature requirements, the signal to background ratio is enhanced by initially requiring events to have $N_{\mathrm{jet}} \geq 2$ in order to exploit the presence of the second light quark from the VLQ production, creating an "associated" forward jet. Then, an optimization of simple rectangular cuts is performed with five (six) angular separation variables as discriminating variables in the CC (NC) channel. The figure of merit $S/\sqrt{S+B}$, where B is the number of expected events according to the MC background estimate, and S is the number of expected events according to the 400 GeV VLQ Monte Carlo, is maximized using TMVA [9] to obtain

[1] ATLAS uses a right-handed coordinate system with the z-axis along the beam pipe. The x-axis points to the center of the LHC ring, and the y-axis points upward. The pseudorapidity is dened in terms of the polar angle θ as $\eta = ln(tan(\theta/2))$

the numerical values for the cuts placed on these variables, the quantitative details of which are provided in Ref. [7]. The sensitivity obtained for the spectrum of masses is not found to be greatly increased if the numerical values of these cuts are varied as a function of signal VLQ mass.

3. Search Results

After the optimization is performed using a Monte Carlo driven estimate of the background composition, the qualitative understanding of the background is confirmed by a comparison to data as in Figure 1(a) and Figure 1(b). However, the final background is parametrized by fitting the reconstructed invariant mass of the four-vector reconstructed from the W or Z boson and the highest *eta* jet, after the full selection. The functional form used to perform this fit is

$$f(m; p_{0,1,2,3}) = p_0 \cdot \frac{(1-x)^{p_1}}{x^{p_2+p_3 \cdot \ln(x)}}, \tag{1}$$

where x is the reconstructed VLQ mass m_Q (in units of 7 TeV) and $p_{0,1,2,3}$ are four free parameters. This fit is performed on a binned distribution, where the binning is chosen to reflect the reconstructed width of the VLQ signal. This background estimation is used to initially perform search using the BumpHunter algorithm as described in Ref. [11] to search all mass windows for the largest excess in data above the smooth background hypothesis as in Figure 1(c) and Figure 1(d). In both channels, no significant deviation is found from the smooth background hypothesis, the most deviant region being illustrated in Figure 1(c) and Figure 1(d) by the vertical blue lines.

4. Interpretation of Results

Since these results are consistent with a background-only hypothesis, Bayesian limits are set on $\sigma(pp \to qQ) \times \mathrm{BR}(\mathrm{Q} \to \mathrm{W/Z + q})$ for the VLQ benchmark signal. These limits, shown in Figure 2, are generated for signal mass points between 400 GeV and 2 TeV using the fitted background estimation described above, with systematic uncertainties from the background fit and signal modelling integrated into the likelihood function with nuisance parameters for each.

From this interpretation, the determined limits on cross section times branching ratio for the CC (NC) channel are found to range from 3 (1.9) pb at 400 GeV to 0.03 (0.07) pb at 2.0 TeV. For the nominal benchmark model considered here, the observed (expected) 95% confidence level lower limits on the VLQ mass are obtained at 1120 GeV (1160 GeV) and 1420 GeV (1570 GeV) for VLQs of charge -1/3 and +5/3 in the CC channel, and at 1080 GeV (1090 GeV) in the NC channel.

5. Conclusion

A search for heavy VLQs has been performed in 4.64 fb^{-1} of pp collision data taken in 2011 by the ATLAS experiment at a center-of-mass energy of 7 TeV. No evidence of VLQs is observed in the reconstructed VLQ candidate mass distributions, and limits are set on the production cross section for signal masses between 400 GeV and 2 TeV. For this benchmark model of VLQs with charge 2/3 and -1/3, these are the strongest limits to date and furthermore, for the charge +5/3 VLQ, which decays to a W boson and a light quark, these are the first limits on such a model.

References

[1] B. A. Dobrescu and C. T. Hill, Phys. Rev. Lett. **81** (1998) 2634-2637.
[2] N. Arkani-Hamed, A. Cohen, E. Katz, and A. Nelson, JHEP 0207 (2002) 034.
[3] A. Atre, M. Carena, T. Han, and J. Santiago, Phys. Rev. **D79** (2009) 054018.
[4] ATLAS Collaboration, Phys. Lett. **B712** (2012) 2239.
[5] CDF Collaboration, *Search for single production of heavy quarks*, CDF-note-10261.

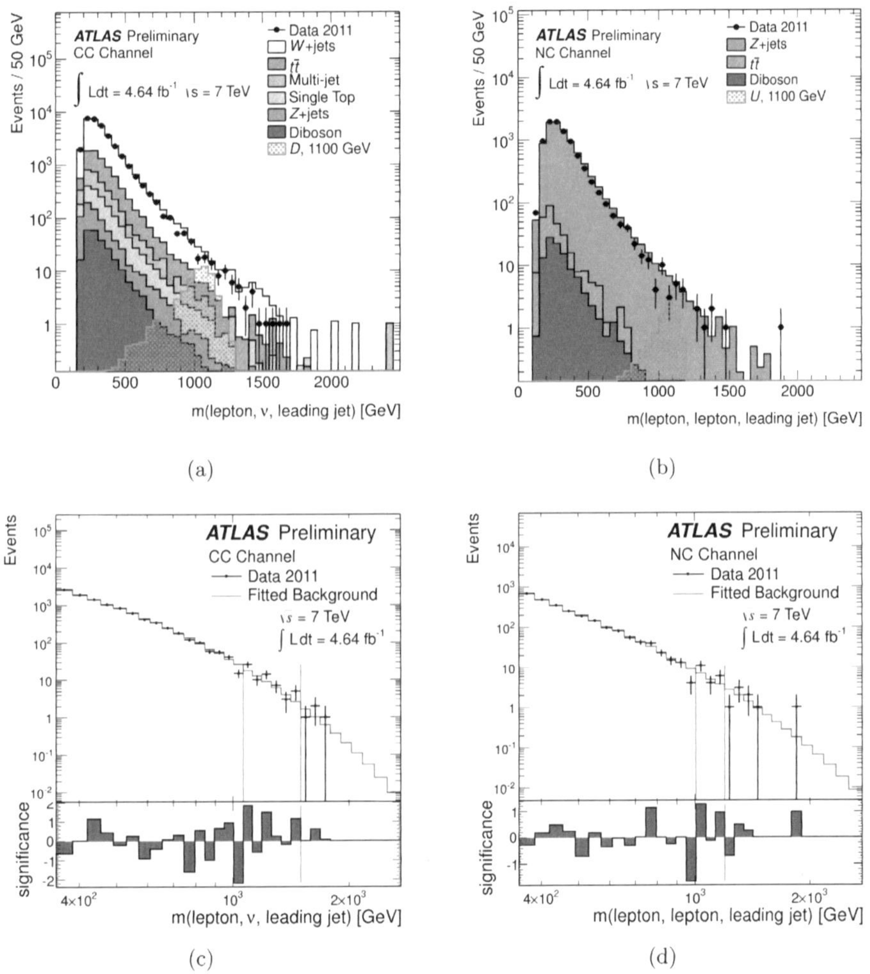

Figure 1. From Ref. [7], figures (a) and (b) show the comparison of the Monte Carlo estimated backgrounds for the combined electron and muon channels for the CC (a) and NC (b) channel. A 1.1 TeV VLQ signal, normalized to the nominal cross section, is overlaid in each case. Figures (c) and (d) show the results from the search using the BumpHunter algorithm for combined electron and muon channels for the CC (c) and NC (d) channels showing the local deviation of the data from the smooth background hypothesis and the most significant deviation and corresponding region bounded by vertical blue lines.

[6] D0 Collaboration, V. M. Abazov et al., Phys. Rev. Lett. **106**, 081801 (2011).
[7] ATLAS Collaboration, ATLAS-CONF-2012-137, *https://cds.cern.ch/record/1480628* (2012).
[8] ATLAS Collaboration, 2008 JINST 3 S08003.
[9] Hoecker et al., PoS **ACAT** (2007) 040, arXiv:physics/0703039.
[10] M. Cacciari, G. P. Salam, and G. Soyez, JHEP 04 (2008) 063, arXiv:0802.1189.
[11] G. Choudalakis, arXiv:1101.0390.

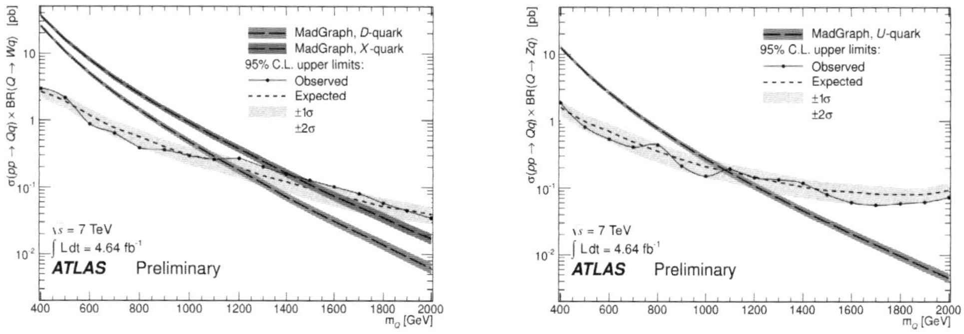

(a) (b)

Figure 2. From Ref. [7], the expected and observed 95% C.L. upper limits on $\sigma(pp \to qQ) \times BR(Q \to Wq)$ where Q is a D or X (a) and on $\sigma(pp \to qU) \times BR(U \to Zq)$ (b) calculated in the CC and NC channels, respectively. The nominal theoretical predictions are overlaid.

Journal of Physics: Conference Series **452** (2013) 012055 doi:10.1088/1742-6596/452/1/012055

Measurement of the top-quark mass using 7 TeV proton-proton collisions with the ATLAS detector

R.Moles-Valls[1] on behalf of the ATLAS Collaboration

[1]Instituto de Física Corpuscular, IFIC (CSIC-UVEG), Valencia, Spain.

E-mail: Regina.Moles@ific.uv.es

Abstract. The top quark is the heaviest known elementary particle. Its mass is intriguingly close to the electroweak symmetry breaking scale. Therefore its accurate determination is an important aspect of the LHC physics program. A summary of the present measurements of the top-quark mass with the ATLAS detector is presented here. The data used in those analyses amounts up to 4.7/fb of proton-proton collisions at 7 TeV with the Large Hadron Collider. The top-quark mass has been measured in different event topologies, namely all-hadronic, lepton plus jets and dilepton channel. The kinematic approach to measure the top-quark mass uses template methods for the all-hadronic and lepton plus jets channels, and the m_{T2} variable, that relates the transverse mass of the two leptons, is used in the dilepton channel.

1. Top-quark mass introduction

The top quark was discovery in 1995 at Tevatron in Chicago, USA [1][2]. It plays an important role due to its large contribution to radiative corrections and high sensitivity to physics beyond the standard model (SM). The top-quark mass is a fundamental parameter of the SM. After its discovery many methods have been developed to measure its mass with a high precision (< 1GeV). Nowadays, the most precise top-quark mass measurement has been provided by the combination of the CDF and D0 experiments: $m_{top} = 173.2 \pm 0.9$ GeV [3].

The LHC (*Large Hadron Collider*) [4] operated at pp centre of mass energy of $\sqrt{s} = 7$ TeV in 2011. During this period ATLAS (*A Toroidal LHC Apparatus*)[5] collected an integrated luminosity of data with all detector subsystems operational of 4.7/fb. The top quark at LHC is mainly produced in pairs through gluon-gluon fusion processes. Once produced, the top quark decays almost exclusively to a W boson and a b-quark (BR>99%). Its lifetime is of the order of 10^{-25}s, therefore the top quark decays before hadronizing. Here, measurements from using the invariant mass of the top-quark decay products are reported. The $t\bar{t}$ events can be divided in three channels depending of the final state objects: all-hadronic channel, lepton plus jets channel and the dilepton channel (lepton=e,μ).

2. ATLAS object reconstruction and selection for top mass analysis

To achieve a good reconstruction of the objects involved in the top-mass measurements (electrons, muons, jets and missing transverse energy (E_T^{miss})) the detector has to be fully operational. All the analyses have used the standard ATLAS reconstruction and selection. Electron candidates are defined as a deposit of energy in the electromagnetic (EM) calorimeter associated with a reconstructed track of the Inner detector (ID). The candidates are selected if they have a transverse energy E_T>25GeV and pseudorapidity $|\eta|<2.5$ excluding the region of $1.37<|\eta|<1.52$. Stringent selection cuts in the calorimeter and tracker are required to provided a good separation between electrons and jets (fake electrons). Moreover electron isolation is necessary to suppress the QCD multi-jet background. The muon candidate is reconstructed with an algorithm that combines track segment from muon chambers and from the ID and refitted them as a single track with a tight quality. The muon candidates are required to have p_T>20 GeV and $|\eta|$<2.5. An isolation cut is used to suppress the background origination from heavy quarks flavour decay. Moreover an overlap between muon and jet is also applied to avoid wrong muon identification. The E_T^{miss} is defined as the event momentum imbalance in the transverse plane of the beam axis, in this plane the momentum conservation is expected. An imbalance indicates the transit of undetectable particles, as for example neutrinos. The E_T^{miss} has been computed with the information of all objects in the event: electrons, muons, jets, soft jets and calorimeter cell out term (this term takes into account the energy not associated with any object). The jets are clustered using Anti-Kt algorithm with R=0.4 as cone size. The constituents of the calorimeter jets are clusters of calorimeter cells. The EM shower energy deposit in the calorimeter is accounted correctly. In addition, a calibration at hadronic scale must be applied. This calibration includes corrections depending of the E and η of the jet at EM scale as well as jet energy scale (JES) corrections based on Monte Carlo (MC) studies that have been validated with data. The jets are selected with p_T>25GeV and $|\eta|$<2.5 respect to the primary vertex. In order to reduce the impact of the pile-up, a cut in the probability that the jet stems from the hard scatter vertex is applied. Furthermore a jet quality criterion is imposed to remove jets not associated to the real energy deposits (hardware problems or cosmic-ray showers). Jets originating form b-quarks are identified using different b-tagging algorithms with a tagging efficiency higher than 70%.

3. ATLAS top-mass analyses

Basically two techniques have been used to perform the mass measurements in ATLAS: the templates methods and the calibration curves. The template methods obtain the top-mass value confronting simulated distributions with their real countpartners. On the other hand, the calibration curves extract the top-mass from a curve that describes the dependence of an experimental observable versus m_{top}. The top-mass measurement obtained for both methods corresponds to the mass definition used in the MC generator. The difference between the MC mass and the pole mass of the top quark is expected of the order of 1 GeV [6]. The ATLAS analysis has been developed in different channel topologies:

3.1. All-hadronic channel

The top-mass measurement in the all-hadronic channel has been done with a template method using 2.04/fb [7]. The selection of full hadronic events is based on a χ^2 equation that measures the compatibility of the jets with a particular assignment of the $t\bar{t}$ decay products:

$$\chi^2 = \frac{(m_{j_1 j_2} - m_W)^2}{\sigma_W^2} + \frac{(m_{j_1 j_2 b_1} - m_t)^2}{\sigma_t^2} + \frac{(m_{j_3 j_4} - m_W)^2}{\sigma_W^2} + \frac{(m_{j_3 j_4 b_2} - m_t)^2}{\sigma_t^2} \qquad (1)$$

These terms constrain the W mass and the top mass with a σ_w=10.2 GeV and σ_t=17.4 GeV respectively. Both resolutions have been determined using MC information. Only the events with a χ^2 lower than 8 are kept for the measurement. In addition, some extra cuts as number of b-jets equal to two and the rejection of events with E_T^{miss} contribution are required to get a purer sample. One of the most important systematic errors for the top-mass measurement is the not well know energy of the jets (JES). To minimize this effect, a ratio of the known W mass and the W mass candidate is calculated event by event and the light jets are corrected by this scale factor. Finally the top-mass is obtained by fitting templates of the three-jet mass combination. The templates for signal have been obtained from five MC $t\bar{t}$ samples generated with a m_{top} between 160-190 GeV. Background contribution has been estimated from data. The final reconstructed mass can be observed in figure 1. This yields a top-quark mass of m_{top} = 174.9 ±2.1 (stat)±3.8(sys) GeV. The systematic errors have been analyzed in detail (table 1).

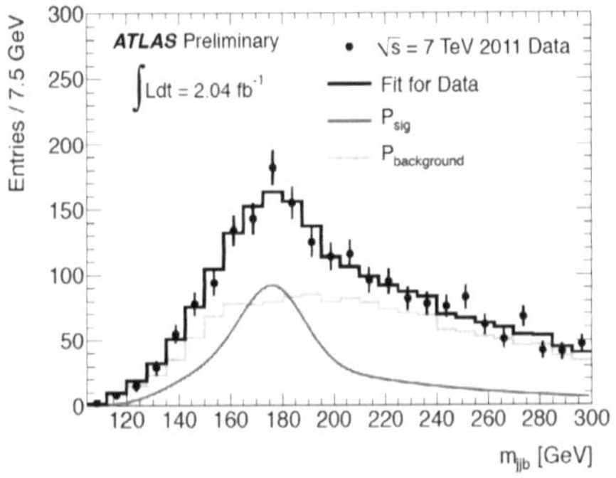

Figure 1. Distribution of m_{jjb} observed in data (points) . The top-quark signal is superimposed on the distribution from multijet background [7].

Source	Uncertainty [GeV]	Source	Uncertainty [GeV]
Method	0.4	Jet energy scale	2.1
Template statistics	0.9	b-jet energy scale	1.4
MC generator	0.5	b-tag efficiency SF	0.3
ISR/FSR	1.7	Jet energy resolution	0.3
PDF	0.6	Jet reco. efficiency	0.2
Backg. Modelling	1.9		
Total systematic uncertainty			**3.8**

Table 1. Table of systematic uncertainty contributions in the measurement of m_{top} in all hadronic channel [7].

3.2. Dilepton channel

A measurement of the top quark mass using the transverse mass variable m_{T2} has been performed using 4.7/fb [8]. The m_{T2} ($m_{T2}(m_{invisible}) = \min\{\max[m_T(m_{invisible}, p_T^{(1)}), m_T(m_{invisible}, p_T^{(2)})]\}$) represents a lower bound on the parent's particle mass and is useful in pair production events with two undetected particles. The b-jets from top quarks are selected by requiring a $p_T > 45$ GeV and $m_{T2} < 200$ GeV. The top decay objects assignment has been done using the multivariate optimization analyses. The dependence of the mean m_{T2} versus eight MC samples generated at different m_{top} have been used to build the calibration curve. Finally, the mean value of m_{T2} distribution measured in data, figure 2, is used to obtain the m_{top} from the calibration curve. This yields a $m_{top} = 175.2 \pm 1.6(stat)^{+3.8}_{-2.8}(syst)$ GeV. The systematic error sources that contribute to the measurement are summarized in table 2.

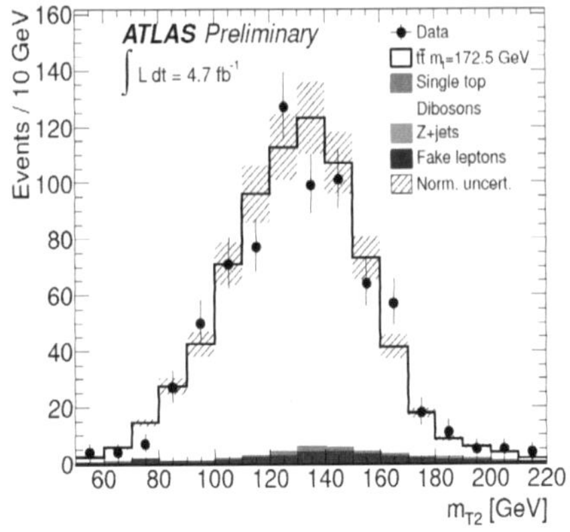

Figure 2. The m_{T2} distribution of the selected events [8].

Source	Uncertainty [GeV]	Source	Uncertainty [GeV]
$t\bar{t}$ generator model	-1.3/+1.3	jet energy scale	-1.4/+1.6
Parton shower	-0.9/+0.9	b-jet energy scale	-1.2/+1.5
Color reconnection	-1.2/+1.2	b-tag efficiency SF	-0.5/+0.5
ISR/FSR	-0.5/0.5	Jet energy res.	-0.1/+1.2
PDF	-0.1/0.1	Leptons	-0.1/0.2
Calibration Curve	-0.3/+0.3	E_T^{miss}	-0.1/0.1
Fake modelling	-0.3/0.3	b-tagging	-0.4/+0.3
Underlying event	-0.2/0.2		
Total Syst. uncertainty			**-3.3/+3.5**

Table 2. Table of systematic uncertainty contributions in the measurement of m_{top} in the dilepton channel [8].

3.3. Lepton+jets channel

In this channel, two templates methods have been developed to measure m_{top} using 1.04/fb.

The 1d-analysis [9] uses a mass ratio between the reconstructed top-mass and the hadronic W mass ($R_{32}=m_{top}/m_w$). The selection of the top decay objects is done using a kinematic fit that exploits the full event topology as well as b-tagging information. The R_{32}, that is sensitive to the top mass variations and helps in the JES reduction, is used to derive the templates at different top masses. The top mass measurement obtained from 1d template is m_{top}=174.4±0.9(stat)±2.8(syst).

The 2d-analysis [9] determines simultaneously the m_{top} and the jet scale factor (JSF) using the m_{top} and m_w distributions. The JSF, calculated using the differences between the predicted m_W and the observed in data, reduces the impact of the JES in the measurement. The top jet triplet candidate is constructed from any b-jet together with any light-jet pair with m_W within 50-110 GeV. The jet triplet with maximum p_T is chosen as top quark candidate. The selected W light-jet pair is calibrated at parton level using a χ^2 equation and the corrections are applied only to the m_{top} parameter to reduce the JES. The templates are constructed as a function of top-quark mass in the range of 160 GeV - 190 GeV and the input value for the JSF is in the range of 0.9-1.1. The unbinned likelihood fit is used to perform a top-mass measurement m_{top}=174.5±0.6(stat)±2.3(sys). Detailed systematic information can be found in table 3. Figure 3 shows the top-mass measurement in the e+jets channel.

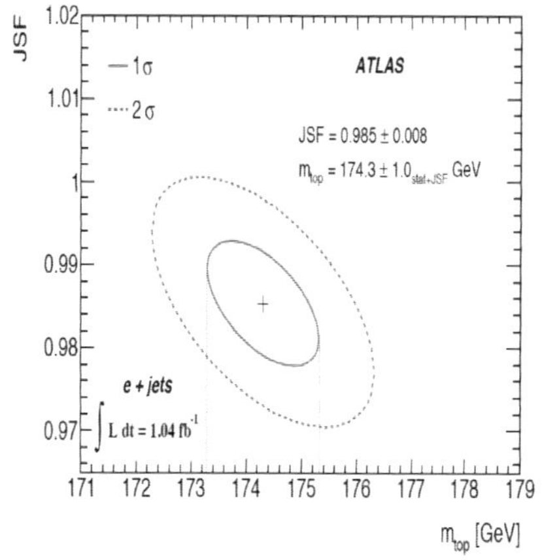

Figure 3. The correlation of the measured top-quark mass, and the jet energy scale factor (JSF) for the e+jets channel [9].

Source	Uncertainty [GeV]	Source	Uncertainty [GeV]
Data statistics	0.61	Proton PDF	0.10
Jet energy SF	0.43	W+jets modelling	0.38
Method calibration	0.07	QCD modelling	0.33
Signal MC generator	0.33	Jet energy scale	0.66
Hadronization	0.15	b-jet energy scale	1.58
Pileup	<0.05	b-tagging efficiency	0.29
Underlying event	0.59	Jet energy resolution	0.07
Color reconnection	0.55	Jet reco. efficiency	<0.05
ISR/FSR	1.01	E_T^{miss}	0.13
Total Syst. uncertainty			**2.31**

Table 3. Systematic uncertainty contributions in the measurement of m_{top} in the lepton plus jets channel with the 2d-template. The uncertainties are combined for electron and muon channel [9].

4. Conclusions

A summary of the present top-mass measurement in ATLAS for the different channels has been reported.

References

[1] CDF Collaboration 1995 Observation of the top quark production in $p\bar{p}$ collision *Phys. Rev. Lett.* **34** 2626-31
[2] D0 Collaboration 1995 Observation of the top quark *Phys. Rev. Lett.* **74** 2632-2637
[3] Tevatron electroweak working group for the CDF and D0 Coll. 2011 Combination of CDF and D0 results on the top mass of the top quark using up to 5.8fb^{-1} of data arXiv:1107.5255 [hep-exp]
[4] Evnas L and Bryant P (editors) 2009 LHC Machine *J. Instrum* **3** SO8001
[5] ATLAS Collaboration 2009 The ATLAS experiment at the CERN Large Hadron Collider *J. Instrum* **3** SO8003
[6] Buckley A et al. 2011 General-purpose event generators for LHC physics *Phys. Rep.* **504** 145
[7] ATLAS Collaboration 2012 Determination of the Top Quark Mass with Template Method in the All-Hadronic Decay Channel using 2.04 fb^{-1} of ATLAS Detector ATLAS-CONF-2012-030
[8] ATLAS Collaboration 2012 Top-quark mass measurement in the eμ channel using the m_{T2} variable at ATLAS. ATLAS-CONF-2012-082
[9] ATLAS Collaboration 2012 Measurement of the Top Quark Mass with the Template Method in the $t\bar{t} \rightarrow lepton + jets$ channel using ATLAS Data *Eur. Phys.J.* **C72** 2046

Measuring the Top Quark Mass using Kinematic Endpoints

Benjamin Nachman, on behalf of the CMS Collaboration
Laboratory for Elementary Particle Physics, Cornell University
E-mail: bnachman@cern.ch

Abstract. We report on a simultaneous measurement of the top quark, W boson, and neutrino masses in $t\bar{t}$ dilepton decays with 4.98fb^{-1} of data from the CMS experiment. The analysis is based on endpoint determinations in kinematic distributions. In addition to the unconstrained fit for three masses, the neutrino and W boson masses can be constrained to standard values; in the maximally constrained fit, the top quark mass is found to be $M_t = 173.9 \pm 0.9\ (\text{stat})\ ^{+1.2}_{-1.8}\ (\text{syst})\ \text{GeV}$.

1. Introduction

In many proposed extensions of the Standard Model, there is a new $\mathbb{Z}_2$ symmetry, like R parity in Supersymmetry (SUSY), under which the new physics is odd and the known particles are even. New physics particles are thus produced in pairs and decays must terminate in a stable particle. Such particles are candidates for dark matter and escape direct observation in the CMS detector [1]. We have developed a model independent method to simultaneously extract all unknown masses in a decay chain. With the production of pairs of neutrinos, dileptonic $t\bar{t}$ resembles the generic new physics topology. Thus, we test this method in data by simultaneously computing the top quark, W boson and neutrino masses. Going beyond the general study of the technique, we also fix the W and neutrino masses to their known values and perform a measurement of the top quark mass. This measurement is competitive in the dilepton channel and does not rely on Monte Carlo simulation. More details are available in Ref. [2].

2. Kinematic Variables - M_{T2}

To combat the problem of underconstrained kinematics, we look at extreme kinematic features of distributions. In particular, endpoints of distributions are determined and related to the underlying masses. This analysis uses the extension of transverse mass called M_{T2} [3]. M_{T2} is the minimum parent mass consistent with observed kinematics. We have three unknowns, M_ν, M_W, and M_t, so we partition the $t\bar{t}$ decay in order to create three mass variables [4]. Two M_{T2} subsystem variables are combined with the bl invariant mass to simultaneously extract the three masses. Figure 1 details the partitioning of the event topology for the M_{T2} variables. In general, the resulting variables depend on the magnitude of the transverse upstream momentum, $\vec{P}_T$ [5]. To remove this dependence, we project all momenta perpendicular to $\vec{P}_T$ and compute $M_{T2\perp}$. From [4], we have analytic formulae that relate distribution endpoints and underlying masses.

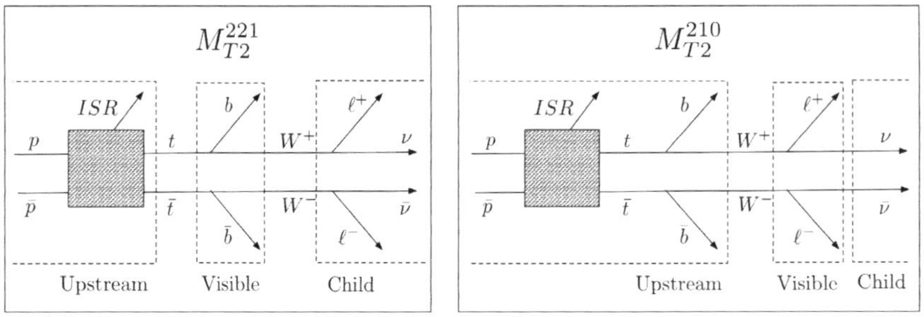

Figure 1. We partition the $t\bar{t}$ decay into subsystems. For a third variable, we use the bl invariant mass, which is in many cases the same as the third possible M_{T2} variable we could form.

3. 2011 Data (4.98/fb)

Monte Carlo (MC) is overlaid on the 2011 data to illustrate the approximate signal and background content of the distributions in Figure 2. There is good agreement, but the signal MC is not used in the reported measurement. From the known values of M_t, M_W, and M_ν the expected endpoints of $M_{T2\perp}^{(210)}$, $M_{T2\perp}^{(221)}$, and m_{bl} are 80.4, 173, and 152.6 GeV (respectively).

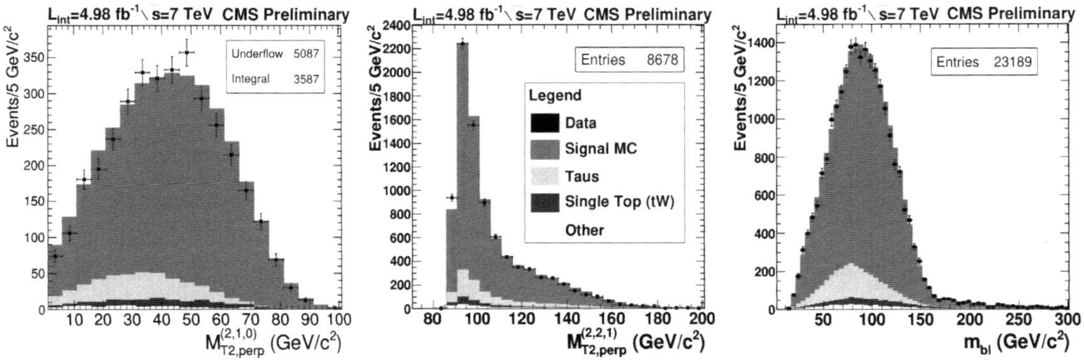

Figure 2. Monte Carlo overlaid on the 2011 data for the three measured kinematic distributions $M_{T2\perp}^{(210)}$, $M_{T2\perp}^{(221)}$, and M_{bl}. The MC is luminosity normalized to the data with NLO cross sections.

4. Fitting Strategy

We perform an unbinned maximum likelihood fit that takes into account event-by-event information such as resolution function shapes. For $\mathbf{M} = (M_\nu^2, M_W, M_t)$[1] and data vector u the likelihood is given by $\mathcal{L}(\mathbf{M}) = \prod_a \prod_{i=1}^{N_{events}} \mathcal{L}_i^a(u_i|\mathbf{M})$ for $a \in \{M_{T2}^{210}, M_{T2}^{221}, M_{bl}\}$ and

$$\mathcal{L}_i^a(x_i|x_{\max}) = \alpha \int S^a(y|x_{\max}^a)\mathcal{R}_i^a(x_i - y)dy + (1-\alpha)B^a(x_i), \tag{1}$$

where α is the signal normalization, $\mathcal{R}$ is a function that incorporates event-by-event resolutions, B is the background estimation and S is the signal shape. This last quantity is approximated as linear decent into the endpoint in a region near the a priori kinematic maximum.

[1] We are only sensitive to the square of the neutrino mass.

5. Background Estimation

The dominant source of background is combinatorial. In particular, the largest contribution to this background is from b jet selection, which is not 100% pure. Non $t\bar{t}$ events do not obey the same bounds as $t\bar{t}$. Figure 3 shows the composition of events in the invariant mass distribution and makes clear that the mis-tag background dominates the events beyond the endpoint (153 GeV). We model this dominant background by using a control sample in which we have inverted our b tagging. We then use a non-parametric estimation (adaptive kernel density estimation) of the shape which enters as B^a in Eq. 6.

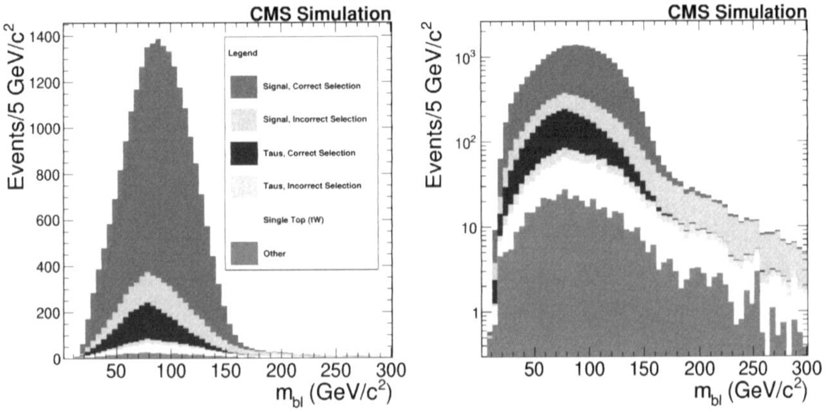

Figure 3. Monte Carlo composition of the M_{bl}. Note that the dominant source of events beyond the endpoint are due to signal events in which the wrong jets were chosen as the b jets.

6. Uncertainty Estimation

The usual methods for statistical uncertainty estimation are not valid because they do not account for fluctuations in the background shape. We thus use bootstrapping to estimate our statistical uncertainty. By recomputing the fit to the control region for each bootstrap sample, we can take into account background fluctuations by taking the central value and width of the distribution of masses evaluated from the ensemble of bootstrapped samples.

We estimate our systematic uncertainties by inserting variations into our fit and re-evaluating the result. Table 1 summarizes the contributions to the total uncertainty. Except for color reconnection, all the variations are applied to the data.

Systematic	Uncertainty (GeV)	Systematic	Uncertainty (GeV)
Jet Energy Scale	$+0.5 \quad -1.4$	Fit Range	± 0.6
Jet Energy Resolution	± 0.5	Background Modeling	± 0.5
Pileup	neligible	Color Reconnection	± 0.6

Table 1. Systematic Uncertainties. Except for color reconnection, the variations used in determining the uncertainties are applied to data. The total uncertainty is $+1.2$, -1.8 GeV.

7. Validation

As a cross check of our analysis, we perform our fit on many pseudo-experiments (PE) of simulated data. For each PE, define Pull $= (m_{\text{true}} - m_{\text{meas}})/\sigma_{\text{meas}}$. If our procedure is unbiased and correctly produces the uncertainty, then the distribution of the pull should be a standard normal. We can also probe potential bias by repeating this exercise for various values of m_{true}. As shown in Figure 4, both of these tests are consistent with an unbiased measurement with correct statistical uncertainty.

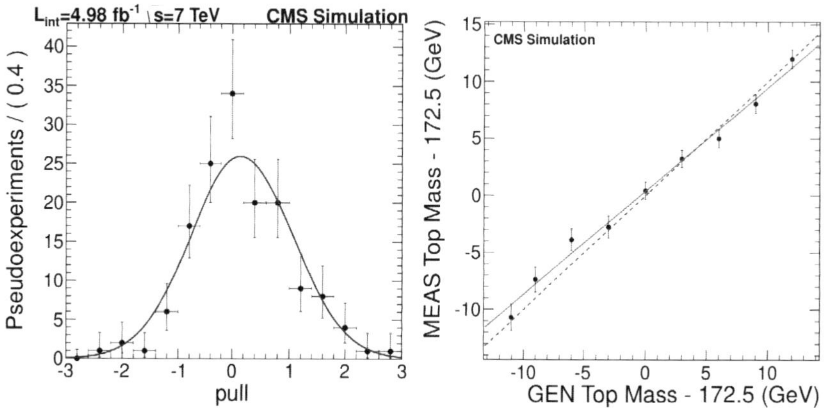

Figure 4. The left plot shows the pull distribution, with a mean of 0.15 ± 0.19 and a standard deviation of 0.92 ± 0.06. The right plot shows the spread in measured top masses as a function of input top mass. The best fit line is shown in red (slope 0.91 ± 0.4, intercept 0.34 ± 0.3) and the line of unit slope is in blue. The blue line agrees with the fit at $\chi^2/\text{ndf} = 10.7/9$.

8. Results

We present three sets of results, corresponding to three levels of constraints on the masses of the neutrino and W boson. The set of masses are shown in Table 8. In the case where both the neutrino and W boson masses are constrained, we compute our most precise top mass measurement: $M_t = 173.9 \pm 0.9$ (stat) $^{+1.2}_{-1.8}$ (syst) GeV. Due to its minimal dependence on MC, this result has many complementary (systematic) uncertainties to other measurements and so can help reduce the world uncertainty. However, in practice this is non-trivial as there is ambiguity on the order of Λ_{QCD} as to what this analysis measures and what traditional MC calibrated approaches compute. As the precision of top mass measurements improves, the top quark community will need to think carefully about the interpretation of the quantities being measured.

	Constraint		
		$m_\nu = 0$	
Fit Quantity	None		$m_W = 80.4$
m_ν^2 (GeV^2)	$-556 \pm 473 \pm 600$	0	(0)
m_W (GeV)	$72 \pm 7 \pm 9$	$80.7 \pm 1.1 \pm 1$	(80.4)
m_t (GeV)	$163 \pm 10 \pm 11$	$174.0 \pm 0.9 \pm 2$	$173.9 \pm 0.9^{+1.2}_{-1.8}$

Table 2. Results for the various levels of constraints. Uncertainties are first statistical and then systematic.

9. Acknowledgements

This work was partly under NSF support, PHY-0970024.

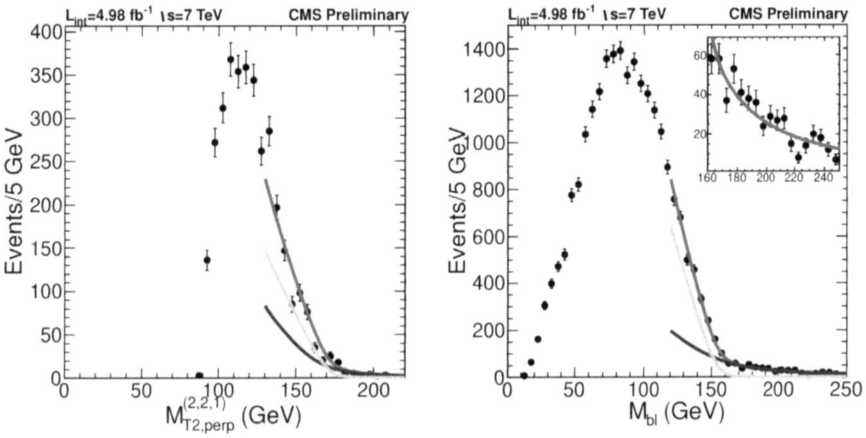

Figure 5. The fit plot showing the data and fit function for the case in which the neutrino and W boson masses are fixed. The displayed fit and data are from a particular bootstrapped sample. The blue line is the background estimation from the control sample, the green line is the signal fit function and the red is their sum. The top right of the m_{bl} plot shows the endpoint region to illustrate the fit to the data.

References

[1] CMS Collaboration, The CMS experiment at the CERN LHC, JINST 3 (2008) S08004.
[2] CMS Collaboration, Mass determination in the $t\bar{t}$ system with kinematic endpoints, CMS Physics Analysis Summary CMS-PAS-TOP-11-027 (2012).
[3] C. G. Lester and D. J. Summers, Measuring masses of semi-invisibly decaying particle pairs produced at hadron colliders, *Physics Letters B* 463 (1999), no. 1, 99-103.
[4] M. Burns, K. Kong, K. T. Matchev et al., Using Subsystem MT2 for Complete Mass Determinations in Decay Chains with Missing Energy at Hadron Colliders, *JHEP* 0903 (2009) 143.
[5] P. Konar, K. Kong, K. T. Matchev et al., Superpartner Mass Measurement Technique using 1D Orthogonal Decompositions of the Cambridge Transverse Mass Variable MT2, *Phys.Rev.Lett.* 105 (2010) 051802.

Journal of Physics: Conference Series **452** (2013) 012057 doi:10.1088/1742-6596/452/1/012057

Search for the Standard Model Higgs boson in association with top quarks and decaying to $b\bar{b}$ at $\sqrt{s}$ = 7 TeV using the ATLAS detector

Olaf Nackenhorst on behalf of the ATLAS Collaboration

II. Institute of Physics, Georg-August-Universität Göttingen, Friedrich-Hund-Platz 1, D–37077 Göttingen

E-mail: `olaf.nackenhorst@cern.ch`

Abstract. A search for a Higgs boson produced in association with a pair of top quarks and decaying into a pair of b quarks is presented. The analysis uses an integrated luminosity of 4.7 fb^{-1} of pp collisions at a center-of-mass energy of $\sqrt{s}$ =7 TeV collected in 2011 with the ATLAS detector at the LHC. The search is focused on the semileptonic decay of the $t\bar{t}$ system and combines nine different topologies given by the jet and b-tagged jet multiplicities of the event. A kinematic reconstruction of the Higgs boson mass is performed in the signal enhanced region, which becomes the primary discriminant variable between signal and background. Background-dominated samples are exploited to constrain the leading systematic uncertainties affecting the background prediction. No significant excess of events above the background expectation is observed. For a Higgs boson with a mass of 125 GeV, an observed (expected) 95% confidence upper limit of 13.1 (10.5) times the Standard Model cross section is obtained.

1. Introduction

One of the major goals of the physics program at the LHC is the search for the Standard Model (SM) Higgs boson. Already wide mass ranges of a possible SM Higgs boson have been excluded by LEP, Tevatron and the two LHC experiments ATLAS and CMS [1, 2, 3], allowing the SM Higgs boson only within the mass window of $m_H \approx [122 - 127]$ eV.

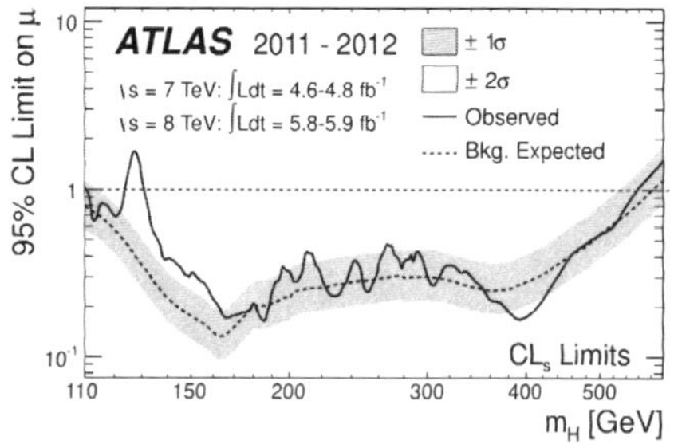

Figure 1. The observed (solid) 95% CL upper limit on the signal strength as a function of m_H and the expectation (dashed) under the background-only hypothesis. The dark and light shaded bands show the one and two sigma uncertainties on the background-only expectation [6].

In addition, an excess of events has been observed which is consistent with a Higgs boson in the mass range of $125 - 127$ eV with a local significance of $5.9\,\sigma$ and $5.0\,\sigma$ by ATLAS (Fig. 1) and CMS, respectively [4, 5, 6, 7]. These searches focused on di-bosonic decay modes of the Higgs boson, while no significant excess has been found yet in fermionic decays.

2. $t\bar{t}H$ Signature and Analysis Strategy

For a low-mass Higgs boson, the dominant decay mode is $H \to b\bar{b}$. In order to be able to distinguish this decay from QCD background, for these studies [8] the Higgs boson is required to be produced in association with a $t\bar{t}$ pair which decays semileptonically. This results typically in a final state with an isolated electron or muon with a high transverse momentum, large transverse missing energy and 6 jets, of which 4 jets originate from b quarks. Events are categorized in 13 independent topologies dependent on the number of jets and b-tags of which 4 are defined as signal regions (SR), 5 as background regions (BR) and 4 as control regions (CR). Depending on the topology either the scalar sum of the jet transverse momenta $H_{\mathrm{T}}^{\mathrm{had}}$ or the invariant mass of the pair of jets $m_{b\bar{b}}$ which are not assigned to the $t\bar{t}$ decay are used as discriminating variables in the signal and background enriched regions (Tab. 1). The invariant

Table 1. Overview of the splitting in different topologies and the used discriminating variable.

	0 b-tags	1 b-tags	2 b-tags	3 b-tags	≥ 4 b-tags
4 jets	BR, $H_{\mathrm{T}}^{\mathrm{had}}$	BR, $H_{\mathrm{T}}^{\mathrm{had}}$	BR, $H_{\mathrm{T}}^{\mathrm{had}}$		
5 jets	CR	CR	BR, $H_{\mathrm{T}}^{\mathrm{had}}$	SR, $H_{\mathrm{T}}^{\mathrm{had}}$	SR, $H_{\mathrm{T}}^{\mathrm{had}}$
≥ 6 jets	CR	CR	BR, $H_{\mathrm{T}}^{\mathrm{had}}$	SR, $m_{b\bar{b}}$	SR, $m_{b\bar{b}}$

mass $m_{b\bar{b}}$ is obtained by a kinematic event reconstruction of the $t\bar{t}$ decay using the maximum likelihood approach which allows for asymmetric energy resolutions [9]. This helps to reduce the combinatorial background as well as it improves the energy resolution of the $m_{b\bar{b}}$ system. No assumptions on the Higgs boson candidate are made. Hypothesis testing is performed in the SR and BR using a profile likelihood fit of the predicted distributions of the discriminants to the data. This allows for constraining the systematic uncertainties by including them as nuisance parameters. The events in the SR provide most information for the limit setting on a Higgs boson production cross section, while the ones in the BR are important to constrain leading systematic uncertainties such as those arising from the jet energy calibration, b-tagging calibration or the modeling of the background. The samples in the CR are used to validate extrapolations across topologies of an improved background prediction resulting from the fit.

3. Post-Fit Event Yields

The post-fit event yields for each considered topology under the signal-plus-background hypothesis (assuming m_H = 125 GeV and SM signal contribution) for the combined electron and muon sample including statistical and constrained systematic uncertainties are summarized in Table 2.

Table 2. Summary of the post-fit event yields in each of the topologies considered, corresponding to the combined e+jets and μ+jets channels. The quoted uncertainties are the sum in quadrature of statistical and total systematic uncertainties on the yields taking correlations into account [8].

	4 jets, 0 b-tags	4 jets, 1 b-tags	4 jets, ≥ 2 b-tags	5 jets, 2 b-tags	5 jets, 3 b-tags
$t\bar{t}H(125)$	0.20 ± 0.03	1.1 ± 0.1	3.0 ± 0.2	2.7 ± 0.2	2.3 ± 0.1
Total bkg.	40200 ± 280	21240 ± 200	15040 ± 150	6640 ± 80	915 ± 24
Data	40209	21248	15066	6653	878

	5 jets, ≥ 4 b-tags	≥ 6 jets, 2 b-tags	≥ 6 jets, 3 b-tags	≥ 6 jets, ≥ 4 b-tags
$t\bar{t}H(125)$	0.74 ± 0.04	3.4 ± 0.2	4.0 ± 0.2	2.2 ± 0.1
Total bkg.	45 ± 3	3360 ± 80	634 ± 19	62 ± 5
Data	41	3340	676	65

The fractional contributions of the various backgrounds to the total background prediction are visualized in pie charts for the different topologies in Figure 2. In the signal region where at least 5 jets and at least 3 b-tags are required the sample is dominated by the $t\bar{t}$ background. While in the 3 b-tag bin this is mainly $t\bar{t}$+light jets, in the 4 b-tag bin the contribution coming from $t\bar{t}$+heavy-flavour decays becomes more important. Also the fraction of multi-jet background increases in the signal region, thus a good modeling of these backgrounds is crucial for this analysis.

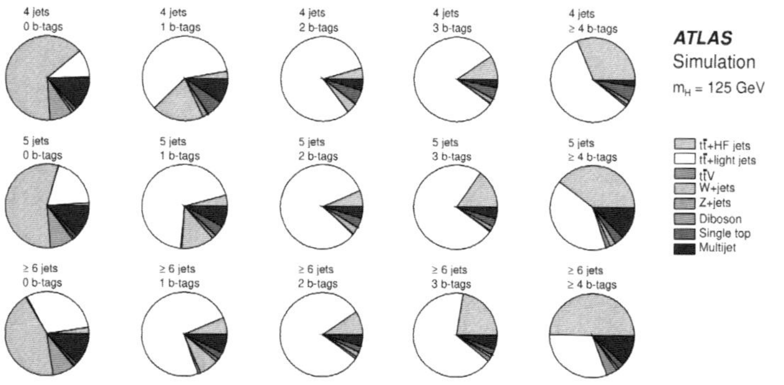

Figure 2. A series of pie charts showing the fractional contributions of the various backgrounds to the total background prediction in this analysis. Each row shows the plots for a specific jet multiplicity (4, 5, 6), and the columns show the b-tagged jet multiplicity (0, 1, 2, 3, 4) [8].

4. Fits to Data

A comparison between data and simulation is shown in Figure 3 for $H_{\mathrm{T}}^{\mathrm{had}}$ in a $t\bar{t}$ dominated region and for $m_{b\bar{b}}$ in a signal enriched region before and after fit to data under the signal-plus-background hypothesis assuming m_H = 125 eV. As a consequence of including background regions and choosing discriminants sensitive to the main systematic uncertainties the latter are significantly constrained and the fit-corrected MC predictions agree much better with the data. In the signal regions the largest contributing uncertainty is the one on the heavy flavor fractions of the $t\bar{t}$ background due to the lack of NLO predictions. This uncertainty is conservatively estimated to be 50% which can be constrained to 30% by means of the fit.

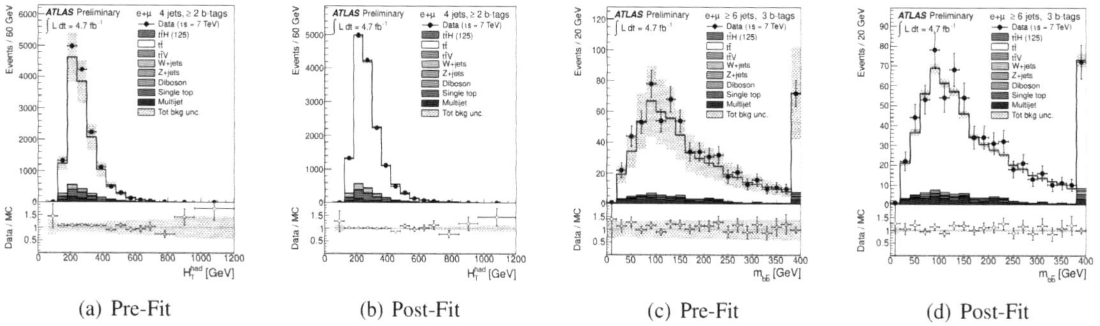

(a) Pre-Fit (b) Post-Fit (c) Pre-Fit (d) Post-Fit

Figure 3. Comparison between data and simulation for the final discriminant variables used in the combined e+jets and μ+jets channels. $H_{\mathrm{T}}^{\mathrm{had}}$ (a, b) is used in 4 jets and ≥ 2 b-tags and $m_{b\bar{b}}$ (c, d) in ≥ 6 jets, 3 b-tags. Both variables are shown before (a, c) and after (b, d) fitting of the nuisance parameters to data under the signal-plus-background hypothesis (assuming m_H=125 GeV). The bottom panel displays the ratio between data and background prediction. The shaded area represents the total background uncertainty [8].

All fitted nuisance parameters are found to have values within the 1σ interval defined by the original uncertainties assigned to the different systematic sources. A number of nuisance parameters are significantly constrained by the fit. The channels with 4 jets play an important role in constraining overall normalisation uncertainties on W+jets, multijet and $t\bar{t}$ background, as well as the b-tagging efficiency. The distribution of the $t\bar{t}$ background across the 0, 1 and 2 b-tags multiplicities provides a powerful constraint on uncertainties arising from the b-tagging efficiency. The 4, 5 and $\geq$ 6 jets channels with 2 b-tagged jets are almost pure in $t\bar{t}$+jets background and provide an important constraint on $t\bar{t}$ modeling uncertainties. Finally, the consideration of the 4 channels with 3 and $\geq$ 4 b-tagged jets allows the uncertainty on the $t\bar{t}$+heavy-flavour fraction to be constrained.

5. Cross Section Limits and Conclusions

Observed and expected 95% confidence level (CL) upper limits on the Higgs boson production cross section relative to the SM are derived as a function of m_H in the range between 110 and 140 GeV and shown in Figure 4. No significant excess of events above the background expectation is observed. For a Higgs boson with a mass of 125 eV, an observed (expected) 95% confidence upper limit of 13.1 (10.5) times the SM cross section is obtained. The most significant systematic uncertainties are $t\bar{t}$+heavy-flavour fractions, tagging efficiencies, multijet background normalisation and jet energy scale. Removing these systematic uncertainties improves the median sensitivity by 38% while removing all systematic uncertainties improves the median sensitivity by 45%.

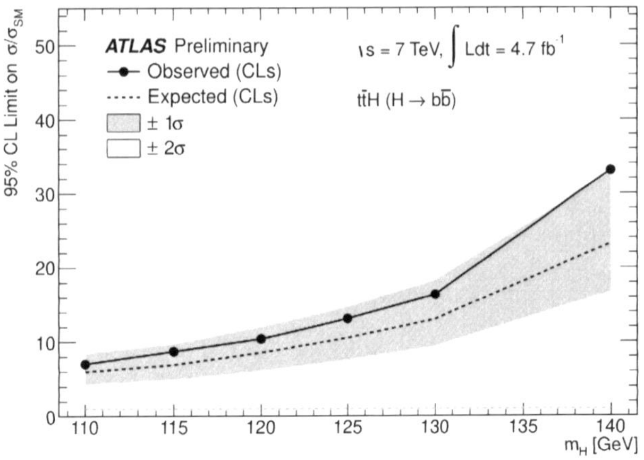

Figure 4. Observed and expected (median, for the background-only hypothesis) 95% CL upper limits on $\sigma(t\bar{t}H) \times \mathrm{BR}(\mathrm{H} \rightarrow \mathrm{b\bar{b}})$ relative to the SM prediction, as functions of m_H. The points are joined by straight lines for better readability [8].

References

[1] ATLAS Collaboration, *Phys. Rev.* D **86** 032003 (2012)
[2] CMS Collaboration, *Phys. Lett.* B **710** 26 (2012)
[3] CDF and D0 Collaborations, arXiv:1207.0449v2 [hep-ex] (2012)
[4] ATLAS Collaboration, *JINST* **3** S08003 (2008)
[5] CMS Collaboration, *JINST* **3** S08004 (2008)
[6] ATLAS Collaboration, *Phys. Lett.* B **716** 1 (2012)
[7] CMS Collaboration, *Phys. Lett.* B **716** 30 (2012)
[8] ATLAS Collaboration, ATLAS-CONF-2012-135 (2012)
[9] ATLAS Collaboration, arXiv:1203.5755v2 [hep-ex] (2012)

Journal of Physics: Conference Series **452** (2013) 012058 doi:10.1088/1742-6596/452/1/012058

Top Quark Charge Asymmetry Measurement with ATLAS at the LHC

Kate Shaw (on behalf of the ATLAS Collaboration)
INFN Gruppo Collegato di Udine and ICTP, Italy
E-mail: kate.shaw@cern.ch

Abstract. In the Standard Model top-antitop pair production is foreseen to be charge asymmetric in quark-antiquark annihilation at NLO. Although the dominant $t\bar{t}$ production mechanism in *pp* collisions at the LHC is gluon-gluon fusion, QCD predicts a small excess of top quarks produced at higher absolute rapidities than antitops. The measurement of the top quark charge asymmetry is therefore an important test of QCD and is also sensitive to new physics. We present measurements of this asymmetry using data corresponding to an integrated luminosity of 1.04 fb^{-1} and 4.70 fb^{-1} of *pp* collisions at $\sqrt{s}$ = 7 TeV collected by the ATLAS detector in the single lepton and dilepton channels respectively.

1. Introduction

Studies of the top quark, being the heaviest known elementary particle with a mass close to the electroweak scale, can provide important tests of the Standard Model (SM) and also promising probes of new physics.

At Leading Order (LO, O(α_s^2)) top-antitop pair production is predicted to be symmetric under charge conjugation, the tops have identical angular distributions. However next-to-leading order (NLO, O(α_s^3)) corrections introduce a charge asymmetry predicting that top (antitop) quarks are emitted preferentially in the direction of the incoming quark (anti-quark). This is a consequence of the interference between the $q\bar{q} \to t\bar{t}$ Born-level and box diagrams which lead to a positive charge asymmetry and the interference between initial and final state radiation contributions ($q\bar{q} \to t\bar{t}g$) leading to a negative charge asymmetry. Similarly the $qg \to t\bar{t}q$ process is asymmetric due to interference between amplitudes which have a relative sign difference under the exchange of the top and antitop. Finally top-antitop production by gluon-gluon fusion ($gg \to t\bar{t}$) is charge symmetric.

In proton-antiproton collisions at the Tevatron top-quark pairs are predominantly produced by quark-antiquark annihilation. CDF [1] and D0 [2] Collaborations measured the charge asymmetry as a forward-backward asymmetry A_{FB} resulting in a 2 - 3 σ excess over SM expectations both in inclusive and differential measurements as a function of the $m_{t\bar{t}}$ variable.

However in proton-proton collisions at the LHC top pairs are predominantly produced by (charge symmetric) gluon-gluon fusion. Furthermore there is no A_{FB} symmetry, instead QCD predicts a small excess of centrally produced antitop quarks, while top quarks are produced at higher absolute rapidities.

The top-based charge asymmetry A_C measured at the LHC is defined as

$$A_C^{t\bar{t}} = \frac{N(\Delta|y| > 0) - N(\Delta|y| < 0)}{N(\Delta|y| > 0) + N(\Delta|y| < 0)} \quad (1)$$

where $\Delta|y| \equiv |y_t| - |y_{\bar{t}}|$ is the difference of the absolute values of the top and antitop rapidities ($|y_t|$ and $|y_{\bar{t}}|$), and N is the number of events where $\Delta|y|$ is positive or negative. Using MC@NLO [6], A_C is theoretically calculated to be ~ 0.6 %.

In addition, a lepton-based asymmetry $A_C^{\ell\bar{\ell}}$ defined as:

$$A_C^{\ell\bar{\ell}} = \frac{N(\Delta|\eta| > 0) - N(\Delta|\eta| < 0)}{N(\Delta|\eta| > 0) + N(\Delta|\eta| < 0)} \quad (2)$$

where $\Delta|\eta| \equiv |\eta_{l^+}| - |\eta_{l^-}|$ is the difference of the absolute values of the positively and negatively charged lepton pseudorapidities and N is the number of events where $\Delta|\eta|$ is positive or negative.

In this article, the top-based and lepton-based asymmetry measurements are presented using data corresponding to 1.04 fb^{-1} and 4.70 fb^{-1} of pp collisions at $\sqrt{s} = 7$ TeV collected by ATLAS in 2011 in the single lepton [4] and dilepton [5] channels.

2. Top charge asymmetry in single lepton channel

Top quarks decay almost 100% of the time to a W-boson and b-quark. In top-antitop pair production in the single lepton channel one W-boson decays hadronically into two quarks and the other decays leptonically into a lepton and a neutrino giving the $t\bar{t} \rightarrow bl\nu bjj$ decay. The $t\bar{t}$ final state is then characterised by an isolated lepton (e or μ) with relatively high p_T, missing transverse momentum (E_T^{miss}) arising from the neutrino, two b-quark jets and two light quark jets from the hadronically decaying W-boson. Selected events are required to have at least one primary vertex with at least 5 associated tracks and exactly one electron with transverse momentum, $p_T > 25$ GeV, or one muon with $p_T > 20$ GeV. To reject multi-jet background in the μ channel $E_T^{miss} > 20$ GeV and $E_T^{miss} + m_T(W) > 60$ GeV are required, where $m_T(W) = \sqrt{2p_T^{\ell}p_T^{\nu}(1 - cos(\phi^{\ell} - \phi^{\nu}))}$ is the W-boson transverse mass, while in the e channel stricter cuts are required, $E_T^{miss} > 35$ GeV and $m_T(W) > 25$ GeV, due to the higher level of multijet background. Finally at least four jets (reconstructed with the anti-k_T algorithm with a 0.4 radius parameter in η-ϕ plane) with $p_T > 25$ GeV and $|\eta| < 2.5$ are required where at least one jet is tagged as originating from a b-quark.

The dominant backgrounds in this channel are multijets and W+jets which are estimated with dedicated data driven methods. Other minor backgrounds like single top production, Z+jets and diboson (WW, WZ, ZZ), are estimated using Monte Carlo (MC) simulated samples.

The $t\bar{t}$ system is reconstructed using a kinematic likelihood fit which relies on a parameterization of the energy and momentum resolution. The hadronically decaying top quark is reconstructed from three of the selected jets and the leptonically decaying top quark from the lepton, the remaining jet and the E_T^{miss}. A likelihood method is used to correctly assign measured jets to partons. The top and W-decay are modeled via Breit-Wigner function.

These measured asymmetries are distorted by event selection bias and detector effects. In order to compare them with theoretical predictions a Bayesian unfolding procedure is performed using the experimental asymmetries to reveal the four-vectors of the top quark before decay. The relation between a true distribution T_j and the reconstructed distribution S_i after detector simulation and event selection is

$$S_i = \sum_j R_{ij} T_j \quad (3)$$

where R_{ij} is the response matrix defined as the probability to observed an event in bin i when it is expected to be in bin j, and is calculated using MC events generated with MC@NLO. By inverting R_{ij} using Bayesian iterative unfolding, T_j is obtained.

The unfolding was performed after background subtraction for both the inclusive measured distribution of $\Delta|y|$, and the measured distribution $\Delta|y|$ as a function of the reconstructed $t\bar{t}$ invariant mass $m_{t\bar{t}}$ using two bins separated at $m_{t\bar{t}} = 450$ GeV as motivated by the CDF measurement [3] mentioned in the introduction.

The measured distributions of the top-antitop rapidity difference after the unfolding is shown in Figure 1 for e channel and μ channel. The resultant charge asymmetries are combined taking into account the correlation and measured as $A_C = -0.019 \pm 0.028(stat.) \pm 0.024(syst.)$.

The asymmetry combined in the two channels and evaluated in the two ranges of $m_{t\bar{t}}$ mentioned above, has been measured to be $A_C = -0.052 \pm 0.070(stat.) \pm 0.054(syst.)$ for $m_{t\bar{t}} < 450$ GeV, and $A_C = -0.008 \pm 0.035(stat.) \pm 0.032(syst.)$ for $m_{t\bar{t}} > 450$ GeV.

and is shown on the right in Figure 1. The uncertainties include that from the parton distribution functions and renormalisation and factorisation scales.

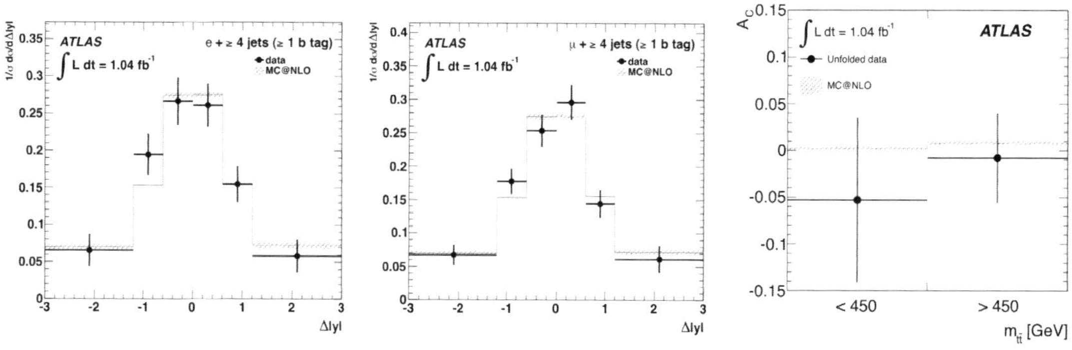

Figure 1. Measured top-antitop rapidity difference $\Delta|y|$ distribution in the single lepton channel after unfolding for e channel (left) and μ channel (center), and the right plot shows the charge asymmetry in the two $m_{t\bar{t}}$ region compared to the prediction from MC@NLO [4].

3. Top charge and lepton-based asymmetries in dileptonic channel

In the dilepton channel, where both W-bosons decay leptonically, producing the $t\bar{t} \to bl\nu bl\nu$ decay, three channels are considered depending on the lepton flavours in the final state: ee, $\mu\mu$ and $e\mu$. Both the top-based and lepton-based asymmetries measurements, as defined in Eq. 1 and Eq. 2 are presented. The dilepton $t\bar{t}$ channel is characterized by two isolated leptons with relatively high p_T, E_T^{miss} arising from the two neutrinos and two b-quark jets. Events are selected requiring at least one primary vertex with at least five associated tracks. Exactly two oppositely charged leptons are required with $E_T > 25$ GeV for electrons or $p_T > 20$ GeV for muons, and in the ee and $\mu\mu$ channels the $E_T^{miss} > 60$ GeV, and in the $e\mu$ channel the sum of lepton and jet E_T should be larger than 130 GeV. In only the ee and $\mu\mu$ channels the dilepton invariant mass is required to be different to the Z-boson mass by at least 10 GeV $|m_{\ell\ell} - m_Z| > 10$ GeV. Finally the event should contain at least two jets with $p_T > 25$ GeV and $|\eta| < 2.5$.

The dominant multijet background is estimated using a dedicated data driven method and background events from single top, Z+jets and diboson are evaluated using MC simulation.

A likelihood method based on the calculation of the leading order matrix element is used to reconstruct the $t\bar{t}$ event topology. For each event the kinematics are solved using the energy-momentum conservation at each vertex of the process and assigned a weight according to a likelihood estimator. The correction to the experimentally measured asymmetries is applied after the background subtraction by using calibration curves obtained by fitting a straight line to the reconstructed asymmetries corresponding to different simulated MC@NLO $t\bar{t}$ samples reweighted to vary the size of the asymmetry at the generator level.

Systematic uncertainties from MC generator and detector modelling are determined, and the asymmetries are recomputed and recalibrated after background subtraction, where the calibration curves are rederived for each systematic source.

The measured top-antitop rapidity difference distribution after corrections is shown in Figure 2 for ee, $e\mu$ and $\mu\mu$ channels. The three channels are combined, taking the correlations among channels into account, to give the lepton and top quark charge asymmetry as $A_C^{\ell} = 0.023 \pm 0.012(stat.) \pm 0.008(syst.)$ and $A_C^{t\bar{t}} = 0.057 \pm 0.024(stat.) \pm 0.015(syst.)$ respectively.

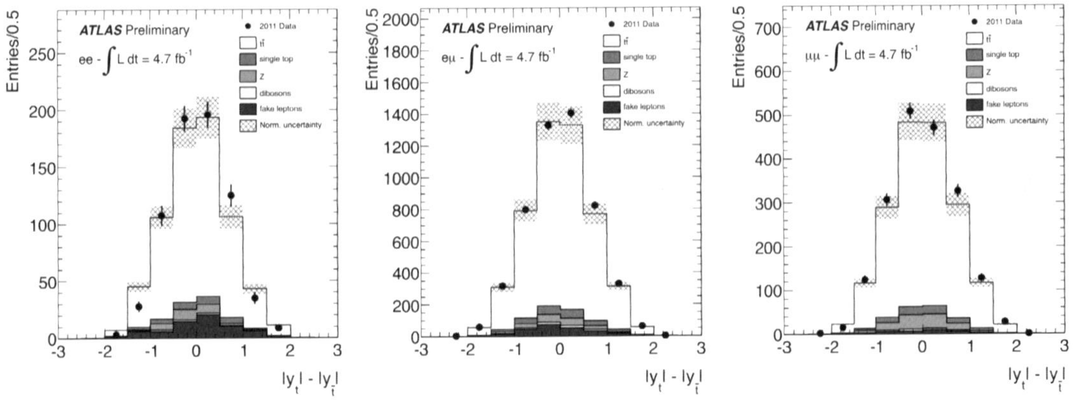

Figure 2. Measured top-antitop rapidity difference $\Delta|y|$ distribution in the dilepton channel after corrections are applied for the ee (left), $e\mu$ (center), and $\mu\mu$ (right) channels [5].

Finally a combination of the top-antitop based asymmetry in the dilepton and single lepton channels gives $A_C^{t\bar{t}} = 0.029 \pm 0.018(stat.) \pm 0.014(syst.)$.

4. Summary

The top quark charge asymmetry measurements performed by ATLAS experiment have been presented for the single lepton and dilepton channel. A lepton-based asymmetry measurement in the dilepton channel has also been presented. All the presented measurements are compatible with the SM predictions.

References

[1] CDF Collaboration, "Forward-backward asymmetry in top quark pair production in $p\bar{p}$ collisions at $\sqrt{s} = 1.69$ TeV", Phys. Rev. Lett. 101, 202001, 2008.

[2] D0 Collaboration, "Forward-backward asymmetry in top quark-antiquark production", Phys. Rev.D 84, 112005, 2011.

[3] CDF Collaboration, "Evidence for a mass dependent forward-backward asymmetry in top quark pair production", Phys. Rev.D 83, 112003, 2011.

[4] ATLAS Collaboration, "Measurement of the charge asymmetry in top quark pair production in pp collisions at $\sqrt{s} = 7$ TeV using the ATLAS detector", Eur. Phys. J. C72 (2012) 2039.

[5] ATLAS Collaboration, "Measurement of the charge asymmetry in dileptonic decays of top quark pairs in pp collisions at $\sqrt{s} = 7$ TeV using the ATLAS detector", ATLAS-CONF-2012-057, 2012.

[6] S. Frixione and B.R. Webber, "Matching NLO QCD computations and parton shower simulations", JHEP 06 (2002), 029, hep-ph/0204244.

AUTHOR INDEX

Institute of Physics
Dirac House, Temple Back
Bristol BS1 6BE UK

proceedings.com